王湘浩文集

王湘浩 著

吉林大学出版社
·长春·

图书在版编目（CIP）数据

王湘浩文集 / 王湘浩著. -- 长春 : 吉林大学出版社，2025. 4. -- ISBN 978-7-5768-4944-8

Ⅰ. N53

中国国家版本馆CIP数据核字第2025519B37号

书　　名：王湘浩文集
WANG XIANGHAO WENJI

作　　者：王湘浩
策划编辑：周　婷
责任编辑：周　婷
责任校对：张鸿鹤
装帧设计：刘　瑜
出版发行：吉林大学出版社
社　　址：长春市人民大街4059号
邮政编码：130021
发行电话：0431-89580036/58
网　　址：http://www.jlup.com.cn
电子邮箱：jldxcbs@sina.com
印　　刷：长春市赛德印业有限公司
开　　本：787mm × 1092mm　1/16
印　　张：31.75
字　　数：480千字
版　　次：2025年4月　第1版
印　　次：2025年4月　第1次
书　　号：ISBN 978-7-5768-4944-8
定　　价：110.00元

纪念王湘浩先生诞辰 ***110*** 周年

数学人生*

我的父亲王桂山，字仙府，靠耕田和卖药为生，在乡村免费行医，时常鼓励我读书，希望我将来当一名教师。叔父早年毕业于天津北洋大学。在父亲和叔父的影响下，1931年我初中毕业后，考取了北洋工学院附属高中。当时北洋工学院的附属高中实际上是预科性质，毕业后可以不经考试升入本科。我自小喜欢数学，小学和中学数学成绩一直很突出，图画、手工课成绩却很差。由于读北洋工学院附属高中需要学机械制图，这使我很难应付，逐渐失去了读工科的兴趣。于是1933年高中毕业时，放弃直升北洋工学院本科的机会和想法，考取了北京大学算学系（数学系）。

* 本文原载于上海教育出版社2005年出版的《科学的道路》。

我有幸进入北京大学数学系学习，真是如鱼得水，喜爱数学的兴趣和“才能”得到发展，成绩遥遥领先，受到老师们的称赞。在三四年级时，还获得每年240元的最高奖学金。

1937年，我从北京大学数学系毕业时，恰值抗日战争爆发，北京大学南迁。我先回到家乡，继而去西安，最后到长沙投奔由北京大学、清华大学、南开大学三校联合成立的长沙临时大学。我在江泽涵教授的帮助下，任临时大学数学系助教，才算结束了流亡生活。1938年春，长沙临时大学迁往昆明，改名西南联合大学。当了两年助教后，1939年当江泽涵教授的研究生，攻拓扑学，1941年毕业后，任西南联合大学讲师。1946年夏到美国普林斯顿大学，在著名代数学家阿廷（E. Artin）指导下攻读学位，1947年夏取得硕士学位，1949年春又取得博士学位，我的博士论文题目是《论格伦瓦尔定理》（*On Grunwald's Theorem*）。1949年6月启程回国，经香港、天津，8月到北京，被北京大学数学系聘为副教授，1950年晋升为教授。在1952年院系调整时，我被分配到东北人民大学数学系任系主任。1955年6月被选聘为中国科学院学部委员。1976年吉林大学计算机科学系成立后，任该系系主任，后又任吉林大学副校长。

1954年，我加入了中国民主同盟，其后曾任吉林省民盟副主任委员、长春市民盟主任委员，民盟中央委员和参议委员，并曾任长春市政协副主席，全国人民代表大会代表。我很感谢国家和人民给予我的荣誉。

在美国普林斯顿大学攻读博士学位期间，我选择了代数学作为研究方向。近世代数中有一个重要命题——迪克逊猜想，这个猜想的证明能彻底阐明有理单纯代数的结构。1931年德国数学家哈塞（H. Hasse）等人使用了类域论方面的重要定理——格伦瓦尔定理证明了这个猜想。这个猜想的证明，在当时的数学界是一件大事。美国著名代数学家阿尔贝特（A. A. Albert）说：线性结合代数的理论，当决定所有有理可除代数的问题得到了解答的时候，也许就达到了它的顶点。

我在研究这个问题时，看出了格伦瓦尔定理的错误，并写出了只有一页半的论文《关于格伦瓦尔定理的反例》。这篇论文使迪克逊猜想又成为悬而未决的问题，从而动摇了有理单纯代数的理论。

1948年底，我在博士论文中纠正了格伦瓦尔定理的错误，将该定理做了推广，并重新证明了迪克逊猜想。当时芝加哥大学数学系主任阿尔贝特看了我的论文后，曾邀请我就这一成果在芝加哥大学作过学术讲演。博士论文中，我只对循环扩张讨论了格伦瓦尔定理。回国后，又对一般的阿贝尔扩张给出了该定理成立的充要条件。

1943年，中山隆（T. Nakayama）和松岛与三（Y. Matuushima）证明了局部域上单纯代数交换子群等于其幺模子群。我利用自己所推广的格伦瓦尔定理证明了上述两群在代数数域情形下仍相等；而且在一般域情形下，当指数无平方因子时，两群也相等。在最一般情形下会是怎样呢?这一问题在以后兴起的代数K理论和代数群论中很重要。在苏联，这个问题称为田中–阿廷（Tanaka–Artin）问题，实际上，阿廷并未具体提出上述问题，我印象是在我的上述论文中提出的。这个问题受到了国际上一些同行的重视，但直到20世纪90年代初仍只得到一些部分结果。

马斯模定理的证明中使用了格伦瓦尔定理，因后者所含错误，马斯模定理原证已不成立。爱区勒（M. E. Erchler）曾经不用格伦瓦尔定理证明了模定理，但论证非常复杂。我在博士论文中曾用我修改的格伦瓦尔定理给出一个证明，后来又给出了一个不用格伦瓦尔定理的非常简单的证明。

1955至1957年，我得到了柯特半单循环的亚直接和表示，并讨论了与此相关的拟赋值环问题。

前期，我在代数学上开展过一些研究工作和教学工作，但后来从国家建设需要出发于1958年便开始了电子计算机和控制论方面的研究。

50年代末，多值逻辑的一个重要问题——函数完备性问题，引起了各国学者的注意，苏联的亚布隆斯基在1958年解决了三值逻辑的完备性问题，对于一般多值逻辑仅给出了一些零星结果。60年代初，我提出了解决这一问题的思想，即利用“保n项关系”的方法解决n值逻辑的完备性问题。根据我的想法，我的学生终于在1964年解决了这一问题。虽然这一结果没有发表，但它比国际上公认的解决这一问题的罗森贝格定理整整早了6年。1963年，我曾提出过多值逻辑中缺值函数的结构问题，并取得一定成果，这一问题后来也由我的学生完全解决了。

60年代初，我在自动机理论方面开展了研究工作，引进了圈环的概念并解决了非奇异线性内动机的分析问题。1990年，又研究解决了该论文中提出的因子分解问题。

1959年，结合当时系内研制计算机的工作，我开出了电子计算机课，并于1971年与吉林大学的同事们共同研制出吉林省第一台台式电子计算机，在通化市投入生产。

1977年，我提出了要在国内开展人工智能研究的建议，并在1980年受教育部委托，在吉林大学举办了全国性的人工智能讨论班，随之成立了全国高校人工智能研究会。同时在定理机器证明方面进行了研究，提出了更一般的广义归结方法和归结方法中的取因子问题。我和同事们在归结方法的研究中，提出了一些有用的改进策略，并且在计算机代数的研究中，提出了代数方程实根分离的一种较好的算法。

1952年我到吉林大学后，立即将主要精力投入建立数学系的工作。在建系过程中，与系里其他负责人共同采取了多方面的措施，在较短的时间内，使吉林大学的数学系和计算机科学系在建立和发展方面达到了预期的要求，得到了国内的承认和重视。

（1）重视教学。1952年建系初期，数学系只有14名教师，却同时需要给全校开出19门课，我在数学系共开出过20几门课程。因此，水平较高的教师经常需要同时开出两三门课。在我主持数学系和计算机科学系的30多年中，系里师生们一直高度重视教学和教材建设。在教学上对青年人的要求是严格的，建立了吉林大学数学系教学高质量和严要求的传统。

（2）重视实际。数学系和计算机科学系的同事们在考虑系的发展方向时，面临两种选择，是主要以自己的学术研究为方向搞起来，以求达到更高水平呢，还是更加重视国家所急需的方向呢?我们选择了后者。50年代中期，我在数学系着重发展了微分方程和计算数学两个方向，而我自己在1958年又搞起了计算机和控制论的研究。从实践结果看来，50年代我们数学系抓住微分方程、计算数学、计算机科学这三个方向，是没有错的。由于大家共同努力和各校同行的支持，到60年代初期，吉林大学数学系已在国内名列前茅，到70年代后期，吉林大学计算机科学系得以建立。

（3）重视人才，重视青年。建系初期，为了提高年轻教师的学术水平，系里的年长教师组织年轻人参加讨论班，年轻人边读文献边做年长教师给的题目，有些题目本来是年长教师已经做完的，为了培养年轻教师，让他们再做，并让他们发表。因此，1954年毕业的学生立即指导55届学生写论文。就用这种方法，在东北人民大学（吉林大学前身）第一期学报上，数学系就发表20多篇论文，在国内形成影响。为了使微分方程这个学科发展起来，1954年从东北工学院聘请了人所未识的王柔怀。在基础数学方面，1956年从中国科学院聘请了学识丰富的孙以丰。为了将计算数学搞起来，1958年请来了这方面的苏联专家。

在建系过程中，当时系里尽可能容纳各种不同意见，发挥各种人才的作用，看到各种人的优点（包括当时一些不得意的人的优点）。我们系里的同事们尽量做到不存私心，每届毕业生留校时，都把学习最好的学生安排到最重要的方向上。在社会各界的支持下，吉林大学数学系很快成长为专业较齐全、在国内有影响的系。这些成绩的取得，一靠国家的教育方针，二靠国内外同行的支持，三靠系里同事们的人品、见识和积极努力。我对此深感欣慰。

▶ 王湘浩院士（1915—1993）

▶ 王湘浩院士，20 世纪 70 年代摄于北京

▶ 王湘浩院士，1982 年摄于办公室

▶ 王湘浩院士

▶ 王湘浩院士在研读文献

▶ 王湘浩院士

▶1937 年，王湘浩大学毕业学位照

▶1937 年，北京大学数学系师生合影，前排左三为王湘浩

▶1936 年，北京大学数学系师生合影，右一为王

▶西南联大数学系师生合影，前排右一为王湘浩

▼西南联大数学系师生合影，右一为王湘浩

1946 年，王湘浩
在美国普林斯顿大学读书时的
学籍卡证件照

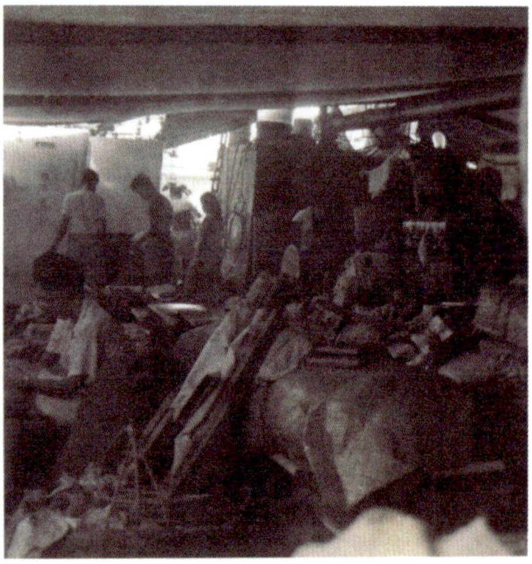

1949 年，王湘浩从香港回到北京所乘坐的
英国货轮统舱内照片

1949 年 8 月，北京大学数学系师生游北海公园合影，前排左四为王湘浩

▶1949 年 8 月，王湘浩在北京留影

▶1955 年，钱学森院士来校视察
左一钱学森院士；左二唐敖庆院士
右二王湘浩院士

▶1955 年 12 月，著名科学家钱学森来校同匡亚明校长及我校部分教授合影
从左至右依次为王湘浩、余瑞璜、钱学森、匡亚明；右一芶清泉、右二龚依群

▶1958 年，王湘浩院士在京剧《空城计》中饰演孔明

▶1959 年 3 月，系领导和苏联专家合影，前排左一为王湘浩院士

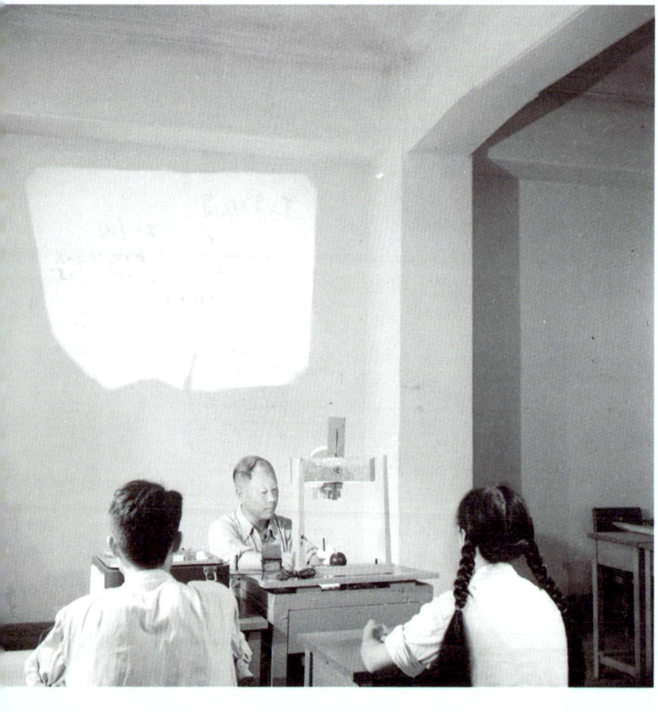

▶1959 年，王湘浩院士在使用新教具

▶1961 年，王湘浩院士在指导学生

▶1961 年，吉林大学毕业典礼上王湘浩院士代表教师讲话

▶1963 年，王湘浩院士在指导学生做毕业论文

▶20 世纪 60 年代，王湘浩院士在上课

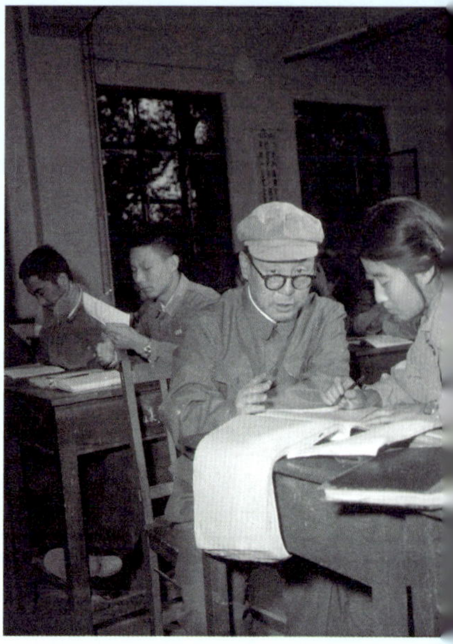

▶20 世纪 70 年代，王湘浩院士在辅导学生

▶1972 年，王湘浩院士在给数学系师生讲解计算机原理

▶1971 年，王湘浩院士（右一）在通化无线电厂研制台式计算机

▶王湘浩院士在讲课

▶1972 年，王湘浩院士（右一）指导南仁东研制台式计算机

▶1972 年，王湘浩院士指导教师研制电子计算机

▶1972 年，数学系台式计算机制成后在通化无线电厂投产，前排居中者为王湘浩院士

▶ 1975 年，王湘浩院士在学术会上作报告

▶ 1978 年 3 月 29 日，出席全国科学大会的 93 位学部委员与中国科学院领导欢聚在北京科学会堂前留下了这张令人难忘的合影，倒数第三排右二为王湘浩院士

▶ 王湘浩院士同两个硕士生讨论学术问题，
左一为王湘浩院士；中间为学生付凝；右一为学生黄秉超

▶ 王湘浩院士在计算机系考核研究生

▶王湘浩院士和刘叙华老师讨论学术论文

▶王湘浩院士在给研究生讲课

▶王湘浩院士在备课

▶1979年，王湘浩在民主党派恢复组织活动大会上发表讲话

▶ 王湘浩院士与计算机系领导及部分教师座谈

▶ 1979 年 8 月，率领国家教委代表团出席日本东京第六届国际人工智能会议，前排左三为王湘浩院士

中国計算机学会

▶（照片左段，右起）

王湘浩　教授，吉林大学计算机系主任，中科院院士，中国计算机学会副理事长。
慈云桂　教授，国防科大计算机系主任，中科院院士，中国计算机学会副理事长。
张效祥　研究员，104 机技术负责人，中科院院士，中国计算机学会理事长。
莫根生　教授级高工，中科院计算所八室，第十五研究所副总工程师。
范新弼　研究员，中科院计算所五室。
张梓昌　研究员，航天 706 所所长，中国计算机学会副理事长。
夏培肃　研究员，中科院计算所，中科院院士。
徐家福　教授，南京大学计算机软件所所长，中国计算机学会副理事长。
郑守淇　教授，西安交通大学计算机系。
李润斋　中科院计算所副所长，科技部科技情报研究所所长。

（照片右段，左起）

阎沛霖　科技教育家，中科院计算所首任所长。
蒋士騛　研究员，主持 109 乙机研发，中科院计算所八室。
吴几康　研究员，中科院计算所，中国计算机学会副理事长。
何绍宗　中科院计算所办公室主任，微电子学所所长，中国计算机学会秘书长。
姚锡珊　中科院计算所高级工程师，104 计算机研制结构组副组长。
张志浩　教授，成都电子科技大学（成都电讯工程学院）计算机系主任。
萨师煊　教授，中国人民大学经济信息管理系主任。
何志均　教授，浙江大学计算机系主任。
欧阳轵能　北京电子管厂总工程师，中国软件与计算机服务公司总经理。
虞浦帆　研究员，华东计算所二室，1965 年国家荣誉“发明”奖获奖者。

▶1986 年，王湘浩院士在大连与时任吉林大学校长伍卓群亲切交谈

▶1987 年 9 月，王湘浩院士（右一）祝贺陈建华博士答辩通过

▶1987 年 9 月，陈建华通过博士论文答辩后对导师表示感谢，左一为王湘浩院士

▶1987 年 9 月，陈建华博士答辩委员会专家合影，
右三为王湘浩院士

▶1987 年，王湘浩院士（中）与博士郑方青和马志方
在吉林大学计算机系资料室讨论学术问题

▶1988 年，王湘浩院士（右二）
与刘叙华、姚玉川、董文权教授等
参加答辩会

▶1988 年，1985 级硕士研究生毕业合影，前排右二为王湘浩院士

▶ 1988 年，王湘浩院士与秘书武艳茹在交谈

▶ 1989 年，王湘浩院士（左一）
与陆汝钤院士、董韫美院士等
参加郑方青、马志方和黄祥喜博士论文答辩会

▶ 1988 年，王湘浩院士在查阅资料

▶ 1989 年，王湘浩院士（右三）同计算机系的领导和教师见面

▶ 1989 年，王湘浩院士（中）与王柔怀（左）、江泽坚教授（右）在一起

▶1990 年，王湘浩院士（前排左四）与知识工程方向的师生合影

▶1991 年 4 月，程明德院士（左）、江泽涵院士（中）和王湘浩院士（右）三位老先生在一起

▶1991 年 4 月，王湘浩院士（右）与老师江泽涵院士合影

▶王湘浩院士和两个孙女在一

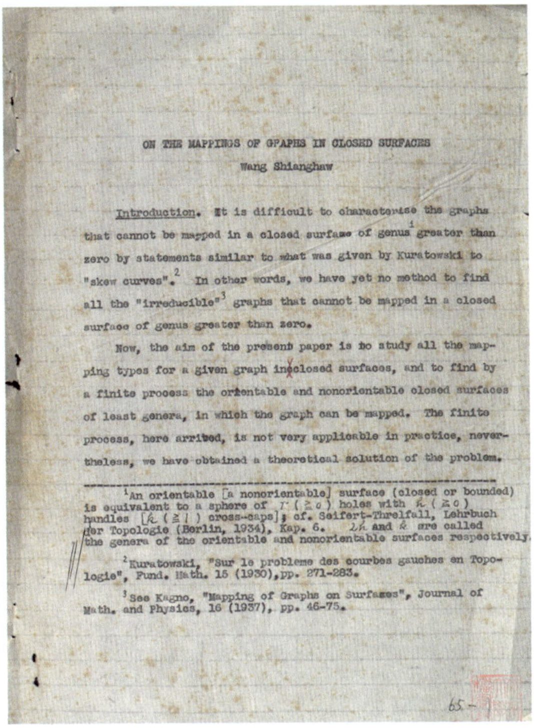

ON THE MAPPINGS OF GRAPHS IN CLOSED SURFACES

Wang Shianghaw

Introduction. It is difficult to characterise the graphs that cannot be mapped in a closed surface of genus[1] greater than zero by statements similar to what was given by Kuratowski to "skew curves".[2] In other words, we have yet no method to find all the "irreducible"[3] graphs that cannot be mapped in a closed surface of genus greater than zero.

Now, the aim of the present paper is to study all the mapping types for a given graph in closed surfaces, and to find by a finite process the orientable and nonorientable closed surfaces of least genera, in which the graph can be mapped. The finite process, here arrived, is not very applicable in practice, nevertheless, we have obtained a theoretical solution of the problem.

[1] An orientable [a nonorientable] surface (closed or bounded) is equivalent to a sphere of $r\ (\geq 0)$ holes with $h\ (\geq 0)$ handles [$k\ (\geq 1)$ cross-caps]; cf. Seifert-Threlfall, Lehrbuch der Topologie (Berlin, 1934), Kap. 6. $2h$ and k are called the genera of the orientable and nonorientable surfaces respectively.

[2] Kuratowski, "Sur le probleme des courbes gauches en Topologie", Fund. Math. 15 (1930), pp. 271-283.

[3] See Kagno, "Mapping of Graphs on Surfaces", Journal of Math. and Physics, 16 (1937), pp. 46-75.

65-1

▶ 1941 年，王湘浩撰写的硕士毕业论文

On the Mapping of Graphs in Closed Surfaces（首页）

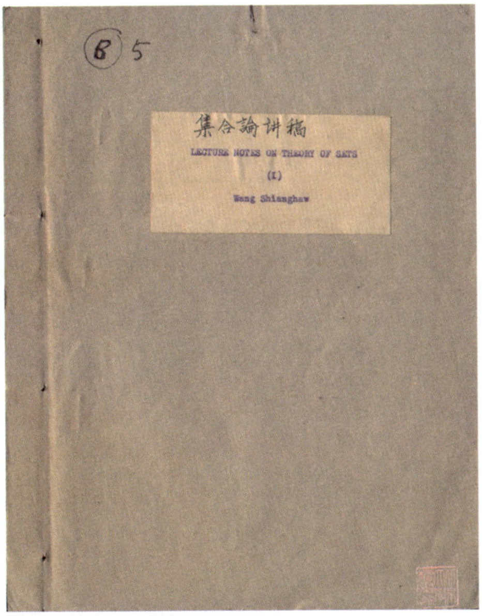

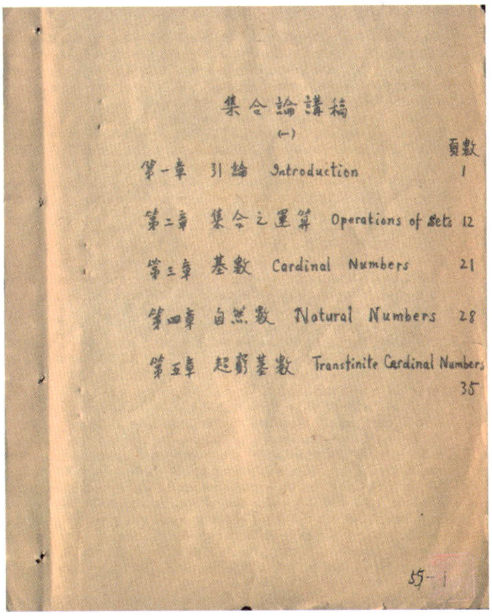

集合論講稿

(一)

55-1

20 世纪 40 年代，王湘浩在西南联合大学担任数学系专任讲师时撰写的《集合论讲稿》手稿（封皮和首页）

APPENDIX VI

Fellowships Awarded to Chinese Students

Two-year fellowships awarded by Department of State for advanced study in the United States under the administration of the China Institute in America

Name*	Field	University
Yat-san Chung	Literature and drama	Yale
Chih-chien Wu	Literature	Yale
Fu-ning Li	Literature	Yale
Ching-tsun Loo	Mathematics	Brown
Hao Wang	Philosophy	Brown
Shiang-haw Wang	Mathematics	Brown
Kwan-wai So	American Government and history	Wisconsin
Shu-chin Yang	Economics	Wisconsin
Chen-hwa Yang	Surgery	Michigan
Zau Foo	Education	Colorado
Liang Hsu	Library science	Denver
Po-chen Lin	Literature	Wesleyan
Ching-hsi Chao	Economics	Washington
Sung-chiao Chao	Geography	Clark
Kuan-hen Lin	Surgery	Long Island College of Medicine
Lan-chang Chiang	Surgery and radiology	Long Island College of Medicine
Shou-cheng Fu	Chemistry	Johns Hopkins
Hua-kang Chou	Pediatrics	Minnesota
Chih-lung Pu	Entomology	Minnesota

▶1946 年，王湘浩
获得美国国务院奖学金记录

Annals of Mathematics
Vol. 49, No. 4, October, 1948

A COUNTER-EXAMPLE TO GRUNWALD'S THEOREM

By Shianghaw Wang

(Received July 26, 1948)

In a recent research, the author of this note found that Grunwald's Theorem on the existence of cyclic extensions over an algebraic number field was not always true. On the other hand, he has obtained a proof of the theorem under appropriate restrictions. Here, a counter-example will be given and existing proofs analyzed. The positive result will be published elsewhere.

Let K be a cyclic extension over an algebraic number field F, of degree l^s, l being a prime, and let Ω be the intermediate field of degree l. Suppose the finite prime $\mathfrak{p}$ in F does not divide l and is ramified in Ω. Then it must be completely ramified in K. Since any $\mathfrak{p}$-adic unit congruent to 1 mod $\mathfrak{p}$ is an $l^{s\text{th}}$ power, the $\mathfrak{p}$-conductor of K over F is exactly $\mathfrak{p}$. Consequently, the multiplicative group of the residue-class field of $\mathfrak{p}$ is homomorphic to the Galois-group of K through the norm-residue symbol. It follows $N\mathfrak{p} - 1$ is divisible by l^s.

Now, let F be the rational field. We shall prove the following statement:

If K is cyclic of degree 2^s over F, $s \geqq 3$, then the prime 2 either is completely ramified in K or has at least two distinct prime factors.

This furnishes us our counter-example. For instance, the statement implies that there exists no cyclic extension of degree 8 over F, for which the local extension over the 2-adic field is unramified of the same degree.

To prove the statement, consider the quadratic field $\Omega = F(\sqrt{m})$ in K. Assume for the sake of argument that the statement is wrong. Then 2 remains prime in $F(\sqrt{m})$. It follows that m is of the form $8n + 5$. So, there exists a prime p which divides m and is not of the form $8n + 1$. Since p is ramified in Ω, we infer $p - 1$ must be divisible by $2^s \geqq 8$. This is a contradiction.

To the author's knowledge, there exist in the literature two proofs of Grunwald's Theorem, namely, the original proof in *Crelle Journal* 169 and G. Whaples' proof in *Duke Mathematical Journal* 9. In the latter, the proof of a fundamental lemma reads in p. 467 as follows: "... since α_1 is an nth power at p_0, p_0 splits completely in $K_1 | k$ ... both p_0, p_1 split completely in K_2," where $K_2 = k(\alpha_2^{1/n})$. This was a mistake. For, the irreducible equation of $\alpha_2^{1/n}$ over k must be fixed before the symbol $\alpha_2^{1/n}$ has any meaning and can be adjoined to k, because $x^n - \alpha_2 = 0$ is in general reducible. But then there is no reason why the nth roots of α_2 at p_0 and p_1 should satisfy the equation we choose. Of course, we may choose as our equation the one which the nth root of α_2 at say p_0 satisfies. Still, we cannot assure that *both* p_0 and p_1 have split factors in K_2, not to say that they split completely in the same.

Now, let us examine the original proof. It was based on *Hilfssatz* 2 in *Mathematische Annalen* 107. The statement proved by induction on pp. 152–3 in

part B of the proof of the *Hilfssatz* was ambiguous. We may interpret it in two different ways (we do it only for $l = 2$):

(A) Let ϵ_i, ρ_j, π_k be units, ideal powers, local primes in $k(i)$. Then

$$\prod \epsilon_i^{x_i} \rho_j^{y_j} \pi_k^{z_k} i^z \underset{(\overline{2^s})}{=} 1$$

in K_{s+2} implies $x_i \equiv y_j \equiv z_k \equiv 0 \pmod{2^s}$.

(B) Let ϵ_i, ρ_j, π_k be things in k (which does not contain i). Then

$$\prod \epsilon_i^{x_i} \rho_j^{y_j} \pi_k^{z_k} (-1)^z \underset{(\overline{2^s})}{=} 1$$

in K_{s+2} and not all of x_i, y_j, $z_k \equiv 0 \pmod{2^s}$ imply that there exists in $k(i)$ some relation

$$(1) \qquad \prod \epsilon_i^{x_i} \rho_j^{y_j} \pi_k^{z_k} i^z \underset{(\overline{2})}{=} 1,$$

where not all of x_i, y_j, $z_k \equiv 0 \pmod 2$.

The first interpretation provides us no knowledge at all about the things in k. So, we may disregard it. If we interpret it as in (B), we must prove later that relation (1) implies $x_i \equiv y_j \equiv z_k \equiv 0 \pmod 2$, not merely that

$$\prod \epsilon_i^{x_i} \rho_j^{y_j} \pi_k^{z_k} (-1)^z \underset{(\overline{2})}{=} 1$$

in $k(i)$ implies $x_i \equiv y_j \equiv z_k \equiv 0 \pmod 2$. But the former statement is false. Let k be the rational field. Then the things ϵ_i, ρ_j do not exist. Taking $z_k = z = 1$, $\pi_k = 2$, we have

$$2 \cdot i = (1 + i)^2 \underset{(\overline{2})}{=} 1.$$

Princeton University

▶1948 年，王湘浩在《数学年刊》上发表的著名学术论文《关于格伦瓦尔德定理的反例》

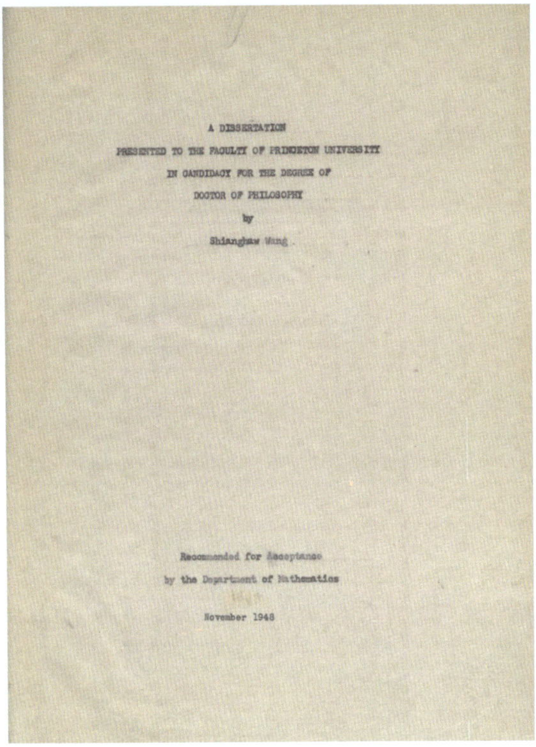

A DISSERTATION
PRESENTED TO THE FACULTY OF PRINCETON UNIVERSITY
IN CANDIDACY FOR THE DEGREE OF
DOCTOR OF PHILOSOPHY
by
Shianghaw Wang

Recommended for Acceptance
by the Department of Mathematics
November 1948

▶ 1948 年 11 月，王湘浩
提交给普林斯顿大学的博士论文（首页）

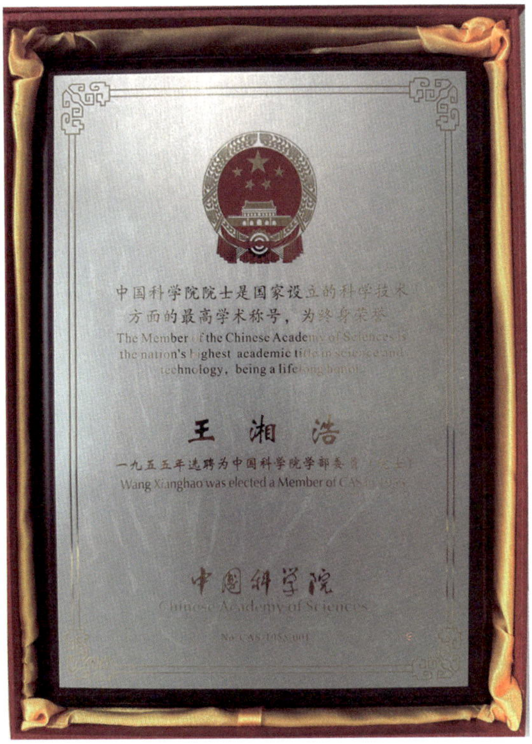

▶ 1955 年，王湘浩的中国科学院学部委员
（院士）聘书

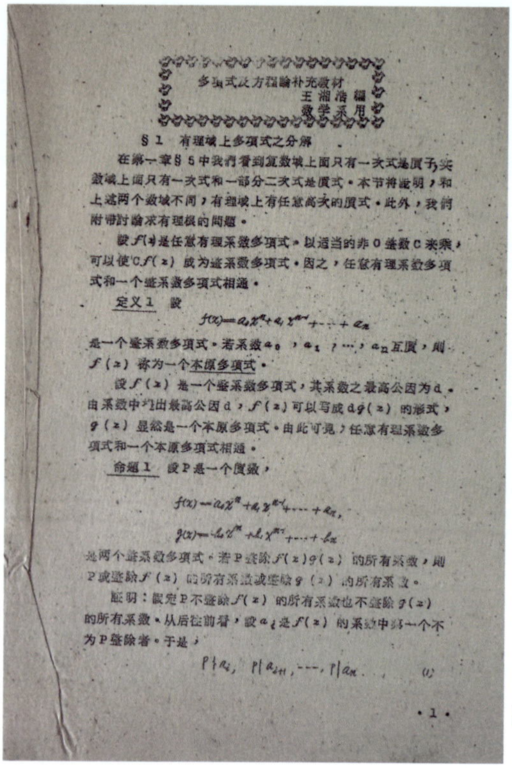
多項式及方程論补充教材
王湘浩编
数学系用

§1 有理域上多項式之分解

在第一章§5中我们看到复数域上面只有一次式是质子，实数域上面只有一次式和一部分二次式是质式。本节将説明，和上述两个数域不同，有理域上有任意高次的质式。此外，我们附带討論求有理根的問題。

設$f(x)$是任意有理系数多項式。以适当的非0整数C来乘，可以使$Cf(x)$成为整系数多項式。因之，任意有理系数多項式和一个整系数多項式相通。

定义1 設

$$f(x)=a_0x^n+a_1x^{n-1}+\cdots+a_n$$

是一个整系数多項式。若系数$a_0, a_1, \cdots, a_n$互质，則$f(x)$称为一个本原多項式。

設$f(x)$是一个整系数多項式，其系数之最高公因为d。由系数中提出最高公因d，$f(x)$可以写成$dg(x)$的形式，$g(x)$显然是一个本原多項式。由此可見，任意有理系数多項式和一个本原多項式相通。

命題1 設P是一个质数，

$$f(x)=a_0x^n+a_1x^{n-1}+\cdots+a_n,$$

$$g(x)=b_0x^m+b_1x^{m-1}+\cdots+b_m$$

是两个整系数多項式。若P整除$f(x)g(x)$的所有系数，則P或整除$f(x)$的所有系数或整除$g(x)$的所有系数。

証明：假定P不整除$f(x)$的所有系数也不整除$g(x)$的所有系数。从后往前看，設a_i是$f(x)$的系数中第一个不为P整除者。于是，

$$P\nmid a_i,\ P\mid a_{i+1},\cdots,P\mid a_n. \qquad (1)$$

·1·

▶ 20世纪60年代，王湘浩院士为数学系编写的补充教材（首页）

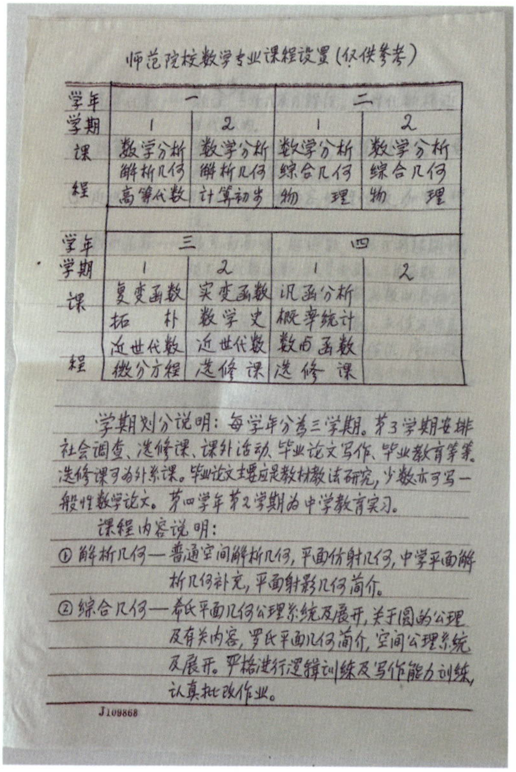
师范院校数学专业课程设置（仅供参考）

学年	一		二	
学期	1	2	1	2
课程	数学分析 解析几何 高等代数	数学分析 解析几何 计算初步	数学分析 综合几何 物理	数学分析 综合几何 物理

学年	三		四	
学期	1	2	1	2
课程	复变函数 拓扑 近世代数 微分方程	实变函数 数学史 近世代数 选修课	泛函分析 概率统计 数与函数 选修课	

学期划分说明：每学年分为三学期。第3学期安排社会调查、选修课、课外活动、毕业论文写作、毕业教育等等。选修课可为外系课。毕业论文主要应是教材教法研究，少数亦可写一般性数学论文。第四学年第2学期为中学教育实习。

课程内容说明：

① 解析几何——普通空间解析几何，平面仿射几何，中学平面解析几何补充，平面射影几何简介。

② 综合几何——希氏平面几何公理系统及展开，关于圆的公理及有关内容，罗氏平面几何简介，空间公理系统及展开。严格进行逻辑训练及写作能力训练，认真批改作业。

J109868

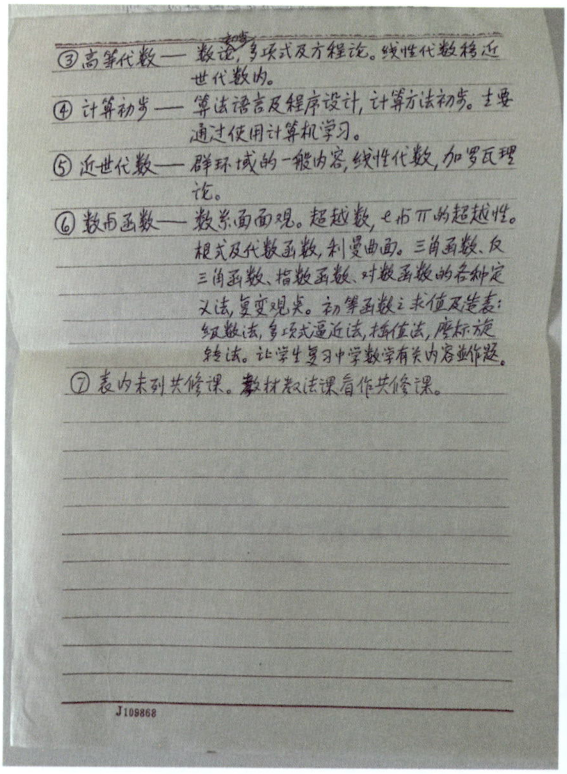
③ 高等代数——数论初步，多项式及方程论。线性代数移近世代数内。

④ 计算初步——算法语言及程序设计，计算方法初步。主要通过使用计算机学习。

⑤ 近世代数——群环域的一般内容，线性代数，加罗瓦理论。

⑥ 数与函数——数系面面观。超越数，e与π的超越性。根式及代数函数，黎曼曲面。三角函数、反三角函数、指数函数、对数函数的各种定义法，复变观点。初等函数之求值及造表：级数法，多项式逼近法，插值法，座标旋转法。让学生复习中学数学有关内容并作题。

⑦ 表内未列共修课。教材教法课看作共修课。

J109868

▶ 王湘浩院士对师范院校数学专业课程设置的建议手稿

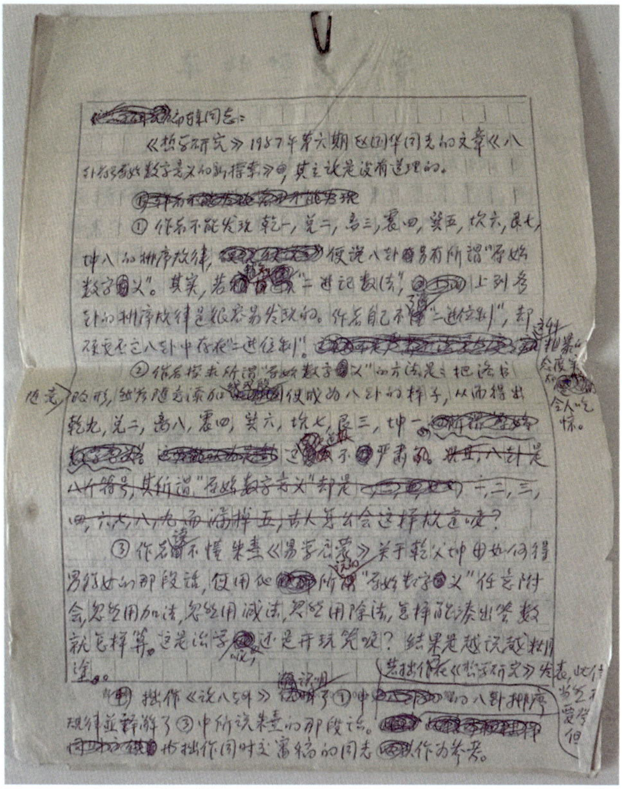

……同志：

《哲学研究》1987年第六期赵国华同志的文章《八卦符号原始数字意义的新探索》，其主要论点是没有道理的。

① 作者不能发现乾一、兑二、离三、震四、巽五、坎六、艮七、坤八的排序规律，……假说八卦另有所谓"原始数字意义"。其实，若用"二进记数法"，上列各卦的排序规律是很容易发现的。作者自己不懂"二进位制"，却硬说古之八卦中存在"二进位制"！

② 作者探求所谓"原始数字意义"的方法是：把诸卦的形，然后随意添加……使成为八卦的样子，从而得出乾九、兑二、离八、震四、巽六、坎七、艮三、坤一，……这不严肃。……

③ 作者不懂朱熹《易学启蒙》关于乾父坤母如何得男得女的那段话，任用他所谓"原始数字意义"任意附会，忽然用加法，忽然用减法，忽然用除法，怎样能凑出答数就怎样算。这是治学还是开玩笑呢？结果是越说越糊涂。

④ 拙作《说八卦》说明了①中……的八卦排序规律并解释了③中所说朱熹的那段话。……附拙作，同时交审稿的同志……作为参考。

▶ 王湘浩院士对《八卦符号原始数字意义的新探索》的意见手稿

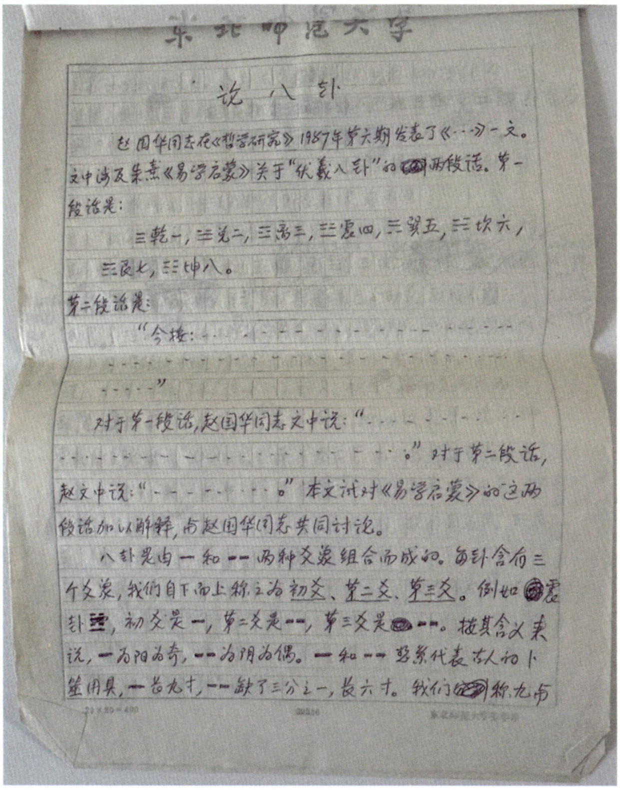

东北师范大学

说八卦

赵国华同志在《哲学研究》1987年第六期发表了《……》一文。文中涉及朱熹《易学启蒙》关于"伏羲八卦"的两段话。第一段话是：

☰乾一，☱兑二，☲离三，☳震四，☴巽五，☵坎六，☶艮七，☷坤八。

第二段话是：

"今按：……"

对于第一段话，赵国华同志文中说："……。"对于第二段话，赵文中说："……。"本文试对《易学启蒙》的这两段话加以解释，与赵国华同志共同讨论。

八卦是由⚊和⚋两种爻象组合而成的。每卦含有三个爻象，我们自下而上称之为初爻、第二爻、第三爻。例如震卦☳，初爻是⚊，第二爻是⚋，第三爻是⚋。按其含义来说，⚊为阳为奇，⚋为阴为偶。⚊和⚋好象代表古人的卜筮用具，⚊长九寸，⚋缺了三分之一，长六寸。我们称九寸

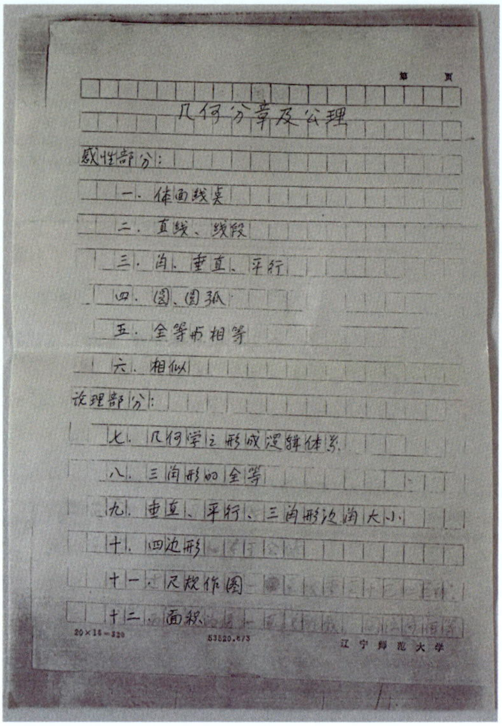
第 页

几何分章及公理

感性部分:

一. 体面线点

二. 直线、线段

三. 角、垂直、平行

四. 圆、圆弧

五. 全等和相等

六. 相似

说理部分:

七. 几何学之形成逻辑体系

八. 三角形的全等

九. 垂直、平行、三角形边角大小

十. 四边形

十一. 尺规作图

十二. 面积

20×16=320 53520.873 辽宁师范大学

王湘浩院士对中学几何内容的建议手稿

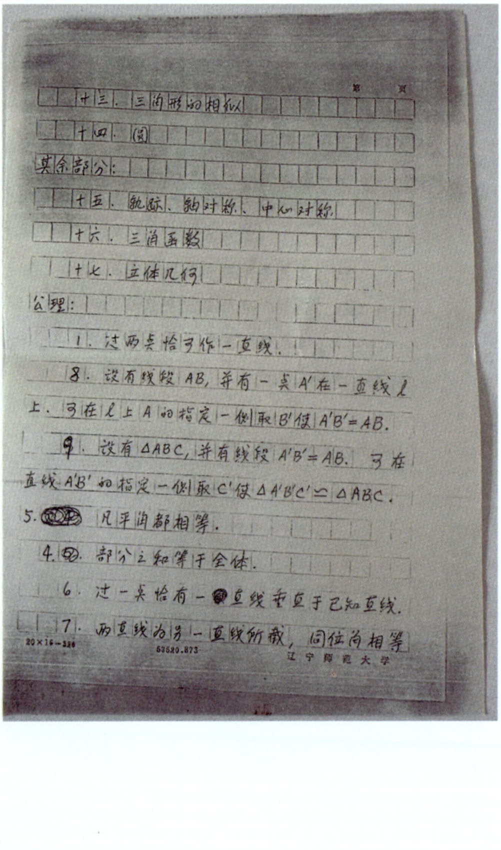
第 页

十三. 三角形的相似

十四. 圆

其余部分:

十五. 轨迹、轴对称、中心对称

十六. 三角函数

十七. 立体几何

公理:

1. 过两点恰可作一直线.

8. 设有线段 AB, 并有一点 A' 在一直线 l 上. 可在 l 上 A' 的指定一侧取 B' 使 $A'B'=AB$.

9. 设有 $\triangle ABC$, 并有线段 $A'B'=AB$. 可在直线 $A'B'$ 的指定一侧取 C' 使 $\triangle A'B'C' \cong \triangle ABC$.

5. 凡平角都相等.

4. 部分之和等于全体.

6. 过一点恰有一直线垂直于已知直线.

7. 两直线为另一直线所截, 同位角相等

20×16=320 53520.873 辽宁师范大学

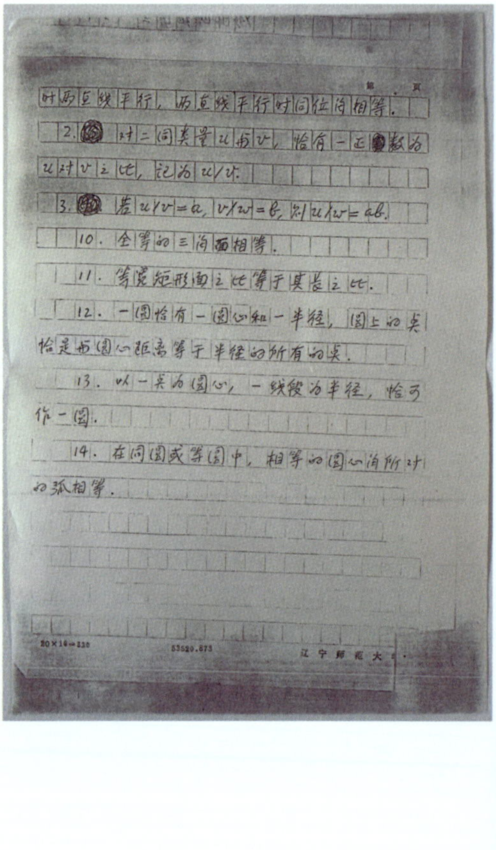
第 页

时两直线平行, 两直线平行时同位角相等.

2. 对二同类量 u 和 v, 恰有一正数为 u 对 v 之比, 记为 u/v.

3. 若 $u/v=a$, $v/w=b$, 则 $u/w=ab$.

10. 全等的三角面相等.

11. 等宽矩形面之比等于其长之比.

12. 一圆恰有一圆心和一半径, 圆上的点恰是和圆心距离等于半径的所有的点.

13. 以一点为圆心, 一线段为半径, 恰可作一圆.

14. 在同圆或等圆中, 相等的圆心角所对的弧相等.

20×16=320 53520.873 辽宁师范大学

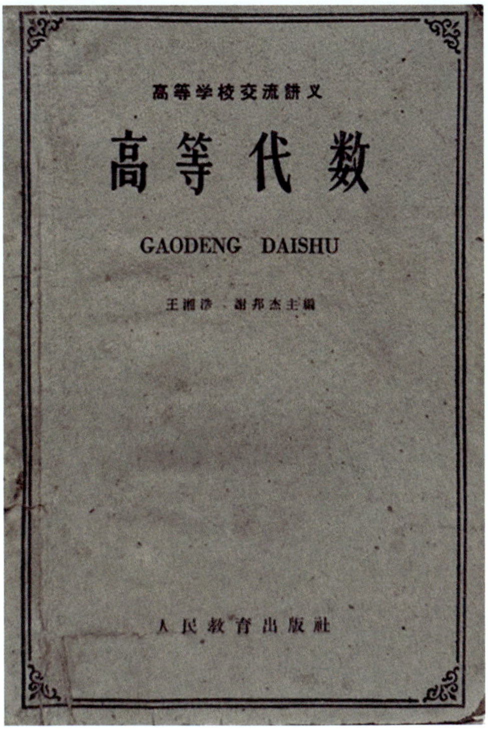

▶1961 年，王湘浩院士与谢邦杰合作编写的《高等代数》教材

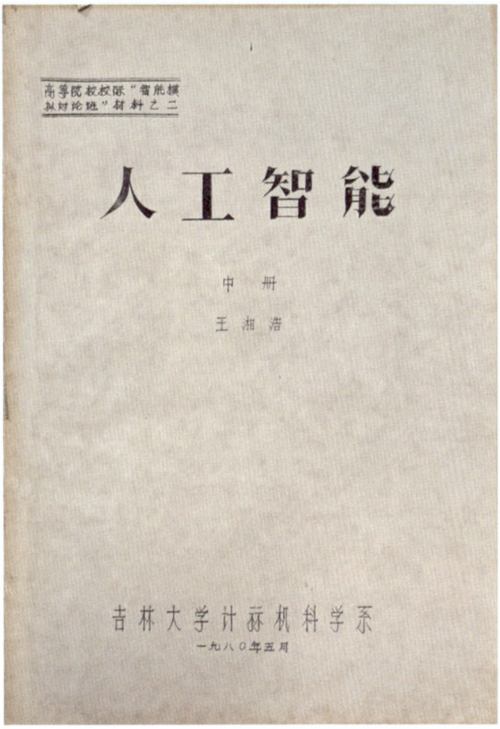

▶1980 年 5 月，王湘浩院士为高等院校校际“智能模拟讨论班”编写的教材

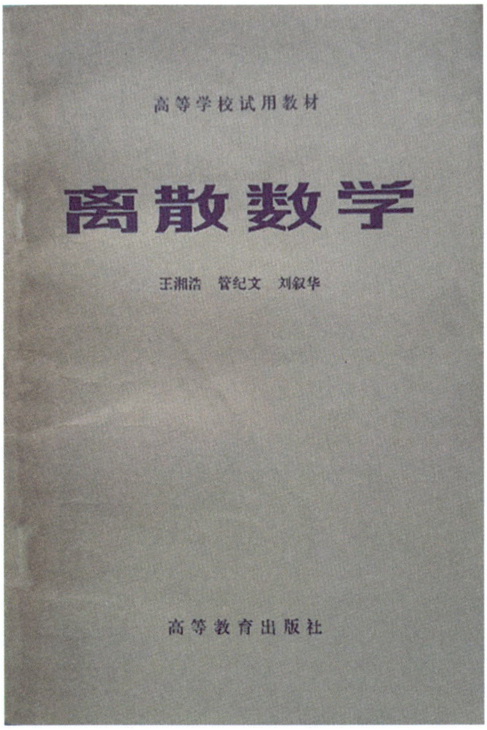

▶1983 年，王湘浩院士与管纪文、刘叙华合作编写的《离散数学》教材

▶1985 年，王湘浩院士与杨荫华合作编写的《线性代数》教材

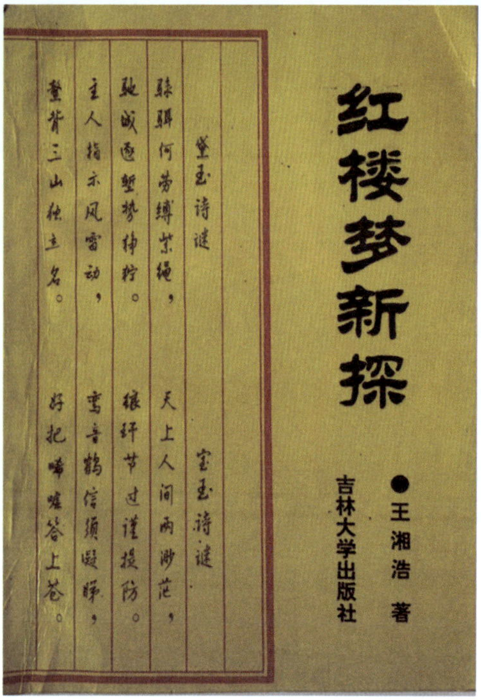

▼1993年，王湘浩院士所著的《红楼梦新探》

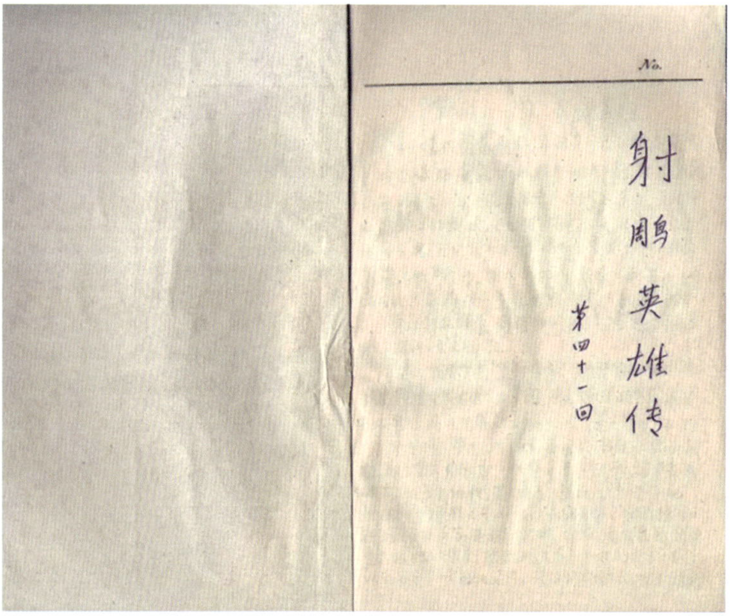

▶王湘浩先生热爱中国文学，
近代文学中，
他特别喜欢金庸先生的作品，
读过《射雕英雄传》后，
王先生觉得书中人物结局不够
于是动笔续写了第四十一回
（原著为四十回）

▶ 1990 年 12 月，王湘浩院士因从事高校科技工作四十年受到国家教委表彰

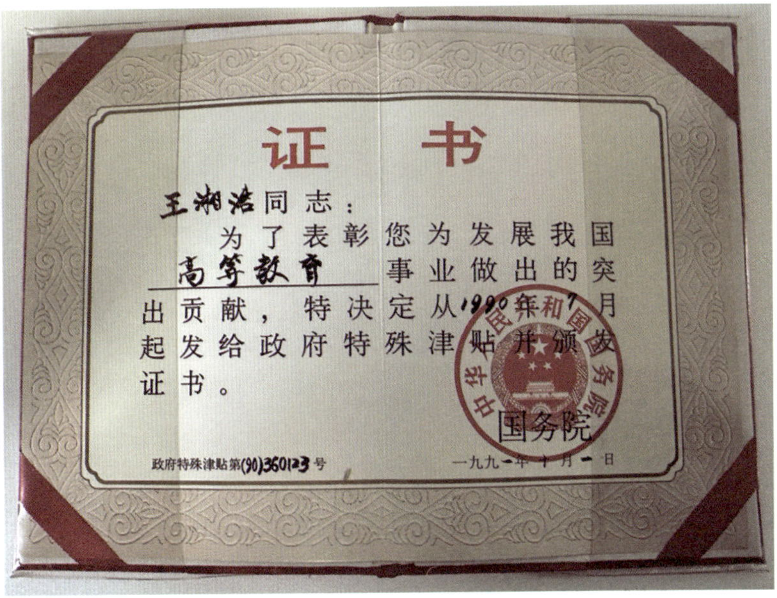

▶ 1991 年，王湘浩院士获得国务院政府特殊津贴

先生向北　精神永存

张　希

中国科学院院士

在王湘浩老师诞辰 110 周年之际，吉大师生校友策划编辑出版《王湘浩文集》，内含王老师生前发表的学术类、科普类文章，各类评论、信函，以及诗歌、红学研究等文学创作。“遍历”其中，那些深邃的理论、精妙的证明和独到的观点跃然纸上，既是对王湘浩老师的深情致敬和怀念，也为后来者在科学之路上勇攀高峰，在教育之路上见贤思齐提供了“时空坐标”和“学习语料”。

上世纪 50 年代初，一大批学术大师，一路向北，云集长春。王湘浩老师也是众多前辈先贤中杰出的一位。王老师曾在美国普林斯顿大学攻读博士学位，青年时期就展现出很高的学术品位。他的博士论文很短，博士研究期间的代表作只有一页半纸，却发现并修正了格伦瓦尔德定理的错误，之后又找到了格伦瓦尔德定理的正确叙述和证明，被称作“格伦瓦尔德－王”定理，铸就了基础数学的经典。在上世纪 80 年代末的一次学校学位工作会上，王老师曾批评当时的部分博士论文像“手册”，记录下做了什么，不像论文，没有论起来，创新性不够强。可见学问大小并不取决于论文的厚度和论文的数量，而在于研究成果的价值，有时“少即是多”。这一点既要让学生们耳濡目染、了然于胸，也有必要请教师们身体力行、率先垂范。

王老师 1949 年博士毕业后，立刻返回百废待兴的祖国，任教北京大学数学系，彰显高尚的家国情怀。1952 年全国高校院系调整，他来到东北人民大学（如今的吉林大学）开创了数学系，并担任首任系主任。建系之初，王湘浩老师在“深耕教师已有学术研究”和“更加重视国家急需方向”之间，毅然选择了后者，在保持代数、分析教学和科研强项的同时，力主发展微分方程和计算数学两个专业，既满足国家的急需，又强调了数学系理论和应用并重的特点，成为吉大较早开展“有组织科研”的范例。

王湘浩老师高度重视师资队伍建设，以“海纳百川”的胸怀和“三顾茅庐”的耐心，四方物色学术带头人，先后从东北工学院和中国科学院数学研究所请来了王柔怀老师和孙以丰老师，分别建立微分方程方向和拓扑学方向，既为数学系特色发展、跻身一流奠定了人才基础，也成为吉大请才敬才的一段佳话。时至今日，“与其引人才，不如请人才”早已深入人心，成为学校人才强校的行动自觉。每每听到数学界的前辈们回忆王柔怀老师、孙以丰老师的卓越成就，更感当年时代背景下，王湘浩老师的战略眼光、博大胸怀和“请”才分量。

严谨务实、锐意创新、敢为人先同样是王湘浩老师的鲜明品格。王老师早在1958年就开展了电子计算机和控制论方面的研究，1976年创建了吉大计算机系并担任首任系主任，1977年在国内最早倡导开展人工智能研究，1980年受教育部委托，在吉大举办了全国性的人工智能讨论班，随后倡导成立了全国高校人工智能研究会，表现出敏锐的学术远见、超常的前瞻思维和大胆的开拓精神。

王湘浩老师在1977年9月指出：“可以预料，到本世纪末，计算机的应用将更为广泛……计算机将能够识别印刷的、甚至手写的文字，能够直接听取人的语言。人工智能的研究将有重大突破，计算机将能够进行某些类似思维的活动，将直接与人进行对话，回答人们提出的问题，协助人们做某些推理和判断。”王老师的“先见之明”令人钦佩。时至今日，人工智能技术深刻影响着人类社会未来发展，成为研究热点。一些科研人员抓住并解决热点领域的关键问题，取得里程碑式的成果。但是一味跟踪热点，可能最后热闹都是别人的，自己却什么也没有。此外，AI算法和强大算力的基础依然是架构、函数、优化等数学理论和方法，基础研究仍不可或缺、任重道远。

王湘浩老师热爱生活、多才多艺，他唱京剧，做诗填词，还研究红学。他曾以黄鹤乡笔名将研究成果发表于《红楼梦学刊》等期刊，出版《红楼梦新探》一书。该书外形上是以探佚学为主，但这是以科学精神从事文史哲方面的研著，研于求实，用数学思想和逻辑在《红楼梦》纷繁的结构关系中推导出他的见解和结论。看来，人文、艺术和科学在求真求美中是相通的。

传承是最好的纪念。吉林大学在数学、计算机、软件、人工智能等相关学院设立了王湘浩奖学金，在人工智能和网络空间安全专业开设了王湘浩班，发起了王湘浩人工智能杰出学者系列讲座，旨在缅怀先生、教育后人，并期待大家在追忆往昔基础上，不断反思和超越。王湘浩老师虽已离开我们32年，但他的精神与众多前辈先贤的精神一道，共同铸就成“听党召唤、至诚报国，扎根东北、兴教育才”的北上精神，激励一代代吉大人求实创新、励志图强，在建设教育强国、科技强国、人才强国的历史进程中不断展现新作为、作出新贡献。

王湘浩先生以严谨的治学精神和跨学科的前瞻视野，在数学与计算机科学领域，留下了不可磨灭的印记。他的学术遗产和教育理念，值得我们认真地学习、传承与发扬！

张景中 2025年4月10日

张景中，计算机科学家、数学家，中国科学院院士。

纯粹数学家们都把能在普林斯顿大学数学系编辑的《数学年刊》（*Annals of Mathematics*）上刊登一篇论文看成一生中无上的荣誉，原因就是它发表的都是质量上乘的创造性成果，是数学界最有分量的杂志之一，相当于英国的《自然》或美国的《科学》在自然科学领域的影响力。有人曾戏称，《数学年刊》上发表的一篇文章就可能成为著名大学数学系终身教职的敲门砖。中国的代数学家王湘浩（1915—1993）20世纪40年代末在美国普林斯顿大学跟随代数大师阿廷（E.Artin，1898—1962）读博士时，发现已经用了近二十年的Grunwald定理不对，并举出了反例，连专门讲授这个定理的导师都未能发现。王湘浩以这项结果写成了或许是历史上最短的数学博士论文。不久后他又找到了Grunwald定理的正确叙述和证明，其主要结果后来被称为Grunwald-Wang定理。他的这两项结果分别发表在1948年和1950年的《数学年刊》上。1955年，四十岁的王湘浩就成为中国科学院的院士（当时称为学部委员）。

汤涛，计算数学家，中国科学院院士、欧洲科学院院士、发展中国家科学院院士。

序

陆汝钤
中国科学院院士

今年5月5日是王湘浩院士诞辰110周年的日子，很高兴在这个日子即将到来时见到了《王湘浩文集》的预印本，触动了我从认识王先生开始的无尽回忆。

我首次见到王湘浩先生就是在他举办的1979年计算机科学研讨会上。这个会名为计算机科学研讨会，实际上人工智能是一个主要的方面。我们中科院数学所有两个人被邀请参加，就是吴文俊先生和我。在这次会议之前，吴先生在几何定理机械化证明方面已经有了很高的成就，所以王湘浩先生请吴先生参会是很自然的。而我却是一个普通的年轻人，而且还从未接触过人工智能，受邀参加这个会议是我的莫大光荣。在会议气氛的鼓舞下，特别是学习了王先生和他的学生所做的机器证明研究工作之后，我才逐渐走进人工智能领域。为此，我一直非常感谢王先生。

王先生不仅在会上，而且据报道在两年之前的1977年就明确提出要推动人工智能研究，在当时提出这个主张意义非常大。国际上人工智能这个名词第一次出现是在1956年美国达特茅斯学院的一个科学家闭门会议上，从那时以来到1977年，已经过去了长长的21年，应该说这是中国人民，特别是中国学者苦等人工智能概念的21年，才等来了王湘浩先生这位"吹哨人"。是王先生的一声哨响，推动了人工智能在中国大地上迅速传遍人心。所以仅仅从那个时间点来看，也能看出王先生能够率先在当时科技还相对落后的中国打出人工智能的旗子，是很不容易的。

同时，王湘浩先生做了大量实际工作。1980年，他在吉林大学举办了国内第一次全国性的人工智能学术会议——首届全国高校系统人工智能研讨会。会后王先生又建立了国内最早的人工智能组织：全国高校系统人工智能研究会并亲自担任会长，以此大力推动国内人工智能研究，培养中国人工智能人才。此后全国形势发展很快，首届全国人工智能学术大会召开，首届中国人工智能学会也成立了。

我记得我们在当时还做了一件事。鉴于已经有联合人工智能各个研究方向的国际会议IJCAI（International Joint Conference on Artificial Intelligence），我们和国内一些学者，包括王湘浩先生的弟子刘叙华老师，在国内发起组织CJCAI会议，就是把

IJCAI 的 I（International）改成 C（China），以便把国内的人工智能会议，包括王湘浩先生的高校人工智能研讨会联合起来。可惜的是没有能坚持下来，只开了大概一两届就停止了。

至于培养人工智能人才，王湘浩先生的贡献可大了。他为国家，也为吉林大学，培养了一支成绩斐然的科研和教师队伍。就我比较熟悉的早期领军人物就包括管纪文、刘叙华、刘大有、孙吉贵、欧阳丹彤等教授，后来的骨干肯定更多。他们都传承了王湘浩先生的科学人生理念。

谈及王湘浩先生的学术贡献，首先当然是数学上的贡献。先生主攻是代数学。他在年轻时就因在博士论文中指出了格伦威尔定理的错误而扬名国际代数学界。其背景是德国数学家用格伦威尔定理“证明”了迪克逊猜想。迪克逊猜想很重要，这个猜想的证明能彻底阐明有理单纯代数的结构。它的证明不成立就像一颗擎天大树塌了一样，引起代数界极大震动。幸好先生不仅是一个错误理论的破坏者，还是一个正确理论的建设者。他在博士论文中反倒是真正证明了格伦威尔定理和它的推广。此后，先生又继续在此基础上扩充和推广成果。考虑到我们主要关注先生根据国家需要，从 1958 年以后转行到计算机科学方向的成就，此处就不再深入探讨先生在数学方面的功绩了。感兴趣的读者可以参阅本书收录的先生“数学人生”一文。

除了数学之外，王湘浩先生和他的团队在计算机科学方面也做了许多工作。我比较了解的是在人工智能领域，但了解得可能不够全面。粗粗地说，主要是在两个方面。一是机器定理证明，二是专家系统开发。在机器定理证明方面王先生和他的团队主要研究基于归结法的逻辑定理证明。有关这方面的主要成果被归纳在专著《定理机器证明》（刘叙华、姜云飞，科学出版社，1987）和《基于归结方法的自动推理》（刘叙华，科学出版社，1994）中。核心贡献者除王先生本人外，还包括刘叙华、孙吉贵、欧阳丹彤等王先生弟子。主要成果包括：1. 经典逻辑基于归结方法的定理机器证明；2. 布尔算子模糊逻辑及其归结形式推理；3. 模态逻辑的归结推理方法；4. 非单调逻辑的形式推理。其中 2，3，4 是非经典逻辑。此外，还包括 1982 年王先生和刘叙华老师提出的广义归结方法（非子句归结法）。

在专家系统开发方面，根据现有资料可知，王湘浩先生指导的学生付凝开发了根据目前已有信息可能是中国最早的专家系统之一：滋养叶细胞疾病专家咨询系统，发表在沈阳自动化所第二届人工智能和模式识别会议文集中。该系统经过严格的实验室专家双盲数据测试，准确率在 90% 以上。此外，管纪文、刘大有、付凝还在 1983 年研制成功“朱仁康中医皮科诊治专家系统”，文章发表在《计算机学报》上。徐鹰做过基于证据理论的专家系统，至少在国际会议发过两篇文章。更为成功的还有王先生指导刘大有等 1992 年开发的 PRES 油气资源评价专家系统。该系统在国际上首次以三维状态进行了油气运聚半定量评价，获得了国家科技进步三等奖和教育部科技进步二等奖。当然这些不是全部。刘大有老师等后来还开发了农业专家系统。在此过程中开发、应用和推进了多种专家系统技术。欧阳丹彤则开发了智能诊断与智能规划专家系统，上述工作有部分在 2008 年的“*AI in China: A Survey*”（*IEEE Intelligent Systems*, 2008, 23(6)）和中科院高技术发展报告“人工智能发展近况”中有介绍。王湘浩先生对中国人工智能的历史贡献是永垂千秋的。

为了更好地体现老科学家们对中国人工智能的贡献，这里我还想提一下中国的人工智能三老。就是吉林大学的王湘浩先生、中国科学院数学研究所的吴文俊先生和浙江大学的何志均先生。这三位老先生都是中国人工智能发展初期的主要开拓者。其中吴文俊先生的主要研究领域是拓扑学，但是他在 1977 年开辟了一个新的方向，在当年发表的论文中给出了初等几何和微分几何的数学定理机械化证明，其理论基础是代数几何的簇理论。国内外数学家很快跟进，并把他的理论应用于机器人操作演算。为此在国际上获得高度评价并于 1997 年获得了“Herbrand 自动推理杰出成就奖”。何志均先生是浙江大学教授，他的主要研究方向是电子信息工程，但是何老先生对人工智能独有偏爱。他组织学生开展了具有多媒体特色，并结合计算机图形学、计算机辅助设计的人工智能研究，指导学生开发出了花布图案自动设计软件、计算机动画、丝绸印染 CAD/CAM 系统。并继王湘浩先生之后，在 1981 年组织举办了又一次全国性的人工智能学术会议，在浙江大学建立了一支强大的人工智能研究队伍。这三位老先生对中国人工智能的贡献是我们不能忘记的。

最后回到王湘浩先生，就说两点。先生首先是一个爱国者，1949 年从美国博士毕业回到北京大学数学系做教授，就已经有很好的发展，但是后来当国家需要的时候，他毫不犹豫地到东北去，到当时比较艰苦的地方去，并且在那里做出了丰功伟业。其次，他除了做科学研究以外，在文学和艺术方面还有自己非常独特的研究甚至才艺。比如他研究《红楼梦》，而且在分析其中人物性格，以及他们的性格如何影响了贾府败落以后他们的命运的基础上，对于八十回以后的《红楼梦》失落部分内容做出具有自己独到见解的推测。他的专著《红楼梦新探》出版以后被红学大师周汝昌赞为“君书动我心”。连专家都“动心”了，可见先生红学研究的独到之处。但先生的多才还不限于此。许多人在回忆录中称赞先生在京剧、桥牌、诗词、楹联等方面都有很深的造诣，不乏亲自登台献艺的佳话。本书收集的先生照片中有一张是王先生羽扇纶巾，饰演孔明坐城楼、观山景的精彩场景。可见先生不仅是一个难得的人才，而且还是天才、全才，甚至怪才。不得不让人心悦诚服。

2025 年 4 月 13 日

回忆王湘浩老师*

伍卓群

王湘浩老师是我国杰出的数学家、计算机科学家和教育家，中国科学院学部委员（院士）。他曾任第三届、第五届至第八届全国人大代表，民盟中央委员，吉林省民盟副主任委员，国务院学位委员会计算机学科评议组组长，中国计算机学会副理事长以及其他许多学术兼职和社会兼职。在吉林大学，他是首批一级教授，长期担任数学系和计算机科学系的系主任，并曾任副校长。他为我国的科技教育事业，为吉林大学的发展和建设，竭忠尽智，奉献终生，创造了光辉的业绩，作出了彪炳史册的贡献。

王湘浩老师已经离开我们二十七年了。每当回想起他在代数学和计算机科学领域作出的重要贡献，缅怀他在吉林大学创建数学系和计算机系的特殊功勋时，往事一幕一幕地在我头脑中闪现。他在教学和科研上的深邃眼光，他的崇高思想和品格，他的严谨求实而又富于创造性的治学精神，他的音容笑貌，是那么清晰，那么亲切地嵌在我们这些老学生的记忆中。

王湘浩老师早年毕业于北京大学数学系，后曾在西南联合大学任教。1946 年夏，他被选送到美国普林斯顿大学深造，在著名代数学家阿廷的指导下，不到三年就以一篇优异的代数学论文获得博士学位。王老师在代数学研究中的一大贡献就是纠正了类域论中一个重要定理——格伦瓦尔定理的错误，并给出了该定理成立的充要条件，从而重新证明了有理单纯代数理论中著名的狄克逊猜想，这一被前人认为已获得证明的猜想由于格伦瓦尔定理发现有错误而重新变成疑案。经王老师修正后的重要结果世称格伦瓦尔一王定理。王老师的工作挽救了有理单纯代数理论的这一危机，震动了当时的国际代数学界。定理证明的传奇过程也一直为数学史家们所津津乐道，半个世纪后

* 本文系伍卓群先生为纪念王湘浩先生诞辰 105 周年《红楼梦新探》再版时所作的代序。

还成为2002年在北京举行的国际数学家大会的一个研讨会的主题。由于王老师在代数学方面的突出贡献，他于1955年成为首批中国科学院学部委员，当时在数学界是最年轻的一位。王老师在学术上勇于创新，早在1958年，他就洞察到电子计算机这一尖端技术对我国发展的重要性，毅然决定逐步离开自己熟悉并已获得重大成就的领域，带领一部分青年教师和学生在吉林大学开启计算机科学的研究。1977年，王湘浩老师在国内最早提出开展人工智能的研究。后来他受教育部委托，首次举办了全国人工智能讨论班，并随之成立了全国人工智能研究会。他在国际上人工智能正处在低潮时，在国内率先倡导人工智能的研究，充分体现了他在学术上的远见卓识。王老师作为中国人工智能研究的奠基者，也成为国内学术界的共识。

1952年秋，全国高校实行院系调整，当时在北京大学任教的王湘浩老师，响应国家的号召，来到当时生活还比较艰苦的长春，在极其困难的条件下为东北人民大学（吉林大学的前身）创建了数学系。1976年，王老师又带领数学系与物理系的一部分人，在吉林大学创办了计算机科学系。王老师在担任两系的系主任期间，以强烈的事业心和具有远见卓识的办学思想，带领全系教师，致力于全系发展和各项基础建设。这里我仅以数学系的创建为例来回忆一下王老师的办学思想和实践。

王湘浩老师既重视教学，又重视科研，一贯实施“两个中心”的办学思想，但始终坚持把教学放在首位。他对教学的要求非常严格。既严格要求学生，也严格要求教师。不管谁在教学上出了问题都不轻易放过，对学术水平高的教师也不例外。对青年教师则要求他们首先过好教学关，进而努力争取教学与科研双优。他自己以身作则，担任繁重的教学任务，并认真对待教学的每一个环节。作为系主任，他既抓保证课堂教学效果的教学方法，更抓能从根本上提高教学质量的教师水平。他把编写高质量的教材摆在十分重要的位置并身体力行。在他的领导下，经过不太长的时间就写出了一批在国内有影响的教材，其中有多部教材后来获得国家和国家教委的优秀教材奖。与此同时，王老师十分重视科研，在教学走上正规的前提下多次适时地推动科研。重视科研这一点，在今天看来是尽人皆知的常识，但在那个年代是非常难得的。

王湘浩老师从建系一开始就十分重视学科建设和教师队伍建设。数学系建系初期，全系只有属于基础数学的代数和分析这两个学科方向。20世纪50年代中期，他抓住两个难得的机会，先后从东北工学院和中国科学院数学研究所引进王柔怀老师和孙以丰老师，分别建立起微分方程和拓扑学这两个新的学科方向。稍后王老师又抓住苏联专家来我校的机遇，配备骨干力量发展计算数学专业，使其成为全国最早创办的计算数学专业之一。经过他卓有成效的运筹，不到十年时间，数学系就发展成国内数学方面学科门类比较齐全的一个系。由于各种原因，那时的吉林大学很难从外地引进优秀人才。他只好把主要注意力放在优秀毕业生的选留上，想方设法把他们培养成各个学科的骨干和新的学术带头人。与此同时，对原有教师队伍不断进行调整，有进有出，有增有减，以优化结构，提高整体水平。

以王湘浩老师为首的老一辈，从建系之日起，就率先垂范，在系内树立勤奋、严谨、求实的学风；坚持艰苦奋斗，团结合作，反对门户之见；对教师的教学和学生的学习，一贯从严要求。作为系主任，他胸怀宽阔，眼光远大，又善于听取各种不同的意见，善于团结具有各种不同性格和不同特点的人一道共事，他思考和处理问题，任何时候都着眼于系的全局。他晚年和我闲谈时，不止一次笑着对我说过："我总是把最优秀的年轻人才投放到最需要发展的学科，而不是拢在自己的门下。"所有这些使得作为系主任的王湘浩老师赢得了全系师生的一致敬仰和爱戴，他的这些表现对于优良系风和学风的形成起着极其重要的作用。

正是在这一系列正确的办学思想指导下，经过以王湘浩老师为核心的全系人员的努力，数学和计算机两个起点原本不高的系，迅速崛起，很快就发展成为全国数学和计算机科学这两个学科的重要教学科研基地之一。1981 年，我国开始建立学位制，吉林大学有 12 个专业被评为首批博士点，数学系的基础数学和计算数学，计算机科学系的计算机软件都在其列。1987 年，吉林大学有 7 个博士点被进一步评定为全国的首批重点学科，上述 3 个专业又全部入选，从比例上占全校的近半数。王老师在吉林大学这样的全国重点大学先后担任数学和计算机两系的首任系主任，创办出两个高水平的一级学科，在两个一级学科中能有三个专业入选首批全国重点学科，这在全国也是不多见的。

我年已九旬，当回顾自己的人生旅程时，有几位老师对我的事业取向和学术发展有着重要影响，王湘浩老师就是其中必须提及的一位。我的大学前两年是在东北工学院念的，那里的数学系水平不高。1952 年 8 月，全国实行院系调整，我们全班被调整到东北人民大学数学系。该系虽是新建，但有从清华和北大等校调来的水平较高的教师。系主任王湘浩老师亲自给我们讲授几何基础和域论等课。他水平很高，思想深刻，却又平易近人，朴实无华。我毕业后，他决定把我留下，并安排我立即担任物理系数学物理方法的主讲，学生就是我的下两班同学。我们上讲台之前，都经过了认真的准备，并通过试讲等方式接受有经验教师的指导，因此教学效果从一开始就都比较好。记得我试讲那一次，系里大部分教师都坐在台下，我紧张地讲完后，慈祥的王湘浩老师说了一句："我看还可以。"而严厉的江泽坚老师则说："我看还不行。"我理解，说"我看还可以"是对我的一种鼓励，而说"我看还不行"则体现了教学上应有的严格要求。王主任（那时我们都这样称呼）从全系的学科布局出发，决定让我跟随王柔怀老师搞常微分方程，以后又根据需要安排我跟随王柔怀老师转攻偏微分方程。他先是任命我为教学秘书，后又任命我为系主任助理，协助他管系里的教学。"文革"前夕我被内定为副系主任，只是由于"文革"的到来没有来得及公布。他转到计算机系后，我们的接触就少得多了，但这丝毫不影响我对他的崇敬，他在我的心目中始终是一位令人仰慕的、德高望重的忠厚长者。

1993 年，我和他同被选为第八届全国人大代表。他表示希望会议期间与我住在一个屋，我当然非常高兴。那一次的人代会从 3 月 15 日到 4 月 1 日共开了十多天。会议期间，我们几乎无话不谈：家事、校事、国事、天下事。他的身体很虚弱，步履维艰，总拄着手杖，出入都由我陪护。人代会后他还要参加国家教委召开的评学部委员（院士）的会。会前有几天空隙，他决定到他的长子王强家去住。4 月 2 日，王强来接他，我把他送上汽车。万万没有想到，那竟是永诀。他很快检查出患有肝癌。得到这个不幸的消息后，我（时任吉林大学校长）立即派人到大连（他在辽宁师范大学兼职，当时正住在大连）去安排治疗。然而病情迅速恶化。当我得到病危通知于 5 月 5 日晨赶到大连时，迎来的竟是令人心碎的噩耗：他老人家已于前一天晚上与世长辞。我急忙前去拜见师母，然后与他的子女商定，连夜将遗体运回长春。在长春举行了庄严肃穆的遗体告别。我怀着对老人无限崇敬的心情，写下了题为《难得京城同起止，谈今论古如昨夕》的吊唁文章，追忆在人代会期间与他相处的日日夜夜。与此同时，为表敬仰并寄托哀思，我还填了一首《满江红》。这首词后来王强把它刻在王老师的墓碑背面，成为永久的纪念。

王湘浩老师献身我国科学和教育事业，为数学和计算机科学在吉林大学的发展运筹帷幄、呕心沥血，作出了不可磨灭的历史贡献，他的名字成为吉林大学师生心中的丰碑。他的学术成就和建系功勋，他的崇高思想和品格，永远为我们所景仰。以他为核心的老一辈科学家在数学和计算机科学领域开创的事业，所树立的优良传统、正确的办学思想和纯正的系风学风，都应该得到继承和发扬光大。为纪念他作出的特殊贡献，吉林大学在数学楼和计算机楼都为他树立了塑像。吉林大学也在数学、计算机、软件等相关学院设立了王湘浩奖学金，我一直担任这个奖学金的管理委员会主任一职。

在追忆王湘浩老师事迹，感受他的人格魅力的同时，我们还会想起他的许多被传颂的趣事。王老师多才多艺，业余生活中兴趣广泛。他研究红楼梦，唱京戏，做诗填词，打桥牌，样样都有一些功夫。王老师闲时做过一些旧诗，填过一些词，写过一些楹联和散曲。其中有一首《沁园春》，将大量计算机语言，纳入这一词牌严格的格律框架，描绘当年调试计算机获得成功的情景。“文革”期间闲无事，他还曾用数学语言编写过一篇《微分三字经》。他爱唱京戏，1958 年还曾在一次全校文艺会演中登台表演，扮演的角色是诸葛亮。同样有趣的是，他还曾用数学语言编写过一篇戏词。他常邀我和其他同事在一起打桥牌，有时甚至通宵达旦。他对《红楼梦》很熟，也很有研究，晚年他曾将研究红楼梦的心得整理成文，以黄鹤乡的笔名发表在《红楼梦学刊》和《红楼梦杂志》等刊物上，颇引起红学界的重视。他对许多问题都有自己的见解，例如他通过论证，断言史湘云最后嫁给了贾宝玉。由他研究红楼梦的一些文章汇编而成的《红楼梦新探》一书，曾于 1993 年 12 月由吉林大学出版社出版，受到著名红学家周汝昌的高度评价。

今年 5 月 5 日是王湘浩老师诞辰 105 周年。为表达对王老师的纪念和敬意，吉林大学计算机科学与技术学院此次再版《红楼梦新探》一书，并汇集了王老师的若干照片。我十分感谢计算机学院为王老师所做的这些事情，应孟繁二同志（计算机学院党委书记）之邀，特意回忆整理了王老师的事迹，以及我同他的交往，是为代序。

于 2020 年 5 月

目录

科普文章……………………………

其他文章……………………………

附录……………………………………

学术文章

On the Mappings of Graphs in Closed Surfaces*

Introduction. It is difficult to characterise the graphs that cannot be mapped in a closed surface of genus① greater than zero by statements similar to what was given by Kuratowski to "skew curves". ② In other words, we have yet no method to find all the "irreducible"③ graphs that cannot be mapped in a closed surface of genus greater than zero.

Now, the aim of the present paper is to study all the mapping types for a given graph inclosed surfaces, and to find by a finite process the orientable and non-orientable closed surfaces of least genera, in which the graph can be mapped. The finite process, here arrived, is not very applicable in practice, nevertheless, we have obtained a theoretical solution of the problem.

The method followed in studying the types is presented in Ⅱ. Suppose the given graph G is mapped through a topological mapping M in a closed surface S, $G' = M(G)$. Cutting S along G', S falls into n pieces S_1, S_2, …, S_n. Each S_i is a bounded surface and therefore can be characterised by its orientability, genus g_i and number of holes r_i. Thus the n surfaces, into which S is broken up, may be described precisely by

(1)$\overleftrightarrow{S}_1(r_1, g_1)$, $\overleftrightarrow{S}_2(r_2, g_2)$, …, $\overleftrightarrow{S}_m(r_m, g_m)$, $S_{m+1}(r_{m+1}, g_{m+1})$, …, $S_n(r_n, g_n)$, where $\overleftrightarrow{S}_i$ denotes the orientability of S_i.

---

* 本文为王湘浩 1941 年撰写的硕士毕业论文。

① An orientable [a non-orientable] surface (closed or bounded) is equivalent to a sphere of $r(\geqslant 0)$ holes with $h(\geqslant 0)$ handles [$k(\geqslant 1)$ cross-caps]; cf. Seifert-Threlfall, Lehrbuch der Topologie (Berlin, 1934), Kap. 6. $2h$ and k are called the genera of the orientable and non-orientable surfaces respectively.

② Kuratowski, "Sur le probleme des courbes gauches en Topologie", Fund. Math. 15 (1930), pp. 271—283.

③ See Kagno, "Mapping of Graphs on Surfaces", Journal of Math. and Physics, 16 (1937), pp. 46—75.

Now, the holes of S_i correspond to a set of cycles of G' and, hence, to a set $\mathfrak{S}_i$ of cycles in G. Corresponding to the two opposite orientations of S_i, $i \leqq m$, there are two opposite orientations of $\mathfrak{S}_i$. Denote $\sum_{i=1}^{n} \mathfrak{S}_i$ by $\mathfrak{M}$ and $\mathfrak{S}_i$, $i \leqq m$, associated with the two opposite orientations by $\overleftrightarrow{\mathfrak{S}}_i$. Thus from a mapping M of the graph G in a closed surface S, we get α) a set $\mathfrak{M}$ of cycles in G, β) a division of $\mathfrak{M}$ in subsets $\overleftrightarrow{\mathfrak{S}}_1$, …, $\overleftrightarrow{\mathfrak{S}}_m$, $\mathfrak{S}_{m+1}$, …, $\mathfrak{S}_n$, γ) an integer g_i associated with $\mathfrak{S}_i$, even for $i \leqq m$ and positive for $i > m$.

The question arises as how to characterise these sets $\mathfrak{M}$ and their divisions into subsets β). This question is solved in Ⅰ. The conditions for $\mathfrak{M}$ are Mi) and Mii) (p. 13), and the condition for the divisions β) is F) (p. 33).

The main theorem may be stated as follows: Let it be given a set $\mathfrak{M}$ of cycles in G with properties Mi) and Mii) and a certain division of $\mathfrak{M}$ into subsets β) with the property F). On associating arbitrarily to each $\mathfrak{S}_i$ a nonnegative integer g_i, even for $i \leqslant m$ and positive for $i > m$, there exists, topologically speaking, one and only one closed surface S and one and only one mapping M, so that the curve $M(G)$ cuts the closed surface into n bounded surfaces, which are characterised by (1). The genus of S is

$$(2) \qquad \chi + \sum_{i=1}^{n} r_i - 2n + 2 + \sum_{i=1}^{n} g_i, \ \chi = \textit{characteristic of } G,$$

and S is orientable if and only if $m = n$ and the set $\{\overleftrightarrow{\mathfrak{S}}_1, \dots, \overleftrightarrow{\mathfrak{S}}_n\}$ can be oriented so that opposite orientations are induced in every arc of G.

Starting from (2) it is very easy to discuss the orientable and non-orientable closed surfaces of least genera, in which G can be mapped. A finite process for the determination of these surfaces is given in Ⅰ.

Ⅰ. MAPPING SETS AND MAPPING FORMS

1. Proper and general cyclic arrangements. A finite set of $n(\geqq 3)$ elements is

said to form a proper cyclic arrangement, when there is a relation of consecutivity for certain pairs of elements in the set with the following properties:

C i) if a is consecutive to l , then $a \neq l$;

C ii) if a is consecutive to l , then l is consecutive to a ;

C iii) each element is consecutive to just two elements;

C iv) if the set is divided in any way into two non-empty disjunct subsets, there exist a and l , one in each subset, such that a is consecutive to l.

When a is consecutive to l , we may say by C ii) that they are consecutive or that they are consecutive elements.

A set of two elements or of a single element is said to form a proper cyclic arrangement, if, in the first case, the two elements are defined as consecutive to each other, and, in the second case, the single element is defined as consecutive to itself.

If in a set of $n(\geqslant 3)$ elements, there is a relation of consecutivity with only the properties C i) , C ii) , and C iii) but not necessarily the property C iv) , then the set falls into a number of subsets each of which forms a proper cyclic arrangement. This separation into subsets is unique.

A proper cyclic arrangement of $n(\geqslant 1)$ elements may obviously be represented in the form

(1) $$a_1 a_2 \cdots a_n,$$

where a_i and a_{i+1}, i being taken mod n , are consecutive.

Also obviously, from the same set of $n(\geqslant 3)$ elements there may be defined $\dfrac{\underline{|n-1}}{2}$ different proper cyclic arrangements.

Let n ($\geqslant 3$). In (1), there are $2n$ ordered pairs of consecutive elements, namely, (a_i, a_{i+1}) and (a_{i+1}, a_i) . Two such pairs are said to be consecutive, if the first element of one pair is the same as the second element of the other, and if the two pairs are not formed with the same two elements. For example, (a_i, a_{i+1}) and (a_{i+1}, a_i) are consecutive. Obviously, this relation of consecutivity defined in the set of $2n$ ordered pairs satisfies C i), C ii), and C iii). It can be easily verified that this set falls into two subsets of n pairs each, forming two proper cyclic arrangements, namely,

(2) $$(a_1, a_2)(a_2, a_3)\cdots(a_{n-1}, a_n)(a_n, a_1),$$
$$(a_1, a_n)(a_n, a_{n-1})\cdots(a_3, a_2)(a_2, a_1).$$

They are called the two opposite orientations of (1). The proper cyclic arrangement of ordered pairs $(a, l)(l, a)$ is called the orientation of the proper cyclic arrangement al, and the proper cyclic arrangement (a, a) is called the orientation of the proper cyclic arrangement a. An oriental proper cyclic arrangement is an arrangement associated with one of its orientations. The two oppositely oriented proper cyclic arrangements of (1) for $n \geqslant 3$ may be represented respectively by

$$\overrightarrow{a_1 a_2 \cdots a_n},$$
$$\overleftarrow{a_1 a_2 \cdots a_n}.$$

The single oriented proper cyclic arrangement of the proper cyclic arrangement al or a is denoted by $\overrightarrow{al}$ or $\overrightarrow{a}$ respectively.

Suppose that, together with a proper cyclic arrangement N as (1), there is a set N' of $n'(\geqslant 1, \leqslant n)$ elements. Suppose that there is defined a function f in N such that $f(a_i)$ is a unique element of N', and that every element of N' is the image of at least one element of N, *i.e.*, $f(N) = N'$. We say that N and f define a general cyclic arrangement of N', or that a general cyclic arrangement of N' is defined with reference to N and f. Two general cyclic arrangements of N', defined with reference to N_1 and f_1, and N_2 and f_2 respectively, are regarded as the same, when there is a correspondence between N_1 and N_2, which is one-to-one and preserves consecutivity, such that corresponding elements in N_1 and N_2 have the same image in N' under f_1 and f_2. Denote $f(a_i)$ by a_i'. The general cyclic arrangement of N' defined with reference to (1) and f may obviously be represented in the form

(3) $$a'_1 a'_2 \cdots a'_n.$$

If an element of N' is the image of just k elements of N, it will be said to appear just k times in (3). For two given elements a' and l', if, for $a' \neq l'$, the equations

$$a'_i = a', \quad a'_{i\pm 1} = l',$$

and if, for $a' = l'$, the equations

$$a'_i = a', \quad a'_{i+1} = l',$$

are satisfied for just k is, $1 \leqslant i \leqslant n$, then a' and l' are said to be consecutive just k times in (3).

The "images" of consecutive elements, orientations, and consecutive ordered pairs in the orientations of N will be called consecutive elements, orientations, and consecutive ordered pairs in the orientations of (3) respectively.

To orient a set $\mathfrak{M}$ of proper or general cyclic arrangements is to orient each and every arrangement of $\mathfrak{M}$. The oriented $\mathfrak{M}$ is denoted by $\overrightarrow{\mathfrak{M}}$. The totality of the orientations of the arrangements of $\overrightarrow{\mathfrak{M}}$ is called the orientation of $\mathfrak{M}$ in $\overrightarrow{\mathfrak{M}}$. The totality of ordered pairs in the orientations of the arrangements of $\mathfrak{M}$ is denoted by $O(\overrightarrow{\mathfrak{M}})$. Two orientations of $\mathfrak{M}$, in which the orientations of each and every arrangement (except these containing one or two elements) are opposite, are said to be opposite.

2. Proper and general linear arrangements. A finite set of $n(\geqslant 2)$ elements is said to form a proper linear arrangement, when in the set there is a relation of consecutivity for certain pairs of elements with the following properties:

L i) same as C i);

L ii) same as C ii);

L iii) there exist just two elements each of which is consecutive to just one element, while any of the remaining elements is consecutive to just two elements;

L iv) same as C iv).

A proper linear arrangement of a single element is the same as the set of the single element.

A proper linear arrangement of $n(\geqslant 1)$ elements may obviously be represented in the form

(1) $$(a_1 a_2 \cdots a_n),$$

Where a_i and a_{i+1}, $i=1, \cdots, n-1$, are consecutive.

Let $n \geqslant 2$. In (1), there are $2(n-1)$ ordered pairs of consecutive elements, namely, (a_i, a_{i+1}) and (a_{i+1}, a_i), $i=1, \cdots, n-1$. The two proper linear arrangements

(2) $$((a_1, a_2)(a_2, a_3)\cdots(a_{n-1}, a_n)),$$

$$((a_n, a_{n-1})\cdots(a_3, a_2)(a_2, a_1)).$$

are called the two opposite orientations of (1). An oriented proper linear arrangement is an arrangement associated with one of its orientations. The two oppositely oriented proper linear arrangements of (1) may be represented respectively by

$$\overrightarrow{(a_1 a_2 \cdots a_n)},$$

$$\overleftarrow{(a_1 a_2 \cdots a_n)}$$

We are not interested in any definition of orientations of a proper linear arrangement of a single element.

Definitions of general linear arrangements, of oriented set of proper or general linear arrangements, etc., are similar to those for cyclic arrangements.

If a subset M of the set of elements in a proper cyclic arrangement N forms a proper linear arrangement through the definition of consecutivity for N ④, it is called a proper linear arrangement in the proper cyclic arrangement. Let the general cyclic arrangement be defined with reference to N' and f. The image of a proper linear arrangement in N is a general linear arrangement. It is called a general linear arrangement in N'.

---

④ If M equals the whole set, one pair of consecutive elements in N must be defined as not consecutive in M, provided that the number of elements in N is not 2.

3. Cycles and paths. If a graph G has an arc with coincident endpoints or two distinct arcs with the same endpoints, we may replace G by a subdivision G' without these kinds of arcs. For our purpose, G' plays the same role as G. Hence, throughout the paper, we suppose that in the graphs any arc has two distinct endpoints and any two distinct arcs have at most one common endpoint.

The degree of a vertex of G is defined to be the number of arcs of G at this vertex. Vertices of degrees 1, 2, and not less than 3 are called free vertices, ordinary vertices, and singular vertices respectively. As a rule, the number of arcs of G is denoted by α^1, the number of vertices by α^0, the number of free, ordinary, and singular vertices by α_1^0, α_2^0, and α_S^0 respectively. Thus

$$\alpha^0 = \alpha_1^0 + \alpha_2^0 + \alpha_S^0$$

The number $\alpha^1 - \alpha^0$ is denoted by χ and called the characteristic of G.

A cycle in G is a general cyclic arrangement of arcs of G

(1) $$a_1 a_2 \cdots a_n$$

with the following properties: *i*) a_i and a_{i+1}, i being taken mod n, have at least one common endpoint; *ii*) either all the arcs are equal and n is even, or not all the arcs are equal and whenever

$$a_{i+1} \neq a_i, \quad a_i = a_{i+1} = \cdots = a_{i+k-1}, \quad a_{i+k-1} \neq a_{i+k},$$

$$k \geqslant 1, \textit{ the subscripts being taken mod } n$$

then the common endpoint of a_{i+1} and a_i is distinct from or the same as that of a_{i+k-1} and a_{i+k} according as k is odd or even.

Obviously, a general cyclic arrangement (1) of arcs of G is a cycle if and only if it is of the form

(1) $$\overline{A_1 A_2} \, \overline{A_2 A_3} \cdots \overline{A_n A_1},$$

where $\overline{A_i A_{i+1}}$ is an arc in G with the vertices A_i and A_{i+1} as endpoints.

A path in G is a general linear arrangement of arcs of G

(2) $$(a_1 a_2 \cdots a_n)$$

with the following properties: $i)$ a_i and a_{i+1}, $i=1, \cdots, n-1$, have at least one common endpoint; $ii)$ whenever

$$a_{i-1} \neq a_i, \ a_i = a_{i+1} = \cdots = a_{i+k-1}, \ a_{i+k-1} \neq a_{i+k},$$

$$k \geqslant 1, \ 2 \leqslant i \leqslant n-1,$$

then the common endpoint of a_{i-1} and a_i is distinct from or the same as that of a_{i+k-1} and a_{i+k} according as k is odd or even.

Obviously, a general linear arrangement (2) of arcs of G is a path if and only if it is of the form

(2)′ $$(\overline{A_1A_2}\,\overline{A_2A_3}\cdots\overline{A_nA_{n+1}}).$$

If all the arcs in (2) are equal and n is odd, then A_1 and A_{n+1} are the endpoints of the single arc in (2). If not all the arcs in (2) are equal, suppose

$$a_1 = a_2 = \cdots = a_k, \ a_k \neq a_{k+1}, \ a_{n-k} \neq a_{n-k+1}, \ a_{n-k+1} = a_{n-k+2} = \cdots = a_n.$$

Then, if k is even, A_1 is the common endpoint of a_k and a_{k+1}; if k is odd, A_1 is the endpoint of a_k not common with a_{k+1}. Similarly for A_{n+1}. Hence, if not all the arcs in (2) are equal or if n is odd, then A_1 and A_{n+1} are uniquely determined vertices A and B, i. e., in any represtentation (2)′ of the path, A_1 and A_{n+1} must be either A and B respectively or B and A respectively. A and B are called the endpoints of the path; they may be distinct or coincident. Endpoints of a path (2) where n is even and all of whose arcs are equal are not defined.

A general linear arrangement in a cycle (see last paragraph in § 2) is a path; it is called a path in the cycle.

The cycle (1)′ is said to be simple if the arcs $\overline{A_1A_2}$, …, $\overline{A_nA_1}$ on the one hand and the vertices A_1, …, A_n on the other are all distinct. The path (2)′ is said to be simple, if the arcs $\overline{A_1A_2}$, …, $\overline{A_nA_{n+1}}$ on the one hand and the vertices A_1, …, A_{n+1} on the other are all distinct, except that the endpoints A_1, A_{n+1} may be coincident.

A broken-arc in G is a subgraph of G consisting of the arcs and vertices in a simple path in G, whose endpoints are not ordinary vertices of G but whose remaining vertices (if there are any) are. The endpoints of the path are called the / endpoints of the broken-arc. A circle in G is a subgraph of G consisting of the arcs and vertices in a simple cycle in G, all of whose vertices are ordinary vertices of G. It can be easily seen that, in a broken-arc, the given simple path is the only simple path in G with the endpoints same as those of the broken-arc. We shall call this path the simple path in the broken-arc. Similarly, we have the simple cycle in a circle.

The set of arcs in G can be divided in a unique manner into non-empty disjoint subsets the arcs and vertices in each of which form a broken-arc or a circle. These are all the broken-arcs and circles in G. Let the number of broken-arcs be β'. Then obviously

$$\chi = \beta' - \alpha_s^0 - \alpha_1^0$$

As a general cyclic arrangement, a cycle can be oriented.

Let (a, l), $a \neq l$, be an ordered pair in the orientation of an oriented cycle $\vec{C}$. We say that $\vec{C}$ induces an orientation of the arc a, which is represented by regarding the common endpoint of a and l as the terminal point of a; also we say that $\vec{C}$ induces an orientation of the arc l, which is represented by regarding the common endpoint of a and l as the initial point of l. Since an arc may appear more than once in a cycle, opposite orientations may be induced in a by $\vec{C}$ Thus, when the oriented cycle $\vec{C}$ is

$$\overrightarrow{\overline{A_1A_2}\ \overline{A_2A_3}\cdots\overline{A_3A_2}\ \overline{A_2A_n}\ \overline{A_nA_1}},\quad A_1 \neq A_3,\quad A_3 \neq A_n,$$

an induced orientation of $\overline{A_2A_3}$ is the orientation of the oriented arc $\overrightarrow{A_2A_3}$ in virtue of the pair $(\overline{A_1A_2}, \overline{A_2A_3})$, and another induced orientation of $\overline{A_2A_3}$ is the orientation of the oriented arc $\overline{A_3A_2}$ in virtue of the pair $(\overline{A_3A_2}, \overline{A_2A_n})$. Opposite orientations are therefore induced in $\overrightarrow{A_2A_3}$. If (a, a) is an ordered pair in the orientation of $\vec{C}$, then we say that $\vec{C}$ induces opposite orientations in a.

The orientations induced in an arc a by an oriented set of cycles $\overrightarrow{\mathfrak{M}}$ are the orientations induced in a by the oriented cycles in $\overrightarrow{\mathfrak{M}}$ containing a.

Similarly, we define orientations induced in an arc by an oriented path or by an oriented set of paths.

4. Mapping sets; Their orientability. All the graphs considered from the present section to § 7 inclusive are connected.

Definition. A set $\mathfrak{M}$ of cycles in a graph G is said to be a mapping set of G if

M i) each arc of G appears just twice in $\mathfrak{M}$ (i. e. , each arc of G appears just once in each of two cycles or twice in one cycle of $\mathfrak{M}$);

M ii) a proper cyclic arrangement of all the arcs of G at each and every vertex can be defined, thus forming a set $\mathfrak{A}$ of proper cyclic arrangements, so that any two consecutive arcs in $\mathfrak{A}$ (i. e. , in some arrangement of $\mathfrak{A}$) are consecutive arcs in $\mathfrak{M}$ (i. e. , in some cycle of $\mathfrak{M}$).

Definition. A mapping set $\mathfrak{M}$ of G is said to be orientable if $\mathfrak{M}$ may be oriented coherently, i. e. , if $\mathfrak{M}$ may be oriented so that $\overrightarrow{\mathfrak{M}}$ induces opposite orientations in every arc of G .

Theorem 1. Let $\mathfrak{M}$ be a mapping set of G . Then

i) $\overline{QP}$ is consecutive (in $\mathfrak{M}$) to itself, if and only if one of its endpoints Q , P is free.

ii) If P is ordinary, $\overline{QP}$ is consecutive to the other arc at P .

iii) If P is singular, $\overline{QP}$ is consecutive to just two other arcs at P .

iv) Two consecutive arcs are consecutive just twice, if they are distinct and their common endpoint is ordinary; otherwise, they are consecutive just once.

Proof. i). Let P be free. Only one proper cyclic arrangement of the single arc at P can be defined and in this arrangement $\overline{QP}$ is defined as consecutive to itself. Hence, $\overline{QP}$ is consecutive to itself in $\mathfrak{M}$. This may also be proved only by the definition of a cycle. If $\overline{QP}$ were not consecutive to itself, let $(a\ \overline{QP}\ l)$ be a path in $\mathfrak{M}$ (i. e. , in some cycle of $\mathfrak{M}$). Then $a \neq \overline{QP}$, $\overline{QP} \neq l$. But by the definition of a cycle, the endpoint common to a and $\overline{QP}$ cannot be the same as that common to $\overline{QP}$ and l , which means that one of a and l would be an arc at P and P would not be free.

Conversely, let both P and Q be not free. If $\overline{QP}$ were consecutive to itself, then by M i), $\overline{QP}$ would appear in only one cycle C of $\mathfrak{M}$. Either C would be of the form $\overline{QP}$ $\overline{PQ}$, or there would exist a and l such that ($a\overline{QP}\,\overline{PQ}l$) is a path in C. In the first case, $\overline{QP}$ would be consecutive in $\mathfrak{M}$ to no other arcs than itself. In the second case, the common endpoint of a and $\overline{QP}$ would be the same as that common to $\overline{QP}$ and l (by definition of a cycle). Let this common endpoint be Q. Then $\overline{QP}$ would be consecutive in $\mathfrak{M}$ to no other arcs at P than itself. In both cases, a proper cyclic arrangement of all the arcs at P as required in M ii) could not be defined.

ii), iii), iv) Let $a=\overline{QP}$ and P be ordinary. Let the other arc at P be l. If a is consecutive in $\mathfrak{M}$ to itself, the cycle in which a appears cannot be of the form aa by M ii). Let ($caad$) be a path in this cycle. Then $c\neq a$, $a\neq d$ by M i). The common endpoint of c and a is the same as that of a and d. This common endpoint must be P, since Q is free by i). Hence $c=d=l$, and a, l are consecutive just twice in $\mathfrak{M}$. If a is not consecutive to itself, let (cad) be a path in $\mathfrak{M}$. Then $c\neq a$, $a\neq d$. The common endpoint of c and a and that of a and d are different. Hence one of c and d must be l. Since a appears just twice in $\mathfrak{M}$, a and l are consecutive just twice in $\mathfrak{M}$. This proves ii) and the first half of iv).

Now, let $a=\overline{QP}$ and P be singular. If a is consecutive to itself, the cycle in which a appears cannot be of the form aa by M ii). Let ($caad$) be a path in this cycle. Then $c\neq a$, $a\neq d$. The common endpoint of c and a and that of a and d must be the same point P. Now, c cannot $=d$, for, otherwise, a proper cyclic arrangement of all the arcs at P as required in M ii) could not be defined. Hence, a in consecutive to just two other arcs at P and each of these two arcs is consecutive to a just once. If a is not consecutive to itself, let (cad) be a path in $\mathfrak{M}$. Then $c\neq a$, $a\neq d$. One and only one of c and d, say c, is an arc at P. Since a appears just twice in $\mathfrak{M}$, we have just one other path ($c'ad'$). Let c' be an arc at P. By M ii), $c'\neq c$. Hence a is consecutive to just two other arcs at P and each of these two arcs is c consecutive to a

just once in $\mathfrak{M}$. This and the fact that an arc cannot be consecutive to itself more than once in $\mathfrak{M}$ (by M *i*)) proves *iii*) and the second half of *iv*).

Theorem 2. For a given mapping set $\mathfrak{M}$ of G only one set $\mathfrak{A}$ as required in M *ii*) can be defined. Any two consecutive arcs in $\mathfrak{M}$ are consecutive in $\mathfrak{A}$.

Proof. If a vertex P is free or ordinary, only one proper cyclic arrangement of all the arcs at P can be defined. Let P be singular. By Theorem 1, *iii*), we can define the two consecutive arcs of an arc a at P in only one way, and, therefore, only one proper cyclic arrangement of all the arcs at P as required in M *ii*) can be defined. Hence, only one $\mathfrak{A}$ as required in M *ii*) can be defined.

Let a, l be consecutive in $\mathfrak{M}$. If $a=l$, one endpoint of a is free by Theorem 1 *i*). But in $\mathfrak{A}$, a is consecutive to itself in the proper cyclic arrangement of all the arcs at P, which contains only one element. If $a \neq l$, let their common endpoint be P. If P is ordinary, in $\mathfrak{A}$, a and l are consecutive in the proper cyclic arrangement of all the arcs at P, which contains only two elements. If P is singular, a is consecutive in $\mathfrak{M}$ to just two other arcs at P, and, in $\mathfrak{A}$, a is consecutive to also just two other arcs in the proper cyclic arrangement of all the arcs at P. Hence, a, l are also consecutive in $\mathfrak{A}$ in this case.

The unique $\mathfrak{A}$ is said to belong to $\mathfrak{M}$.

Theorem 3. Let $\mathfrak{M}$ be a mapping set of G and $\mathfrak{A}$ belong to $\mathfrak{M}$. $\mathfrak{M}$ is orientable, if and only if $\mathfrak{M}$ and $\mathfrak{A}$ can be oriented so that $O(\overrightarrow{\mathfrak{M}})=O(\overrightarrow{\mathfrak{A}})$. When $O(\overrightarrow{\mathfrak{M}})=O(\overrightarrow{\mathfrak{A}})$, $\overrightarrow{\mathfrak{M}}$ is coherently oriented.

Proof. Let $\mathfrak{M}$ be orientable and $\overrightarrow{\mathfrak{M}}$ be coherently oriented. Consider any vertex P. Let A be the arrangement at P in $\mathfrak{A}$. If P is free, and $A=a$, there is the ordered pair (a, a) in $O(\overrightarrow{\mathfrak{M}})$. On orienting A, we have also the pair (a, a) in the orientation of $\overrightarrow{A}$. If P is ordinary, let $A=al$. If l appears in $\mathfrak{M}$ so that $(alla)$ is a path in $\mathfrak{M}$, we have in $O(\overrightarrow{\mathfrak{M}})$ the pairs (a, l) and (l, a). If l appears in $\mathfrak{M}$ so that (ald) and (ald'), d, d', distinct from l, are paths in $\mathfrak{M}$, we have also (a, l),

(l, a) in $O(\overrightarrow{\mathfrak{M}})$, since $\overrightarrow{\mathfrak{M}}$ induces opposite orientations in l. Hence, in both cases, we have (a, l), (l, a) in $O(\overrightarrow{\mathfrak{M}})$. On orienting A, we have also these pairs in the orientation of $\overrightarrow{A}$. If P is singular, let $A = a_1 a_2 \cdots a_n$, $n \geqslant 3$. In $O(\overrightarrow{\mathfrak{M}})$, we have (a_1, a_2) or (a_2, a_1) but not both, since a_1, a_2 are consecutive just once in $\mathfrak{M}$. Suppose (a_1, a_2) is in $O(\overrightarrow{\mathfrak{M}})$. (a_2, a_3) must also be in $O(\overrightarrow{\mathfrak{M}})$. For, if a_2 appears in $\mathfrak{M}$ so that $(a_1 a_2 a_2 a_3)$ is a path in $\mathfrak{M}$, (a_2, a_3) is in $O(\overrightarrow{\mathfrak{M}})$. If a_2 appears in $\mathfrak{M}$ so that $(a_1 a_2 a')$ and $(a_3 a_2 a'')$, a', a'' distinct from a_2, are paths in $\mathfrak{M}$, the pair (a_2, a_3) must be in the orientation of the cycle in which $(a_3 a_2 a'')$ is a path, since $\overrightarrow{\mathfrak{M}}$ induces opposite orientations in a_2. Similarly, the pairs (a_3, a_4), ⋯, (a_n, a_1) are in $O(\overrightarrow{\mathfrak{M}})$. Let us orient A so that (a_1, a_2) is in the orientation of $\overrightarrow{A}$. If, however, (a_2, a_1) is in $O(\overrightarrow{\mathfrak{M}})$, then (a_3, a_2), ⋯, (a_1, a_n) are in $O(\overrightarrow{\mathfrak{M}})$. In this case, we shall orient A so that (a_2, a_1) and therefore (a_1, a_n), (a_n, a_{n-1}), ⋯, (a_3, a_2) are in the orientation of $\overrightarrow{A}$. When all the arrangements in $\overrightarrow{\mathfrak{A}}$ are oriented in this manner, we obtain an $\mathfrak{A}$ such that $O(\overrightarrow{\mathfrak{A}}) = O(\overrightarrow{\mathfrak{M}})$.

Conversely, suppose $\overrightarrow{\mathfrak{M}}$ and $\overrightarrow{\mathfrak{A}}$ are oriented so that $O(\overrightarrow{\mathfrak{M}}) = O(\overrightarrow{\mathfrak{A}})$. Let a be any arc and P one of its endpoints. If P is free, the pair (a, a) is in $O(\overrightarrow{\mathfrak{M}})$. If P is ordinary and l is the other arc at P, the pairs (a, l), (l, a) are in $O(\overrightarrow{\mathfrak{A}})$ and therefore in $O(\overrightarrow{\mathfrak{M}})$.. If P is singular and the arrangement at P in $\mathfrak{A}$ is $lac\cdots$, the pairs (a, l), (c, a) or (l, a), (a, c) are in $O(\overrightarrow{\mathfrak{A}})$ and therefore in $O(\overrightarrow{\mathfrak{M}})$. In all cases, $\overrightarrow{\mathfrak{M}}$ induces opposite orientations in a. Hence, $\overrightarrow{\mathfrak{M}}$ is oriented coherently.

Theorem 4. Let l be a broken-arc in G and $\mathfrak{M}$ be any mapping set of G. The simple path in l "appears just twice" as a path in $\mathfrak{M}$.

Proof. Let the simple path in l be

(1) $$(a_1 a_2 \cdots a_n)$$

For $n=1$, the theorem is a consequence of M i). Let $n > 1$. Suppose that $(ca_1 d)$ is a path in $\mathfrak{M}$. From the proof of Theorem 1, one of c and d must be a_2, because the common endpoint of a_1, a_2 is ordinary. Let $d = a_2$. Since both endpoints of a_2 is ordinary (when $n>2$), a_2 is consecutive to a_1 and a_3 and only to these arcs. We have then the path $(ca_1a_2a_3)$ in $\mathfrak{M}$. Similarly to these arcs. We have then the path $(ca_1a_2a_3)$ in $\mathfrak{M}$. Similarly we have the paths $(ca_1a_2a_3a_4)$, $\cdots$, $(ca_1a_2\cdots a_n)$, *and therefore* $(a_1a_2\cdots a_n)$ in $\mathfrak{M}$. Since a_1 appears just twice in $\mathfrak{M}$, $(a_1a_2\cdots a_n)$ appears just twice as a path in $\mathfrak{M}$.

As a direct application of Theorem 4, we shall determine all the mapping sets of the graph G which is a single broken-arc. Let the simple path in l be (1). Then (1) appears just twice as a path in any mapping set of G. Since a_n has one of its endpoints free, we have the path $(a_1a_2\cdots a_na_n\cdots a_2a_1)$ and hence the cycle

$$a_1a_2\cdots a_na_n\cdots a_2a_1$$

in any mapping set. But every arc of l appears already just twice in this cycle, and, consequently, the graph has only one mapping set containing this cycle as its only element. This mapping set is obviously orientable.

Let G be a circle the simple cycle in which is

$$a_1a_2\cdots a_n \tag{2}$$

By similar reasonings as used in the proof of Theorem 4 and in the above paragraph, it is readily shown that G has only one orientable mapping set containing (2) twice as its elements[⑤], and only one nonorientable mapping set containing

$$a_1a_2\cdots a_naa_1a_2\cdots a_n$$

as its only element.

5. Determination of all the mapping sets of a graph. The problems whose solutions are the objects of the present and the next sections have been solved in the last section for a graph which is a single broken-arc or a circle. Accordingly, we assume G is neither a single broken-arc nor a circle in the present and the next

---

⑤ Here we make the convention that a set of cycles may contain on cycle twice as its elements and, when orienting the set, we may orient these "two" elements independently. It is clear that these queer elements appear in a mapping set, only if G is a circle.

sections, unless the contrary is indicated. Thus, $\alpha_s^0 \neq 0$, because G is connected. Let the arcs of G be a_1, $\cdots$, $a_{\alpha'}$.

Defining arbitrarily a proper cyclic arrangement A_P of all the arcs at each and every vertex P of G, we obtain a set $\mathfrak{A}$ of proper cyclic arrangements.

Consider any broken-arc l. Let the simple path in l be

$$(a_i \cdots a_j),$$

and the endpoints of l be P and Q, which are endpoints of a_i and a_j respectively. At least one of P and Q must be singular. First, suppose one of P and Q is free; we may assume that it is Q. a_j is consecutive to itself in A_Q and a_i is consecutive in A_P to just two other arcs $a_{j'}$ and $a_{j''}$. Form the paths:

(1) $$(a_{j'}a_i \dots a_j a_j),\ (a_{j''}a_i \dots,\ a_j a_j).$$

Those two paths are distinct, and are said to form a canonical pair for l. Now, suppose both P and Q are singular; they may be distinct or coincident. In other cases, a_i is consecutive in A_P to just two other arcs $a_{j'}$ and $a_{j''}$, and $a_{j'}$ is consecutive in A_Q to just two other arcs $a_{i'}$ and $a_{i''}$. Form the paths:

(2) $$(a_{j'}a_i \dots a_j a_{i'}),\ (a_{j''}a_i \dots,\ a_j a_{i''}),$$

(3) $$(a_{j'}a_i \dots a_j a_{i''}),\ (a_{j''}a_i \dots,\ a_j a_{i'}).$$

It can be easily proved that the four paths are distinct. Each of the pairs (2) and (3) is called a canonical pair for l. Choose one pair arbitrarily and hold it fast. Thus, we have two such paths for each broken-arc and therefore a set $\mathfrak{P}$ of $2\beta'$ such distinct paths, called a canonical set (of paths) for the given $\mathfrak{A}$ of G.

A path (aa'), where a and a' are consecutive in some A_P, P being singular or free, is called a link. For example, the paths $(a_{j'}a_i)$, $(a_j a_{i'})$ "in" the path $(a_{j'}a_i \dots a_j a_{i'})$ in $\mathfrak{P}$ are links. We shall prove that a link (aa') "appears just twice" in $\mathfrak{P}$. First, suppose P is free. Then $a = a'$, and (aa') appears once in each of the paths for the broken-arc containing a, and appears in no other paths. Secondly, suppose P is singular. Let the broken-arcs containing a and a' be l and l' respectively. Obviously,

(aa') appears only in the paths for l and l'. If $l \neq l'$, (aa') appears once in a unique path for l and a second time in a unique path for l'. If $l = l'$, (aa') appears either once in each of the two paths for l, or twice in a unique path for l. Hence, in every case, (aa') appears just twice in $\mathfrak{P}$.

Two distinct paths in $\mathfrak{P}$ are defined as consecutive, if they have a link in common. A path in $\mathfrak{P}$ is defined as consecutive to itself, if it contains a link twice, or, in other words, if the two links in it are the same. If a link appears twice in a path in $\mathfrak{P}$, it appears in no other paths. Hence, if a path is consecutive to itself, it is consecutive to no other paths. If a path $\mathcal{L}$ is consecutive to only one path $\mathcal{L}'$ distinct from $\mathcal{L}$, then the two links in $\mathcal{L}$ must be the two links in $\mathcal{L}'$, and $\mathcal{L}'$ is therefore consecutive to the path $\mathcal{L}$ only. Finally, since a path contains at most two distinct links, it is consecutive to at most two paths. Denote the totality of the paths each of which is consecutive to itself by $\mathfrak{P}_0$ and that of the paths each of which is consecutive to only one path distinct from itself by $\mathfrak{P}_1$. Set $\mathfrak{P}_2 = \mathfrak{P} - \mathfrak{P}_1 - \mathfrak{P}_0$. Then any two paths in different sets $\mathfrak{P}_0$, $\mathfrak{P}_1$, and $\mathfrak{P}_2$ are not consecutive. The set $\mathfrak{P}_0$ falls into subsets each of which forms a proper cyclic arrangement of a single element; the set $\mathfrak{P}_1$ falls into subsets each of which forms a proper cyclic arrangement of two elements; and the set $\mathfrak{P}_2$ falls into subsets each of which forms a proper cyclic arrangement of at least three elements, because the relation of consecutivity in $\mathfrak{P}_2$ has the properties O_i), O_{ii}), O_{iii}).

Any of these proper cyclic arrangements is of the form

(4) $$(a_{j_n} a_{i_1} \cdots a_{j_1} a_{i_2})(a_{j_1} a_{i_2} \cdots a_{j_2} a_{i_3}) \cdots (a_{j_{n-1}} a_{i_n} \cdots a_{j_0} a_{i_1}),\quad n \geqslant 1$$

Corresponding to each arrangement (4), let us form a general cyclic arrangement

(5) $$a_{i_1} \cdots a_{j_1} \cdots a_{i_2} \cdots a_{j_2} \cdots a_{i_n} \cdots a_{j_n}$$

(5) is obviously a cycle. Let the totality of these cycles be $\mathfrak{M}$. I say that $\mathfrak{M}$ is a mapping set, (5) is constructed by "linking" together the simple paths in certain broken-arcs, each of which is in the "middle" of a path in φ. But the simple paths in

each broken-arc appears in the middle of just two paths in β , namely, the two paths of the canonical pair for the broken-arc. Hence, the simple path injeach broken-arc appears just twice as a path in the cycles of $\mathfrak{M}$. Consequently, each arc of G appears just twice in $\mathfrak{M}$ and M_i) is satisfied. To show that M_{ii}) is also satisfied, we shall prove that $\mathfrak{A}$ is just what is required in M_{ii}) for $\mathfrak{M}$ to be a mapping set. It is obvious from the construction that any two consecutive arcs in β are consecutive in $\mathfrak{M}$ (and conversely). It remains to show that any two consecutive arcs in $\mathfrak{A}$ are consecutive in β. But this follows from the construction of β .

Hence, given $\mathfrak{A}$, each β determines in this manner a unique mapping set $\mathfrak{M}$, to which the given $\mathfrak{A}$ belongs. we shall denote this $\mathfrak{M}$ by $[\mathfrak{A}, \beta]$.

Now, let $\mathfrak{M}$ be any mapping set of G . Together with $\mathfrak{M}$, there is given the $\mathfrak{A}$ belonging to $\mathfrak{M}$. We shall prove that a unique β for the given $\mathfrak{A}$ can be constructed such that $[\mathfrak{A}, \beta]=\mathfrak{M}$. By Theorem 4, a cycle in a mapping set must be of the form (5) where $(a_{i_k} \cdots a_{j_k})$ are simple pathe in certain broken-arcs. If β is to be constructed so that $[\mathfrak{A}, \beta]=\mathfrak{M}$, in β there must be a proper cyclic arrangement of the form (4), so must contain the paths $(a_{j_0} a_{i_1} \cdots a_{j_1} a_{i_2})$, $\cdots$, $(a_{j_{n-1}} a_{i_n} \cdots a_{j_1} a_{i_1})$. Hence form these paths from the cycles of the form (5) in the given $\mathfrak{M}$, and denote the totality of these paths by β . Since the simple path $(a_i \cdots a_j)$ in any broken-arc l appears as a path in $\mathfrak{M}$ just twice, there are just two paths

(6) $$(a_{j'} a_i \cdots a_j a_{i'}),\ (a_{j''} a_i \cdots a_j a_{i''})$$

in β corresponding to l . From the construction of these paths, a_i is consecutive to $a_{j'}$ and $a_{j''}$, and a_j to $a_{i'}$ and $a_{i''}$ in $\mathfrak{M}$, and therefore in $\mathfrak{A}$ in virtue of Theorem 2. Hence, from definition, the paths (6) form a canonical pair for l . That $[\mathfrak{A}, \beta]$ is the given $\mathfrak{M}$ follows from the construction of β . Hence,

Theorem 5. For any given mapping set $\mathfrak{M}$ or G , to which $\mathfrak{A}$ belongs, a unique β can be constructed such that $[\mathfrak{A}, \beta]=\mathfrak{M}$.

Thus, by constructing all possible β and then all possible β , we obtain by the above process all the mapping sets or G. Now, if $d_1, \ldots, d_{a_s^0}$ are the degrees of the singular vertices of G, there are $\prod_{i=1}^{d_s^0} \frac{\lfloor\underline{d_i - 1}}{2}$ ways to construct $\mathfrak{A}$. Let $\mathfrak{A}$ be held fast. The two paths for l may be constructed in two ways or one according as l has no or has a free endpoint. But in G there are as many broken-arcs having free endpoints as there are free vertices. Thus, for fixed $\mathfrak{A}$, we have $2^{\beta' - a_1^0}$ ways to construct β. Consequently, the total number of possible β and therefore the number of mapping sets of G is

$$2^{\beta^1 - a_1^0} \prod_{i=1}^{a_s^0} \frac{\lfloor\underline{d_i - 1}}{2} = 2^{\beta' - a_1^0 - a_s^0} \prod_{i=1}^{a_s^0} \lfloor\underline{d_i - 1}$$

Refering to a formula in ζ 3, we have

Theorem 6. The number of mapping sets of G is $2^a \prod_{i=1}^{a_s^0} d_i - 1$.

6. Number of orientable mapping sets of a graph. Let $\mathfrak{M} = [\mathfrak{A}, \beta]$. Let

(1) $$(a_{j_n} a_{i_1} \cdots a_{j_1} a_{i_2})(a_{j_1} a_{i_2} \cdots a_{j_2} a_{i_3}) \cdots (a_{j_{n-1}} a_{i_n} \cdots a_{j_n} a_{i_1})$$

where $n \geqslant 1$, be a typical proper cyclic arrangements of paths in β, and

(2) $$a_{i_1} \cdots a_{j_1} a_{i_2} \cdots a_{j_2} \cdots a_{i_n} \cdots a_{j_n}$$

the corresponding cycle in $\mathfrak{M}$. Suppose (2) be oriented arbitrarily. Obviously, the paths in (1) may be oriented so that the set of ordered pairs in the given orientation of (2) is equal to the set of ordered pairs in the orientations of the paths in (1). Hence, to any given orientation of $\mathfrak{M}$, we can orient β so that $O(\vec{\beta}) = O(\overrightarrow{\mathfrak{M}})$. Thus, in virtue of Theorem 3, we have proved the necessary part of the following theorem:

Theorem 7. A necessary and sufficient condition for $\mathfrak{M}$ to be orientable is that β and $\mathfrak{A}$ can be oriented so that $O(\vec{\beta}) = O(\overrightarrow{\mathfrak{A}})$.

To show that the condition is sufficient, let $\vec{\beta}$ and $\overrightarrow{\mathfrak{A}}$ be such that $O(\vec{\beta}) = O(\overrightarrow{\mathfrak{A}})$. It suffices to prove that $\mathfrak{M}$ can be oriented so that $O(\overrightarrow{\mathfrak{M}}) = O(\vec{\beta})$. Suppose the oriented path $\overrightarrow{a_{j_n} a_{i_1} \cdots a_{j_1} a_{i_2}}$ is in $\vec{\beta}$. The oriented path of $(a_{j_1} a_{i_2} \cdots a_{j_2} a_{i_3})$ in $\vec{\beta}$ must be $\overrightarrow{(a_{j_1}a_{i_2}\cdots a_{j_2}a_{i_3})}$ but not $\overleftarrow{(a_{j_1}a_{i_2}\cdots a_{j_2}a_{i_3})}$. For, if $a_{j_1}\neq a_{i_2}$, the pair (a_{j_1}, a_{i_2}) is in $O(\vec{\beta})$ and therefore in $O(\vec{\mathfrak{A}})$. Consequently, the pair (a_{i_2}, a_{j_1}) is not in $O(\vec{\mathfrak{A}})$ and therefore not in $O(\vec{\beta})$. If $a_{j_1}=a_{i_2}$, then $a_{i_1}=a_{j_2}$, because two broken-arcs having an arc in common must be the same. The paths $(a_{j_n}a_{i_1}\cdots a_{j_1}a_{i_2})$ and $(a_{j_1}a_{i_2}\cdots a_{j_2}a_{i_3})$ are therefore the paths (1), ζ 5, for a certain broken-arc in β. Hence, $a_{j_n}\neq a_{i_1}$, $a_{j_2}\neq a_{i_1}$, $a_{j_n}\neq a_{i_3}$. Since the pair $(a_{j_n}, a_{i_1})=(a_{j_n}, a_{j_2})$ is in $O(\vec{\beta})$ and therefore in $O(\vec{\mathfrak{A}})$, the pair (a_{i_3}, a_{j_2}) is not in $O(\vec{\mathfrak{A}})$ and therefore not in $O(\vec{\beta})$. But the orientation of $\overleftarrow{(a_{j_1}a_{i_2}\cdots a_{j_2}a_{i_3})}$ contains the pairs (a_{i_2}, a_{j_1}) and (a_{i_3}, a_{j_2}). Similarly, the oriented paths $\overrightarrow{(a_{j_{k-1}}a_{i_k}\cdots a_{j_k}a_{i_{k+1}})}$, $k=3,\cdots,n$ are in $\vec{\beta}$. The pairs in the orientations of these oriented paths are in the orientations of $\overrightarrow{a_{i_1}\cdots a_{j_1}a_{i_2}\cdots a_{j_2}\cdots a_{i_n}\cdots a_{j_n}}$. Hence, $\mathfrak{M}$ can be oriented so that $O(\vec{M})=O(\vec{\beta})$.

Consider a fixed $\mathfrak{A}$ and a fixed $\vec{\mathfrak{A}}$. We shall prove that a unique β can be constructed having the property that it can be oriented so that $O(\vec{\beta})=O(\vec{\mathfrak{A}})$. Let l be any broken-arc and $(a_i\cdots a_j)$ the simple path in l. If one of the endpoints P and Q of l is free, we construct the canonical pair for l as in ζ 5; the two paths for l can be oriented so that the pairs in their or iontations are in $O(\vec{\mathfrak{A}})$. Suppose that both P and Q are singular, and that $a_{j'}$, $a_{j''}$ are the arcs consecutive to a_i in A_p and $a_{i'}$, $a_{i''}$ the arcs consecutive to a_j in A_Q. Suppose that, in $\vec{\mathfrak{A}}$, A_P and A_Q are so oriented that the pairs $(a_{j'}, a_i)$, $(a_i, a_{j''})$, $(a_{i''}, a_j)$, $(a_j, a_{i'})$ are in $O(\vec{\mathfrak{A}})$. The paths for l in β cannot be $(a_{j'}a_i\cdots a_ja_{i''})$, $(a_{j''}a_i\cdots a_ja_{i'})$. For any orientation of $(a_{j'}a_i\cdots a_ja_{i''})$ contains an ordered pair not in $O(\vec{\mathfrak{A}})$. Let $(a_{j'}a_i\cdots a_ja_{i'})$, $(a_{j''}a_i\cdots a_ja_{i''})$ be the paths for l. We see the ordered pairs in the orientations of $\overrightarrow{(a_{j'}a_i\cdots a_ja_{i'})}$ and $\overleftarrow{(a_{j''}a_i\cdots a_ja_{i''})}$ are in $O(\vec{\mathfrak{A}})$. Thus, we get a unique β which can be oriented so that $O(\vec{\beta})\subset O(\vec{\mathfrak{A}})$, and, from the fact that any two consecutive arcs in $\mathfrak{A}$ are consecutive in β, $O(\vec{\beta})=O(\vec{\mathfrak{A}})$. Hence, any $\vec{\mathfrak{A}}$ determines a unique orientable mapping set $[\mathfrak{A}, \beta]$ such that β

can be oriented so that $O(\vec{\beta})=O(\vec{\mathfrak{A}})$.

It is obvious that if $\vec{\mathfrak{A}}$ and $(\mathfrak{A})'$ are oppositely oriented, they determine the same orientable mapping set. But, if $\vec{\mathfrak{A}}$ and $(\vec{\mathfrak{A}})'$ are neither the same nor oppositely oriented, the orientable mapping sets determined by them are actually different. For, $\vec{\mathfrak{A}}$ and $(\vec{\mathfrak{A}})'$ being not oppositely oriented, there exists an arrangement A_p (P singular) that is oriented in the same way in $\vec{\mathfrak{A}}$ and $(\vec{\mathfrak{A}})'$; and, $\vec{\mathfrak{A}}$ and $(\vec{\mathfrak{A}})'$ being not the same, there exists an arrangement A_Q (Q singular) that is oriented oppositely in $\vec{\mathfrak{A}}$ and $(\vec{\mathfrak{A}})'$. Since G is connected, there is a simple path L in G with endpoints P and Q. Suppose that $P_0(=p)$, P_1, $\cdots$, P_{n-1}, $P_n(=Q)$ are all the singular vertices of G which are "on" L and are in successive order along L from P to Q. Then, there is a unique P_i such that A_{P_i} is oriented in the same way in $\vec{\mathfrak{A}}$ and $(\vec{\mathfrak{A}})'$, but $A_{P_{i+1}}$, $\cdots$, A_{P_n} are oriented oppositely in $\vec{\mathfrak{A}}$ and $(\vec{\mathfrak{A}})'$. Write P' and Q' for P_i and P_{i+1} respectively. Let the "subpath" of L having P' and Q' as endpoints be L': $a_i \cdots a_j$, P' being an endpoint of a_i and Q' an endpoint of a_j. Since L' is simple and passes through no singular vertices except the endpoints, it is the simple path in some broken-arc l. Let the arcs consecutive in A'_P to a_i be $a_{j'}$ and $a_{j''}$ and the arcs consecutive in A'_Q to a_j be $a_{i'}$ and $a_{i''}$. Let the ordered pairs $(a_{j'}, a_i)$, $(a_i, a_{j''})$, $(a_{i''}, a_j)$, $(a_j, a_{i'})$ be in $O(\vec{\mathfrak{A}})$. Then the ordered pairs $(a_{j'}, a_i)$, $(a_i, a_{j''})$, $(a_{i'}, a_j)$, $(a_j, a_{i''})$ are in $O((\vec{\mathfrak{A}})')$. The paths for l in the β determined by $\vec{\mathfrak{A}}$ are $(a_{j'}a_i \cdots a_j a_{j''})$, $(a_{j''}a_i \cdots a_j a_{i''})$ and those in the β' determined by $(\vec{\mathfrak{A}})'$ are $(a_{j'}a_i \cdots a_j a_{i''})$, $(a_{j''}a_i \cdots a_j a_{i'})$. The orientable mapping sets $[\mathfrak{A}, \beta]$ and $[\mathfrak{A}, \beta']$ are therefore different.

Now, an $\mathfrak{A}$ has $2^{\alpha_s^0}$ different orientations in which two by two are opposite. Hence, there are

$$\frac{1}{2}2^{\alpha_s^0}\prod_{i=1}^{\alpha_s^0}\frac{\lfloor d_i-1}{2}=\frac{1}{2}\prod_{i=1}^{\alpha_s^0}\lfloor d_i-1$$

Orientable mapping sets of G. This is

Theorem 8. The number of orientable mapping sets of G in $\frac{1}{2}\prod_{i=1}^{\alpha_s^0} \underline{|d_i-1}$.

The number X is in general greater than -1. $X=-1$, if and only if G is a tree. This, together with Theorem 6 and the results at the end of ζ 4, gives

Theorem 9. *A* graph has no nonorientable mapping sets, when and only when it is a tree.

7. Another method of determining the mapping sets. Let a graph G be given. We can find a sequence of connected graphs G_1, ⋯, $G_{\alpha'}(=G)$, where G_i is a subgraph of G_{i+1} obtained from G_{i+1} by letting fall a single arc, and where G_1 contains only a single arc. If we can determine all the mapping sets of G_{i+1} from those of G_i, all the mapping sets of G can then be determined step by step starting from the single mapping set of G_1. Accordingly, we need only to discuss how all the mapping sets of a graph G can be determined from those of G' obtained by letting fall a single arc of G.

We shall call for convenience a free vertex P the common endpoint of the arc at P and the arc itself. Thus, if a and a' are consecutive in a mapping set, we may speak of the common endpoint of a and a', no matter $a=a'$ or not.

Let G' be obtained from G by letting fall the arc $a=PQ$. Let the arcs of G' be a_1, ⋯, $a_{\alpha'}$. Consider a mapping set $\mathfrak{M}'$ of G'. From $\mathfrak{M}'$, we can construct mapping sets of G in the following manner. If a has free endpoint, say Q, in G, then Q is not a vertex of G'. Let a_i, a_l be consecutive arcs in some cycle $c'=a_i\cdots a_l$ of $\mathfrak{M}'$, whose common endpoint is P. Form the cycle $c=aaa_i\cdots a_l$. I say $\mathfrak{M}=\mathfrak{M}'-C'+C$ is a mapping set of G. M_i) is obviously satisfied. Let $\mathfrak{A}'=\{A'_R\}$ belong to $\mathfrak{M}'$, and $A'_P=a_i\cdots a_l$. From $A_P=aa_i\cdots a_l$ and $A_Q=a$. Then $\mathfrak{A}=\mathfrak{A}'-A'_P+A_P+A_Q$ is evidently what is required in M_{ii}) for $\mathfrak{M}$ to be a mapping set. Now, suppose neither P nor Q is free in G. Let a_i, a_l be consecutive arcs in some cycle c'_1 of $\mathfrak{M}'$, whose common endpoint is P, and a_j, a_m be consecutive arcs in same cycle c'_2 of $\mathfrak{M}'$, whose common endpoint is Q. First, suppose $C'_1 = C'_2 = a_i a_{i'} \cdots a_{j'} a_j \cdots a_m a_{m'} \cdots a_{l'} a_l$. From $C = aa_i a_{i'} \cdots a_{j'} a_j aa_l a_{l'} \cdots a_{m'} a_m$, $C_1 = aa_i a_{i'} \cdots a_{j'} a_j$, $C_2 = aa_l a_{l'} \cdots a_{m'} a_m$. I say both $\mathfrak{M}_1 = \mathfrak{M}' - C'_1 + C$ and $\mathfrak{M}_2 = \mathfrak{M}' - C'_1 + C_1 + C_2$ are mapping sets of G. M_i) is obviously satisfied. Let $\mathfrak{A}' = \{A'_R\}$ belong to $\mathfrak{M}'$, and $A'_P = a_i \cdots a_l$, $A'_Q = a_j \cdots a_m$. Form $A_P = aa_i \cdots a_l$, $A_Q = aa_j \cdots a_m$. Then $\mathfrak{A} = \mathfrak{A}' - A'_P - A'_Q + A_P + A_Q$ is what is required in M_{ii}) for $\mathfrak{M}_1$ and $\mathfrak{M}_2$ to be mapping sets. Secondly, suppose $C'_1 \neq C'_2$ and $C'_1 = a_i a_{i'} \cdots a_{l'} a_l$, $C'_2 = a_j a_{j'} \cdots a_{m'} a_m$. From $C_1 = aa_i a_{i'} \cdots a_{l'} a_l aa_j a_{j'} \cdots a_{m'} a_m$, $C_2 = aa_i a_{i'} \cdots a_{l'} a_l aa_m a_{m'} \cdots a_{j'} a_j$. Then both and are mapping sets of G. The proof is the same as in the first case.

Let $\{\mathfrak{M}\}_{\mathfrak{M}'}$ be the totality of mapping sets of G which can be obtained from $\mathfrak{M}'$ in the above described manner, and let $\{\mathfrak{M}\}_{G'} = \sum_{\mathfrak{M}' \lambda G'} \{\mathfrak{M}\}_{\mathfrak{M}'}$. We shall prove that $\{\mathfrak{M}\}_{G'}$ is the totality of all mapping sets of G. Let $\mathfrak{M}$ be any mapping set of G. If a has a free vertex in G, say Q, then a appears in $\mathfrak{M}$ in some cycle $C = caa_i \cdots a_l$. Form $C' = a_i \cdots a_l$. Then $\mathfrak{M}' = \mathfrak{M} - C + C'$ is a mapping set of G'. M_i) is obviously satisfied. Let $\mathfrak{A} = \{A_R\}$ belong to $\mathfrak{M}$ and $A_P = ca_i \cdots a_l$. Form $A'_P = a_i \cdots a_l$. Then $\mathfrak{A}' = \mathfrak{A} - A_P - A_Q + A'_P$ is what is required in M_{ii}) for $\mathfrak{M}$ to be a mapping set. Suppose neither P nor Q is free in G. First, suppose a appears in $\mathfrak{M}$ in only one cycle $C = aa_i a_{i'} \cdots a_{j'} a_j aa_l a_{l'} \cdots a_{m'} a_m$ [or $C = aa_i a_{i'} \cdots a_{l'} a_l aa_j a_{j'} \cdots a_{m'} a_m$], where a_i, a_l are arcs at P and a_j, a_m arcs at Q. From $C' = a_i a_{i'} \cdots a_{j'} a_j a_m a_{m'} \cdots a_{l'} a_l$ [or $C'_1 = a_i a_{i'} \cdots a_{l'} a_l$, $C'_2 = a_{j'} a_j \cdots a_m a_{m'}$. Then $\mathfrak{M}' = \mathfrak{M} - C + C'$ [or $\mathfrak{M}' = \mathfrak{M} - C + C'_1 + C'_2$] is a mapping set of G'. M_i) is obviously satisfied. Let $\mathfrak{A} = \{A_R\}$ belong to and $A_P = aa_i \cdots a_l$, $A_Q = aa_j \cdots a_m$. Form $A'_P = a_i \cdots a_l$, $A'_Q = a_j \cdots a_m$. Then $\mathfrak{A}' = \mathfrak{A} - A_P - A_Q + A'_P + A'_Q$ is what is required in M_{ii}) for $\mathfrak{M}'$ to be a mapping set. Secondly, suppose a appears in $\mathfrak{M}$ in two distinct cycles $C_1 = aa_i a_{i'} \cdots a_{j'} a_j$ and $C_2 = aa_l a_{l'} \cdots a_{m'} a_m$, where a_i, a_l are arcs at P and a_j, a_m arcs at Q. Form the cycle $C' = a_i a_{i'} \cdots a_{j'} a_j a_m a_{m'} \cdots a_{l'} a_l$. Then $\mathfrak{M}' = \mathfrak{M} - C_1 - C_2 + C'$ is a mapping set of G'. The proof is the same as in the first case. In all cases, $\mathfrak{M}$ can be obtained from $\mathfrak{M}'$ by the process in the preceeding paragraph.

Consequently, $\mathfrak{M}$ is in $\{\mathfrak{M}\}_{\mathfrak{M}'}$.

This method of determining the mapping sets has the following advantage. Let G' be any connected subgraph of G . The sequence G_1, …, $G_{\alpha'}$ may be chosen so that $G_{\alpha'(G')}=G'$. If all the mapping sets of G' are known, the mapping sets of G may be found by starting from these of G' . Thus, we may find once for all the mapping sets of number of graphs which or whose subdivisions are often subgraphs of others, thus shortening the process of determining the mapping sets of a suitable given graph. Examples or these graphs are the complete four points (whose mapping sets are tabulated in Ⅲ), the complete five points, and the graph obtained by joining with an arc each two vertices one in each of two sets of three vertices, etc.

If G is a tree, then, in any sequence G_1, …, $G_{\alpha'}$, the arc in G_{i+1} but not in G_i has a free endpoint in G_{i+1}. From the method of constructing the mapping sets of G_{i+1} from those of G_i, and from the fact that G_1 has only one mapping set which consists of only one cycle, we have immediately the following

Theorem 10. Any mapping set of a tree contains only a singledcycle.

8. Mappine forms. Genera of mapping forms and of graphs. we shall now drop the assumption that the graph considered is connected. Let a graph G have the components G_1, G_2, …, G_r : $G=G_1+G_2+\cdots+G_r (r \geqq 1)$. Then $\sum_{i=1}^{r}\mathfrak{M}_i$ where $\mathfrak{M}_i$ is a mapping set of G_i, is called a mapping set of G .

A set $\mathfrak{S}$ of cycles, when associated with two opposite orientations, wilI be denoted by $\overleftrightarrow{\mathfrak{S}}$. To orient $\overleftrightarrow{\mathfrak{S}}$ is to associate one of the two opposite orientations in $\overleftrightarrow{\mathfrak{S}}$ to $\mathfrak{S}$ to form an oriented set $\vec{\mathfrak{S}}$. Finally, a set $\mathfrak{S}$ or $\overleftrightarrow{\mathfrak{S}}$, when associated with a nonnegative integer g , will be denoted by $\mathfrak{S}+g$ or $\overleftrightarrow{\mathfrak{S}}+g$ respectively. When $g=0$, we shall write $\mathfrak{S}$ or $\overleftrightarrow{\mathfrak{S}}$ instead of $\mathfrak{S}+0$ or $\overleftrightarrow{\mathfrak{S}}+0$.

Definition. A set

(1) $\{\overleftrightarrow{\mathfrak{S}}_1+g_1, \cdots, \overleftrightarrow{\mathfrak{S}}_m+g_m, \overleftrightarrow{\mathfrak{S}}_{m+1}+g_{m+1}, \cdots, \overleftrightarrow{\mathfrak{S}}_n+g_n\}$

where $n \geqq 1$, $0 \leqq m \leqq n$, $g_i \equiv 0 \pmod 2$ for $i \leqq m$, $g_i>0$ for $i>m$, and

$\mathfrak{S}_1$, …, $\mathfrak{S}_n$ are jointly exhaustive, disjunct, nonempty subsets of a mapping set $\mathfrak{M} = \sum_{i=1}^{r} \mathfrak{M}_i$ of G ⑥, is called a (mapping) form of G belonging to $\mathfrak{M}$, provided

F) When the set {$\mathfrak{S}_1$, …, $\mathfrak{S}_n$} is divided in any way into two disjunct nonempty subsets, there exist an $\mathfrak{S}_i$ in one subset and an $\mathfrak{S}_j$ in the other and there exists an $\mathfrak{M}_t$ in {$\mathfrak{M}_1$, …, $\mathfrak{M}_r$} such that $\mathfrak{S}_i \cdot \mathfrak{M}_t \neq 0$ and $\mathfrak{S}_j \cdot \mathfrak{M}_t \neq 0$.

A mapping set $\mathfrak{M}$ of a connected graph G will be considered as a special form of G, belonging to this $\mathfrak{M}$, in which $m = n =$ *the number of cycles in* $\mathfrak{M}$, each cycle is associated with its two opposite orientations, and $g_i = 0$

As a generalization of orientability of a mapping set, we have

Definition. The form (1) is said to be orientable, if $m = n$ and if the set {$\overleftrightarrow{\mathfrak{S}}_1$, …, $\overleftrightarrow{\mathfrak{S}}_n$} can be oriented $\overleftrightarrow{\mathfrak{S}}_1$, …, $\overleftrightarrow{\mathfrak{S}}_n$ can be oriented so that opposite orientations are induced in every arc of G.

Definition. The number

$$X + r - 2n + \alpha + \sum_{i=1}^{n} g_i,$$

where r is the number of cycles in $\mathfrak{M}$, is called the genus of the form (1).

Hence the genus of a mapping set of α^2 cycles of a connected graph is

$$-\alpha^0 + \alpha^1 - \alpha^2 + \mathcal{L}.$$

Now, for given G, $n \leqq \alpha^1(G)$. Thus, the genera of all the forms of G has a minimum.

Definition. The minimum of the genera of all the orientable [nonorientable] mapping forms of G is called the orientable [nonorientable] genus of G.

9. Determination of the genera of a graph. Theorem 11. Let G be a connected graph, of which the orientable and non-orientable genora are h and k reapectively. All the orientable forms of genus h of G are orientable mapping sets of G. When $k \leqq h$,

---

⑥ If a component of G, say G_i, is a circle, one cycle may appear twice as elements of $\mathfrak{M}_i$. If such a cycle is present, it is considered to appear twice as elements of $\mathfrak{M}$, and twice as elements of $\mathfrak{S}_1$, …, $\mathfrak{S}_n$

all the nonorientable forms of genus k of G are nonorientable mapping sets of G. When $k>h$, then $k=h+1$ and if, in addition, G is not a tree, it has a nonorientable mapping set of genus k.

Proof. Suppose that $\mathfrak{F}=\{\overset{\leftrightarrow}{\mathfrak{S}}_1+g_1, \cdots, \overset{\leftrightarrow}{\mathfrak{S}}_n+g_n\}$ is an orientable form of G, belonging to the mapping set $\mathfrak{M}$ of α^2 cycles of G, and that the genus of $\mathfrak{F}$ is the orientable genus h of G. I say $\mathfrak{F}=\mathfrak{M}$, $\mathfrak{M}$ being considered as a special form of G. For, if $\mathfrak{F}\neq\mathfrak{M}$, then either $n<\alpha^2$ or some $g_i\neq 0$, and the genus of $\mathfrak{M}$ would be less than that of $\mathfrak{F}$. Moreover, since $\mathfrak{F}$ is orientable, $\mathfrak{M}$ must be orientable. The genus of $\mathfrak{F}$ could then not be the minimun h. Hence, any orientable form of genus is simply an orientable mapping set.

Suppose $k\leqq h$. Let $\mathfrak{F}=\{\overset{\leftrightarrow}{\mathfrak{S}}_1+g_1, \cdots, \overset{\leftrightarrow}{\mathfrak{S}}_m+g_m, \overset{\leftrightarrow}{\mathfrak{S}}_{m+1}+g_{m+1}, \cdots, \overset{\leftrightarrow}{\mathfrak{S}}_n+g_n\}$ be a nonorientable form of genus k belonging to the mapping set $\mathfrak{M}$ of α^2 cycles. I say $\mathfrak{F}=\mathfrak{M}$. For, if $\mathfrak{F}\neq\mathfrak{M}$, either $n<\alpha^2$ or some $g_i\neq 0$, and the genus of $\mathfrak{M}$ would be less than k. But since $k\leqq h$, there exists no form, orientable or nonorientable, of genus less than k. Hence, any nonorientable form of genus k is simply a nonorientable mapping set.

Now, suppose $k>h$ and G is not a tree, Let

$$\mathfrak{M}=\{c_1, c_2, c_3, \cdots, c_{\alpha^2}\}$$

be an orientable mapping set of genus h. Then $\alpha^2>1$. For, if G is a circle, from the end of §4, the only orientable mapping set of G has two elements. If G is not a circle, there is a simple cycle in G. It is easy to see that we can define an $\vec{\mathfrak{A}}$ which determines as in §6 an orientable mapping set in which the simple cycle is an element. This orientable mapping set has obviously more than one element. Since $\mathfrak{M}$ has the maximun number of elements, $\alpha^2>1$.

Now, I say that C_1 must have an arc in common with another cycle in $\mathfrak{M}$. For, if other wise, let the totality of arcs in C_1 be M_1 and the totality of arcs in $C_2\cdots$, C_{α^2} be $\mathfrak{M}_2$. Then $M_1\cdot M_2=0$. Since G is connected, there exist a_1 *in* M_1 and a_2 in M_2

which have a common endpoint P .

Consider the arrangement A_p in $\mathfrak{A}$ belonging to $\mathfrak{M}$. From M and $M_1 \cdot M_2 = 0$, tho arcs consecutive to a_1 *in* A_p would be in M_1, and so would be their consecutive arcs in A_p , otc. Consequently, all the arcs at P , and, in particular, a_2, would be in M_1. This contradiction shows that C_1 have an arc, say $\overline{PQ}$, in common with another cycle, say C_2.

Suppose $C_1 = \overline{PQQR} \cdots \overline{SQ}$, $C_2 = \overline{PQQV} \cdots \overline{VP}$. Form

the cycle $C = \overline{PQ}\,\overline{QR} \cdots \overline{SQ}\,\overline{PQ}\,\overline{QV} \cdots \overline{VP}$. Then $\mathfrak{M}' = \{C_1, C_3, \cdots, C_n\}$ is obviously a nonorientable mapping set. The genus of $\mathfrak{M}'$ is $h+1$. Since $h > k$, and k is the minimum of the geners of all the nonorientable forms, $k = h+1$, the genus of $\mathfrak{M}'$.

To show that, when G is a tree, we still have $k = h+1$, let us determine the genera of a tree. From Theorems 9 and 10, any mapping set of a tree is orientable and consists of a single cycle. Hence for any mapping set of a tree, $X = -1$. $\alpha^2 = 1$, and therefore the genus is 0. From the above results, $h = 0$. A nonorientable form of a tree must be of the form $\{c + g\}$, the genus of which is g . since $g \geqq 1$, then $k = 1$. Consequently $k = h+1$.

From Theorem 11 and the last paragraph in its proof, we have

Theorem 12. The orientable genus of a connected graph is equal to the minimum of the genera of all the orientable mapping sets of G . If G is not a tree, the nonorientable genus of G is equal to the minimim of the genera of all the nonorientable mapping sets of G ; and, if G is a tree, its nonorientable genus is 1.

From Theorem 12, the genera of connected graph can be determined directly from its mapping sets without any construction of forms. The genera of a disconnected graph can be determined from those of its components in virtue of the following theorem:

Theorem 13. Let G be a graph whose components are G_1, $\cdots$, G_y . Let h_i and k_i

be the orientable and nonorientable genera of G_i respectively. Then the orientab le and nonorientable genera of G are respectively $h = h_1 + \cdots + h_y$ and k = the minimum of $x_1 + \cdots + x_y$ where x_i is h_i *or* k_i and at leaat one x_i is k_i.

Proof. For a connected graph the theorem is obvioue. We shall prove by mathematical induction the theorem for $G = G_1 + \cdots + G_y$, on the supposition that it is true for any graph of $y-1$ components, and, in particular, for $G' = G_1 + \cdots + G_{y-1}$.

We shall prove only the formula for k, The proof of the fomula for h is simpler and will follow by analogy.

Let

$$\mathfrak{F} = \{\overleftrightarrow{\mathfrak{S}}_1 + g_1, \cdots, \overleftrightarrow{\mathfrak{S}}_m + g_m, \overleftrightarrow{\mathfrak{S}}_{m+1} + g_{m+1}, \cdots, \overleftrightarrow{\mathfrak{S}}_{m+n} + g_{m+n}\}$$

be a nonorientable mapping form of G, belonging to the mapping set $\mathfrak{M} = \mathfrak{M}_1 + \cdots + \mathfrak{M}_y$, and its gon us is the nonorientable genus k of G:

$$x + r - \alpha(m+n) + \alpha + \sum_{i=1}^{m+n} g_i = k$$

Suppose that the sets G_i is in $\mathfrak{F}$ are so enumerated that

$$\mathfrak{S}_1, \cdots, \mathfrak{S}_{m'}, \mathfrak{S}_{m+1}, \cdots, \mathfrak{S}_{m+n'}$$

are all those $\mathfrak{S}_i$ is which do contain cycles in $\mathfrak{M}_y$. Divide each of these sets into 2 disjunct subsets $\mathfrak{S}_i{}'$ and $\mathfrak{S}_i{}''$, the former consisting exclusively of cycles in $\mathfrak{S}_y$ and the latter cycles in $\mathfrak{S}'$, Consider the set

$$\{\mathfrak{S}'_1, \cdots, \mathfrak{S}'_{m'}, \mathfrak{S}'_{m+1}, \cdots, \mathfrak{S}'_{m+n'}\}$$

By definition, every element of this set is a nonempty set of cycles and their sum is $\mathfrak{M}_y$. Because $\mathfrak{S}_y$ is connected, this set has the property F) with respect to $\{\mathfrak{M}_y\}$, i. e. , F) as described above except that $\{\mathfrak{M}_1, \cdots, \mathfrak{M}_y\}$ is replaced by $\{\mathfrak{M}_y\}$.

Let $\mathfrak{S} = \mathfrak{S}''_1, \cdots, \mathfrak{S}''_{m'}, \mathfrak{S}''_{m+1}, \cdots, \mathfrak{S}''_{m+n'}$ and form the second set

(3) $$\{\mathfrak{S}_1, \mathfrak{S}_{m'+1}, \cdots, \mathfrak{S}_m, \mathfrak{S}_{m+n'+1}, \cdots, \mathfrak{S}_{m+n}\}$$

$\mathfrak{S}$ is not empty, because of the property F) of the form $\mathfrak{F}$. Moreover, the sum of the elements in (3) is the mapping set $\mathfrak{M}' = \mathfrak{M}_1 + \cdots + \mathfrak{M}_{r-1}$ of G'. From the proporty F) of $\mathfrak{F}$ we shall proove as follows that the set (3) has also the proporty F) with respect to $\{\mathfrak{M}_1, \cdots, \mathfrak{M}_{r-1}\}$ of G'. Let this set be divided in any way into two disjunct nonempty subsets S_1 and S_2, and suppose that the subset containing $\mathfrak{S}$ is S_1. Then $S'_1 = S_1 - \mathfrak{S} + \{\mathfrak{S}_1, \cdots, \mathfrak{S}_{m'}, \mathfrak{S}_{m+1}, \cdots, \mathfrak{S}_{m+n'}\}$ and S_2 are two jointly exhaustive, disjunct, nonempty subsets of $\{\mathfrak{S}_1, \cdots, \mathfrak{S}_{m+n}\}$, the set in $\mathfrak{F}$. Because of the property F) of $\mathfrak{F}$ there exist $\mathfrak{S}_i$ in S'_1, $\mathfrak{S}_j$ in S_2, and $\mathfrak{M}_t$ such that $\mathfrak{S}_i \cdot \mathfrak{M}_t \neq 0$, $\mathfrak{S}_j \cdot \mathfrak{M}_t \neq 0$. If $i = 1, \cdots, m', m+1, \cdots,$ or $m+n'$, then $t \neq r$, since S_2 contains no cycles in G_r. In the first case, there exist $\mathfrak{S}_i$ in S_1, $\mathfrak{S}_j$ in S_2 and $\mathfrak{M}_t$ such that $\mathfrak{S}_i \cdot \mathfrak{M}_t \neq 0$, $\mathfrak{S}_j \cdot \mathfrak{M}_t \neq 0$. In the second case, there exist $\mathfrak{S}$ ($\supset \mathfrak{S}''_i$) in S_1, G_j in S_2 and $\mathfrak{M}_t$ such that $\mathfrak{S} \cdot \mathfrak{M}_t \supset \mathfrak{S}''_i \cdot \mathfrak{M}_t = \mathfrak{S}_i \cdot \mathfrak{M}_t \neq 0$. This proves that (3) has the property F) with respect to $\{\mathfrak{M}_1, \cdots, \mathfrak{M}_{r-1}\}$

Now from $\mathfrak{F}$ and the sets (2) and (3) respectively we shall construct the form

$$\mathfrak{F}_r = \{\overleftrightarrow{\mathfrak{S}}'_1 + g_1, \cdots, \overleftrightarrow{\mathfrak{S}}'_{m'} + g_{m'}, \overleftrightarrow{\mathfrak{S}}'_{m+1} + g_{m+1}, \cdots, \overleftrightarrow{\mathfrak{S}}'_{m+n'} + g_{m+n'}\}$$

of G_y belonging to $\mathfrak{M}_y$ and the form

$$\mathfrak{F}' = \{\overleftrightarrow{\mathfrak{S}}_1, \overleftrightarrow{\mathfrak{S}}_{m'+1} + g_{m'+1}, \cdots, \overleftrightarrow{\mathfrak{S}}_m + g_m, \mathfrak{S}_{m+n'+1} + g_{m+n'+1}, \cdots, \mathfrak{S}_{m+n} + g_{m+n}\}$$

of G' belonging to $\mathfrak{M}'$ so that at least one of them is nonortentable. The opposite orientations of $\overleftrightarrow{\mathfrak{S}}_j$ in $\mathfrak{F}'$, $j = m'+1, \cdots, m$, are those in $\mathfrak{F}$. In order to complete the construction it remains to define the opposite orientations of $\overleftrightarrow{\mathfrak{S}}'_i$, $i = 1, \cdots, m'$, and $\overleftrightarrow{\mathfrak{S}}$. If $n = 0$, the opposite orientations in each $\overleftrightarrow{\mathfrak{S}}'_i$ is defined to be the subsets of the opposite orientations in $\overleftrightarrow{\mathfrak{S}}'_i$. If $\{\overleftrightarrow{\mathfrak{S}}'_1, \cdots, \overleftrightarrow{\mathfrak{S}}'_{m'}\}$ is now nonorientable, define arbitrarily the opposite orientations in $\overleftrightarrow{\mathfrak{S}}$. If, however, $\{\overleftrightarrow{\mathfrak{S}}'_1, \cdots, \overleftrightarrow{\mathfrak{S}}'_{m'}\}$ is orientable, it has two coherent orientations. In each of these orientations, there is a well defined $\overleftrightarrow{\mathfrak{S}}'_i$ and hence a well defined $\vec{\mathfrak{S}}_i$ (*an orionted* $\overleftrightarrow{\mathfrak{S}}_i$) [7]. The totality of the orientations of the cycles in $\vec{\mathfrak{S}}_i$, but not in $\vec{\mathfrak{S}}'_i$, defines an orientation of $\mathfrak{S}$. The two

---

⑦ If the two orientations in $\vec{\mathfrak{S}}'$ are equal, we choose an arbitrary orientation in $\overleftrightarrow{\mathfrak{S}}_i$ as the orientation determined by $\vec{\mathfrak{S}}'_i$.

orientations so defined are opposite and will be defined to be those in $\overleftrightarrow{\mathfrak{S}}$, since the given form $\mathfrak{F}$ is nonorientable and $\{\overleftrightarrow{\mathfrak{S}}_1, \cdots, \overleftrightarrow{\mathfrak{S}}'_{m'}\}$ is orientable, $\{\overleftrightarrow{\mathfrak{S}}, \overleftrightarrow{\mathfrak{S}}_{m'+1}, \cdots, \overleftrightarrow{\mathfrak{S}}_m\}$ must be nonorientable. Finally, if $n > 0$, define $\overleftrightarrow{\mathfrak{S}}'_i$, $\overleftrightarrow{\mathfrak{S}}$ arbitrarily. At least one of $\mathfrak{F}_y$ and $\mathfrak{F}'$ is also nonorientable.

Let the characteristics of G_y, G' be X_y, X', and the numbers of elements in $\mathfrak{M}_y$, $\mathfrak{M}'$ be r_y, r' respectively. Then the genera of $\mathfrak{F}_y$ and $\mathfrak{F}'$ are respectively

$$g(\mathfrak{F}_y) = x_y + r_y - \alpha(m' + n') + \alpha + \sum_{i=1}^{m'} g_i + \sum_{i=1}^{n'} g_{m+i},$$

$$g(\mathfrak{F}') = x' + r' - \alpha(m + n - m' - n') + \sum_{i=m'+1}^{m} g_i + \sum_{i=n'+1}^{n} g_{m+i}$$

Now, from $x = x_y + x'$, $r = r_y + r'$, and from the values of $g(\mathfrak{F}_y)$, $g(\mathfrak{F}')$, we have immediataly $g(\mathfrak{F}_y) + g(\mathfrak{F}') = k$. From the hypothesis of induction, if $\mathfrak{F}'$ is orientable, $g(\mathfrak{F}') \geqq h_1 + \cdots + h_{y-1}$ and $g(\mathfrak{F}_y) \geqq k_y$; and, if $\mathfrak{F}'$ is nonorientable, $g(\mathfrak{F}_y) \geqq \min(x_1 + \cdots + x_{y-1})$, where x_i is h_i or k_i and at least one x_i is k_i, and $g(\mathfrak{F}_y) \geqq \min(h_y, k_y)$. Hence

(4) $k \geqq \min(x_1 + \cdots + x_{y-1})$, where x_i is h_i or k_i and at least one x_i is k_i.

Now to complete the proof of our theorem, we shall prove (4) with the inequality sign reversed. Let $min(x_1 + \cdots + x_{y-1})$, where x_i is h_i or k_i and at least one x_i is k_i, be $x'_1 + \cdots + x'_y$. Since our theorem is true for G_y and is supposed to be true for G', there are a form

$$\mathfrak{F}_y^* = \{\overleftrightarrow{\mathfrak{S}}_1^{**} + g_1^{**}, \cdots, \overleftrightarrow{\mathfrak{S}}_{p''}^{**} + g_{p''}^{**}, \overleftrightarrow{\mathfrak{S}}_{p''+1}^{**} + g_{p''+1}^{**}, \cdots, \overleftrightarrow{\mathfrak{S}}_{q''}^{**} + g_{q''}^{**}\}$$

of G_y of genus $g(\mathfrak{F}_y) = x'_y$, which belongs to the mapping set $\mathfrak{M}_y^*$ and which is orientable or nonorientable according as x'_y is h_y or k_y; and a form

$$\mathfrak{F}^* = \{\overleftrightarrow{\mathfrak{S}}_1^* + g_1^*, \cdots, \overleftrightarrow{\mathfrak{S}}_{p'}^* + g_{p'}^*, \overleftrightarrow{\mathfrak{S}}_{p'+1}^* + g_{p'+1}^*, \cdots, \overleftrightarrow{\mathfrak{S}}_{q'}^* + g_{q'}^*\}$$

of G' of genus $g(\mathfrak{F}^*) = x'_1 + \cdots + x'_{y-1}$, which belongs to the mapping set $\sum_{i=1}^{y-1} \mathfrak{M}_i^*$ and which is orientable or nonorientable according as none or at least one of x'_i, $i = 1, \cdots, y-1$, is k_i.

It is obvious that one of $\mathfrak{F}^*$, $\mathfrak{F}_y^*$ is nonorientable. Let $\mathfrak{S}^* = \mathfrak{S}_1^* + \mathfrak{S}_1^{**}$. Consider the set

(5) $$\{\mathfrak{S}^*, \mathfrak{S}_2^*, \cdots, \mathfrak{S}_{q'}^*, \mathfrak{S}_2^{**}, \cdots, \mathfrak{S}_{q''}^{**}\}.$$

It has the property F). For, let it be divided in any way into two disjunct nonempty subsets S_1 and S_2, and suppose that the subset containing $\mathfrak{S}^*$ is S_1. Then at least one of $S_2^* = S_2 \cdot \{\mathfrak{S}_2^*, \cdots, \mathfrak{S}_{q'}^*\}$, $S_2^{**} = S_2 \cdot \{\mathfrak{S}_2^{**}, \cdots, \mathfrak{S}_{q''}^{**}\}$ is nonempty. If $S_2^{**} \neq 0$, there are $\mathfrak{S}^*$ in S_1 and certain $\mathfrak{S}_i^{**}$ in S_2 and $\mathfrak{M}_y^*$ such that $\mathfrak{S}^* \cdot \mathfrak{M}_y^* \neq 0$, $\mathfrak{S}_i^{**} \cdot \mathfrak{M}_y^* \neq 0$. If $S_2^* \neq 0$, then $S_1^* = S_1 \cdot \{\mathfrak{S}_2^*, \cdots, \mathfrak{S}_{q'}^*\} + \mathfrak{S}_1^*$ and S_2^* are two jointly exhaustive, disjinct, nonempty subsets of $\{\mathfrak{S}_1^*, \cdots, \mathfrak{S}_{q'}^*\}$. From the property F) of $\mathfrak{F}^*$ there are $\mathfrak{S}_i^*$ in S_1^* and $\mathfrak{S}_j^*$ in S_2^* and an $\mathfrak{M}_t^*$ $(t < y)$ such that $\mathfrak{S}_i^* \cdot \mathfrak{M}_t^* \neq 0$, $\mathfrak{S}_j^* \cdot \mathfrak{M}_t^* \neq 0$. If $\mathfrak{S}_i^* \neq \mathfrak{S}_1^*$, $\mathfrak{S}_i^*$ is in S_1 and $\mathfrak{S}_j^*$ in S_2. If $\mathfrak{S}_i^* = \mathfrak{S}_1^*$, there are $\mathfrak{S}^*$ in S_1 and $\mathfrak{S}_j^*$ in S_2 such that $\mathfrak{S}^* \cdot \mathfrak{M}_t^* \supset \mathfrak{S}_1^* \cdot \mathfrak{M}_t^* \neq 0$, $\mathfrak{S}_j^* \cdot \mathfrak{M}_t^* \neq 0$. This proves that (5) has the property F).

Now define a form $\mathfrak{F}_1$ of G as follows. If some g_i^* or g_i^{**} is not zero, let

$$\mathfrak{F}_1 = \{\overleftrightarrow{\mathfrak{S}}_2^*, \cdots, \overleftrightarrow{\mathfrak{S}}_q^*, \overleftrightarrow{\mathfrak{S}}_2^{**}, \cdots, \overleftrightarrow{\mathfrak{S}}_{q''}^{**}, \mathfrak{S}^* + \sum_{i=1}^{q'} g_i^* + \sum_{i=1}^{q''} g_i^{**}\},$$

Where $\overleftrightarrow{\mathfrak{S}}_{p'+1}^*, \cdots, \overleftrightarrow{\mathfrak{S}}_{q'}^*, \overleftrightarrow{\mathfrak{S}}_{p''+1}^{**}, \cdots, \overleftrightarrow{\mathfrak{S}}_{q''}^{**}$ are assigned arbitrarily. If all g_i^* and g_i^{**} are zero, and therefore, $q' = p'$ and $q'' = p''$,

Let

$$\mathfrak{F}_1 = \{\overleftrightarrow{\mathfrak{S}}^*, \overleftrightarrow{\mathfrak{S}}_2^*, \cdots, \overleftrightarrow{\mathfrak{S}}_p^*, \overleftrightarrow{\mathfrak{S}}_2^{**}, \cdots, \overleftrightarrow{\mathfrak{S}}_{p''}^{**}\}.$$

where one orientation in $\overleftrightarrow{\mathfrak{S}}^*$ is the sum of one orientation in $\overleftrightarrow{\mathfrak{S}}_1^*$ and one orientation in $\overleftrightarrow{\mathfrak{S}}_1^{**}$, and the other orientation in $\overleftrightarrow{\mathfrak{S}}^*$ is the sum of the remaining orientations in $\overleftrightarrow{\mathfrak{S}}_1^*$ and $\overleftrightarrow{\mathfrak{S}}_1^{**}$. In either case, $\mathfrak{J}_1$ is nonorientable and therefore of a genus $g(\mathfrak{J}_1) \geqq k$. But, as in the proof of (4), it is obvious that $g(\mathfrak{J}_1) = g(\mathfrak{J}_1^*) + g(\mathfrak{J}^*) = x'_1 + \cdots + x'_{y-1} + x'_y$. Hence $x'_1 + \cdots + x'_y \geqq k$.

This result together with (4) establishes the formula for k.

Ⅱ. CONNECTION BETWEEN MAPPINGS AND MAPPING FORMS

10. From a mapping to a form. Let G be a graph, whose components are G_1, …, G_r. Since a mapping of G corresponds naturally to a mapping of any subdivision G^* of G and from Theorem 4 a mapping set [a mapping form] of G corresponds naturally to a mapping set [a mapping form] of G^*, we shall in the sequel make no distinction between mappings and mapping forms of G and those of a subdivision, and shall accordingly assume that G has been subdivided as fine as we please.

Let G be mapped through a (topological) mapping M in a closed surface S. Divide S into a simplicial complex of which $G' = M(G)$ (G is supposed to have been subdivided sufficiently fine) is a subcomplex. S may be considered as obtained through identification of pairs of sides of the triangles in it. Let us form the identification in two steps: first those pairs which after identification become arcs not of G', and, secondly, those pairs which after identification become arcs of G'.

Through the first step, we get a complex consisting of n bounded surfaces S_1, …, S_n as components, and, through the second step, we get S anew.

Let S_1, …, S_m, $0 \leqq m \leqq n$, be orientable and S_{m+1}, …, S_n nonorientable, and let S_i be of r_i holes and of genus g_i.

As subcomplexes of $\sum S_i$, the rims of S_1, …, S_n are circles[8], and identification of pairs of arcs of these circles gives G'. Let these circles be K_{ij}, $j = 1, \cdots, r_i$, $i = 1, \cdots, n$, and the simple cycle in K_{ij} be C'_{ij}. Denote by F the second identification which is considered as a continuous mapping defined on $\sum S_i$:

$$F(\sum S_i) = S,\quad F(\sum K_{ij}) = G',\quad F(\textit{an arc of } \sum K_{ij}) = \textit{an arc of } G'.$$

---

[8] To avoid circles of two arcs, we assume that if a component of G is a single broken-arc, it has been subdivided so as to contain at least two arcs.

Since the image of an arc in C'_{ij} through F is an arc of G', C'_{ij} and $M^{-1}F$ define a general cyclic arrangement C_{ij} of arcs of G, which is obviously a cycle. Denote $\{C_{i1}, \cdots, C_{ir_i}\}$ by $\mathfrak{S}_i$ and $\sum \mathfrak{S}_i$ by $\mathfrak{M}$.

Let the totality of cycles in $\mathfrak{M}$ that are cycles in the component G_t of G be $\mathfrak{M}_t$: $\mathfrak{M} = \sum_{t=1}^{y} \mathfrak{M}_t$. Then, $\mathfrak{M}_t$ is a mapping set of G_t. For, since there are just two arcs in $\sum K_{ij}$, corresponding to any arc of G through $M^{-1}F$, Mi) is satisfied. To show that Mii) is also satisfied, consider a vertex P of G. Since S is a manifold, the triangles and areas on S incident with $P' = M(P)$ can be arranged as follows:

$$\Delta_1,\ \alpha'_1,\ \Delta_2,\ \alpha'_2,\ \cdots,\ \Delta_e,\ \Delta'_e,$$

so that Δ_u and Δ_{u+1} have the common side α_u, u taken mod e. Let the arcs of G' which appear among $\alpha'_1, \cdots, \alpha'_e$ be $\alpha'_{u_1}, \cdots, \alpha'_{u_d}$, $1 \leqq u_1 < \cdots < u_d \leqq e$, and $\alpha'_{u_r} = M(\alpha_{u_\sigma})$. Form the proper cyclic arrangement

(1) $$\alpha_{u_1},\ \alpha_{u_2},\ \cdots,\ \alpha_{u_d}.$$

Now, the triangles $\Delta_{u_\upsilon+1}, \cdots, \Delta_{u_\sigma+1}$, are connected together along the arcs $\alpha'_{u_\upsilon+1}, \cdots, \alpha'_{u_{\sigma+1}-1}$ through the first identification. Hence, of the two pairs of arcs $F^{-1}(\alpha'_{u_\upsilon})$ and $F^{-1}(\alpha'_{u_{\upsilon+1}})$, the two arcs of the two triangles $F^{-1}(\Delta'_{u_\sigma+1})$ and $F^{-1}(\Delta'_{u_{\sigma+1}})$ are consecutive in some cycle C'_{ij} and therefore α_{u_υ} and $\alpha_{u_{\sigma+1}}$ are consecutive in $\mathfrak{M}_t$. (1) is thus what is required at P in M ii) for $\mathfrak{M}_t$ to be a mapping set. Since $\mathfrak{M}_t$ is a mapping set of G_t, $\mathfrak{M}$ is a mapping set of G.

The set $\{\mathfrak{S}_1, \cdots, \mathfrak{S}_n\}$ has the property F. For, let it be divided in any way into two disjunct nonempty subsets. This division corresponds to a division of $\{S_1, \cdots, S_n\}$ into two disjunct nonempty subsets. Since S is connected, there exists S_i and S_j one in each subset and G_t such that a certain arc in $K_{i1}, \cdots, K_{ir_i}$ and a certain arc in $K_{j1}, \cdots, K_{jr_j}$ correspond to the same arc of $M(G_t)$ through F. Thus, $\mathfrak{S}_i \cdot \mathfrak{M}_t \neq 0$, $\mathfrak{S}_j \cdot \mathfrak{M}_t \neq 0$.

Finally, when $i \leqq m$, two opposite orientations of $\{C'_{i1}, \cdots, C'_{ir_i}\}$ are induced by (the two coherent orientations of the triangles in) S_i. These orientations define two opposite orientations of $\mathfrak{S}_i$ through $M^{-1}F$, thus giving an $\overleftrightarrow{\mathfrak{S}}_i$.

Consequently, we get a form of G

$$\mathfrak{F}=\{\overleftrightarrow{\mathfrak{S}}_1+g_1,\ \cdots,\ \overleftrightarrow{\mathfrak{S}}_m+g_m,\ \mathfrak{S}_{m+1}+g_{m+1},\ \cdots,\ \mathfrak{S}_n+g_n\}.$$

Again, G, M, and S are supposed given. Now, let S_1^*, $\cdots$, $S_{n^*}^*$ be a set of bounded surfaces, of which S_1^*, $\cdots$, $S_{m^*}^*$ are orientable and $S_{m^*+1}^*$, $\cdots$, $S_{n^*}^*$ nonorientable. $0\leqq m^*\leqq n^*$. Let S_i^* be of r_i^* holes and of genus g_i^*, and the rims of the r_i^* holes of S_i^* be divided into circles K_{ij}^*, $j=1$, $\cdots$, r_i^*. Suppose that the following condition is fulfilled by an identification F^* of pairs of arcs in $\sum K_{ij}^*$:

(2) $F^*(\sum S_i^*)=S$, $F^*(\sum K_{ij}^*)=G'$, F^*(*an arc of* $\sum K_{ij}^*$)=*an arc of* G'.

Then, the set

$$\mathfrak{F}^*=\{\overleftrightarrow{\mathfrak{S}}_1^*+g_1^*,\ \cdots,\ \overleftrightarrow{\mathfrak{S}}_{m^*}^*+g_{m^*}^*,\ \overleftrightarrow{\mathfrak{S}}_{m^*+1}+g_{m^*+1}^*,\ \cdots,\ \mathfrak{S}_{n^*}^*+g_{n^*}^*\}$$

formed in the same way as $\mathfrak{F}$ above, is a mapping form of G and equal to $\mathfrak{F}$. For, it is obvious that any two points in $F^*(S_i^*)-G'$ can be connected by a curve not containing points in G', but any point in $F^*(S_i^*)-G'$ and any point in $F^*(S_j^*)-G'$, $i\neq j$, cannot be connected by such a curve. Hence, $F^*(S_i^*)-G'$, $i=1$, $\cdots$, n^*, are the domains into which G' divides $S-G'$. Similarly, $F(S_i)-G'$, $j=1$, $\cdots$, n, are also these domains. Thus, $n=n^*$, and S_j, S_i^* can be rearranged so that $F^*(S_i^*)=F(S_i)$, $i=1$, $\cdots$, n. Consequently, a simplex in the simplicial division of S above considered, which is a simplex in $F(S_i)$, but not of G' may be carried through F^{*-1} into a simplex in S_i^*. These simplexes together with those in $\sum\limits_j K_{ij}^*$ form a simplicial division of S^*, which is isomorphic to the complex S_i and in the isomorphism two corresponding arcs in the rima of S_i and S_i^* correspond to the same area of G' through F^* and F. Hence, the set $\mathfrak{F}^*$ is the same as $\mathfrak{F}$ and is therefore a form of G.

Thus, the form $\mathfrak{F}$ may be considered as determined from any set of bounded surfaces S_1^* , …, $S_{n^*}^*$, and an arbitrary identification F^* with the property (2). $\mathfrak{F}$ is said to belong to M .

Now, the characteristic of S_i is $g_i + r_i - \pi$ [9]. Through the identification, $2\alpha^1(G) - \alpha^0(G)$ vertices and $\alpha^1(G)$ arcs are diminished. Hence, the characteristic of S is

$$\sum_{i=1}^{n}(g_i + r_i - 2) + 2\alpha^1(G) - \alpha^0(G) - \alpha^1(G) \text{ or } X(G) + \sum_{i=1}^{n} r_i - 2u + \sum_{i=1}^{n} g_i \text{ , and}$$

its genus is $X(G) + \sum_{i=1}^{n} r_i - 2u + 2 + \sum_{i=1}^{n} g_i$ or equal to the genus of $\mathfrak{F}$.

If $\mathfrak{F}$ is orientable, then $m = n$ and the set $\{\overleftrightarrow{\mathfrak{S}}_1, \cdots, \overleftrightarrow{\mathfrak{S}}_n\}$ is orientable. Corresponding to a coherent orientation of $\{\overleftrightarrow{\mathfrak{S}}_1, \cdots, \overleftrightarrow{\mathfrak{S}}_n\}$, we have for every i an orientation of $\{C'_{i1}, \cdots, C'_{ir_i}\}$ induced by a coherent orientation of the triangles in S_i . Through F , we obtain a corresponding orientation of the triangles in S . This orientation induces opposite orientations in each arc of G' because the chosen orientation of $\{\overleftrightarrow{\mathfrak{S}}_1, \cdots, \overleftrightarrow{\mathfrak{S}}_n\}$ is coherent, and induces opposite orientations in each arc of $S - G'$ because the orientation of the triangles in S_i is coherent. Consequently, S is orientable. Conversely, it is also easy to see that, if S is orientable, so must $\mathfrak{F}$ be.

Hence,

Theorem 14. The form belonging to a mapping M of G in a closed surface S has the same genus and orientability as S .

11. From a form to a mapping, Let

$$\mathfrak{F} = \{\overleftrightarrow{\mathfrak{S}}_1 + g_1, \cdots, \overleftrightarrow{\mathfrak{S}}_m + g_m, \mathfrak{S}_{m+1} + g_{m+1}, \cdots, \mathfrak{S}_n + g_n\}$$

be a form of G belonging to the mapping set $\mathfrak{M} = \sum_{t=1}^{y} \mathfrak{M}_t$. Suppose that $\mathfrak{S}_i$ contains r_i cycles.

Corresponding to $\mathfrak{S}_i$ take a bounded surface S_i of r_i holes and of genus g_i , which is orientable, when and only when $i \leqq m$. Set up an arbitrary one-to-one correspondence between the rims of the r_i holes of s_i and r_i cycles in $\mathfrak{S}_i$. When a cycle

---

⑨ See Seifert-Threlfall, loc. cit.

C_{i1}, $j=1$, …, r_i in $\mathfrak{S}_i$ consists of a certain number of arcs (one arc appearing in the cycle twice will be counted twice in this number), divide the corresponding rim in S_i into a circle K_{ij} of the same number of arcs and denote the simple cycle in K_{ij} by C'_{ij}. Let us choose functions f_{ij}., $j=1$, …, r_i, $i=1$, …, n, such that C'_{ij}. and f_{ij}. define the corresponding cycles C_{ij} in $\mathfrak{S}_i$, and such that, when $0 \leqq m$, the two opposite orientations of $\{C'_{i1}, \cdots, C'_{ir_i}\}$ induced by S_i define the two opposite orientations in $\overset{\leftrightarrow}{\mathfrak{S}}_i$ (ξ1). Let the functions f_{ij} be summarized under the name f.

Now, through f each arc in $\{C'_{11}, \cdots, C'_{nr_n}\}$ has as the unique image an arc in $\mathfrak{M}$. The common endpoint of two consecutive arcs in $\{C'_{11}, \cdots, C'_{nr_n}\}$ will be said to have as the unique image the common endpoint (ξ7) of the two corresponding consecutive arcs in $\mathfrak{M}$. Two arcs or vertices in $\sum K_{ij}$ will be said to be equivalent, when they have the same image. From the property M i) of a mapping set and the function f, the number of arcs in $\sum K_{ij}$ is even and the arcs are equivalent by pairs. Through the identification of each pair of equivalent arcs so that the two pairs of equivalent endpoints are identified, there results from $\sum S_i$ a complex S consisting of a number of closed surfaces as components, and from $\sum K_{ij}$ a graph G' in S. Denote the identification by F. Then

(1) $F(\sum S_i)=S$, $F(\sum K_{ij})=G'$, $F(an\ arc\ of\ K_{ij})=an\ arc\ of\ G'$.

I declare that G is mapped topologically by a mapping M in S so that G' is isomorphic to G:

$$M(G)=G'.$$

Obviously, it suffices to prove that all the equivalent vertices in $\sum K_{ij}$ are identified. Through F. Consider the equivalent vertices which have any vertex P (of G_t, say) as image. If P is free or ordinary, this is obvious. Suppose that P is singular, and that the proper cyclic arrangement at P as required in M ii) for $\mathfrak{M}_t$ be

$$\overline{PQ_1}\ \overline{PQ_1}\cdots\overline{PQ_d}.$$

Then in $\mathfrak{M}_t$ there are the pairs of consecutive arcs:

(2) $\overline{Q_1P}$, $\overline{PQ_2}$; $\overline{Q_2P}$, $\overline{PQ_3}$; …; $\overline{Q_{d-1}P}$, $\overline{PQ_d}$; $\overline{Q_dP}$, $\overline{PQ_1}$.

Let the pair of consecutive arcs in $\{C'_{11}, \cdots, C'_{nr_n}\}$ having the pair $\overline{Q_iP}$, $\overline{PQ_{i+1}}$ as image be $\overline{Q''_{i+1}P'_i}$, $\overline{P'_iQ'_{i+1}}$. Then, corresponding to (2), there are the pairs of consecutive arcs

(3) $\overline{Q''_1P'_1}$, $\overline{P'_1Q'_2}$; $\overline{Q''_2P'_2}$, $\overline{P'_2Q'_3}$; …; $\overline{Q''_{d-1}P'_{d-1}}$, $\overline{P'_{d-1}Q'_d}$; $\overline{Q''_dP'_d}$, $\overline{P'_dQ'_1}$.

These areas are distinct, since the pairs in (2) are distinct.

Now, in (3), there are already two arcs $\overline{P'_iQ'_{i+1}}$, $\overline{P'_{i+1}Q''_{i+1}}$ have the arc $\overline{PQ_{i+1}}$ in (2) as image, so there is no arc that is not in (3) and has any arc in (2) as image. Hence, P'_1, $P'_2\cdots$, P'_d are all the equivalent vertices having P as image. Now, the identification of $\overline{P'_iQ'_{i+1}}$ and $\overline{P'_{i+1}Q''_{i+1}}$ makes P'_i and P'_{i+1} coincide. Consequently, the vertices P'_1, $P'_2\cdots$, P'_d are identified through F.

S must be connected and hence is a single closed surface. In fact, if otherwise, we should be able to divide the set of bounded surfaces into two disjunct nonempty subsets so that the image of any bounded surface in the one subset and that of any bounded surface in the other belong to different components of S. This division would correspond to a division of the set $\{\mathfrak{S}_1, \cdots, \mathfrak{S}_n\}$ into two disjunct nonempty subsets. There would exist then by F) $\mathfrak{S}_i$ and $\mathfrak{S}_j$ one in each subset and $\mathfrak{M}_t$ such that $\mathfrak{S}_i \cdot \mathfrak{M}_t \neq 0$, $\mathfrak{S}_j \cdot \mathfrak{M}_t \neq 0$. G_t would be mapped in different components of S. But, since G_t is connected, this cannot be the case.

Now, from (1), the bounded surfaces S_1, …, S_n and F have the property (2) in §10 with the stars dropped. The form belonging to M may then be considered as determined from S_1, …, S_n and F. But the form thus determined is obviously $\mathfrak{J}$ from the constructions of S_1, …, S_n and F.

Hence,

Theorem 15. Any form of G belongs to a mapping of G in a closed surface.

12. Orientable and nonorientable surfaces minimal for a group. In Kagol' s terminology, a closed surface S is said to be minimal for a group G, if G can be mapped in S but cannot be mapped in any closed surface of orientability same as S but of genus less than that of S. From Theorems 14 and 15, we have immediately

Theorem 16. The genus of an orientable [a nonorientable] surface minimal for G is equal to the orientable [nonorientable] genus of G.

If the formbelong to the mapping M of a connected graph G in S is in particular a mapping set, then $M(G)$ divides S into surface elements. Hence, there follows from Theorems 13, 14, and 15

Theorem 17. When a connected graph G is mapped through any mapping M in an orientable surface minimal for it, $M(G)$ divides the surface into surface elements. If the nonorientable genus of G is less than or equal to its orientable genus, then, when it is mapped through any mapping M in a nonorientable surface minimal for it, $M(G)$ divides the surface into surface elements. If the nonorientable genus of G is greater (hence, greater by 1) than its orientable genus and if G is not a tree, we can define a mapping M of G in any nonorientable surface minimal for it so that $M(G)$ divides the surface into surface elements.

13. Types of mappings. Definition. A mapping M of G in S and a mapping M' of G in S' are said to be of the same type if a topological transformation T of S on S' exists such that for any point P in G

(1) $$TM(P)=M'(P).$$

Theorem 18. A mapping M of G in S and a mapping M' of G in S' are of the same type if and only if the forms belonging to M and M' are the same.

Proof. Suppose that M and M' are of the same type. Then there exists a topological transformation T of S on S' such that (1) holds. Let $\mathfrak{F}$ be the form belonging to M and be considered as determined from a simplicial division of S as in §10. Through T, the simplicial division of S is carried into an isomorphic simplicial division of S', and, from (1), this isomorphism induces an isomorphism of $M(G)$ and $M'(G)$ in which two corresponding arcs or vertices correspond to the same arc or vertex of G through M^{-1} and M'^{-1}. From the process of determining the form belonging to a mapping given in § 10, we see immediately that the form belonging to M' is the same as $\mathfrak{F}$.

Conversely, suppose that the forms belonging to M and M' are the same

$$\mathfrak{F}=\{\overset{\leftrightarrow}{\mathfrak{S}}_1+g_1, \cdots, \overset{\leftrightarrow}{\mathfrak{S}}_m+g_m, \mathfrak{S}_{m+1}+g_{m+1}, \cdots, \mathfrak{S}_n+g_n\}.$$

To prove that M and M' are of the same type, we assume the truth of the following lemma, which is intuitionally obvious and not difficult to be proved:

<u>Lemma</u>. Let S^* and S^{**} be two homeomorphic bounded surfaces and let the rims of the holes or S^* and S^{**} be $K_1^*, \cdots, K_r^*$ and $K_1^{**}, \cdots, K_r^{**}$. Suppose that t_i is an arbitrary topological transformation of K_i^* on K_i^{**}, and, when S^* and S^{**} are orientable, suppose in addition that these transformations carry an "orientation" of $\{K_1^*, \cdots, K_r^*\}$ induced by S^* into an "orientation" of $\{K_1^{**}, \cdots, K_r^{**}\}$ induced by S^{**}. Then, there exists a topological transformation t of S^* on S^{**} that in identical with t_i in K_i^*.

Now, let $\mathfrak{F}$, as the form belonging to M [M'], be obtained from the bounded surfaces $S_1, \cdots, S_n$ [$S_1', \cdots, S_n'$], the rims of whose holes are divided into circles K_{ij} [K_{ij}'], $j=1, \cdots, r_i$, $i=1, \cdots, n$, the simple cycles in which are C_{ij} [C_{ij}'], $j=1, \cdots, r_i$, $i=1, \cdots, n$, and from the identification F [F'] such that

$$F\left(\sum S_i\right)=S,\ F\left(\sum K_{ij}\right)=M(G),\ F\left(an arc of \sum K_{ij}\right)=an arc of\ M(G).$$

$$\left[F'\left(\sum S_i'\right)=S',\ F'\left(\sum K_{ij}'\right)=M'(G),\ F'\left(an\ arc\ of\ \sum K_{ij}'\right)=an\ arc\ of\ M'(G).\right]$$

Suppose that $S_1, \cdots, S_n$ are obtained from a simphicial division of S as in § 10. Since both S_i and S_i' are of r_i holes, of genus g_i, and orientable if and only if $i \leqq m$, they are homeomorphic. Further, since C_{ij} and C_{ij}' define through $M^{-1}F$ and $M'^{-1}F'$ respectively the same cycle in $\mathfrak{S}_i$, we can define an isomorphism t_{ij} of K_{ij} on K_{ij}' such that if Q is any point in K_{ij}

(2) $$M'^{-1}F't_{ij}(Q)=M^{-1}F(Q),$$

and, since both the orientations of $\{C_{i1}, \cdots, C_{ir_i}\}$ induced by S_i and these of $\{C'_{i1}, \cdots, C'_{ir_i}\}$ induced by S'_i define the orientations in $\overleftrightarrow{\mathfrak{S}}_i$ for $i \leqq m$, the isomorphisms $t_{i1}, \cdots, t_{ir_i}$ may be chosen so as to carry an orientation of $\{K_{i1}, \cdots, K_{ir_i}\}$ induced by S_i into an orientation of $\{K'_{i1}, \cdots, K'_{ir_i}\}$ induced by S'_i. Hence, by the Lemma, there exists a topological transformation t_i of S_i on S'_i that is identical with t_{ij} in K_{ij}. t_i carries the simphicial division of S_i into a simphicial division of S'_i and t_i may thus be considered as an isomorphism of the complex S_i on the complex S'_i. On account of (2), the isomorphisms $t_1, \cdots, t_n$ thus defined carry arcs or vertices in $\sum K_{ij}$ to be identified through F to arcs or vertices in $\sum K'_{ij}$ to be identified through F'. Thus, performing the identifications F and F'. we obtain from $\sum S_i$ and $\sum S'_i$ isomorphic simphicial divisions of S and S' respectively. In the isomorphism, an arc or a vertex of $M(G)$ corresponds to an arc or a vertex of $M'(G)$ and conversely, any two corresponding arcs or vertices of $M(G)$ and $M'(G)$ correspond to the same arc or vertex of G through M^{-1} and M'^{-1}. Denote the isomorphism of S on S' by T. Then, if P is any point in

$$TM(P)=M'(P),$$

and M and M' are of the same type.

From this theorem, we see that a form may be considered as a model of a type of mappings. Thus, when the form $\mathfrak{F}$ belongs to M, we may say that M is <u>of the type</u> $\mathfrak{F}$.

14. <u>Categories and subcategories of mappings.</u> In this section, we fix our attention on mappings of G in a given closed surface S.

Let M, M' be of types

$$\mathfrak{F}=\{\overleftrightarrow{\mathfrak{S}}_1+2h_1, \cdots, \overleftrightarrow{\mathfrak{S}}_m+2h_m, \mathfrak{T}_1+k_1, \cdots, \mathfrak{T}_n+k_n\},$$

$$\mathfrak{F}'=\{\overleftrightarrow{\mathfrak{S}}_1'+2h_1', \cdots, \overleftrightarrow{\mathfrak{S}}_{m'}'+2h_{m'}', \mathfrak{T}'_1+k_1', \cdots, \mathfrak{T}_{n'}'+k_{n'}'\}$$

respectively. Let δ_i, t_j, $\delta_{i'}'$, $t_{j'}'$ be the number of cycles in $\mathfrak{S}_i$, $\mathfrak{T}_j$, $\mathfrak{S}_{i'}'$, $\mathfrak{T}_{j'}'$, respectively, M and M' are said to be the same category, if $m = m'$, $n = n'$, and $\mathfrak{S}_1$, …, $\mathfrak{T}_n$, $\mathfrak{S}_1'$, …, $\mathfrak{T}_{n'}'$ can be rearranged so that $\delta_i = \delta_i'$, $i = 1$, …, m, $t_j = t_j'$, $j = 1$, …, n. The category is denoted by $[(S_1 \cdots S_m)(t_1 \cdots t_n)]$. Only when S is orientable, we write also $[S_1 \cdots S_m]$ instead of $[(S_1 \cdots S_m)()]$.

M and M' are said to be of the same subcategory, if $m = m'$, $n = n'$, and $\mathfrak{S}_1$, …, $\mathfrak{T}_n$, $\mathfrak{S}_1'$, …, $\mathfrak{T}_{n'}'$ can be rearranged so that $\overleftrightarrow{\mathfrak{S}}_i = \overleftrightarrow{\mathfrak{S}}_i'$, $i = 1$, …, m, $\mathfrak{T}_j = \mathfrak{T}_j'$, $j = 1$, …, n. The subcategory is denoted by $[\overleftrightarrow{\mathfrak{S}}_1, \cdots, \overleftrightarrow{\mathfrak{S}}_m, \mathfrak{T}_1 \cdots, \mathfrak{T}_n]$.

It is obvious that a category falls into a number of subcategories, and a subcategory falls into a number of types.

Theorem 19. Let $r = \sum_{i=1}^{m} \delta_i + \sum_{j=1}^{n} t_j$, $x =$ characteristic of G, $g =$ genus of S, and $c = g + 2(m + n) - x - r - 2$. The number of types in the subcategory $[\overleftrightarrow{\mathfrak{S}}_1, \cdots, \overleftrightarrow{\mathfrak{S}}_m, \mathfrak{T}_1 \cdots, \mathfrak{T}_n]$. is $\binom{m + c/2 - 1}{c/2}$ provided $n = 0$, $\binom{c-1}{c-n}$ provided $m = 0$, and

$$\sum_{a=0}^{\left[\frac{c-n}{2}\right]} \binom{m-a-1}{a} \binom{c-2a-1}{c-2a-n}$$

provided $m \neq 0 \neq n$.

Proof. Let the $\mathfrak{F}$ at the beginning of the present sections represent a type in the subcategory. Then

$$x + r + 2 - 2(m+n) + 2\sum_{i=1}^{m} h_i + \sum_{j=1}^{n} k_j = g, \text{ or}$$

$$2\sum_{i=1}^{m} h_i + \sum_{j=1}^{n} k_j = c$$

The problem is merely to find the number of solutions of the equation

$$2\sum_{i=1}^{m} x_i + \sum_{j=1}^{n} y_j = c$$

in nonnegative x_i' s and positive y_i' s. Each solution corresponds to a type and conversely.

If $n=0$, we have

$$\sum_{i=1}^{m} x_i = \frac{c}{2},$$

which has $\binom{m+c/2-1}{c/2}$ solutions. If $m=0$, we have

$$\sum_{j=1}^{n} y_j = c,$$

which has $\binom{c-1}{c-n}$ solutions. Finally, if $m \neq 0 \neq n$, the original equation may be replaced by the simultaneous equations

$$\sum_{i=1}^{m} x_i = a, \quad \sum_{j=1}^{n} y_i = c-2a,$$

$$0 \leqq a \leqq \left[\frac{c-n}{2}\right].$$

Now, the first equation has for fixed a $\binom{m+a-1}{a}$ solutions, and the second equation $\binom{c-2a-1}{c-2a-n}$ solutions. The simultaneous equations have then for fixed a $\binom{m+a-1}{a}\binom{c-2a-1}{c-2a-n}$ solutions, and the original equation has in the total

$$\sum_{a=0}^{\left[\frac{c-n}{2}\right]} \binom{m+a-1}{a}\binom{c-2a-1}{c-2a-n}$$

solutions.

It is obvious from this theorem that two subcategories in the same category are composed of the same number of types.

Ⅲ. AN EXAMPLE: MAPPINGS OF THE COMPLETE FOUR POINTS

15. Mapping sets of the complete four points. We shall apply the previous results to the investigation of the mappings of the complete four points, the graph of four vertices A, B, C, D, each two of which are joined by an arc.

The graph is connected and of characteristic 2. The degree of each vertex is 3.

Hence, it has $\frac{1}{2}\cdot 2^4=8$ orientable and $2^2\cdot 2^4-8=56$ nonorientable mapping sets. By means of the method described in § 5 or § 7, we find all its mapping sets, of which the following is a complete list (where $ABCDA$ for example, represents the cycle $\overline{AB}\,\overline{BC}\,\overline{CD}\,\overline{DA}$).

Orientable mapping sets

7 of two cycles

ABCDBACBDA, ACDA　ACDBCABDA, ABCDA

ACBDCABCDA, ABDA　ABDCBACDA, ACBDA

ABDCADBCDA, ABCA　ABCADCBDA, ABDCA

ABCADBACDA, BCDB

1 of four cycles

ABCA, ABDA, ACDA, BCDB

Nonorientable mapping sets

28 of a single cycle

ABCABDACDBCDA　ABDCADBCABCDA

ABCABDCBDACDA　ABDCADCBACBDA

ACBACDBCDABDA　ABCDACBDCABDA

ACBACDABDCBDA　ABCDBACDACBDA

ABCADBACDBCDA　ABDACDBCABCDA

ABCADBCDBACDA　ABDCBACBDACDA

ABCADCBDCABDA　ABCDACDBCABDA

ABCADCABDCBDA　ABCDBACBDACDA

ABCDACDBACBDA　ACBDCABCDABDA

ABCDBCABDACDA　ABDCBACDACBDA

ABDCABCDACBDA　ABCABDCADBCDA

ACBDCBACDABDA　ABCABDCADCBDA

ABCDBCADCABDA　ABDCABCADBCDA

ABCDBCADBACDA　ABDCABCADCBDA

21 of two cycles

ABCABDA, ACDBCDA　ABCDACBDA, ABDCA

ABCABDCBDA, ACDA　ABCDA, ACBDCABDA

ACBACDBCDA, ABDA　ABCDBACDA, ACBDA

ACBACDA, ABDCBDA　ABDA, ACDBCABCDA

ABCA, ABDACDBCDA　ABDCBACBDA, ACDA

ABCA, ABDCBDACDA　ABCABDACDA, BCDB

ABCDA, ACDBACBDA　ACBACDABDA, BCDB

ABCDBCABDA, ACDA　ABCADBCDA, ABDCA

ABDCABCDA, ACBDA　ABCA, ABDCADCBDA

ACBDCBACDA, ABDA　ABCADCABDA, BDDB

ABCDBCA, ABDACDA

7 of three cycles

ABCA, ABDA, ACDBCDA　ABCABDA, ACDA, BCDB

ABCA, ABDCBDA, ACDA　ACBACDA, ABDA, BCDB

ABCDBCA, ABDA, ACDA　ABCA, ABDACDA, BCDB

ABCDA, ACBDA, ABDCA

From the list, we see that maximum number of cycles in orientable mapping sets is 4 and that in nonorientable mapping sets is 3. Thus, an orientable surface minimal for the graph is of genus $2-4+2=0$ and is therefore a sphere; a nonorientable surface minimal for the graph is of genus $2-3+2=1$ and is therefore a projective plane. These results are trivial, because we do know that the graph can be mapped in a sphere and a projective plane. But it is interesting to discuss the types of mappings in a given (closed) surface.

16. Types of mappings of the complete four points. Let S be a closed surface of genus g. In how many ways can we map the complete four points in S, or, more

precisely how many types of mappings of the graph in S are there? Now, Theorem 19 gives us the number of types in a subcategory. If we know the number of categories and the number of subcategories in every category, we shall obtain the required number.

If S is orientable, there are 7 categories in general, i. e., provided g is sufficiently large, and, if S is nonorientable, there are 36 categories in general. The following lists gives for every category the number of subcategories in it and the number of types in each subcategory. Also the minimum g for which a category actually exists is given.

S being Orientable ($g=2h$)

Category	Number of subc.	Number of types in each subc.	Minimum h
[2]	7	1	2
[11]	7	h	1
[4]	1	1	3
[13]	4	$h-1$	2
[22]	3	$h-1$	2
[112]	6	$\frac{1}{2}h(h+1)$	1
[1111]	1	$\frac{1}{6}(h+1)(h+2)(h+3)$	0

S being nonorientable

Category	No. of subc.	Number of types in each subc.		Minimum h	
		$g=2k+1$	$g=2k$	$g=2k+1$	$g=2k$
[(1)()]	28	1	—	1	—
[()(1)]	28	1	1	2	2
[(2)()]	49	—	1	—	2
[()(2)]	28	1	1	2	3
[(11)()]	21	—	k	—	1

续 表

Category	No. of subc.	Number of types in each subc.		Minimum h	
		$g=2k+1$	$g=2k$	$g=2k+1$	$g=2k$
[()(11)]	28	$2k-2$	$2k-3$	2	2
[(1)(1)]	56	k	$k-1$	1	2
[(3)()]	28	1	—	2	—
[()(3)]	7	1	1	3	3
[(12)()]	42	k	—	1	—
[()(12)]	21	$2k-3$	$2k-4$	2	3
[(2)(1)]	42	$k-1$	$k-1$	2	2
[(1)(2)]	21	$k-1$	$k-1$	2	2
[(111)()]	7	$\frac{1}{2}(k+1)(k+2)$	—	0	—
[()(111)]	7	$(k-1)(2k-1)$	$(k-1)(2k-3)$	2	2
[(11)(1)]	21	$\frac{1}{2}k(k+1)$	$\frac{1}{2}k(k+1)$	1	1
[(1)(11)]	21	k^2	$k(k-1)$	1	2
[(4)()]	7	—	1	—	3
[()(4)]	1	1	1	3	4
[(13)()]	12	—	$k-1$	—	2
[()(13)]	4	$2k-4$	$2k-5$	3	3
[(3)(1)]	16	$k-1$	$k-2$	2	3
[(1)(3)]	4	$k-1$	$k-2$	2	3
[(22)()]	9	—	$k-1$	—	2
[()(22)]	3	$2k-4$	$2k-5$	3	3
[(2)(2)]	12	$k-1$	$k-2$	2	3
[(112)()]	6	—	$\frac{1}{2}k(k+1)$	—	1
[()(112)]	6	$(k-1)(2k-3)$	$(k-2)(2k-3)$	2	3
[(2)(11)]	12	$k(k-1)$	$(k-1)^2$	2	2
[(11)(2)]	6	$\frac{1}{2}k(k+1)$	$\frac{1}{2}k(k-1)$	1	2
[(12)(1)]	24	$\frac{1}{2}k(k+1)$	$\frac{1}{2}k(k-1)$	1	2
[(1)(12)]	12	$k(k-1)$	$(k-1)^2$	2	2

续 表

Category	No. of subc.	Number of types in each subc.		Minimum h	
		$g=2k+1$	$g=2k$	$g=2k+1$	$g=2k$
[()(1111)]	1	$\frac{2}{3}k(k-1)(2k-1)$	$\frac{1}{3}(k-1)(2k-1)(2k-3)$	2	2
[(111)(1)]	4	$\frac{1}{6}(k+1)(k+2)(k+3)$	$\frac{1}{6}k(k+1)(k+2)$	0	1
[(1)(111)]	4	$\frac{1}{6}k(k+1)(4k-1)$	$\frac{1}{6}k(k-1)(4k+1)$	1	2
[(11)(11)]	6	$\frac{1}{3}k(k+1)(k+2)$	$\frac{1}{6}k(k+1)(2k+1)$	1	2

I shall merely indicate the method of calculation by considering the categories [2] and [(2)()]. Here, for both [2] and [(2)()], $r=2$, $m=1$, $n=0$. Hence, $c=g+2(1+0)-2-2-2=g-4$. In order that [2] or [(2)()] may exist, c must be even and nonnegative and hence g must be even and $\frac{g}{2}\geqq 2$. By Theorem 19, each subcategory in [2] or [(2)()] is equal to $\binom{1+c/2-1}{c/2}=1$ type. Now, let $\mathfrak{M}$ be a mapping set of two cycles. A subcategory in [2] or [(2)()] is of the form $[\overleftrightarrow{\mathfrak{S}}]$. From $\mathfrak{M}$, we have just one way to form $\mathfrak{S}$ and hence just two ways to form $\overleftrightarrow{\mathfrak{S}}$. If $\mathfrak{M}$ is nonorientable, each $\{\overleftrightarrow{\mathfrak{S}}\}$ is nonorientable, and hence $\mathfrak{M}$ gives two subcategories in [(2)()]. If $\mathfrak{M}$ is orientable, one $\{\overleftrightarrow{\mathfrak{S}}\}$ is orientable and another $\{(\overleftrightarrow{\mathfrak{S}})'\}$ is nonorientable, and hence $\mathfrak{M}$ gives one subcategory in [2] and another in [(2)()]. But we have 21 nonorientable and 7 orientable mapping sets of two cycles. Consequently, [2] is composed of $7\times 1=7$ subcategories, and [(2)()], of $21\times 2+7\times 1=49$ subcategories.

The total number of types in each case is now easily calculated. For example, let S be orientable and $h\geqq 3$, The Number is

$$7+7h+1+4(h-1)+3(h-1)+\frac{6}{2}h(h+1)+\frac{1}{6}(h+1)(h+2)(h+3).$$

Thus, we have the following theorems.

Theorem 20. Let S be an orientable closed surface of genus $2h$, $h\geqq 3$.

There are

$$\frac{1}{6}(h^3 + 24h^2 + 113h + 12)$$

types of mappings of the complete four points in S. The graph can be mapped in a sphere, a torus, and a double-torus in 1, 17, and 56 ways respectively.

Theorem 21. Let S be a nonorientable closed surface of genus g. If $g = 2k + 1$ and $k \geqq 3$, there are

$$\frac{1}{6}(40k^3 + 660k^2 + 1664k - 516)$$

types of mappings of the complete four points in S. If $g = 2k$ and $k \geqq 4$, the number of types is

$$\frac{1}{6}(40k^3 + 591k^2 + 977k - 1242).$$

The graph can be mapped in a projective plane, a nonorientable torus, a sphere with 3 cross-caps, with 4 cross-caps, with 5 cross-caps, and with 6 cross-caps in 11, 58, 251, 532, 954, and 1347 ways respectively.

A System of Completely Independent Axioms for the Sequence of Natural Numbers*

1. Peano's five axioms for the sequence of natural numbers run as follows: (1) 1 is a number, (2) Any number a has a unique sequent a', (3) Different numbers have different sequents, (4) 1 is not a sequent, and (5) The principle of mathematical induction.

Now, in a system of axioms, it seems desirable, in the first place, to avoid symbols for particular individuals and symbols for functions, and, secondly, to secure the property of complete independence. In fact, Peano's system can be replaced by an equivalent one which fulfills both requirements. In such a system, the symbol "1" and the function "sequent" will not appear, although they may be introduced afresh through definitions. Besides, the principle of mathematical induction will no more play the rôle of an axiom, because, under the second requirement, it can hardly be stated in a simple and elegant form.

To exhibit such a system is the object of the present note. The axioms, together with a few deductions, will be given in § 2, and their complete independence will be discussed in § 3.

To deduce Peano's Axiom (3), we introduce the concept "the set of numbers from 1 to n." With this concept, the relation "being less than" can be defined, and the theorem on "recurrence definitions" proved. In virtue of this theorem, the definitions of addition and multiplication can be given, and the categoricalness of the

* 本文原载于《符号逻辑杂志》(*The Journal of Symbolic Logic*) 1943 年第 8 卷第 1 期。

present system proved immediately. These developments, however, need not concern us here.

2. Let N be a non-empty set and P a set of ordered pairs of elements of N. By $P(a, b)$ and $\overline{P}(a, b)$ will be meant respectively that (a, b) is in P and that (a, b) is not in P. The set $\{N, P\}$ is called a *progression*, if the following axioms are fulfilled:

AXIOM Ⅰ. *For any a (in N), there exists x such that $P(a, x)$;*

AXIOM Ⅱ. *If $P(a, b)$ and $P(a, c)$, then $b=c$;*

Axiom Ⅲ. *If, for all x, $\overline{P}(x, a)$, and, for all x, $\overline{P}(x, b)$, then $a=b$;*

Axiom Ⅳ. *If A is a non-empty sub-set of N, then there exists x in A such that, for all y in A, $\overline{P}(y, x)$.*

Suppose that $\{N, P\}$ is a progression. We shall hold this progression fixed till the close of the present section, and shall call the elements of N *natural numbers or simply numbers*.

From Axioms Ⅰ and Ⅱ, we see that, for any number a, there is a unique number x such that $P(a, x)$. This number x will be called the sequent of a and denoted by a'. From Axiom Ⅲ and Axiom Ⅳ for $A=N$, we see that there is a unique number x such that, for all y, $\overline{P}(y, x)$. This number x will be denoted by 1.

THEOREM 1 (Principle of mathematical induction). *A set A of numbers is identical with the set N of all numbers, if (i) 1 is in A, and (ii) n' is in A, whenever n is in A.*

Proof. Let B be the set of numbers not in A. Suppose that b is any number in B. By (i), $b \neq 1$. Hence, by Axiom Ⅲ, we have an x such that $x'=b$. But, by (ii), x cannot be in A, and, therefore, must be in B. Thus, for any number b in B, there is an x in B such that $P(x, b)$. By Axiom Ⅳ, B must be empty.

When A is considered as determined by a property, Theorem 1 yields the ordinary method of proof by mathematical induction.

A set S of numbers is called a *section*, if b is in S, whenever a is in S and $b'=a$. A set R of numbers is called a remainder, if a' is in R, whenever a is in R. The set N is obviously a section as well as a *remainder*. The intersection of all sections containing n is still a section and will be denoted by S_n. The intersection of all remainders containing n is still a remainder and will be denoted by R_n.

THEOREM 2. *If a is in R_n and $a \neq n$, then there exists an x in R_n such that $x'=a$.*

Proof. Suppose that the theorem were false. Then there would be an $a \neq n$ in R_n such that, for all x in R_n, $x' \neq a$, whence R_n minus $\{a\}$ would be a remainder containing n. But a proper sub-set of R_n cannot be a remainder containing n. This is a contradiction.

THEOREM 3. *n is not in $R_{n'}$.*

Proof. If n were in $R_{n'}$, then, by Theorem 2, for any number a in $R_{n'}$, there would exist an x in $R_{n'}$ such that $P(x, a)$. This conclusion and the fact that $R_{n'}$, is non-empty contradict Axiom Ⅳ.

THEOREM 4. *1 is in S_n.*

Proof. If 1 were not in S_n, then, for any number a in S_n, there would be an x in S_n such that $P(x, a)$ (by the definition of a section). This conclusion and the fact that S_n is non-empty contradict Axiom Ⅳ.

THEOREM 5. *n' is not in S_n.*

Proof. Let S^* denote S_n minus the intersection of S_n and $R_{n'}$. Then n' is not in S^*. If we can show that $S_n=S^*$, the theorem is proved. Let a be in S^* *and* $b'=a$. Then a and hence b are in S_n. Now, b is not in $R_{n'}$, for, otherwise, $a=b'$ would be in $R_{n'}$. Hence b is in S^* and S^* is a section. But, by Theorem 3, n is in S^*. Thus, S^* cannot be a proper sub-set of S_n, and $S_n=S^*$.

THEOREM 6. *If a is in S_n and $a \neq n$, then a' is in S_n.*

Proof. If a' were not in S_n, then S_n minus $\{a\}$ would be a section containing n.

THEOREM 7. *For any m and n, either m is in S_n, or n is in S_m.*

Proof. Suppose that n is not in S_m. Let A denote S_n minus the intersection of S_n and S_m. Since A is non-empty, we have an a in A such that, for all x in A, $x' \neq a$. By Theorem 4, $a \neq 1$. Hence, we have a number b such that $b' = a$. Since a is in S_n, b is in S_n and hence in S_m (because it is not in A). But, since $b' = a$ is not in S_m, $b = m$ by Theorem 6. Hence, m is in S_n.

THEOREM 8 (Uniqueness of the immediate predecessor). *If $a' = b'$, then, $a = b$.*

Proof. Consider the sections S_a and S_b. By Theorem 7, either a is in S_b, or b is in S_a. By symmetry, we may suppose that b is in S_a. Since $b' = a'$ is, by Theorem 5, not in S_a, b must be equal to a by Theorem 6.

THEOREM 9. *$S_{n'}$ is equal to the union S^* of S_n and $\{n'\}$.*

Proof. Since $S_{n'}$ is a section containing n, S_n and hence S^* are sub-sets of $S_{n'}$.

Now, S^* is a section. For, let a be in S^* and $b' = a$. If $a \neq n'$, then a is in S_n and hence b is in S_n and therefore in S^*. If $a = n'$, then $b = n$ by Theorem 8 and hence b is in S^*. But S^* contains n'. Thus, $S_{n'}$ is a sub-set of S^*.

3. The complete independence of the system of axioms is proved by means of the following sixteen examples. The example with the symbol (+−+−), for instance, fulfills Axioms Ⅰ and Ⅲ but not Axioms Ⅱ and Ⅳ.

(++++). A progression.

(++−+). A progression with a foreign object a and the ordered pair (a, the second element in the progression).

(+−++). A progression with the additional pair (the first element, the third element).

(+−−+). A progression with a foreign object a and the pairs (a, the second element in the progression) and (a, the third element in the progression).

(+++−). Element: a. Pair: (a, a).

(++−−). Elements: a, b, c. Pairs: (a, c), (b, c), (c, c).

(+−+−). Elements: a, b. Pairs: (a, b), (a, a), (b, b).

(+−−−). Elements: a, b, c, d. Pairs: (a, b), (a, a), (b, b), (c, a), (d, b).

(−+++). Element: a. Pair: wanting.

(−+−+). Elements: a, b. Pair: wanting.

(−−++). Elements: a, b, c. Pairs: (a, b), (a, c).

(−−−+). Elements: a, b, c, d. Pairs: (a, b), (a, c).

(−++−). Elements: a, b. Pair: (a, a).

(−+−−). Elements: a, b, c. Pair: (a, a).

(−−+−). Elements: a, b. Pairs: (a, a), (a, b).

(−−−−). Elements: a, b, c, d. Pairs: (a, a), (a, b).

The first four examples are constructed on the supposition that a progression does exist. Now, examples for (++−+), (+−++), and (+−−+) are not progressions. Thus, the question arises, can examples for these cases be constructed by means of a finite number of objects? This question is answered in the negative by the following.

THEOREM 10. *A set $\{N, P\}$, where N is finite, cannot fulfill both Axioms* Ⅰ *and* Ⅳ.

Proof. We shall prove the theorem by mathematical induction. If N is equivalent to (in a one-to-one correspondence with) S_1, which contains merely the element 1, the truth of the theorem is obvious. Assume that the theorem is true when N is equivalent to S_n. Let us prove it for the case that N is equivalent to $S_{n'}$. Suppose, on the contrary, that N were equivalent to $S_{n'}$, but that $\{N, P\}$ fulfilled both Axioms Ⅰ and Ⅳ. By Ⅳ, there would be an element a in N such that, for all x in N, $\overline{P}(x, a)$. Let N^* denote the set N minus $\{a\}$, and P^* the set of all those ordered pairs in P which do not contain a as an element. In virtue of Theorem 9, the set N^* is easily seen to be equivalent to S_n. Since N^* is a sub-set of N, the set $\{N^*, P^*\}$ would fulfill Axiom Ⅳ. Since, for all x in N^*, $\overline{P}(x, a)$, $\{N^*, P^*\}$ would fulfill I also. These contradict the hypothesis of induction.

A Counter-Example to Grunwald's Theorem *

In a recent research, the author of this note found that Grunwald's Theorem on the existence of cyclic extensions over an algebraic number field was not always true. On the other hand, he has obtained a proof of the theorem under appropriate restrictions. Here, a counter-example will be given and existing proofs analyzed. The positive result will be published elsewhere.

Let K be a cyclic extension over an algebraic number field F, of degree l^s, l being a prime, and let Ω be the intermediate field of degree l. Suppose the finite prime $\mathfrak{p}$ in F does not divide l and is ramified in Ω. Then it must be completely ramified in K. Since any $\mathfrak{p}$-adic unit congruent to 1 mod $\mathfrak{p}$ is an $l^{s\,th}$ power, the $\mathfrak{p}$-conductor of K over F is exactly $\mathfrak{p}$. Consequently, the multiplicative group of the residue-class field of $\mathfrak{p}$ is homomorphic to the Galois-group of K through the norm-residue symbol. It follows $N\mathfrak{p}-1$ is divisible by l^s.

Now, let F be the rational field. We shall prove the following statement:

If K is cyclic of degree 2^s *over* F, $s \geqq 3$, *then the prime* 2 *either is completely ramified in* K *or has at least two distinct prime factors.*

This furnishes us our counter-example. Forinstance, the statement implies that there exists no cyclic extension of degree 8 over F, for which the local extension over the 2-adic field is unramified of the same degree.

To prove the statement, consider the quadratic field $\Omega = F(\sqrt{m})$ in K. Assume for the sake of argument that the statement is wrong. Then 2 remains prime in $F(\sqrt{m})$. It follows that m is of the form $8n+5$. So, there exists a prime p which

---

* 本文原载于《数学年刊》(*Annals of Mathematics*) 1948 年第 49 卷第 4 期。

divides m and is not of the form $8n+1$. Since p is ramified in Ω, we infer $p-1$ must be divisible by $2^s \geqq 8$. This is a contradiction.

To the author's knowledge, there exist in the literature two proofs of Grunwald's Theorem, namely, the original proof in *Crelle Journal 169 and G. Whaples' proof in Duke Mathematical Journal 9*. In the latter, the proof of a fundamental lemma reads in p. 467 as follows: "... since α_1 is an nth power at p_0, p_0 splits completely in $K_1 \mid k$ ⋯ both p_0, p_1 split completely in K_2," where $K_2 = k(\alpha_2^{1/n})$. This was a mistake. For, the irreducible equation of $\alpha_2^{1/n}$ over k must be fixed before the symbol $\alpha_2^{1/n}$ has any meaning and can be adjoined to k, because $x^n - \alpha_2 = 0$ is in general reducible. But then there is no reason why the nth roots of α_2 at p_0 and p_1 should satisfy the equation we choose. Of course, we may choose as our equation the one which the nth root of α_2 at say p_0 satisfies. Still, we cannot assure that both p_0 and p_1 have split factors in K_2, not to say that they split completely in the same.

Now, let us examine the original proof. It was based on *Hilfssatz* 2 in *Mathematische Annalen* 107. The statement proved by induction on pp. 152－3 in part B of the proof of the *Hilfssatz* was ambiguous. We may interpret it in two different ways (we do it only for $l=2$):

(A) Let ε_i, ρ_j, π_k be units, ideal powers, local primes in $k(i)$. Then

$$\prod \varepsilon_i^{\pi_i} \rho_j^{y_j} \pi_k^{z_k} i^z \overset{=}{2^s} 1$$

in K_{s+2} implies $x_i \equiv y_j \equiv z_k \equiv 0 (mod\, 2^s)$.

(B) Let ε_i, ρ_j, π_k be things in k (which does not contain i). Then

$$\prod \varepsilon_i^{\pi_i} \rho_j^{y_j} \pi_k^{zk} (-1)^z \overset{=}{2^s} 1$$

in K_{s+2} and not all of x_i, y_j, $z_k \equiv 0 (mod\, 2^s)$ imply that there exists in $k(i)$ some relation

(1) $$\prod \varepsilon_i^{\pi i} \rho_j^{y_j} \pi_k^{zk} i^z \overset{=}{2} 1, \quad \sharp$$

where not all of x_i, y_j, $z_k \equiv 0 (mod\, 2)$.

The first interpretation provides us no knowledge at all about the things in k. So, we may disregard it. If we interpret it as in (B), we must prove later that relation (1) $x_i \equiv y_j \equiv z_k \equiv 0(mod\,2)$, not merely that

$$\prod \varepsilon_i^{\pi i} \rho_j^{y_j} \pi_k^{zk} (-1)^z \overset{=}{2} 1$$

in $k(i)$ implies $x_i \equiv y_j \equiv z_k \equiv 0(mod\,2)$. But the former statement is false. Let k be the rational field. Then the things ε_i, ρ_j do not exist. Taking $z_k = z = 1$, $\pi_k = 2$, we have

$$2 \cdot i = (1+i)^2 \overset{=}{2} 1$$

A Dissertation Presented to the Faculty of Princeton University in Candidacy for the Degree of Doctor of Philosophy*

Recommended for Acceptance

by the Department of Mathematics

November 1948

* 本文为王湘浩1948年11月提交给普林斯顿大学的博士学位论文。该成果（*On Grunwald's Theorem*）后来发表于《数学年刊》（*Annals of Mathematics*）1950年第51卷第2期。

An Abstract of the Doctoral Dissertation

On Grunwald's Theorem

Abstract

It was found by the author that Grunwald's Theorem on the existence of cyclic extensions over an algebraic number field was not always true. The purpose of the dissertation is to exhibit a necessary and sufficient condition for the theorem to hold and to extend the theorem to a more general one in which the cyclic extensions to be constructed contains a cyclic extension previously given.

The method used was essentially that of G. Whaples. The mistake in his main lemma is corrected. In order not to base the theorem on any analytic-number-theoretic statements, a weak form of the theorem on the existence of primes in a generalized arithmetic progression in deduced from the first inequality in class field theory.

The main theorem is as follows:

Let C be cyclic of degree l^r over F, $r \geqq 0$. Let S be a finite set of primes in F, containing all primes ramified in C, and also all evenly-even primes in the special case when i) $l=2$, ii) i, $i \cos 2\pi/2^{t+1} \notin F$, iii) $s \geqq t+l$, iv) there exist no oddly-even primes outside S. To each $\mathfrak{p} \in S$, there is given a cyclic extension $K^{\mathfrak{p}}$ of degree $l^{r_p+S_p}$ over $F_{\mathfrak{p}}$, where $l^{r_{\mathfrak{p}}} = (C^{\mathfrak{p}} : F_{\mathfrak{p}})$. Let $s \geqq \max s_{\mathfrak{p}}$.

Suppose a definite isomorphism of the Galois-group of $K^{\mathfrak{p}}$ over $F_{\mathfrak{p}}$ into a cyclic group (σ) of order l^{r+s} is given so that $\left(\frac{K^{\mathfrak{p}}/F_{\mathfrak{p}}}{\alpha_{\mathfrak{p}}}\right)$ may be regarded as a homomorphic mapping of $\alpha_{\mathfrak{p}}$ into (σ). Furthermore, suppose a definite isomorphism of the Galois-group of C over F onto (σ) mod (σ^{l^r}) is given so that $\left(\frac{C/F}{\mathfrak{a}}\right)$ may be regarded as a homomorphic mapping of $\mathfrak{a}$ onto (σ) mod (σ^{l^r}).

General Hypotheses:

(a) $C^{\mathfrak{p}} \subset K^{\mathfrak{p}}$.

(b) $r_{\mathfrak{p}} > 0$ implies $s_{\mathfrak{p}} = s$.

(c) $\left(\frac{C/F}{\alpha_{\mathfrak{p}}}\right) \equiv \left(\frac{K^{\mathfrak{p}}/F_{\mathfrak{p}}}{\alpha_{\mathfrak{p}}}\right) \qquad (mod(\sigma^{l^r}))$.

Special Hypothesis: In the special case, the following condition should be fulfilled;

(d) $\prod_{\mathfrak{p} \in S} \left(\frac{K^{\mathfrak{p}}/F_{\mathfrak{p}}}{(\sec 2\pi/2^{t+1})^{2^s}}\right) = 1$

Conclusion: Under the above conditions, there exists a cyclic extension K of degree l^{r+s} over F such that

(A) $C \subset K$,

(B) the $\mathfrak{p}$-adic completion of K is $K^{\mathfrak{p}}$ for $\mathfrak{p} \in S$,

(C) $\left(\frac{K/F}{\alpha_{\mathfrak{p}}}\right) \equiv \left(\frac{K^{\mathfrak{p}}/F_{\mathfrak{p}}}{\alpha_{\mathfrak{p}}}\right)$.

As applications of the theorem, the theorem that every simple algebra over an algebraic number field is cyclic and the norm theorem of Hasse-Schilling-Maass are proved.

On Grunwald's Theorem

Introduction

In a note in these *Annals of Mathematics* 49, 1948, the author gave an example showing that Grunwald's Theorem was not always true as it stood. Our purpose here is to exhibit a necessary and sufficient condition for the theorem to hold. Besides, we shall extend the theorem to a more general one in which.

Ⅰ. the cyclic extension K to be constructed over F shall contain a given cyclic extension C,

Ⅱ. it shall have certain desirable local properties relative to a given cyclic extension $\widetilde{C}$.

The extended theorem will be applied in a separate paper to prove a statement concerning the commutator group of a simple algebra over an algebraic number field.

In Grunwald's original proof of his theorem, the cyclic extension K constructed had the special property that, apart from the given primes, only a single prime in F was ramified in K, provided that the given degree $(K:F)$ was a prime power. This property of K was sometimes very convenient. However, as we have seen in the note mentioned above, the proof did not apply when, roughly speaking, $(K:F)$ was even. We shall therefore neglect this point of the theorem altogether, and adopt the more flexible method of G. Whaples. For it is this method that renders the extended theorem easy to handle and enables us to avoid the use of the theorem on the existence of primes in a generalized arithmetic progression.

It turns out that the theorem that every simple algebra over an algebraic number field is cyclic, which is by far the most important application of Grunwald's Theorem, still holds. By a special device, the norm theorem Hasse-Schilling-Maass can also be proved without much difficulty.

Best thanks are to Professor Emil Artin for his kind directions and helpful suggestions throughout the preparation of this thesis.

1. Cyclotomic Extensions

Let F be an algebraic number field of finite degree over the rational field R. l being a rational prime, ζ_s will always denote a primitive $l^{s\,th}$ root of unity.

If l is odd, $F(\zeta_s)$ is cyclic over F, of degree a divisor of $l^{s-1}(l-1)$. When $l = 2$, $F(\zeta_s) = F(e^{2\pi i/2^s})$ over F is either cyclic or is the direct product of two cyclic components. For our purpose, it is important to determine these components explicitly.

First, consider the field $R(e^{2\pi i/2^s})$, assuming $s \geqq 3$. Since

$$\cos 2\pi/2^s = \frac{1}{2}(e^{2\pi i/2^s} + e^{-2\pi i/2^s}),$$

$R(e^{2\pi i/2^s})$ contains the number $\cos 2\pi/2^s$. The formula

$$\cos 2\pi/2^v = 2(\cos 2\pi/2^{v+1})^2 - 1 \tag{1}$$

shows recursively that the subfield $R(\cos 2\pi i/2^s)$ contains the numbers $\cos 2\pi/2^v$, $v \leqq s$, and the formula

$$\sin 2\pi/2^v \cos 2\pi/2^v = \frac{1}{2}\sin 2\pi/2^{v-1}$$

shows recursively that it contains also the numbers $\sin 2\pi/2^v$ and in particular $\sin 2\pi/2^s$. So, it follows from

$$e^{2\pi i/2^s} = \cos 2\pi/2^s + i\sin 2\pi/2^s$$

that $R(e^{2\pi i/2^s})$ is the composed of $R(i)$ and $R(\cos 2\pi/2^s)$. Now, $R(\cos 2\pi/2) = R(\cos 2\pi/4) = R$, and $(\cos 2\pi/2^{v+1})^2 \in R(\cos 2\pi/2^v)$ by (1), which show that $R(\cos 2\pi/2^s)$ is of degree at most 2^{s-2} over R. Since $R(e^{2\pi i/2^s})$ is of degree 2^{s-1} over R and is the direct product of a quadratic field and a cyclic field of degree 2^{s-2}, we conclude that $R(e^{2\pi i/2^s})$ is the direct product of $R(i)$ and $R(\cos 2\pi/2^s)$ and that $R(\cos 2\pi/2^s)$ is cyclic of degree 2^{s-2} over R .

We shall assume once for all that F contains the number $\cos 2\pi/2^t$ but not $\cos 2\pi/2^{t+1}$, where t is naturally $\geqq 2$.

Now, suppose that $F(e^{2\pi i/2^s})$ is not cyclic over F . Since F contains $\cos 2\pi/2^t$, $F(i)$ contains the number $e^{2\pi i/2^t}$. Since $F(i)$ is certainly cyclic over F , we have $s \geqq t+1$.

Consider the field $R(\cos 2\pi/2^t) \subset F$. Over this field, $R(e^{2\pi i/2^s})$ is the direct product of $R(\cos 2\pi/2^t, i)$ and $R(\cos 2\pi/2^s)$, with the generating automorphisms τ and σ respectively:

$$\tau^2 = 1,\ \sigma^{2^{s-t}} = 1.$$

It can be easily verified that any subgroup of the group (σ, τ) is of the form (σ^{2^n}, τ) , (σ^{2^n}) , (τ) or $(\sigma^{2^n}\tau)$. Since $F(e^{2\pi i/2^s})$ is by assumption not cyclic over F , the field $F \cap R(e^{2\pi i/2^s})$ corresponds necessarily to a group of the form (σ^{2^n}, τ) . If $n > 0$, $F \cap R(e^{2\pi i/2^s})$ would contain $\cos 2\pi/2^{t+1}$. So, $F \cap R(e^{2\pi i/2^s})$ is exactly $R(\cos 2\pi/2^t)$, and $F(e^{2\pi i/2^s})$ over F is the direct product of $F(i)$ and $F(\cos 2\pi/2^s)$ with the generating automorphisms σ and τ respectively.

$F(e^{2\pi i/2^s})$ contains exactly three quadratic fields over F , namely,

$$F_\sigma = F(i) = F(e^{2\pi i/2^t}),$$

$$F_{\sigma^2,\tau} = F(\cos 2\pi/2^{t+1}),\ F_{\sigma\tau} = F(i\cos 2\pi/2^{t+1}).$$

Let $\mathfrak{p} \nmid 2$ be a prime in F . If $\mathfrak{p}$ is finite, it is unramified in $F(e^{2\pi i/2^s})$, and therefore $F_\mathfrak{p}(e^{2\pi i/2^s})$ is cyclic over $F_\mathfrak{p}$. So, $F_\mathfrak{p}$ contains at least one of the three numbers $e^{2\pi i/2^t}$, $\cos 2\pi/2^{t+1}$, $i\cos 2\pi/2^{t+1}$. If $\mathfrak{p}$ is infinite, $F_\mathfrak{p}$ contains of course $\cos 2\pi/2^{t+1}$. For $v \geqq t+1$, we have

$$(\sec 2\pi/2^{t+1})^{2^v} = \left(\frac{1}{\cos 2\pi/2^{t+1}}\right)^{2^v}$$

$$= \left(\frac{1}{i\cos 2\pi/2^{t+1}}\right)^{2^v}$$

$$= \left(\frac{2}{1+e^{2\pi i/2^t}}\right)^{2^v}.$$

It is well-known that, if ζ_t is any primitive $2^{t\,th}$ root of unity, then $(1-\zeta_t)^{2^{t-1}}=2$ up to a unit. But $-e^{2\pi i/2^t}$ is a primitive $2^{t\,th}$ root of unity. Hence,

$$(\sec 2\pi/2^{t+1})^{2^v}=2^{2^v-2^{v-t-1}}\varepsilon,$$

ε being a unit.

An *even* prime is one which divides 2. When i, $i\cos 2\pi/2^{t+1}\notin F$, an even prime $\mathfrak{p}$ is said to be *oddly-even*, if

$$(F_{\mathfrak{p}}(e^{2\pi i/2^s}):F_{\mathfrak{p}})=(F(e^{2\pi i/2^s}):F)$$

for any s, i.e., if $F_{\mathfrak{p}}$ does not contain any of the three numbers i, $\cos 2\pi/2^{t+1}$, $i\cos 2\pi/2^{t+1}$; otherwise, $\mathfrak{p}$ is said to be *evenly-even*.

We summarize our results in

Lemma 1. If $F(e^{2\pi i/2^s})$ is not cyclic over F, then $s\geqq t+1$, and i, $i\cos 2\pi/2^{t+1}\notin F$. The converse also holds. In these circumstances, $F(e^{2\pi i/2^s})$ is the direct product of $F(i)$ and $F(\cos 2\pi/2^s)$, and is that of $F(i\cos 2\pi/2^{t+1})$ and $F(\cos 2\pi/2^{t+1})$ as well. For any prime $\mathfrak{p}$ in F, which is not oddly-even, $F_{\mathfrak{p}}$ contains a $2^{t\,th}$ root of $(\sec 2\pi/2^{t+1})^{2^v}$, $v\geqq t+1$. The number $(\sec 2\pi/2^{t+1})^{2^v}$ in divisible only by even primes.

Examples. The prime 2 is oddly-even in R. Consider the field $R(\sqrt{7})$. We have $\sqrt{2}$, i, $\sqrt{-2}\notin R(\sqrt{7})$. Since -7 is a square in the 2-adic field R_2, $R_2(\sqrt{7})=R_2(i)$. So, there is an evenly-even prime in $R(\sqrt{7})$ but no oddly-even primes.

Let θ be a root of the equation

$$x^3+x+8=0.$$

We have $\sqrt{2}$, i, $\sqrt{-2}\notin R(\theta)$. Now, in R_2.

(2) $$x^3+x+8=(x+8\varepsilon)(x^2-8\varepsilon x+1+64\varepsilon^2),$$

ε being a unit in R_2. A root of the quadratic factor in (2) is $4\varepsilon+i\sqrt{1+48\varepsilon^2}$.

Since $1+48\varepsilon^2$ is a square in R_2, this quadratic factor defines the field $R_2(i)$. So, there are two even primes in $R(\theta)$, one of which is oddly-even and the other evenly-even.

2. Necessary Conditions

Let C be cyclic of degree l^r over F, l being a prime. When is it possible to construct an extension K of degree l^s over C, which remains cyclic over F? Suppose such an extension exists. Let $\mathfrak{p}$ be a prime in F, and consider the relations of the fields $K^{\mathfrak{p}}$, $C^{\mathfrak{p}}$, $F_{\mathfrak{p}}$. Let Z be the decomposition field of $\mathfrak{p}$ in K. Then either C is contained in Z or Z is a proper subfield of C. In the first case, $C^{\mathfrak{p}}=F_{\mathfrak{p}}$ and $(K^{\mathfrak{p}}:C^{\mathfrak{p}})$ might be any divisor of l^s. In the second case, $C^{\mathfrak{p}}\neq F_{\mathfrak{p}}$ and $(K^{\mathfrak{p}}:C^{\mathfrak{p}})$ must be maximal, i. e., l^s. So, we have obtained a necessary condition for the existence of K:

Lemma 2. If K is cyclic of degree l^{r+s} over F and if C is the subfield of K of degree l^r over F, then, for any prime $\mathfrak{p}$ in F, $(C^{\mathfrak{p}}:F_{\mathfrak{p}})>1$ implies $(K^{\mathfrak{p}}:C^{\mathfrak{p}})=l^s$.

Suppose i, $i\cos 2\pi/2^{t+1}\notin F$. Let C be cyclic of degree 2^r over F, $r\geqq 0$, and let K be an extension of C, which is cyclic of degree 2^{r+s} over F, $s\geqq t+1$. Let S be a finite set of primes in F, containing all even-primes and all primes ramified in C.

For $\mathfrak{p}\notin S$, there exists by Lemma 1 an element $\alpha_{\mathfrak{p}}$, in F_p such that

$$(3)\qquad \alpha_{\mathfrak{p}}^{2^s}=(\sec 2\pi/2^{t+1})^{2^s}.$$

It follows that $\alpha_{\mathfrak{p}}$ is a unit of $F_{\mathfrak{p}}$. So,

$$\left(\frac{K^{\mathfrak{p}}/F_{\mathfrak{p}}}{\alpha_{\mathfrak{p}}}\right)$$

is an element in the inertia group of $K^{\mathfrak{p}}$ over $F_{\mathfrak{p}}$. Since C is unramified outside S, the order of the inertia group is at most 2^s. Hence,

$$\left(\frac{K^{\mathfrak{p}}/F_{\mathfrak{p}}}{(\sec 2\pi/2^{t+1})^{2^s}}\right)=\left(\frac{K^{\mathfrak{p}}/F_{\mathfrak{p}}}{\alpha_{\mathfrak{p}}}\right)^{2^s}=1.$$

Since

$$\prod_{\mathfrak{p}}\left(\frac{K^{\mathfrak{p}}/F_{\mathfrak{p}}}{(\sec 2\pi/2^{t+1})^{2^s}}\right)=1,$$

we have

(4)
$$\prod_{\mathfrak{p}\in S}\left(\frac{K^{\mathfrak{p}}/F_{\mathfrak{p}}}{(\sec 2\pi/2^{t+1})^{2^s}}\right)=1.$$

Now, suppose $C=F$. If $\mathfrak{p}\in S$ is not oddly-even, there exists an element $\alpha_{\mathfrak{p}}$ in $F_{\mathfrak{p}}$ such that (3) holds. So, $(\sec 2\pi/2^{t+1})^{2^s}$ is a norm in $K^{\mathfrak{p}}$. If $\mathfrak{p}$ is oddly-even, $(\sec 2\pi/2^{t+1})^{2s}$ is not always a norm in $K^{\mathfrak{p}}$. But its square is certainly one, because $(\sec 2\pi/2^{t+1})^{2^s}\in F$. So, in case $C=F$, (4) means plainly that the number of oddly-even primes $\mathfrak{p}$, for which $(\sec 2\pi/2^{t+1})^{2^s}$ is not a norm in $K^{\mathfrak{p}}$, is even.

We have proved

Lemma 3. Suppose i, $i\cos 2\pi/2^{t+1}\notin F$. Let C be cyclic of degree 2^r over F and let S contain all even primes in F and all primes ramified in C. Then, a necessary condition for the existence of an extension K of C, which is cyclic of degree 2^{r+s} over F, $s\geqq t+1$, and whose $\mathfrak{p}$-adic completion is a given extension $K^{\mathfrak{p}}$ over $F_{\mathfrak{p}}$ for $\mathfrak{p}\in S$ is that the norm-residue symbols could be fixed so that (4) is satisfied. When $C=F$, this condition is equivalent to that the number of oddly-even primes $\mathfrak{p}$ in F, for which $(\sec 2\pi/2^{t+1})^{2^s}$ is not a norm in $K^{\mathfrak{p}}$, should be even.

We observe that, when $(K^{\mathfrak{p}}:F_{\mathfrak{p}})<2^s$, $(\sec 2\pi/2^{t+1})^{2^s}$ is always a norm in $K^{\mathfrak{p}}$. When $(K^{\mathfrak{p}}:F_{\mathfrak{p}})=2^s$, it is a norm in $K^{\mathfrak{p}}$ if and only if the number

$$(\sec 2\pi/2^{t+1})^2=\frac{2}{1+\cos 2\pi/2^t}$$

is a norm in the quadratic intermediate field.

3. Main Lemma

Lemma 4. Let Δ be normal over F and suppose that $(\Delta:F)$ is a power of l. Let σ be an automorphism of Δ over F, and denote by $\bar{\sigma}$ the set of all conjugates of σ in the

Galois-group of Δ over F. Then, for a certain m prime to l, there exist infinitely many primes $\mathfrak{p}$ in F, for which

$$\left(\frac{\Delta/F}{\mathfrak{p}}\right)=\bar{\sigma}^{m}.$$

In particular, if Δ is the direct product of the cyclic extensions E_1, …, E_u, E_{u+1}, …, E_v, $0 \leqslant u \leqslant v$, then there exist infinitely many primes in F, which remain prime in E_1, …, E_u and split completely in E_{u+1}, …, E_v.

PROOF. We shall first construct a cyclic extension E of degree l over F which is independent with Δ. Let p be a rational prime which divides $2^{l^n}-1$. Then, since the period of 2 mod p is a power of l, we have $l \mid (p-1)$. It can be easily seen that $2^{l^n}-1$ and

$$(2^{l^{n+1}}-1)/(2^{l^n}-1)=(2^{l^n})^{l-1}+\cdots+2^{l^n}+1$$

are relatively prime. So, there exist infinitely many p for which $l \mid (p-1)$. If ζ is a primitive p^{th} root of unity, $R(\zeta)$ contains a cyclic extension of degree l over R. Hence, there exist infinitely many cyclic extensions of degree l over R. From this, we infer that there exists a cyclic extension E of degree l over F independent with Δ. Write $\Delta^*=\Delta\times E$.

Let τ be a generating automorphism of E over F, and let F^* be the fixed field in Δ^* of the automorphism $\sigma\tau$. Then, Δ^* is cyclic over F^*. By the first inequality in class field theory, there exist infinitely many primes in F^*, which remain prime in Δ^*. So, for a certain c prime to l, there exist infinitely many primes $\mathfrak{P}$ in F^*, for which

$$\left(\frac{\Delta^*/F^*}{\mathfrak{P}}\right)=(\sigma\tau)^{c}=\sigma^{c}\tau^{c}.$$

Again, there are infinitely many of these primes $\mathfrak{P}$ for which the corresponding $\mathfrak{p}'s$ in F are unramified in Δ^* and $f(\mathfrak{P}/\mathfrak{p})$ is constant. We have

$$\left(\frac{\Delta^*/F^*}{\mathfrak{P}}\right)\in\left(\frac{\Delta^*/F}{\mathfrak{p}}\right)^{f(\mathfrak{P}/\mathfrak{p})}$$

Since τ^c is a generating automorphism of E over F, $f(\mathfrak{P}/\mathfrak{p})$ is prime to l.

Let

$$d \cdot f\left(\frac{\mathfrak{P}}{\mathfrak{p}}\right) \equiv 1(mod(\Delta : F)).$$

Then,

$$\left(\frac{\Delta^*/F}{\mathfrak{p}}\right) = \bar{\sigma}^{cd}\tau^{cd}.$$

So,

$$\left(\frac{\Delta/F}{\mathfrak{p}}\right) = \sigma^{cd},$$

proving the first part of the lemma with $(m = cd)$.

For the second part, we take $\sigma = \sigma_1, \cdots, \sigma_u$, σ_j being a generating automorphism of E_j over F, $j = 1, \cdots, u$.

Now, let $\widetilde{C}$ be cyclic of degree a power of l over F, which is independent with $F(\zeta_s)$. Let C be cyclic of degree l^r over F, $r \geqq 0$, $C = \widetilde{C}$ or not.

We introduce the following notations, S being any finite set of primes in F:

α = numbers($\neq 0$) in F,

$\mathfrak{a}$ = idèles in F,

$\alpha_{\mathfrak{p}}$ = elements($\neq 0$) in $F_{\mathfrak{p}}$,

$\mathfrak{a}_S$ = idèles with arbitrary S-components and all other components 1,

$\mathfrak{b}^S$ = idèles with S-components 1 and all other components local units,

$\mathfrak{g}$ = group of idèles in F, to which C is class field (or idèles in this group),

$\mathfrak{g}_S = \mathfrak{a}_S \cap \mathfrak{g}$.

$(\sec 2\pi/2^{t+1})_S{}^{2^\nu}$ = idèle with S-components $(\sec 2\pi/2^{t+1})^{2^\nu}$ and all other components 1.

Lemma 5 (Main Lemma). Let S be a finite set of primes in F, which contains all primes ramified in C, and which contains also all evenly-even primes in the special case when i) $l = 2$, ii) i, $i\cos 2\pi/2^{t+1} \notin F$, iii) $s \geqq t+1$, iv) there exists no oddly—even prime outside S. Then, there exists in the general case a finite set T of primes in F outside S such that, for $v = 1, 2, \cdots, s$,

$$\mathfrak{g}_S \cap \alpha \mathfrak{g}^{l^v} \mathfrak{b}^{S+T} = \mathfrak{g}_S{}^{l^v}. \tag{5}$$

In the special case, (5) should be replaced by the statement that any idèle in $\mathfrak{g}_S \cap \alpha \mathfrak{g}^{2^v} \mathfrak{b}^{S+T}$ is either of the form $\mathfrak{g}_S^{2^v}$ or of the form

$$(\sec 2\pi/2^{t+1})_S{}^{2^v} \mathfrak{g}_S^{2^v}.$$

The primes in T may be chosen so as to remain prime in $\widetilde{C}$, except that, in a special case which will be explained in the proof, an oddly-even prime must be included.

PROOF. Let Γ be the maximal abelian extension of $F(\zeta_s)$, having l as an exponent and unramified outside $S + lp_\infty$. Let E_1, E_2, $\cdots$ be all cyclic subfields of Γ over $F(\zeta_s)$. Consider the composed field $\Delta_j = E_j \widetilde{C}(\zeta_s)$ over $F(\zeta_s)$. Δ_j is either identical to $\widetilde{C}(\zeta_s)$ or is the direct product of the latter and E_j. In both cases, there exists by Lemma 4 a prime $\mathfrak{P}_j$ in $F(\zeta_s)$ outside S, which remains prime in E_j and in $\widetilde{C}(\zeta_s)$. Let $\mathfrak{p}_j$ be the prime in F divisible by $\mathfrak{P}_j$. We put temporarily $T = \{\mathfrak{p}_1, \mathfrak{p}_2, \cdots\}$ and shall enlarge it later. The primes in T remain prime in $\widetilde{C}$.

Let $\mathfrak{g}_s$ be an idèle belonging to the left side of (5), i. e.,

$$\mathfrak{g}_S = \alpha \mathfrak{g}^{l^v} \mathfrak{b}^{S+T}. \tag{6}$$

We wish to show that (6) implies all solutions of

$$x^{l^v} - \alpha = 0 \tag{7}$$

are in F ζ_s. Suppose on the contrary that this is not the case. Then (7) defines a proper cyclic extension F (ζ_s, $\sqrt[l^v]{\alpha}$) of F (ζ_s). Since, by (6), $l^v \mid or\ d_{\mathfrak{p}}\alpha$ for $\mathfrak{p} \notin S$, an easy estimation of the local different shows that this extension is unramified outside $S + l_{p_\infty}$ over $F(\zeta_s)$. So, it has a subfield E in common with Γ, say $E = E_j$. Since, by (6), α is an $l^{v\,th}$ power at $\mathfrak{p}_j$, it is an $l^{v\,th}$ power at $\mathfrak{P}_j$, which means $\mathfrak{P}_j$ splits completely in F (ζ_s, $\sqrt[l^v]{\alpha}$) and hence in E_j, contrary to the choice of $\mathfrak{P}_j$.

We now distinguish the two cases:

I. F (ζ_s) is cyclic over F or there exist oddly-even primes outside S. When l is odd and when F (ζ_s) $\neq$ F (ζ_1), F (ζ_s) is the direct product of two fields W_1 and W_2 such that (W_1 : F) $<l$, (W_2 : F) is a power of l. Choose a prime $\mathfrak{p}^*$ outside S, which remains prime both in W_2 and in $\widetilde{C}$. When $l=2$, when F (ζ_s) $\neq$ F and when F (ζ_s) is cyclic over F, we choose a prime $\mathfrak{p}^*$ outside S, which remains prime in F (ζ_s) and in $\widetilde{C}$. When F (ζ_s) is not cyclic over F, we take as $\mathfrak{p}^*$ one of the oddly-even primes outside S. This is the only case in which we have no control over the prime selected and it might not remain prime in $\widetilde{C}$. Adjoining $\mathfrak{p}^*$ to T, we wish to show that α of (6) is an $l^{v\,th}$ power in F.

Let θ be a solution of (7) determined as follows: When F (ζ_s) =F (ζ_1), no $\mathfrak{p}^*$ was chosen. In this case, we take as θ any solution of (7). In all other cases, (6) implies that θ is an $l^{v\,th}$ power at $\mathfrak{p}^*$. So, there exists $\theta \in F_{\mathfrak{p}^*}$ such that $\theta^{l^v}=\alpha$. This element θ may be regarded as an algebraic quantity over F.

Let

$$x^n+\cdots\pm a_n=(x-\theta)(x-\eta'\theta)\cdots$$

be the irreducible polynomial of θ over F, where η', $\cdots$ are $l^{v\,th}$ roots of unity. Since we have proved that all solutions of (7) are in $F(\zeta_s)$, $F(\theta)\subset F(\zeta_s)$.

If $F(\zeta_s)=F$, θ is already in F. When $l=2$ and when $F(\zeta_s)\neq F$, we have, by the choice of $\mathfrak{p}^*$,

$$(F_{\mathfrak{p}^*}(\zeta_s):F_{\mathfrak{p}^*})=(F(\zeta_s):F).$$

Should $n>1$, $\mathfrak{p}^*$ would have a split factor in $F(\theta)\subset F(\zeta_s)$, which is a contradiction. So, there remains only the case $l\neq 2$. We prove first that $n<l$. If $F(\zeta_s)=F(\zeta_1)$, this is trivial. Suppose $F(\zeta_s)\neq F(\zeta_1)$. Since $F(\theta)\subset F(\zeta_s)$, $n\geqq l$, implies the field $\Omega=F(\theta)\cap W_2$ is a proper extension of F. But $\mathfrak{p}^*$ splits completely in the cyclic field $F(\theta)$. So, $\mathfrak{p}^*$ splits completely in Ω and therefore cannot remain prime in W_2, a contradiction. Consequently $n<l$. We have $a_n=\eta\,\theta^n$, η

being an $l^{v\,th}$ root of unity. Since $n < l$, there exists u, v such that $un + vl^v = 1$. Hence, F contains the number

$$a_n^u \alpha^v = \eta^u \theta^{un+vl^v} = \eta^u \theta,$$

which is an $l^{v\,th}$ root of α.

Since α is an $l^{v\,th}$ power in F, (6) may be transformed into

$$\mathfrak{g}_S = \mathfrak{g}'^{l^v} \mathfrak{b}^{S+T}$$

This equation implies $(\mathfrak{g}')_\mathfrak{p}$ is a unit for $\mathfrak{p} \notin S$. Since C is unramified outside S, $(\mathfrak{g}')_\mathfrak{p}$ is in the group $\mathfrak{g}$ for $\mathfrak{p} \notin S$. Hence $\mathfrak{g}_S{}'$ is in the group $\mathfrak{g}$. We have

$$\mathfrak{g}_\zeta = \mathfrak{g}_S{}'^{l^v}$$

proving the lemma in the present case.

Ⅱ. Special case. In this case, $F(\zeta_s)$ over F is the direct product of $F(i)$ and $F(cos 2\pi/2^s)$ and is also that of $F(icos 2\pi/2^{t+1})$ and $F(cos 2\pi/2^s)$. Let $\mathfrak{p}_1^*$ be a prime outside S, which remains prime in $\widetilde{C}$ and in $F(cos 2\pi/2^s)$ but which splits in $F(icos 2\pi/2^{t+1})$. And let $\mathfrak{p}_2^*$ be a prime outside S, which remains prime in $\widetilde{C}$ and in $F(cos 2\pi/2^s)$ but which splits in $F(i)$. We adjoin $\mathfrak{p}_1^*$ and $\mathfrak{p}_2^*$ to T.

If α is a $2^{v\,th}$ power in F, the lemma can be proved as in the last part of Ⅰ. Let us therefore assume α is not a $2^{v\,th}$ power in F.

Now, α is a $2^{v\,th}$ power at $\mathfrak{p}_1^*$: $\theta^{2^v} = \alpha$, $\theta \in F_{\mathfrak{p}_1^*}$. Since p_1^* splits into exactly two prime factors in $F(\zeta_s)$ and since $F(\theta) \subset F(\zeta_s)$, the irreducible polynomial of θ over F is of degree 1 or 2. In the first case, α would be a $2^{v\,th}$ power in F. Since $i \notin F_{\mathfrak{p}_1^*}$, the irreducible equation of θ over F must be of the form

$$(x-\theta)(x+\theta)=0,\ \theta^2 \in F.$$

Since $\mathfrak{p}_1^*$ splits in $F(icos 2\pi/2^{t+1}) = F(i\sec 2\pi/2^{t+1})$ but not in $F(cos 2\pi/2^{t+1})$ nor in $F(i)$, we have $\theta \in F(i\sec 2\pi/2^{t+1})$. As both θ^2 and $(i\sec 2\pi/2^{t+1})^2$ are in F, θ differs from $i\sec 2\pi/2^{t+1}$ only by a factor in F:

$$\theta = ia \sec 2\pi/2^{t+1},\ a \in F.$$

So,

(8) $$\alpha=\theta^{2^v}=i^{2^v}(\sec 2\pi/2^{t+1})^{2^v}a^{2^v}.$$

First, suppose $v \leqq t$. Since α is a $2^{v\,th}$ power at $\mathfrak{p}_2^*$: $\xi^{2^v}=\alpha$, $\xi \in F_{\mathfrak{p}_2^*}$, and since $e^{2\pi i/2^t} \in F_{\mathfrak{p}_2^*}$, (8) implies $\sec 2\pi/2^{t+1}=\xi i^{-1}a^{-1}e^{2u\pi i/2^t} \in F_{\mathfrak{p}_2^*}$, contrary to the fact that $\mathfrak{p}_2^*$ remains prime in $F(cos 2\pi/2^s)$. This shows that in case $v \leqq t$ the assumption that α is not a $2^{v\,th}$ power in F is untenable.

Now, let $v \geqq t+1$. (8) gives

$$\alpha=(\sec 2\pi/2^{t+1})^{2^v}a^{2^v}.$$

In (6), the factor a^{2^v} can be thrown into $\mathfrak{g}^{2^v}$. By Lemma 1, $(\sec 2\pi/2^{t+1})^{2^v}$ is a $2^{v\,th}$ power in $F_{\mathfrak{p}}$ for $\mathfrak{p} \notin S$, and is a unit in $F_{\mathfrak{p}}$ for these $\mathfrak{p}$'s because S contains all even primes. Since S contains also all primes ramified in C, the part of $(\sec 2\pi/2^{t+1})^{2^v}$ outside S can be transferred to $\mathfrak{g}^{2^v}$. Hence, (6) becomes

$$\mathfrak{g}_S=(\sec 2\pi/2^{t+1})_S^{2^v}\mathfrak{g}'^{2^v}\mathfrak{b}^{S+T},$$

which may be transformed into

$$\mathfrak{g}_S=(\sec 2\pi/2^{t+1})_S^{2^v}\mathfrak{g}_S'^{2^v}$$

as in I. The lemma is thus completely proved.

For a particular application, we prove

COROLLARY 1. Let $\alpha_0 \in F$ be a fixed number, and let S be a finite set of primes in F. There exists a finite set T of primes outside S such that

(A) $\mathfrak{a}_S \cap \alpha\mathfrak{a}^l\mathfrak{b}^{S+T}=\mathfrak{a}_S^l$

(B) α_0 is an lth power at every $\mathfrak{p} \in T$ with possibly a single exception.

PROOF: If α_0 is an lth power in F, we just disregard it. So, assume α_0 is not an lth power in F. $F(\zeta_1, \sqrt[l]{\alpha_0})$ is then a proper cyclic extension of $F(\zeta_1)$.

For (A), we choose T to consist of the primes $\mathfrak{p}_1$, $\mathfrak{p}_2$, $\cdots$ as in the proof of Lemma 5, with $C=F$, $s=1$. Now, if $E_j \neq F(\zeta_1, \sqrt[l]{\alpha_0})$, $\mathfrak{P}_j$ may be selected to split completely in $F(\zeta_1, \sqrt[l]{\alpha_0})$. α_0 is then an lth power at $\mathfrak{p}_j$. Since at most one of E_1, E_2, $\cdots$ coincides with $F(\zeta_1, \sqrt[l]{\alpha_0})$, (B) is fulfilled.

We state here the following lemma of G. Whaples:

Lemma 6. Let G be a finite abelian group having l^s as an exponent, and G_1 a subgroup of G such that, for all $v \leqq s$,

$$G_1 \cap G^{l^v} = G_1^{l^v}$$

Then G is the direct product of G_1 and some subgroup H of G.

4. Main Theorem

Let C be a cyclic extension of degree $m = l_1^{r_1} l_2^{r_2} \cdots$ over F and let S be a finite set of primes in F. Our problem is to construct a cyclic extension K of degree $n = l_1^{r_1+s_1} l_2^{r_2+s_2} \cdots$ over F, which contains C and whose $\mathfrak{p}$-adic completion is a given extension $K^{\mathfrak{p}}$ for $\mathfrak{p} \in S$. This problem may be reduced to the simpler one in which $m = l^r$, $n = l^{r+s}$. In fact, if the latter problem is solved, all we need to do in the general case is constructing for each l_j a cyclic extension K_j of degree $l_j^{r_j+s_j}$ over F, which contains the l_j-component of C and whose $\mathfrak{p}$-adic completion is the l_j-component of $K^{\mathfrak{p}}$ for $\mathfrak{p} \in S$, and then taking the direct product of K_j, $j=1, 2, \cdots$, as the field K required.

Theorem. Let $\widetilde{C}$ be cyclic of degree a power of l over F, which is independent with $F(\xi_s)$ over F. Let C be cyclic of degree l^r over F, $r \geqq 0$, $C = \widetilde{C}$ or not. Let S be a finite set of primes in F, containing all primes ramified in C, and also all evenly-even primes in the special case when $i) l=2$, $ii) i$, $icos 2\pi/2^{t+1} \notin F$, $iii) s \geqq t+1$, $iv)$ there exists no oddly-even primes outside S. To each $\mathfrak{p} \in S$, there is given a cyclic extension $K^{\mathfrak{p}}$ of degree $l^{r_{\mathfrak{p}}+s_{\mathfrak{p}}}$ over $F_{\mathfrak{p}}$, where $l^{r_{\mathfrak{p}}} = (C^{\mathfrak{p}} : F_{\mathfrak{p}})$. Let $s \geqslant \max s_{\mathfrak{p}}$.

Suppose a definite isomorphism of the Galois-group of $K^{\mathfrak{p}}$ over $F_{\mathfrak{p}}$ into a cyclic group (σ) of order l^{r+s} is given so that

$$\left(\frac{K^{\mathfrak{p}} / F_{\mathfrak{p}}}{\alpha_{\mathfrak{p}}}\right)$$

may be regarded as a homomorphic mapping of $\alpha_{\mathfrak{p}}$ into (σ). Furthermore, suppose a definite isomorphism of the Galois-group of C over F onto (σ) mod (σ^{l^r}) is given so that

$$\left(\frac{C/F}{\mathfrak{a}}\right)$$

may be regarded as a homomorphic mapping of $\mathfrak{a}$ onto (σ) mod (σ^{l^r}).

General Hypotheses;

(a) $C^{\mathfrak{p}} \subset K^{\mathfrak{p}}$.

(b) $r_{\mathfrak{p}} > 0$ implies $s_{\mathfrak{p}} = s$.

(c) $\left(\frac{C/F}{\alpha_{\mathfrak{p}}}\right) \equiv \left(\frac{K^{\mathfrak{p}}/F_{\mathfrak{p}}}{\alpha_{\mathfrak{p}}}\right)$ (mod (σ^{l^r})).

Special Hypothesis: In the special case, the following condition should be fulfilled:

(d) $$\prod_{\mathfrak{p}\in S}\left(\frac{K^{\mathfrak{p}}/F_{\mathfrak{p}}}{(\sec 2\pi\ /\ 2^{t+1})^{2^s}}\right)=1.$$

Conclusion: Under the above conditions, there exists a cyclic extension K of degree l^{r+s} over F such that:

(A) $C \subset K$,

(B) the $\mathfrak{p}$ -adic completion of K is $K^{\mathfrak{p}}$ for $\mathfrak{p} \in S$,

(C) $$\left(\frac{K/F}{\alpha_{\mathfrak{p}}}\right) = \left(\frac{K^{\mathfrak{p}}/F_{\mathfrak{p}}}{\alpha_{\mathfrak{p}}}\right),$$

(D) apart from an eventual oddly-even prime, all primes outside S ramified in K remain prime in $\widetilde{C}$.

Proof. First of all, let us enlarge S to include a prime $\mathfrak{p}_0 \nmid l p_\infty$, which remains prime in C and define $K^{\mathfrak{p}_0}$ as the unramified extension of degree l^{r+s} over $F_{\mathfrak{p}_0}$. Conditions (a) and (b) are satisfied at $\mathfrak{p}_0$.

Observe that (b) is a consequence of (c). Conversely, if (b) is satisfied,

$$\left(\frac{K^{\mathfrak{p}}/F_{\mathfrak{p}}}{\alpha_{\mathfrak{p}}}\right)$$

can always be fixed so that (c) is satisfied. In particular, this can be done for $\mathfrak{p}_0$. In the special case, the enlarged S satisfies still condition (d), because $K^{\mathfrak{p}_0}$ is unramified and because $(\sec 2\pi / 2^{t+1})^{2^s}$ is a unit in $F_{\mathfrak{p}_0}$ and therefore a norm in $K^{\mathfrak{p}_0}$.

We define a mapping of $\mathfrak{a}_S$ by the following equation:

$$\left(\frac{K^S/F_S}{\mathfrak{a}_S}\right)=\prod_{\mathfrak{p}\in S}\left(\frac{K^{\mathfrak{p}}/F_{\mathfrak{p}}}{(\mathfrak{a}_S)_{\mathfrak{p}}}\right).$$

This is clearly a homomorphic mapping of $\mathfrak{a}_S$ onto the group (σ). Let $\mathfrak{h}_S$ be the kernel of this mapping, and let $\beta_0 \in F_{\mathfrak{p}_0}$ be such that

$$\left(\frac{K^{\mathfrak{p}_0} / F_{\mathfrak{p}_0}}{\beta_0}\right)=\sigma.$$

Then $\mathfrak{a}_S=(\beta_0)\mathfrak{h}_S$, where (β_0) is the cyclic group generated by β_0 in $\alpha_{\mathfrak{p}_0}$. By (c), $\mathfrak{h}_S \subset \mathfrak{g}_S$.

Now, choose a set T of primes as in Lemma 5. Write

$$\mathfrak{h}_0=\alpha\mathfrak{h}_S\mathfrak{g}^{l^s}\mathfrak{b}^{S+T}.$$

Then $\mathfrak{h}_0 \subset \mathfrak{g}$. Set $G=\mathfrak{g}/\mathfrak{h}_0$ and $G_1=\mathfrak{g}_S\mathfrak{h}_0/\mathfrak{h}_0$. Since $\alpha\mathfrak{g}^{l^s}\mathfrak{b}^{S+T}\subset\mathfrak{h}_0$, G is finite and has l^s as an exponent. Consider the group $G_1 \cap G^{l^v}$. Any element in this group may be represented by an idèle $\mathfrak{g}_S$ and by an idèle $\mathfrak{g}^{l^v}$ as well, i. e.,

(9) $$\mathfrak{g}_S=\mathfrak{g}^{l^v}\cdot\alpha\mathfrak{h}_S\mathfrak{g}^{l^v}\mathfrak{b}^{S+T},$$

or

$$\mathfrak{g}_S\mathfrak{h}_S^{-1}=\alpha\mathfrak{g}^{l^v}\mathfrak{g}^{l^s}\mathfrak{b}^{S+T}.$$

By Lemma 5, $\mathfrak{g}_S\mathfrak{h}_S^{-1}$ is of the form $\mathfrak{g}_S^{l^v}$ in general. Only in the special case, it may be of the form $(\sec 2\pi / 2^{t+1})_S^{2^v}\mathfrak{g}_S^{2^v}$. By (d),

$$\left(\left(\frac{K^S/F_S}{(\sec 2\pi / 2^{t+1})_S^{2^v}}\right)\right)^{2^{s-v}}=\left(\frac{K^S/F_S}{(\sec 2\pi / 2^{t+1})_S^{2^s}}\right)=1.$$

So,

$$\left(\frac{K^S/F_S}{(\sec 2\pi / 2^{t+1})_S^{2^v}}\right)\in(\sigma^{2^{r+v}}),$$

or

$$(\sec 2\pi / 2^{t+1})_S^{2^v}\in(\beta_0^{2^{r+v}})\mathfrak{h}_S\subset\mathfrak{g}_S^{2^v}\mathfrak{h}_S$$

In all cases, $\mathfrak{g}_S$ of (9) is in the same coset of $\mathfrak{h}_0$ as some $\mathfrak{g}_S^{l^v}$. This proves $G_1 \cap G^{l^v} = G_1^{l^v}$. By Lemma 6, there exists a group $\mathfrak{h}$ such that

(10) $\mathfrak{h}_0 \subset \mathfrak{h} \subset \mathfrak{g}$,

(11) $\mathfrak{g}_S \mathfrak{h} = \mathfrak{g}$,

(12) $\mathfrak{g}_S \cap \mathfrak{h} = \mathfrak{g}_S \cap \mathfrak{h}_0$.

We contend that the class field K of F to the group $\mathfrak{h}$, which is clearly admissible, fulfills all our requirements.

Since $\mathfrak{h} \subset \mathfrak{g}$, (A) is fulfilled. Since $\mathfrak{h} \supset \mathfrak{b}^{S+T}$, K is unramified outside $S+T$. By the choice of T, (D) is fulfilled.

To prove (B), consider the group $\mathfrak{a}_S \cap \mathfrak{h}$. Since $\mathfrak{h} \subset \mathfrak{g}$, $\mathfrak{a}_S \cap \mathfrak{h} = \mathfrak{g}_S \cap \mathfrak{h} = \mathfrak{g}_S \cap \mathfrak{h}_0$ by (12). But, as we have seen, any $\mathfrak{g}_S$ of the form $\mathfrak{h}_0 = \alpha \mathfrak{h}_S \mathfrak{g}^{l^S} \mathfrak{b}^{S+T}$ is of the form

$$\mathfrak{h}_S \mathfrak{g}_S^{l^v} = \mathfrak{h}_S ((\beta_0^{l^r}) \mathfrak{h}_S)^{l^S} = \mathfrak{h}_S .$$

So, $\mathfrak{a}_S \cap \mathfrak{h} = \mathfrak{h}_S$. In particular, for $\mathfrak{p} \in S$, $\alpha_{\mathfrak{p}} \in \mathfrak{h}$ if and only if

$$\left(\frac{K^{\mathfrak{p}} / F_{\mathfrak{p}}}{\alpha_{\mathfrak{p}}}\right) = 1$$

i. e., (B) is fulfilled.

Now,

$$(\beta_0)\mathfrak{h} = (\beta_0)\mathfrak{h}\mathfrak{h}_S = (\beta_0)\mathfrak{h}\mathfrak{a}_S \supset (\beta_0)\mathfrak{h}\mathfrak{g}_S .$$

But $\mathfrak{h}\mathfrak{g}_S = \mathfrak{g}$ by (11), and $(\beta_0)\mathfrak{g} = \mathfrak{a}$. It follows $\mathfrak{a} = (\beta_0)\mathfrak{h}$. Since $\beta_0^{l^{r+s}} \in \mathfrak{h}$, $(\mathfrak{a} : \mathfrak{h}) \leqq l^{r+s}$. But $(K^{\mathfrak{p}_0} : F_{\mathfrak{p}_0}) = l^{r+s}$. So, K is cyclic of degree l^{r+s} over F.

For (C), we define an isomorphism of the Galois-group of K over F onto (σ) so that $\left(\frac{K/F}{\mathfrak{a}_s}\right)$ is mapped to

$$\left(\frac{K^S / F_S}{\mathfrak{a}_S}\right),$$

that is to say, we simply identify the automorphism $\left(\frac{K/F}{\beta_0}\right)$ with σ.

The theorem is therefore proved.

5. Corollaries and Applications

We observe that in the proof of the theorem the local fields $K^{\mathfrak{p}}$ were used only in an indirect manner; all that was really needed was the existence of a kernel $\mathfrak{h}_S$. What we mean by this is that the theorem may be used to prove the existence theorem in local class field theory. For example, let $k=R_2(\eta)$, $\eta\neq 0$, be a 2-adic field and let γ be a subgroup of the group β of all non-zero elements in k such that $(\beta:\gamma)=2^s$. It is required to construct an extension of k, which is the class field of k to the group γ. Let η be of degree n over R_2 and let $g(x)$ be its irreducible polynomial over the same. Let $f(x)$ be a polynomial in R, which is sufficiently near $xg(x)$ at the prime 2 and near $x^{n+1}-3$ at the prime 3. Then $f(x)$ is irreducible in R and has in R_2 a linear factor near x and a factor of degree n near $g(x)$. θ being a root of $f(x)$, the field $F=R(\theta)$ has two even primes $\mathfrak{p}_1$ and $\mathfrak{p}_2$ such that $F_{\mathfrak{p}_1}=k$ and $F_{\mathfrak{p}_2}=R_2$. It follows i, $\sqrt{2}$, $\sqrt{-2}\notin F$ and $\mathfrak{p}_2$ is oddly-even. With $S=\mathfrak{p}_1$, $\mathfrak{h}_S=\gamma$, and with an oddly-even prime outside S, a cyclic extension K of degree 2^s over F can always be constructed such that $K^{\mathfrak{p}_1}$ is the class field of k to the group γ.

Now, in certain applications, what is important is not the exact structure of the local fields but only their degrees. Let $s_{\mathfrak{p}}$ be given integers for $\mathfrak{p}\in S$. Suppose the situation is such as described by (i) — iv) in the theorem, and suppose it is required to construct a cyclic extension K of degree 2^s over F such that $(K^{\mathfrak{p}}:F_{\mathfrak{p}})=2^{s_{\mathfrak{p}}}$ for $\mathfrak{p}\in S$. If the finite $\mathfrak{p}\in S$ is not oddly-even, or if it is but $s_{\mathfrak{p}}<s$, we take as $K^{\mathfrak{p}}$ the unramified extension of degree $2^{s_{\mathfrak{p}}}$ over $F_{\mathfrak{p}}$. If $\mathfrak{p}\in S$ is infinite, we define $K^{\mathfrak{p}}$ in the only way available. If $\mathfrak{p}$ is oddly-even and $s_{\mathfrak{p}}=s$, we take $K^{\mathfrak{p}}=F_{\mathfrak{p}}(\cos 2\pi/2^{t+s})$. We have

$$N_{K^{\mathfrak{p}}F_{\mathfrak{p}}}(\sec 2\pi/2^{t+1})=(-(\sec 2\pi\ /\ 2^{t+1})^2)^{2^{s-1}}=(\sec 2\pi/2^{t+1})^{2^s}.$$

So, $(\sec 2\pi/2^{t+1})^{2^s}$ is a norm in $K^{\mathfrak{p}}$ over $F_{\mathfrak{p}}$ for all $\mathfrak{p}$ in S, and (d) is fulfilled.

Hence, K can be constructed. Combining this with the general case which is simpler, we obtain Corollary 2. Let S be a finite set of primes in F and n any positive integer. To each $\mathfrak{p} \in S$, there is given a positive integer $n_{\mathfrak{p}}$ such that

(a) $n_{\mathfrak{p}} \mid n$,

(b) $n_{\mathfrak{p}} = 1$ when $\mathfrak{p}$ is complex,

(c) $n_{\mathfrak{p}} \leqq 2$ when $\mathfrak{p}$ is real.

Then there exists a cyclic extension K of degree n over F such that, for $\mathfrak{p} \in S$,

$$(K^{\mathfrak{p}} : F_{\mathfrak{p}}) = n_{\mathfrak{p}}.$$

Furthermore, if $n = l^s$, and if $\widetilde{C}$ is independent with $F(\zeta_s)$ over F and is cyclic of degree a power of l, K can be constructed so that, if $\mathfrak{p} \notin S$ is ramified in K, it remains prime in $\widetilde{C}$.

It follows from Corollary 2 that every simple algebra over an algebraic number field is cyclic.

Corollary 3. Let $\alpha_0 \in F$ be a fixed number and let S be a finite set of primes in F. To each $\mathfrak{p} \in S$, there is given a cyclic extension $\Omega^{\mathfrak{p}}$ of degree l or 1 over $F_{\mathfrak{p}}$. Then, there exists a cyclic extension Ω of degree l over F such that

(A) the $\mathfrak{p}$-adic completion of Ω is $\Omega^{\mathfrak{p}}$ for $\mathfrak{p} \in S$,

(B) if T is the set of all primes outside S, where Ω is ramified, then α_0 is an lth power at every $\mathfrak{p} \in T$ with possibly a single exception.

Proof. We choose the set T in the proof of the theorem according to Corollary 1. (B) is fulfilled by the choice of T which contains all primes outside S, where Ω is ramified.

We use Corollary 3 to prove the norm theorem of Hasse-Schilling-Maass:

Let A be a central simple algebra over an algebraic number field F. Then $\alpha_0 \in F$ is a norm in A if and only if it is positive at all infinite primes where A is ramified.

For a proof of the following lemma, see H. Maass, Hamb. Abh. 12, 1938:

Lemma. To prove the norm theorem, it suffices to show that, if $A = D$ is a division algebra, and if $\alpha_0 \in F$ is positive at all infinite primes where D is ramified, and is a local unit at all finite primes where D is ramified, then α_0 is a norm in D .

Proof of the norm theorem. By the lemma, we may prove it only for a division algebra. If the degree of the algebra is 1, the theorem is trivial. Assume as an induction hypothesis that it has been proved for all division algebras of degree $< n$, $n > 1$. Suppose that $\alpha_0 \in D = D_{(n)}$ satisfies thee conditions in the Lemma.

Let $l \mid n$ be a prime. When n is even, we take $l = 2$. Applying Corollary 3, we construct a cyclic extension Ω of degree l over F such that

(A) $\Omega^{\mathfrak{p}}$ is unramified of degree l over $F_{\mathfrak{p}}$ for all finite $\mathfrak{p}$ where D is ramified,

(B) $\Omega^{\mathfrak{p}}$ is complex for all infinite $\mathfrak{p}$ where D is ramified,

(C) $\Omega^{\mathfrak{p}} = F_{\mathfrak{p}}$ for all $\mathfrak{p}$ where α_0 is not a unit,

(D) if T is the set of all primes other than those in (B), where Ω is ramified, then α_0 is an lth power at every $\mathfrak{p} \in T$ with possibly a single exception.

It follows from the local properties of Ω and from the product formula of the norm residue symbol that α_0 is a norm everywhere in Ω. So, there exists an element ω in Ω such that

$$N_{\Omega F}(\omega) = \alpha_0.$$

Now, Ω is clearly a subfield of D. Let D_1 be the centralizator of Ω in D . Then D_1 is ramified at no infinite primes. By the hypothesis of induction, there exists an element β in D_1 such that

$$N_{D_1\Omega}(\beta) = \omega.$$

We have

$$N_{DF}(\beta) = N_{\Omega F}(N_{D_1\Omega}(\beta)) = N_{\Omega F}(\omega) = \alpha_0,$$

proving the norm theorem.

On the Commutator Group of a Simple Algebra *①

Introduction.

Let A be an algebra. If α in A is a commutator: $\alpha=\beta\gamma\beta^{-1}\gamma^{-1}$, clearly, its norm (in any sense) must be unity. It follows that any element in the commutator group must be of norm 1.

Now, consider as an example the algebra D of quaternions over the field R of real numbers. Any element α in D is contained in a quadratic extension of R, which is by necessity a field $R(i)$, isomorphic to the field of complex numbers. If α is of reduced norm 1: $\alpha\bar{\alpha}=1$, it can be written in the form $\cos\theta+i\sin\theta=e^{\theta i}$. Since there exists an element j in D such that $j\xi j^{-1}=\bar{\xi}$ for any ξ in $R(i)$, we have $e^{\theta i}=e^{\frac{1}{2}\theta i}j(e^{\frac{1}{2}\theta i})^{-1}j^{-1}$. This shows any element of reduced norm 1 in D is a commutator.

Nakayama and Matsushima② proved that any element of reduced norm 1 in a p-adic division algebra is a product of at most three commutators. Combining this result with the above simple fact in the quaternion algebra, we may say that any element of reduced norm 1 in a "local" division algebra is in the commutator group. It is but a step to infer the corresponding statement for any "local" simple algebra.

Our main purpose here is to prove that the statement holds "in the large," i. e., the commutator group of the group of all regular elements in a simple algebra A over an algebraic number field k coincides with the group of all elements of reduced norm 1. From this theorem and the norm theorem of Hasse-Schilling-Maass, we infer that the

---

* 本文原载于《美国数学杂志》(*American Journal of Mathematics*) 1950 年第 72 卷第 2 期。

① This investigation of the multiplicative group of a simple algebra was suggested to the author by Professor E. Artin.

② T. Nakayama and Y. Matsushima, "Über die multiplikative Gruppe einer p-adischen Divisionsalgebra," *Proceedings of the Imperial Academy of Japan*, 19, 1943.

factor commutator group of A is isomorphic in a natural way to the group of all elements in k, which are positive at all infinite primes where A is ramified.

Whether the theorem is true for any simple algebra we cannot yet decide. Nevertheless, we shall prove its general validity in the special case where the index of the algebra is square free.

The commutator group of a simple ring which is not a pure division ring is a generalization of the classical special linear group. Apart from certain special cases to be discussed in I below, the question of its structure was settled by Dieudonné③. But the group of a pure division ring is entirely different. The structure problem in this case seems very difficult even when a division algebra over an algebraic number field is concerned.

Ⅰ. Commutator Groups of Simple Rings.

Let A be a simple ring, i. e., a full matrix ring over a division ring D. The set of all regular elements in A forms a group which will be denoted by the same symbol denoting the ring itself.

Suppose A is of degree $m > 1$ over D : $A = M_{(m)} \times D$, and let E_{ij}, $i, j = 1, \cdots, m$, form the usual matrix basis. E being the unit matrix, let $B_{ij}(\lambda) = E + \lambda E_{ij}$, $i \neq j$, and let $E(\alpha) = E + (\alpha - 1) E_{mm}$. Then the set of all $B_{ij}(\lambda)$ generates the commutator group A' of A and any coset of A' in A can be represented by a matrix $E(\alpha)$. The central elements of A or of A' are contained in the center k of D.

Dieudonné proved in [2] that the factor commutator group A/A' is isomorphic to D/D' and the isomorphism can be brought up by the function Δ which maps the coset of $E(\alpha)$ to the coset of α in D/D'. $\Delta(X)$ was called the determinant of the matrix X. The following lemma is an immediate consequence of this result of Dieudonné:

---

③ J. Dieudonné, "Les déterminants sur un corps non commutatif," *Bulletin de la Société Mathématique de France*, vol. 71 (1943).

LEMMA 1. *The matrix* $\alpha_1 E_{11}+\cdots+\alpha_m E_{mm}$ *is in* A' *if and only if the product* $\alpha_1\cdots\alpha_m$ *is in* D'.

We shall call a group simple if any proper normal subgroup is contained in the center. Dieudonné determined the structure of A' by showing that it is simple when $m>2$, or when $m=2$ and the center k of D is not the prime field R_2, R_3, or R_5 of characteristic 2, 3, or 5 respectively. We wish to show that this restriction can be removed. In fact, we shall prove that *the group* A' *is simple, when* $m>2$, *or when* $m=2$ *and* $D\neq R_2$ *or* R_3.

Proof. The general proof given below does not apply when $D=R_5$. For this singular case, see Dickson [3], p. 85.④

Let $A=M_{(2)}\times D$, $D\neq R_2$, R_3, R_5. Let N be a normal subgroup of A, containing a matrix

$$X=\begin{pmatrix}\alpha & \beta\\ \gamma & \delta\end{pmatrix}\notin A'\cap k.$$

Following the method in [2], we shall first construct a matrix of the form $B_{12}(\lambda_0)$ from X and its transforms.

Suppose one of β and γ is not 0. We may assume $\beta\neq 0$. For, if $\gamma\neq 0$, X may be replaced by

$$\begin{pmatrix}0 & 1\\ -1 & 0\end{pmatrix}\begin{pmatrix}\alpha & \beta\\ \gamma & \delta\end{pmatrix}\begin{pmatrix}0 & 1\\ -1 & 0\end{pmatrix}^{-1}=\begin{pmatrix}\delta & -\gamma\\ -\beta & \alpha\end{pmatrix}.$$

Furthermore, we may assume $\alpha=0$, because

$$\begin{pmatrix}-1 & 0\\ -\beta^{-1}\alpha & 1\end{pmatrix}^{-1}\begin{pmatrix}\alpha & \beta\\ \gamma & \delta\end{pmatrix}\begin{pmatrix}1 & 0\\ -\beta^{-1}\alpha & 1\end{pmatrix}=\begin{pmatrix}0 & \beta\\ \gamma_1 & \delta_1\end{pmatrix}. \tag{1}$$

For $\xi\in D$, k_ξ will denote a subfield of D, containing the field $k(\xi)$, and containing at least one element outside k in case $k=R_2$, R_3 *or* R_5.

If $\eta\in k_\gamma$, we have

④ L. E. Dickson, Linear Groups, Leipzig, 1901.

$$(2)\qquad \begin{pmatrix}0 & \beta\\ \gamma & \delta\end{pmatrix}^{-1}\begin{pmatrix}\eta & 0\\ 0 & \eta^{-1}\end{pmatrix}^{-1}\begin{pmatrix}0 & \beta\\ \gamma & \delta\end{pmatrix}\begin{pmatrix}\eta & 0\\ 0 & \eta^{-1}\end{pmatrix}=\begin{pmatrix}\eta^{2} & *\\ 0 & \beta^{-1}\eta^{-1}\beta\eta^{-1}\end{pmatrix}=Y.$$

When $k \neq R_2$, R_3, R_5, we choose η in k such that $\eta^4 \neq 1$. When $k = R_2$, R_3 or R_5, we choose η in k_γ outside k such that $\eta^2 \notin k$. In both cases, Y has the property that either its two diagonal elements are distinct or they do not belong to k.

If $\beta = \gamma = 0$ in the original matrix X, X has already the above property, because it is not in the center of A' by assumption. Hence, N always contains a matrix $\begin{pmatrix}\alpha & \beta\\ 0 & \delta\end{pmatrix}$ where $\alpha \neq \delta$ or $\alpha = \delta$ but $\alpha \notin k$. So, there exists an element $\xi \in D$ for which $\alpha\xi \neq \xi\delta$. We have

$$(3)\qquad \begin{pmatrix}1 & \xi\\ 0 & 1\end{pmatrix}\begin{pmatrix}\alpha & \beta\\ 0 & \delta\end{pmatrix}\begin{pmatrix}1 & \xi\\ 0 & 1\end{pmatrix}^{-1}\begin{pmatrix}\alpha & \beta\\ 0 & \delta\end{pmatrix}^{-1}=\begin{pmatrix}1 & \xi-\alpha\xi\delta^{-1}\\ 0 & 1\end{pmatrix}=B_{12}(\lambda_0),$$

where $\lambda_0 = \xi - \alpha\xi\delta^{-1} \neq 0$.

Now, consider a field k_{λ_0}. Starting with $B_{12}(\lambda_0)$ as the original matrix X, we first obtain a matrix of the form $\begin{pmatrix}0 & \beta\\ \gamma & \delta\end{pmatrix}$ as in (1), β, γ, $\delta \in k_{\lambda_0}$. This time, we choose $\eta \in k_{\lambda_0}$ such that $\eta^4 \neq 1$ and obtain a matrix of the form $\begin{pmatrix}\alpha & \beta\\ 0 & \delta\end{pmatrix}$ as in (2), α, β, $\delta \in k_{\lambda_0}$, $\alpha \neq \delta$. Taking $\xi = (1-\alpha\delta^{-1})^{-1}$ in (3), we obtain the matrix $B_{12}(1)$.

Let $\lambda \in D$ be arbitrary. Starting with $B_{12}(1)$ as our matrix X, we repeat the above process, keeping the coefficients of the matrices inside a field k_λ. Setting $\xi = \lambda(1-\alpha\delta^{-1})^{-1}$ in (3), we obtain $B_{12}(\lambda)$. Since

$$\begin{pmatrix}1 & 0\\ \lambda & 1\end{pmatrix}=\begin{pmatrix}0 & 1\\ -1 & 0\end{pmatrix}\begin{pmatrix}1 & \lambda\\ 0 & 1\end{pmatrix}^{-1}\begin{pmatrix}0 & 1\\ -1 & 0\end{pmatrix}^{-1},$$

we have finally the matrix $B_{21}(\lambda)$. So, $N = A'$ and the statement is proved.

Since, as we shall prove, the commutator group of a simple algebra over an algebraic number field coincides with the group of all elements of norm 1, it is conceivable that the group D/D' might always be isomorphic to a subgroup of the center k and Dieudonné's determinant of a matrix could then be realized by an element in k. As the following example shows, this is not the case:

Let K be the field of all power series over R_2 of the form

$$\sum_{v=r}^{\infty} a_v x^v, \quad a_v = 0 \text{ or } 1, \quad r \lesseqgtr 0.$$

We introduce a second variable y and consider all the double power series

$$\sum_{\mu=s}^{\infty} \alpha_\mu y^\mu, \quad \alpha_\mu \in K, \quad s \lesseqgtr 0.$$

We add and multiply them formally applying the rule $y\alpha = \alpha^\sigma y$, $\alpha \in K$, where σ is the automorphism of K defined by the equation

(4) $$x^\sigma = x + x^2.$$

As can be readily verified, these double power series form a division ring D. Now, in a power series field, an automorphism of infinite period, which can be defined by an equation like (4), leaves fixed only the coefficient field. Since σ is obviously of infinite period, it can be easily seen from the above remark that the center of D is exactly $R_2 = \{0, 1\}$.

Let $\xi = \sum_{\mu=s}^{\infty} \alpha_\mu y^\mu$ be in D, $\alpha_s \neq 0$. We define $|\xi| = y^s$. By the rule of multiplication, we have $|\xi\eta| = |\xi||\eta|$. So, this is a homomorphism of D onto the infinite cyclic group generated by y. Since the latter is abelian, D/D' is certainly not trivial.

Ⅱ. Proof of the Main Theorem.

Let A be a central simple algebra over k. By the norm $N\alpha$ of $\alpha \in A$ we mean the reduced norm $N_{Ak}(\alpha)$ of α over k. If α is in the maximal subfield K of A, then $N\alpha = N_{Kk}(\alpha)$.

Let K be a cyclic maximal subfield over k and let σ be a generating automorphism. Then there exists an element τ in A such that $\tau\xi\tau^{-1} = \xi^\sigma$ for any $\xi \in K$. If $N\alpha = 1$, α

$\in K$, there exists an element ξ in K such that $\alpha=\xi^{\sigma-1}$ by a theorem of Hilbert. So, $\alpha=\tau\xi\tau^{-1}\xi^{-1}$ and we have proved

LEMMA 2. *Any element of norm* 1 *in* A, *that is contained in a cyclic maximal subfield, is a commutator.*

Let D be a central division algebra over k and F a field of degree m over k.

LEMMA 3. *If* $\alpha\in D$ *is in the commutator group* D'_F *of* D_F, *then* α^m *is in* D'.

Proof. Consider the algebra $A=M_{(m)}\times D$. Since F is isomorphic to maximal subfield of $M_{(m)}$, we may regard $D_F=F\times D$ as a subalgebra of A. Hence, $\alpha\in D'_F\subset A'$. By Lemma 1, $\alpha^m\in D'$.

Now, let K be a splitting field of $D=D_{(n)}$, of degree n over k. If $\alpha\in D$ has norm 1 in D, it still has norm 1 in $D_K=M_{(n)}$. But an element of norm 1 in a full matrix algebra is nothing but a matrix of determinant 1. So, $\alpha\in D'_K$. This, together with Lemma 3, proves

LEMMA 4. *If* $N\alpha=1$ *in* $D=D_{(n)}$, *then* $\alpha^n\in D'$.

Let k be a p-adic field and $D=D_{(n)}$ be a central division algebra over k. Then D contains an unramified maximal subfield W. For a suitably chosen generating automorphism σ of W and a suitably chosen prime element π in D, we have

$$D=W+W_{\pi}+\cdots+W_{\pi^{n-1}};\ \pi\xi\pi^{-1}=\xi^{\sigma}\ \text{for}\ \xi\in W,\ \pi^n=p.$$

So, D admits an irreducible representation over W, in which the element $\alpha=\alpha_0+\alpha_1\pi+\alpha_2\pi^2+\cdots+\alpha_{n-1}\pi^{n-1}$, $\alpha_i\in W$, is represented by the matrix

$$\begin{pmatrix} \alpha_0 & \alpha_1 & \cdots & \alpha_{n-1} \\ p\alpha^{\sigma}_{n-1} & \alpha^{\sigma}_0 & \cdots & \alpha^{\sigma}_{n-2} \\ p\alpha_1^{\sigma^{n-1}} & p\alpha_2^{\sigma^{n-1}} & \cdots & \alpha_0^{\sigma^{n-1}} \end{pmatrix}.$$

It can be shown that α is integral if and only if so are α_0, α_1, $\cdots$, α_{n-1}.

Now, let α be a unit in D. Let q be the number of residue-classes of p in k. We have $\alpha_0\equiv\omega\pmod{p}$, ω being a (q^n-1)-th root of unity in W. It follows from the above representation that $N\alpha\equiv N\alpha_0\equiv N\omega\pmod{p}$.

The following lemma is a slight improvement of a result of Nakayama and Matsushima, referred to in the introduction:

LEMMA 5. *Any element* α *of norm* 1 *in* D *is a product of at most two commutators.*

Proof. Since α is of norm 1, it is a unit. Using the notations introduced above, we have $\alpha\equiv\omega\,(\mathrm{mod}\ \pi)$ and $1=N\alpha\equiv N\omega\,(\mathrm{mod}\ p)$. Since $N\omega$ is a (q^n-1)-th root of unity, the latter congruence implies $N\omega=1$. So, $\omega^{1+q+\cdots+q^{n-1}}=1$, or ω is a $(q^n-1)/(q-1)$-th root of unity.

Suppose ω is a primitive $(q^n-1)/(q-1)$-th root of unity. If $v<n$, we have $q^v-1<(q^n-1)/(q-1)$. So, ω is not a (q^v-1)-th root of unity for $v<n$, and therefore generates the residue-class field of π in D (i. e., the residue-class field of p in k by adjoining the class of ω). It follows that α also generates the residue-class field. Consequently, α generates a maximal unramified subfield of D. Since α is in a cyclic maximal subfield, it is a commutator.

If ω is not primitive, let $\omega_1\in W$ be a primitive $(q^n-1)/(q-1)$-th root of unity. Then $\omega_1^{-1}\alpha\equiv\omega_1^{-1}\omega\,(\mathrm{mod}\ \pi)$, and $\omega_1^{-1}\omega$ is primitive. Since both ω_1 and $\omega_1^{-1}\alpha$ are commutators, α is in this case a product of two commutators.

Observe that, if we have proved that any element of norm 1 in D is in D', D being an arbitrary division algebra, we may conclude immediately that any element of norm 1 in $A=M_{(m)}\times D$ is in A'. For, any element $\alpha\in A$ is equivalent to a diagonal matrix $E(\beta)$ mod A'. If $N\alpha=1$, we have $N_{Dk}(\beta)=1$ and $\beta\in D'$. Consequently, $E(\beta)\in A'$ and $\alpha\in A'$.

Let $A=A_{(n)}$ be a central simple algebra over the p-adic field k, and ω_1, ω_2, $\cdots$, ω_{n^2} form a basis. Let $\xi=t_1\omega_1+\cdots+t_{n^2}\omega_{n^2}$ be a general element of A. We valuate A by defining $|\xi|=max(|t_1|,\cdots,|t_{n^2}|)$. This is a valuation of A in a modified sense, inasmuch as the condition $|\xi\eta|=|\xi||\eta|$ for an orthodox valuation is not necessarily satisfied. Nevertheless, it turns A into a complete normed vector space over k, and different bases give rise to topological equivalent spaces. Needless to say, any subalgebra is a linear subspace and therefore, as a point set, is closed in A. Furthermore, if the subalgebra is a field, the valuation introduced above induces in it the same topology as its standard valuation does.

As in any normed vector space, $\xi+\eta$, $\xi-\eta$, $\alpha\xi$ are of course continuous functions of ξ, η. Since the coefficients of $\xi\eta$ are polynomials in those of ξ, η, $\xi\eta$ is also a continuous function of ξ, η. Let $f_\xi(x)=x^n+a_1(t)x^{n-1}+\cdots+a_n(t)$ be the principal polynomial of ξ. Then $a_1(t)$, $\cdots$, $a_n(t)$, as polynomials of t_1, $\cdots$, t_{n^2}, are continuous functions of ξ. If ξ is regular, then $a_n(t)\neq 0$ and ξ^{-1} may be written in the form $\xi^{-1}=-a_n(t)^{-1}(\xi^{n-1}+a_1(t)\xi^{n-2}+\cdots+a_{n-1}(t))$. Consequently, ξ^{-1} is a continuous function of ξ.

LEMMA 6. *If α generates a maximal subfield K of A, then all elements in a sufficiently small neighborhood of α generate maximal subfields which are isomorphic to K.*

Proof. Since there are only finitely many intermediate fields between K and k, and they are closed in K, there exists a neighborhood U_K of α in K, which does not intersect them. Since $f_\alpha(x)$ is irreducible over k, it has no multiple roots. So, there exists a neighborhood U_A of α in A such that, if $\xi\in U_A$, $f_\xi(x)$ will be sufficiently near $f_\alpha(x)$ so that it will have a root θ in U_K on account of Hensel's Lemma. Since θ does not belong to the intermediate fields, it generates K: $K=k(\theta)$. But then $f_\xi(x)$ must be irreducible. Consequently, $k(\xi)$ is a maximal subfield and is isomorphic to $k(\theta)=K$.

Let A be a central simple algebra over the algebraic number field k. Then $A^p=A\times k_p$ is the p-adic completion of A, p being a prime in k. If $\alpha_i\in A^{p_i}$, $i=1$, $\cdots$, t, are given, we can approximate them simultaneously by an element ξ in A. In fact, we need only approximate the coefficients.

LEMMA 7. *Let $\varepsilon>0$ be given and let $\alpha_i\in A^{p_i}$, $i=1$, $\cdots$, t, be elements of* *norm* 1. *There exists an element* $\xi \in A'$ *such that* $|\xi-\alpha_1|^{p_1}<\varepsilon$, …, $|\xi-\alpha_t|^{p_t}<\varepsilon$.

Proof. Since α_i is of norm 1, it is in $(A^{p_i})'$. Let

$$\alpha_i=\beta_{i1}\gamma_{i1}\beta_{i1}^{-1}\gamma_{i1}^{-1}\cdots\beta_{ix_i}\gamma_{ix_i}\beta_{ix_i}^{-1}\gamma_{ix_i}^{-1},\ \beta_{ij},\ \gamma_{ij}\in A^{p_i}.$$

Let η_{ij}, $\xi_{ij}\in A$ be good approximations to β_{ij}, γ_{ij} at p_i and to 1 at $p_{i'}$, $i'=1$, …, t, $i'\neq i$. Then

$$\xi=\prod_{j=1}^{x_1}(\eta_{ij}\xi_{ij}n_{ij}^{-1}\xi_{ij}^{-1})\cdots\prod_{j=1}^{x_t}(\eta_{tj}\xi_{tj}n_{tj}^{-1}\xi_{tj}^{-1})$$

is a good approximation to α_i at p_i, $i=1$, …, t.

THEOREM. *Any element of norm* 1 *in a simple algebra over an algebraic number field is in the commutator group.*

Proof. Let k be the center of the algebra. We prove the theorem in three steps, n being the index of the algebra:

i) $n=1$, a prime. In this case, the assumption that k is an algebraic number field will not be used. By a remark made before, we may prove the theorem only for a division algebra $D_{(l)}$. Let $\alpha_0\in D$ be an element of norm 1. α_0 is contained in a maximal subfield K of D. If D is of characteristic l and K is inseparable over k, α_0 satisfies the equation $\alpha_0{}^l-1=0$ which implies $\alpha_0=1$. So, we may assume K is separable over k. Let E be an extension of K, which is normal over k with the Galois group $G(k)$. Let $G(F)$ be an l-Sylow group of $G(k)$ with the fixed field F. Then $(F:k)=m$ is not divisible by l and D_F remains a division algebra. Consider the maximal subfield $K_F=K\times F$ of D_F and the group of E over K_F be $G(K_F)$. Since $G(K_F)$ is a maximal subgroup in the p-group $G(F)$, it is normal in the same. Hence, K_F is normal and cyclic over F. $\alpha_0\in K_F$ is therefore in D'_F by Lemma 2. Applying Lemma 3, we have $\alpha_0^m\in D'$. But Lemma 4 gives $\alpha_0{}^l\in D'$. Since $(l, m)=1$, $\alpha_0\in D$.

ii) $n=l^s$, $s>1$. We prove the theorem by induction. Accordingly, assume the theorem has been proved for $n = l^{s-1}$. Let α_0 be an element in $D = D_{(l^s)}$, of norm 1.

Let S_0 be a finite set of finite primes in k, which contains

(a) all finite primes where D is ramified,

(b) all even primes,⑤

(c) a prime which splits completely in the l component of $k(\zeta_{s+1})$ over k, ζ_{s+1} denoting a primitive l^{s+1}-th root of unity.

For $p \in S_0$, let K^p be a maximal unramified subfield of D^p, and let θ_p be an element of norm 1, which generates $K^p : K = k_p(\theta_p)$. For example, θ_p may be taken as a primitive $(q^{l^s} - 1)/(q-1)$-th root of unity in K^p, q being the number of residue-classes of p in k. Let $\xi \in D'$ be a sufficiently good approximation to the elements θ_p for all $p \in S_0$. Then $\alpha = \xi\alpha_0$ will be a sufficiently good approximation to the elements θ_p, and will generate on account of Lemma 6 maximal subfields in D^p isomorphic to K^p for all $p \in S_0$. It follows, first of all, that $K = k(\alpha)$ is a maximal subfield of D.

Now, let E, F, $G(k)$, $G(F)$, $G(K_F)$ have the same significance as in i). Passing a composition series of $G(F)$ through $G(K_F)$, we see K_F has a subfield Ω_1, which is cyclic of degree l over F. We are now in the division algebra D_F. Let $\nu_1 = N_{K_F\Omega_1}(\alpha)$.

We shall denote the extension of S_0 in F by S_0 again. Let S^* be the set of all primes in F, which are not relatively prime to v_1 in Ω_1. Since $N_{\Omega_1 F}(v_1) = 1$ and Ω_1 is unramified of degree l at the primes in S_0, $S^* \cap S = 0$. Finally, let $\mathfrak{q} \notin S_0 + S^*$ be a prime which remains prime in Ω_1 and in the l-component of $F(\zeta_s)$ over F.

By (c), Ω_1 is independent with $F(\zeta_{s+1})$ over F. Applying Corollary 2 in [4] with $\widetilde{C} = \Omega_1$, we construct a cyclic extension Γ of degree l^{s+1} over F such that

(A) $(\Gamma^{\mathfrak{p}} : F_{\mathfrak{p}}) = l^{s+1}$ *for* $\mathfrak{p} \in S_0$,

(B) $\Gamma^{\mathfrak{p}} = F_{\mathfrak{p}}$ *for* $\mathfrak{p} \in S^* + \mathfrak{q}$,

---

⑤ S. Wang, "On Grunwald's theorem," to appear in the *Annals of Mathematics*.

(C) if Γ is ramified at any $\mathfrak{p}$ outside $S+S^*+\mathfrak{q}$, $\mathfrak{p}$ remains prime in Ω_1.

Let D_1 be the centralizator of Ω_1 in D_F. It is a division algebra of degree l^{s-1} over Ω_1, and is isomorphic to the division algebra in $D_F \times \Omega_1$. So, if $\mathfrak{p}_\infty$ is a real prime in F, where D_F is ramified, then D_1 is ramified over Ω_1 at $\mathfrak{P}_\infty \mid \mathfrak{p}_\infty$, if and only if $\mathfrak{P}_\infty$ is real, i.e., if and only if Ω_1 is unramified over F at $\mathfrak{p}_\infty$. Let S_∞ be the set of all real primes in F, where D_F is ramified but where Ω_1 is unramified.

Let Ω^* be the subfield of Γ of degree l over F. Let $\overline{S}$ be the set of all primes in F outside $S_0+S^*+\mathfrak{q}$, where Ω^* is ramified over F. Put $S=S_0+S^*+\overline{S}+\mathfrak{q}+S_\infty$.

Let Γ_1 be the subfield of Γ of degree l^s over F. For $\mathfrak{p}\in S$, we define $\Gamma^{*\mathfrak{p}}$ and $(\alpha_{\mathfrak{p}}, \Gamma^{*\mathfrak{p}}/F_{\mathfrak{p}})$ as follows:⑥

(a_1) for $\mathfrak{p}\in S_0+S^*+\overline{S}+\mathfrak{q}$, $\Gamma^{*\mathfrak{p}}=\Gamma_1{}^{\mathfrak{p}}$ and

$$(\alpha_{\mathfrak{p}}, \Gamma^{*\mathfrak{p}}/F_{\mathfrak{p}})=(\alpha_{\mathfrak{p}}, \Gamma_1{}^{\mathfrak{p}}/F_{\mathfrak{p}}),$$

(b_1) for $\mathfrak{p}\in S_\infty$, $\Gamma^{*\mathfrak{p}}$ is complex and $(\alpha_{\mathfrak{p}}, \Gamma^{*\mathfrak{p}}/F_{\mathfrak{p}})$ is defined in the only way possible.

Write $S'=S_0+S^*+\overline{S}+\mathfrak{q}$. When $l=2$ and when $F(\zeta_{s+1})$ is not cyclic over F, we have, by Lemma 3 in [4],

$$\prod_{\mathfrak{p}\in S'}((\sec 2\pi/2^{t+1})^{2^{s-1}}, \Gamma_1{}^{\mathfrak{p}}/F_{\mathfrak{p}})=1. \tag{5}$$

Since $(\sec 2\pi/2^{t+1})^{2^{s-1}}$ is totally positive, (5) implies that

$$\prod_{\mathfrak{p}\in S}((\sec 2\pi/2^{t+1})^{2^{s-1}}, \Gamma^{*\mathfrak{p}}/F_{\mathfrak{p}})=1$$

and that condition (d) in the theorem in [4] is satisfied. Applying that theorem with $\widetilde{C}=\Omega_1$, $C=\Omega$, we construct a cyclic extension Γ^* of degree l^s over F such that

(A_1) $\Omega^*\subset\Gamma^*$,

(B_1) the p-adic completion of Γ^* is $\Gamma^{*\mathfrak{p}}$ for $\mathfrak{p}\in S$,

(C_1) if Γ^* is ramified at any $\mathfrak{p}$ outside S, $\mathfrak{p}$ remains prime in Ω_1.

---

⑥ This is the upper part of the orthodox norm residue symbol. The lower part, consisting of a horizontal line and the prime $\mathfrak{p}$, is omitted for printing convenience.

Consider the field $K_1 = \Omega_1 \times \Gamma^*$. This is a direct product, because $\mathfrak{q}$ remains prime in Ω_1, but splits completely in Γ^* by (B) and (a_1). We contend that v_1 is a norm everywhere in K_1. In fact, at all primes extending those in S_∞, $v_1 = N_{D_1\Omega_1}(\alpha)$ is a norm in D_1 and is therefore positive. At all primes extending those in S^*, K_1 splits completely over Ω_1. Since v_1 is a unit at all finite primes not in S^*, it is a norm everywhere in K_1 except possibly at the finite primes $\mathfrak{P}$ where K_1 is ramified over Ω_1. Let $\mathfrak{p}$ be the prime in F divisible by $\mathfrak{P}$. Then $\mathfrak{p} \in S_0 + \overline{S} +$ *the primes in* (C_1). By the construction of K, Γ, and Γ^*, $\mathfrak{p}$ remains prime in Ω_1. We have

$$(v_1,\ K_1{}^{\mathfrak{P}}/\Omega_{1\mathfrak{P}}) = (N_{\Omega_1{}^{\mathfrak{p}}}F_{\mathfrak{p}}(v_1),\ \Gamma^{*\mathfrak{p}}/F_{\mathfrak{p}}) = (1,\ \Gamma^{*\mathfrak{p}}/F_{\mathfrak{p}}) = 1.$$

Consequently, v_1 is a norm everywhere in K_1, and there exists an element $\beta_1 \in K$ such that $N_{K_1\Omega_1}(\beta_1) = v_1$.

Consider the algebra $A = M_{(l)} \times D_F$ over F. It contains the algebra $A_1 = M_{(l)} \times D_1$ over Ω_1. Since K_1 clearly splits D_1 and is of degree l^s over Ω_1, it is a maximal subfield of A_1. Consider the element

$$\gamma_1 = \begin{pmatrix} 1 & 0 & \cdots & 0 \\ 0 & 1 & \ddots & \vdots \\ \vdots & \ddots & \ddots & 0 \\ 0 & \cdots & 0 & a \end{pmatrix}$$

in A_1. We have

$$N_{A_1\Omega_1}(\gamma_1) = N_{D_1\Omega_1}(\alpha) = v_1 = N_{K_1\Omega_1}(\beta_1) = N_{A_1\Omega_1}(\beta_1),$$

i. e., $N_{A_1\Omega_1}(\gamma_1) = N_{A_1\Omega_1}(\beta_1)$. By the hypothesis of induction, γ_1 differs from β_1 only by an element in $A'_1 \subset A'$.

Let us assume for a moment that it had already been proved that $\beta_1 \in A'$.

We would infer then $\gamma_1 \in A'$. Lemma 1 would give $\alpha \in D_F$. By the same argument as in i), we would conclude $\alpha \in D'$. So, $\alpha_0 \in D'$ and the theorem in the case of prime power indices would have been proved.

To prove that β_1 is actually in A', consider the centralizator A^* of Ω^* in A. It contains of course the element β_1. Now, it is easy to construct a field of degree l^{s-1} over Ω^*, which splits D_F ⑦. So, Ω^* is isomorphic to a subfield Ω_2 of D_F of degree l over F. Let D_2 be the centralizator of Ω_2 in D_F. Then that of Ω_2 in A is $A_2 = M_{(l)} \times D_2$. Since A_2 is conjugate to A^* in A, A_2 contains an element β_2 conjugate to β_1. To prove $\beta_1 \in A'$, it suffices to prove $\beta_2 \in A'$.

Let $v_2 = N_{A_2\Omega_2}(\beta_2)$. Let S^{**} be the set of all primes in F, which are not relatively prime to v_2 in Ω_2. Let S'_∞ be the set of all real primes in F, where D_F is ramified. Put $S = S_0 + S^{**} + S'_\infty + \overline{S}$.

We are going to construct a cyclic extension K_2 of degree l^{s+1} over F, applying the theorem in [4] with $\widetilde{C} = C = \Omega_2$. Since $\mathfrak{q}$ remains prime in the l-component of $F(\zeta_s)$, but splits completely in Ω_2, Ω_2 is independent with $F(\zeta_s)$ over F.

For $\mathfrak{p} \in S$, we define $K_2^{\mathfrak{p}}$ and $(\alpha_{\mathfrak{p}}, K_2^{\mathfrak{p}}/F_{\mathfrak{p}})$ as follows:

(a_2) for $\mathfrak{p} \in S_0 + \overline{S}$, $K_2^{\mathfrak{p}} = \Gamma^{\mathfrak{p}}$ and $(\alpha_{\mathfrak{p}}, K_2^{\mathfrak{p}}/F_{\mathfrak{p}}) = (\alpha_{\mathfrak{p}}, \Gamma^{\mathfrak{p}}/F_{\mathfrak{p}})$,

(b_2) for $\mathfrak{p} \in S^{**}$, $K_2^{\mathfrak{p}} = F_{\mathfrak{p}}$,

(c_2) for $\mathfrak{p} \in S'_\infty$, $K_2^{\mathfrak{p}}$ is complex and $(\alpha_{\mathfrak{p}}, K_2^{\mathfrak{p}}/F_{\mathfrak{p}})$ is defined in the only way possible.

Since $\Omega_2^{\mathfrak{p}} = F_{\mathfrak{p}}$ for $\mathfrak{p} \in S^{**}$, $\Omega_2^{\mathfrak{p}} \subset K_2^{\mathfrak{p}}$ for $\mathfrak{p} \in S^{**}$. As before, we prove that, when $l = 2$ and when $F(\zeta_s)$ is not cyclic over F,

$$\prod_{\mathfrak{p} \in S} ((\sec 2\pi/2^{t+1})^{2^s}, K_2^{\mathfrak{p}}/F_{\mathfrak{p}}) = 1,$$

and that condition (d) is satisfied. Consequently, there exists a cyclic extension K_2 of degree l^{s+1} over F having the following properties:

(A_2) $\Omega_2 \subset K_2$,

(B_2) the $\mathfrak{p}$-adic completion of K_2 is $K_2^{\mathfrak{p}}$ for $\mathfrak{p} \in S$,

(C_2) if K_2 is ramified at any prime $\mathfrak{p}$ outside S, $\mathfrak{p}$ remains prime in Ω_2.

---

⑦ For example, let $\eta \in \Omega^*$ be totally negative and divisible exactly by the first power of $\mathfrak{p}$ for $\mathfrak{p} \in S_0$. Then, $\Omega^*(\eta^{1/a})$, where $a = l^{s-1}$, splits D_F.

Again, we prove that there exists an element γ_2 in K_2 such that $N_{K_2\Omega_2}(\gamma_2)=v_2$.

Since K_2 splits D_F and therefore D_2, it is a maximal subfield of A_2 and therefore of A. We have $N_{A_2\Omega_2}(\gamma_2) = N_{K_2\Omega_2}(\gamma_2) = v_2 = N_{A_2\Omega_2}(\beta_2)$, i.e., $N_{A_2\Omega_2}(\gamma_2) = N_{A_2\Omega_2}(\beta_2)$. By the hypothesis of induction β_2 differs from γ_2 only by an element in $A'_2 \subset A'$. Now,

$$N_{AF}(\gamma_2)=N_{\Omega_2 F}(v_2)=N_{AF}(\beta_2)=N_{AF}(\beta_1)=N_{\Omega_1 F}(v_1)=N_{D_F F}(\alpha)=1,$$

i.e., $N_{AF}(\gamma_2)=1$. But γ_2 is contained in the cyclic maximal subfield K_2 of A. So, γ_2 is a commutator in A and $\beta_2 \in A'$.

iii) $n=l_1^{s_1}l_2^{s_2}\cdots l_r^{s_r}$, general. Let F_i be an extension of degree m_i of k such that $(m_i, l_i)=1$, and that D_{F_i} is of index $l_i^{s_i}$ over F_i. Let $\alpha_0 \in D$ be of norm 1. Then $\alpha_0 \in D'_{F_i}$. By Lemma 3, $\alpha_0^{m_i} \in D'$. But, by Lemma 4, $\alpha_0^n \in D'$. Since $(m_1, m_2, \cdots, m_r, n)=1$, $\alpha_0 \in D'$. The theorem is thus completely proved.

If, in iii), the fields F_i are chosen such that $(m_i, l_i^{s_i})=l_i^{s_i-1}$ and that D_{F_i} is of index l, we obtain $\alpha_0^{n/(l_1 l_2 \cdots l_r)} \in D'$. Since i) was proved for general algebras, we have the following

COROLLARY. *Let A be any simple algebra. Then* $N\alpha = 1$ *in A implies* $\alpha^{n/(l_1\cdots l_r)} \in A'$, $n=l_1^{s_1}l_2^{s_2}\cdots l_r^{s_r}$ *being the index of A. In particular, if the index of A is square free, any element of norm* 1 *in A is in* A'.

Maass 模定理的一个简单证明*

设 A 是代数数域 F 上面的一个正规单纯代数．Maass 模定理可以叙述如下：$a \in F$ 在 A 中是一个模，必要而且只要在所有使 A 分歧的无穷远质点处 a 恒为正．由于 Grunwald 定理中所包含的错误，Maass（1937）的原证不适用．后来，Eichler（1938）曾不用 Grunwald 定理证明模定理，证明中用到关于单纯代数中的理想的某些讨论，而证明本身也比较复杂．本文作者（1950）曾利用一个修改了的 Grunwald 定理给出模定理的另一个证明．本文的目的则是用一个非常简单的推理证明模定理，证明中不用 Grunwald 定理．

条件的必要性易见（Maass 1937）．今证充分性．设 A 对 F 的次数为 n．设 p_i，$i=1, \cdots, m$，为 F 中所有使 A 分歧的有穷质点，并设 $\operatorname{ord} p_i\, a = r_i$，$i=1, \cdots, m$．按照 Maass（1937）的推理，可取 $\beta_i \in F$ 使 $\operatorname{ord} p_i\, \beta_i = 1$；而 β_i 在 p_j，$j \neq i$，处为单位；β_i 在所有使 A 分歧的无穷远质点处为正；此外，β_i 在 A 中是一个模．

命

$$a' = a\beta_1^{1-r_1} \cdots \beta_m^{1-r_m},$$

则 $\operatorname{ord} p_i a' = 1$，$i=1, \cdots, m$，且 a' 在所有使 A 分歧的无穷远质点处为正．显然，只要证明 a' 在 A 中是一个模，则本定理得证．试看扩张 $K = F(\sqrt[n]{(-1)^{n-1}a'})$．$K$ 对于 F 的次数为 n 而且显然分解 A，故 K 可以看作是 A 中的一个极大域．但如此则有

$$N_{AF}(\sqrt[n]{(-1)^{n-1}a'}) = a'.$$

* 本文原载于《东北人民大学自然科学学报》1955 年第 1 期。

参考文献

王湘浩，1950，On Grunwald's Theorem，Ann. of Math. 51，471.

Eichler，M.，1938，Über die Idealklassenzahl hyperkomplexer Systeme，Math. Zeit. 43，481.

Maass，H.，1937，Beweis des Normensatzes in einfachen hyperkomplexen Systemen，Ham. Abh. 12，64.

A Simple Proof of the Norm Theorem of Maass

Abstract

Let A be a normal simple algebra over an algebraic number field F. The norm theorem of Maass states that a necessary and sufficient condition for $a_0 \in F$ to be a norm in A is that it should be positive at all primes where A is ramified. Our aim here is to give a simple proof of this theorem, without the use of Grunwald's Theorem.

The condition is obviously necessary (Maass 1937). We prove the sufficiency. Suppose the degree of A over F is n. Let p_i, $i=1, \cdots, m$, be all the finite primes in F where A is ramified, and let $\mathrm{ord} p_i \alpha = r_i$, $i=1, \cdots, m$. By an argument of Maass (1937), there exists $\beta_i \in F$ such that $\mathrm{ord} p_i \beta_i = 1$, that β_i is a unit at all p_j, $j \neq i$, and that β_i is a norm in A. Clearly, it suffices to show that

$$a' = a\beta_1^{1-r_1} \cdots \beta_m^{1-r_m}$$

is a norm in A. Since $\mathrm{ord} p_i a' = 1$, $i=1, \cdots, m$, so $F(\sqrt[n]{(-1)^{n-1} a'})$ splits A and may be regarded as a maximal field in A. But then

$$N_{AF}(\sqrt[n]{(-1)^{n-1} a'}) = a'.$$

关于 Köthe 半单纯环*

引言

在一般环的理论中，有四种重要的根理想：Baer 根理想，Jacobson 根理想，Brown-McCoy 根理想，Köthe 根理想．对于前三种根理想已经得到了半单纯环的构造定理，就是说，半单纯环可以表为一些较简单的半单纯环的亚直接和（见 Levitzki 1951，Jacobson 1945，Brown and McCoy 1947）．在本文中，我们给出 Köthe 半单纯环的一种亚直接和表示，并且稍为详细地讨论一下交换环的情形．

1．Köthe 半单纯环的亚直接和表示

设 R 是任意环．R 的 Köthe 根理想 K 就是 R 的最大的幂零元素理想．若 $K=(0)$，则 R 说是一个 Köthe 半单纯环．

若在 R 中，两个理想 M，N 之积 $MN=0$ 蕴含 $M=(0)$ 或 $N=(0)$，则 R 叫做一个质环．如果 R 是一个质环且其 Köthe 根理想为 (0)，则 R 说是一个 Köthe 质环．

定理一．　R 的 Köthe 根理想 K 等于所有具下列性质之理想 N 的交集：R/N 是一个 Köthe 质环．

证．　设 R/N 是一个 Köthe 质环．显然 $N\supset K$，盖否则 R/N 中将有幂零元素理想．因之，要证本定理只要证明对于任意 $a\notin K$，必有一个理想 N 存在：$a\notin N$ 且 R/N 是一个 Köthe 质环．既 $a\notin K$，则 (a) 必然不是一个幂零元素理想，因而 (a) 含有一个非幂零元素 b．利用 Zorn 引理，不难说明 R 中有一个不包含 b 的任意方 b^n 的极大理想 N 存在．N 既不包含 b 的任意方，故 $a\notin N$．今试看 R/N 而且用 $\bar{b}$ 代表 R/N 中包含 b 的剩余类．由于 N 的极大性，R/N 的任意非零理想必包含 $\bar{b}$ 的若干方．因为 N 不包含 b 的任意方，故 $\bar{b}$ 非幂零，因而 $\bar{b}$ 的任意方亦非幂零，可见 R/N 中没有非零幂零元素理想，是即 R/N 的 Köthe 根理想为零．设 A，B 是 R/N 的两个非零

* 本文原载于《东北人民大学自然科学学报》1955 年第 1 期。

理想，并设 $A \ni \bar{b}^m$ ，$B \ni \bar{b}^n$ ，如此则 $AB \ni \bar{b}^{m+n}$ ，故 AB 也是一个非零理想. 因之，R/N 是一个 Köthe 质环而定理得证.

任意一组 Köthe 质环的亚直接和显然必是一个 Köthe 半单纯环，故由定理一有

定理二. R 是一个 Köthe 半单纯环，必要而且只要 R 是一组 Köthe 质环的亚直接和.

2. 交换情形

设 R 是一个 Köthe 半单纯环. R 说是对于 Köthe 根理想亚直接不可分，如果，将 R 表为任意一组 Köthe 半单纯环 R_λ 的亚直接和时，必有一个 λ 存在使 R 到 R_λ 的自然同态写像为同构写像.

下面的定理容易证明：

定理三. 设 R 是一个 Köthe 半单纯环. R 对于 Köthe 根理想亚直接不可分，必要而且只要 R 有一个最小的具下列性质之非零理想 P 存在：R/P 是一个 Köthe 半单纯环.

以下只讨论交换环，凡出现的环都是交换环.

对于交换环 R ，Köthe 根理想等于 Baer 根理想. 因之，根据 Levitzi（1951）的结果，我们知道，一个交换环 R 是一个 Köthe 半单纯环必要而且只要 R 可以表为一组无零因子的环的亚直接和. 把定理二用在交换环上我们得到同样的结果而得不出什么新东西. 但是，在定理一的证明中，环 R/N 具有一种特殊性质，这种特殊性质使我们在交换情形下可以得到更进一步的结果.

交换环 R 说是一个拟赋值环，如果 R 中有一个非幂零元素 a 具下列性质：R 的任意非零理想包含 a 的若干方.

定理四. 任意交换的 Köthe 半单纯环可以表为一组对于 Köthe 根理想亚直接不可分环的亚直接和.

证. 根据定理一的证明，我们知道任意交换的 Köthe 半单纯环可以表为一组拟赋值环的亚直接和. 今证任意拟赋值环 R 对于 Köthe 根理想亚直接不可分. 首先，因为 R 中有一个非幂零元素 a ，R 的任意非零理想包含 a 的若干方，故 R 中没有幂零元

素理想，因而 R 是 Köthe 半单纯环. 设 R 表为一组环 R_λ 的亚直和. 于是在 R 中有理想 N_λ，其交集为 (0)，而 $R/N_\lambda \cong R_\lambda$. 假定 N_λ 都不是零理想（即假定 R 到任意 R_λ 的自然同态写像都不是同构写像）. 因为上面所取的 a 非 0，故有 N_λ 存在：$N_\lambda \not\ni a$. 但 N_λ 包含 a 的若干方，故 R_λ 中有非零幂零元素存在，因而 R_λ 不是 Köthe 半单纯环. 这就证明了 R 对于 Köthe 根理想亚直接不可分.

定理五. 交换的 Köthe 半单纯环 R 对于 Köthe 根理想亚直接不可分，必要而且只要 R 是一个拟赋值环.

证. "只要"已证，今证"必要". 设交换的 Köthe 半单纯环 R 不是拟赋值环. R 可以表为一组拟赋值环 R_λ 的亚直接和. 今 R 不是拟赋值环，故 R 不可能与任何 R_λ 同构. 因之，R 对于 Köthe 根理想非亚直接不可分.

现在我们略为研究一下拟赋值环的性质，因而可以看出这样的环和普通的赋值环有某些相似之点.

设 R 是一个拟赋值环. 我们除去 R 是一个域的平凡情形.

现在我们把 R 的任意非零理想算是 0 的一个邻域，而把任意非零理想的任意剩余类算是其中所有元素的邻域. 这样，R 成为一个拓扑空间.

定理六. 在上面引进的拓扑之下，R 成为一个拓扑环.

证. 加法减法乘法的连续性易证. 今证 R 是一个 T_1 空间，这只要证明 R 的所有非零理想的交集为 (0) 即可. 假定 R 的所有非零理想的交集 J 非 (0)，则 J 是 R 的最小非零理想. 试看任意非零元素 $j \in J$. 因为 J 是理想，故 $jJ \subset J$. 但 jJ 是一个非零理想，而 J 是最小的非零理想，故 $jJ \supset J$. 因之，$jJ = J$. 这证明了 J 是一个域. 设 e 是 J 中的单位元素. 于是，a 为 R 的任意元素时，$e(a-ea)=0$. 但 R 中无零因子，故 $a-ea=0$，而 $a=ea \in J$，是即 $R \subset J$. 因之，R 本身是一个域，但这种情形前面已经除去.

据定理三，R 有一个最小的具下列性质之非零理想 P：R/P 中无非零幂零元素.

定理七. P 由所有具下列性质之元素 a 组成：

(1) $$\lim_{n\to\infty} a^n = 0.$$

证. 适合条件 (1) 之元素 a 实际上即适合下列条件之元素 a：R 的任意非零理

想包含 a 的若干方．命 Q 为所有这样的 a 的集合．今证 Q 是一个理想．设 $a \in Q$，$b \in Q$．设 N 为任意非零理想，并设 $a^m \in N$，$b^n \in N$．只要把 $(a-b)^{m+n-1}$ 展开即不难看出 $(a-b)^{m+n-1} \in N$，因之，$a-b \in Q$．设 $a \in Q$，r 任意．则 N 为任意非零理想而 $a^n \in N$ 时，有 $(ar)^n = a^n r^n \in N$．因之，$ar \in Q$．故 Q 是一个理想．次证 R/Q 中无非零幂零元素．事实上，否则有 $r \in R$，$r \notin Q$ 而 r 的若干方 $\in Q$．但如此，则易见 $\lim\limits_{n\to\infty} r^n = 0$ 而 $r \in Q$，矛盾．既 R/Q 中无非零幂零元素，故 $Q \supset P$．今若 $Q > P$，则有 $a \in Q$，$a \notin P$．但 P 为非零理想，故 a 的若干方在 P 内，即 R/P 中有非零幂零元素，矛盾．故 $Q = P$．

定理八．　设 $a \in P$ 而 r 为 R 的任意非零元素．则 r 整除 a 的若干方．特别，若 $a \in P$，$b \in P$ 皆非零，则 a 整除 b 的若干方，b 也整除 a 的若干方．

证．　理想 rR 必包含 a 的若干方，比方 a^n．于是，$r \mid a^n$．

定理八说明 R 中好像只有一个质元素．

定理九．　设 $a \in P$，$a \neq 0$．则

$$R > aR > a^2R > \cdots, \tag{2}$$

而 a^nR 作成 0 的一个完全邻域系．

证．　因 R 的任意非零理想包含 a 的若干方，故 R 的任意非零理想包含某 a^nR．所以，a^nR，$n=1, 2, \cdots$，作成 0 的一个完全邻域系．若在（2）中某两个相邻项相等，比方 $a^nR = a^{n+1}R$，则 $a^nR = a^{n+1}R = a^{n+2}R = \cdots$．但既（2）中各项作成 0 的一个完全邻域系而 R 是一个 T_1 空间，此不可能．

参考文献

Brown and McCoy, 1947, Radicals and Subdirect Sums, Am. J. 69, 46.

Jacobson N., 1945, The Radical and Semi-Simplicity for Arbitrary Rings, Am. J. 67, 300.

Levitzki, J., 1951, Prime Ideals and the Lower Radical. Am. J. 73, 25.

On Köthe Semi-Simple Rings

Abstract

In this note, a subdirect decomposition of a ring of zero Köthe radical is given, and in the commutative case, it is proved that a ring contains no nilpotents if and only if it is a subdirect sum of rings "subdirectly irreducible with respect to the Köthe radical". The structure of these latter rings is investigated and it is shown that they are similar to the valuation rings in the valuation theory.

关于 Cantor 公理*

H. B. Ефимов 在其所著高等几何学旧版中叙述 Cantor 公理如下："设在任意线段 a 上给了线段的无穷叙列 A_1B_1，A_2B_2，…，其中每个后面的都在前面一个的内部；再有设不存在这样的线段，它在所有这些线段的内部．那末在直线 a 上就存在着一个而且只一个点 X，落在所有线段 A_1B_1，A_2B_2 等等的内部"．这一公理事实上等于下列命题："设有线段叙列 A_1B_1，A_2B_2，…，其中 A_nB_n 包含 $A_{n+1}B_{n+1}$．如此，必有一点在所有这些线段之上"．在证明 Archimedes 公理的独立性时，著者引用了 Колмогоров 的一个例子．著者的证明是错误的，Колмогоров 的例子实际上并不适合 Cantor 公理。由于此，著者在第三版中将 Cantor 公理加以修改，修改以后 Коломогоров 的例子就可以用了．

本文的目的是在保持 Cantor 公理的旧有较强的形式下证明 Archimedes 公理的独立性。为此，显然只要造一个非 Archimedes 序域 F 满足下列两个条件：

(C) 设 $A_n \in F$，$B_n \in F$，$n=1, 2, \cdots$，并设 $A_n \leqslant A_{n+1}$，$B_{n+1} \leqslant B_n$，$A_n \leqslant B_n$．如此，必有 $C \in F$ 适合

$$A_n \leqslant C \leqslant B_n.$$

(D) 若 $A \in F$，$B \in F$，则有 $C \in F$ 使

$$A^2 + B^2 = C^2.$$

现在逐步进行如下．

试看所有整序叙列

$$(r_0, r_1, r_\tau, \cdots),$$

其中 τ 经过所有第一二级序数，$r_\tau = -1, 0$，或 1，此外，叙列中至多有可数个项 $r_\tau \neq 0$．命 M 为这些叙列的集合．用字典排列法规定这些序列的大小，这样，M 成为一

---

* 本文原载于《东北人民大学自然科学学报》1956 年第 2 期。

个序集.

对于 M_1 我们证明下列两个引理：

引理一. 对于 M 中任意可数个元素 $u_1, u_2\cdots$，有 $u\in M$ 存在，适合 $u>u_n$，$n=1, 2\cdots$，同样有 $v\in M$ 存在，适合 $v<u_n$，$n=1, 2, \cdots$.

证. 设 $u_n=(r_{n0}, r_{n1}, \cdots, r_{n\tau}\cdots)$。取 $r_\tau=\underset{n}{\mathrm{Max}}\, r_{n\tau}$，命 $u'=(r_0, r_1, \cdots, r_\tau, \cdots)$. 因为至多有可数个 $r_\tau\neq 0$，故 u' 之尾部都是 0，将这些 0 之一改为 1，命所得之叙列为 u，则 $u>u_n$，$n=1, 2, \cdots$. 同样，取 $s_\tau=\underset{n}{\mathrm{Min}}\, r_{n\tau}$，命 $v'=(s_0, s_1, \cdots, s_\tau, \cdots)$，将 v' 尾部某一个 0 改为 -1 而得 v，则 $v<v_n$，$n=1, 2, \cdots$.

引理二. 设 $u_n\in M$，$v_n\in M$，$n=1, 2, \cdots$，并设 $u_n\leqslant u_{n+1}$，$v_{n+1}\leqslant v_n$，$u_n<v_n$. 如此，必有 $w\in M$ 适合 $u_n<w<v_n$.

证. 用反证法，假定这样的 w 不存在. 设

$$u_n=(r_{n0}, r_{n1}, \cdots, r_{n\tau}\cdots),$$

$$v_n=(s_{n0}, s_{n1}, \cdots, s_{n\tau}\cdots).$$

命 $r_0=\underset{n}{\mathrm{Max}}\, r_{n0}$，$s_0=\underset{n}{\mathrm{Min}}\, s_{n0}$，则 $r_0\leqslant s_0$. 若 $r_0<s_0$，仿照引理一证明中的方法，可以取 $u>u_n$ 而使 u 的首项为 r_0. 这样，则 $u_n<u<v_n$ 与反证法所作假定矛盾. 故 $r_0=s_0$. 命 $r_1=\underset{n}{\mathrm{Max}}\, r_{n1}$，$s_1=\underset{n}{\mathrm{Min}}\, s_{n1}$，其中 Max 及 Min 仅对首项为 r_0 之 u_n 及 v_n 而取. 若 $r_1<s_1$ 则如上可得矛盾，故 $r_1=s_1$. 如此类推可定义 r_τ 及 s_τ 而有 $r_\tau=s_\tau$（τ 为极限数时，推理方法基本上也是一样的）. 命 $w'=(r_0, r_1, \cdots, r_\tau, \cdots)$，则 $u_n\leqslant w'\leqslant v_n$. 由反证法所作假定，不等式中必有一个等号对充分大的 n 成立，而由题设只能有一个等号成立. 若对充分大的 n，$u_n<w'=v_n$，只要把 w' 的尾部充分靠后处的一个 0 改为 -1 而得 w，则有 $u_n<w<v_n$，矛盾. 同样，若对充分大为 n，$u_n=w'<v_n$，亦可引出矛盾.

今试看下面的形式线性组合：

$$\sum_{u\in M}\lambda_u u,$$

其中系数 λ_u 为任意整数，而最多当 u 经过 M 的一个整序子集时 $\lambda_u\neq 0$. 命 G 为所有这些线性组合的集合. 规定两个性组合相加只要把对应的系数相加，而且按字典排列法规定这些线性组合的大小，这样，G 成为一个序群.

对于 G，我们证明下面的两个引理：

引理三. 对于 G 中任意可数个元素 α_1，$\alpha_2\cdots$，有 $\alpha\in G$ 存在适合 $\alpha>\alpha_n$，$n=1$，$2\cdots$，同样有 $\beta\in G$ 存在适合 $\beta<\alpha_n$，$n=1$，$2\cdots$.

证. 若 α_1，$\alpha_2\cdots$ 中包含 0，将 0 除去. 设 α_n 的第一个非 0 项为 $\lambda_{n0}u_{n0}$。由引理一，可取 $u<u_{n0}$，$n=1$，$2\cdots$. 取 $\alpha=1u$，$\beta=-1u$，则 α，β 合于本引理中的要求.

引理四. 设 $\alpha_n\in G$，$\beta_n\in G$，$n=1$，$2\cdots$，并设 $\alpha_n\leqslant\alpha_{n+1}$，$\beta_{n+1}\leqslant\beta_n$，$\alpha_n<\beta_n$. 如此，必有 $\gamma\in G$ 适合 $\alpha_n<\gamma<\beta_n$.

证. 假定这样的 γ 不存在. 若对充分大的 n，α_n 都是 0，设 β_n 的第一个非 0 项为 $\mu_{n0}v_{n0}$. 取 $v>v_{n0}$，$\gamma=1v$，则 $\alpha_n<\gamma<\beta_n$，与反证法假定矛盾. 同样，若对充分大的 n，β_n 都是 0，也可推出矛盾. 因之，最多除去有限个 n，可设 $\alpha_n\neq0$，$\beta_n\neq0$. 今设 α_n 的第一个非 0 项为 $\lambda_{n0}u_{n0}$，β_n 的第一个非 0 项为 $\mu_{n0}v_{n0}$. 不难说明，最多除去有限个 n，下述两种情形之一必然出现：

1° u_{n0} 作成一个降列，v_{n0} 作成一个升列，$u_{n0}\geqslant v_{n0}$，而 λ_{n0}，μ_{n0} 皆为正；

2° u_{n0} 作成一个升列，v_{n0} 作成一个降列，$u_{n0}\leqslant v_{n0}$，而 λ_{n0}，μ_{n0} 皆为负.

因为两种情形可以同样讨论，我们只讨论 1°. 若 $u_{n0}>v_{n0}$ 成立，则由引理二，有 $w\in M$ 存在适合 $u_{n0}>w>v_{n0}$. 取 $\gamma=1w$，则 $\alpha_n<\gamma<\beta_n$，矛盾.
故从某一个 N 开始，

$$u_{N0}=v_{N0}=u_{N+1,\ 0}=v_{N+1,\ 0}=\cdots=u_0.$$

设 $\lambda_0=\underset{n\geqslant N}{\mathrm{Max}}\lambda_{n0}$，$\mu_0=\underset{n\geqslant N}{\mathrm{Min}}\mu_{n0}$. 若 $\lambda_0<\mu_0$，命 $\lambda_{n1}u_{n1}$ 为 α_n 的第二个非 0 项. 由引理二，可取 u 适合 $u_0<u<u_{n1}$，$n\geqslant N$. 命 $\gamma=\lambda_0u_0+1u$，则 $\alpha_n<\gamma<\beta_n$，矛盾. 故 $\lambda_0=\mu_0$，而从某 N_1 起，α_n 和 β_n 的第一非 0 项都是 λ_0u_0. 下一步我们看 α_n 和 β_n，$n\geqslant N_1$，的第二非 0 项. 如此类推，我们得到一个 $\gamma=\lambda_0u_0+\cdots$ 适合 $\alpha_n\leqslant\gamma\leqslant\beta_n$. 由反证法假定，不等式中必有一个等号对于充分大的 n 成立. 但这样只要看叙列 $\alpha'_n=\alpha_n-\gamma$，$\beta'_n=\beta_n-\gamma$，由本证开始时所说即可推出矛盾.

现在看下面的广义形式幂级数：

$$\sum_{\alpha\in G}\alpha_{\alpha}x^{\alpha}$$

其中 x 是一个抽象文字，α_{α} 是任意实数，而最多当 α 经过 G 的一个整序子集时 $\alpha_{\alpha}\neq$

0. 命 F 为这些级数的集合. 我们规定，两个级数相加只要把对应的系数相加，相乘只要按照 $x^{\alpha}x^{\beta}=x^{\alpha+\beta}$ 把一个级数的各项乘其余一个级数的各项而后合并同类项，最后，我们用字典排列法规定级数的大小. 这样，F 成为一个非 Archimedes 序域满足条件（D）. 条件（C）可以仿照引理四证明.

参考文献

Н. В. Ефимов，ВысшаЯ геометрия，1949；汉译本：叶菲莫夫，高等几何学，裘光明译，商务，1953.

Н. В. Ефимов，ВысшаЯ геометрия，1953；汉译本：叶菲莫夫，高等几何学，裘光明译，高教出版社，1954.

关于似收敛*

一、引　言

设 F 是一个赋值域，K 是 F 的一个代数扩张，赋值理论的一个非常基本的事实是 F 的赋值可以开拓为 K 的一个赋值. Schilling 在其所著 *The Theory of Valuations* 中采用一种简接方法来论证这一事实，就是说，首先把 F 扩张为一个“完满域”$\overline{F}$，然后作 K 与 $\overline{F}$ 的合成域，再利用 $\overline{F}$ 上面较简单的赋值开拓理论便得到 F 到 K 的一个赋值开拓. 但 Schilling 关于“完满域”存在的论证是建筑在 Kaplansky（1942）关于赋值极大域的理论之上的，而书中 40 页引理 9（见本文定理八）的证明暗中假定了上述赋值开拓的存在. 这样，Schilling 的论证便成为循环推理.

Kaplansky（1942）对于上述引理 9 的证明可以说并没有什么大问题，因为 Krull 的论文（1932）中已经有了上述赋值开拓的存在定理. 应当指出，Krull 对于这个存在定理的证明并不完全对，但可以加以修改，事实上，我们可以修改 Krull 的论证（参看 Chevalley，*Algebraic Functions of One Variable*，第 6 页，定理 1 的证明）证明下面更普遍的存在定理：对于 F 的任意扩张 K（不一定是代数扩张），F 的赋值总可以开拓为 K 的一个赋值.

如果我们希望保持 Schilling 的简接方法，我们最好修改 Kaplansky 的论证系统. 我们发现可以避免上述引理 9 直接证明任意似收敛叙列在 F 的一个适当的直接扩张中有一个似极限（定理七）. 然后，利用似收敛与似极限的简单性质立即推出上述引理 9. 此外，我们还把这个引理扩充到有理式的情形（定理九）. 本文的论证没有用到 Ostrowski（1935）的比较特殊的引理（见 Schilling 书 40 页引理 8）.

二、似极限与似收敛

设 F 是一个赋值域，其赋值记为 V.

---

* 本文原载于《东北人民大学自然科学学报》第 1956 年第 2 期。

设 $\{a_\rho\}$ 是 F 中的一个无最后项的整序叙列. 我们说 $\{a_\rho\}$ 似收敛于 a 或说 $\{a_\rho\}$ 以 a 为似极限，如果从 ρ 的某个值开始，$V(a-a_\rho)$ 严格上升，若对充分大的 $\tau>\sigma>\rho$ 有

(1) $$V(a_\tau-a_\rho)>V(a_\sigma-a_\rho)$$

则 $\{a_\rho\}$ 说是似收敛.

由定义易见在一个似收敛叙列内，由某处起无重复项.

定理一. 若 $\{a_\rho\}$ 有一个似极限 a，则 $\{a_\rho\}$ 似收敛.

证. 设 $\rho\geqslant\rho_0$ 时，$V(a-a_\rho)$ 严格上升. 设 $\tau>\sigma>\rho\geqslant\rho_0$. 于是

$$V(a_\sigma-a_\rho)=V((a-a_\rho)-(a-a_\sigma))=V(a-a_\rho),$$

同样，

$$V(a_\tau-a_\sigma)=V(a-a_\sigma),$$

因为 $V(a-a_\sigma)>V(a-a_\rho)$，故 (1) 成立.

用类似的推理容易证明，若 $\{a_\rho\}$ 似收敛，则对充分大的 $\sigma>\rho$，$V(a_\sigma-a_\rho)$ 只与 ρ 有关. 此外，由上证，若 $\{a_\rho\}$ 似收敛于 a，则对充分大的 $\sigma>\rho$，$V(a-a_\rho)=V(a_\sigma-a_\rho)$

定理二. 若似收敛叙列 $\{a_\rho\}$ 非似收敛于 0，则从某处起，$V(a_\rho)$ 为常量.

证. 因为似收敛叙列 $\{a_\rho\}$ 非似收敛于 0，故一方面对于 $\tau>\sigma>\rho\geqslant\rho_0$，(1) 成立，一方面有 $\sigma'>\rho'\geqslant\rho_0$ 使

$$V(a_{\sigma'})\leqslant V(a_{\rho'}).$$

但如此则当 $\tau>\sigma'$ 时有

$$V(a_\tau-a_{\sigma'})>V(a_{\sigma'}-a_{\rho'})\geqslant V(a_{\sigma'}),$$

即 $V(a_\tau-a_{\sigma'})>V(a_{\sigma'})$. 由此易见

$$V(a_\tau)=V(a_{\sigma'}).$$

定理三. 若 $\{a_\rho\}$ 似收敛于 a 而不似收敛于 0，则对充分大的 ρ，$V(a)=V(a_\rho)$.

证. 设对 $\rho\geqslant\rho_0$，$V(a-a_\rho)$ 严格上升而 $V(a_\rho)$ 保持不变，取 $\rho>\sigma\geqslant\rho_0$，则

$$V(a-a_\rho)>V(a-a_\sigma)=V(a_\rho-a_\sigma)\geqslant V(a_\rho).$$

因之，

$$V(a)=V(a-a_\rho+a_\rho)=V(a_\rho).$$

定理四. 设 $\{a_\rho\}$ 似收敛于 a，$f(x)$ 是一个非常数多项式．这样，则 $\{f(a_\rho)\}$ 似收敛于 $f(a)$．

证．若 $f(x)$ 是一个一次多项式 $cx+d$，则 $f(a)=ca+d$，$f(a)-f(a_\rho)=c(a-a_\rho)$；因为从某处起 $V(a-a_\rho)$ 严格上升，故从该处起 $V(f(a)-f(a_\rho))=V(c(a-a_\rho))$ 严格上升，是即 $\{f(a_\rho)\}$ 似收敛于 $f(a)$．今用归纳法，假定定理对于次数为 $n-1$ 的多项式已证．设 $f(x)$ 的次数为 n．以 $x-a$ 除 $f(x)$ 得

$$f(x)=q(x)(x-a)+f(a).$$

今 $q(x)$ 为 $n-1$ 次，故 $\{q(a_\rho)\}$ 似收敛于 $q(a)$，因而由定理一 $\{q(a_\rho)\}$ 似收敛，于是由定理二从某处起 $V(q(a_\rho))$ 严格上升或保持不变，但 $V(a-a_\rho)$ 终极严格上升，故 $V(q(a_\rho)(a-a_\rho))$ 终极严格上升，是即 $V(f(a)-f(a_\rho))$ 终极严格上升，故 $\{f(a_\rho)\}$ 似收数于 $f(a)$．

定理五. 设 $\{a_\rho\}$ 似收敛于 a，$\dfrac{f(x)}{g(x)}$ 是一个非常数有理式．若 $\{g(a_\rho)\}$ 非似收敛于 0，则 $\left\{\dfrac{f(a_\rho)}{g(a_\rho)}\right\}$ 似收敛于 $\dfrac{f(a)}{g(a)}$．

证．试看

$$\frac{f(x)}{g(x)}-\frac{f(a)}{g(a)}=\frac{g(a)f(x)-f(a)g(x)}{g(a)g(x)}.$$

命

$$h(x)=g(a)f(x)-f(a)g(x).$$

因为 $h(a)=0$，故若 $h(x)$ 为常数，则 $h(x)=0$，与 $\dfrac{f(x)}{g(x)}$ 非常数矛盾，既 $h(a)=0$，故 $\{h(a_\rho)\}$ 似收敛于 0，因之，$V(h(a_\rho))$ 终极严格上升．今 $\{g(a_\rho)\}$ 非似收敛于 0，故 $V(g(a_\rho))$ 终极保持不变，因之，

$$V\left(\frac{f(a_\rho)}{g(a_\rho)}-\frac{f(a)}{g(a)}\right)=V\left(\frac{h(a_\rho)}{g(a)g(a_\rho)}\right)$$

终极严格上升，故 $\left\{\dfrac{f(a_\rho)}{g(a_\rho)}\right\}$ 似收敛于 $\dfrac{f(a)}{g(a)}$．

三、直接扩张

设赋值域 K 是 F 的扩张（其赋值 W 也是 F 的赋值 V 的开拓）. 若二者的值群和赋值剩余域都相同，则 K 叫做 F 的一个直接扩张.

定理六. 设赋值域 K 是 F 的扩张. 若对任意 $a \in K$，$\notin F$，有 F 中一个叙列 $\{a_\rho\}$，$\{a_\rho\}$ 似收敛于 a 而不似收敛于 0，则 K 是 F 的直接扩张.

证. 因为 $\{a_\rho\}$ 似收敛于 a 而不似收敛于 0，故由定理三对充分大的 ρ，$V(a)=V(a_\rho)$，所以 K 的值群和 F 的值群相等，设 a 是 K 中的一个赋值单位. 设对 $\sigma>\rho\geqslant\rho_0$，$V(a)=V(a_\rho)$，$V(a-a_\sigma)>V(a-a_\rho)$. 于是，$V(a-a_\rho)\geqslant 0$，而 $V(a-a_\sigma)>0$. 故模 K 的赋值极大理想，a_σ 和 a 在同一剩余类内，因而 K 的赋值剩余域和 F 的赋值剩余域相同.

定理七. 若 F 中的似收敛叙列 $\{a_\rho\}$ 在 F 中无似极限，则可以作一个直接扩张 K，使 $\{a_\rho\}$ 在 K 中有一个似极限.

证. 甲）设对 F 中的任意多项式 $f(x)$，$V(f(a_\rho))$ 终极保持不变. 于是，对任意有理式 $r(x)=\dfrac{f(x)}{g(x)}\neq 0$ 有 $\rho_0(r)$ 使 $\rho\geqslant\rho_0(r)$ 时，$V(r(a_\rho))$ 保持不变，取抽象文字 t 而作超越扩张 $K=F(t)$. 规定

$$W(r(t))=V(r(a_\rho)),\quad \rho\geqslant\rho_0(r)$$

因为 V 适合赋值的两个条件，易见 W 也适合赋值的两个条件，所以 W 是 K 的赋值而且显然是 V 的开拓. 今由 W 的定义，对充分大的 $\sigma>\rho$，$W(t-a_\rho)=V(a_\sigma-a_\rho)$，故 $W(t-a_\rho)$ 终极严格上升，即 t 为 $\{a_\rho\}$ 的似极限，由定理五及六，K 是 F 的直接扩张.

乙）设对 F 中有的多项式 $f(x)$，$V(f(a_\rho))$ 非终极保持不变，命 $\varphi(x)$ 为这些多项式中次数最低的一个，$\varphi(x)$ 自然不能是一个常数. 其次，$\varphi(x)$ 也不能是一次式. 盖若 $\varphi(x)=cx+d$，$\{ca_\rho+d\}$ 显然为似收敛，既 $V(ca_\rho+d)$ 非终极保持不变，故由定理二 $\{ca_\rho+d\}$ 似收敛于 0，此与 $\{a_\rho\}$ 在 F 中无似极限矛盾，最后，$\varphi(x)$ 必是不可约多项式. 盖若 $\varphi(x)=g(x)h(x)$，而 $g(x)$ 和 $h(x)$ 的次数低于 $\varphi(x)$ 的次数，则 $g(a_\rho)$，$h(a_\rho)$ 终极保持不变，因而 $\varphi(a_\rho)$ 亦终极保持不变，与 $\varphi(x)$ 的性质矛盾. 今作扩张 $K=F(\theta)$，$\varphi(\theta)=0$. 设 $\varphi(x)$ 的次数为 n，于是 K 中任意元素 α 可唯一地

表为 θ 的最多 $n-1$ 次的多项式 $f_{\alpha}(\theta)$. $f_{\alpha}(x)$ 的次数既低于 n，故有 $\rho_0(\alpha)$ 使 $\rho\geqslant\rho_0(\alpha)$ 时，$V(f_{\alpha}(\alpha_{\rho}))$ 保持不变. 规定

$$W(\alpha)=V(f_{\alpha}(\alpha_{\rho})),\ \rho\geqslant\rho_0(\alpha).$$

设 $\alpha\beta=\gamma$. 于是，

$$f_{\alpha}(x)f_{\beta}(x)=q(x)\varphi(x)+f_{\gamma}(x). \tag{2}$$

今 $q(x)$ 的次数小于 n，故 $V(q(a_{\rho}))$ 终极保持不变. 此外，$V(f_{\alpha}(a_{\rho}))$，$V(f_{\beta}(a_{\rho}))$，$V(f_{\gamma}(a_{\rho}))$ 亦终极保持不变，故有 ρ_0 存在使 $\rho\geqslant\rho_0$ 时，$V(q(a_{\rho}))$，$V(f_{\alpha}(a_{\rho}))$，$V(f_{\beta}(a_{\rho}))$，$V(f_{\gamma}(a_{\rho}))$ 都保持不变，但 $V(\varphi(a_{\rho}))$ 非终极保持不变，故有 $\rho_1\geqslant\rho_0$，$\rho_2\geqslant\rho_0$，使

$$V(\varphi(a_{\rho_1}))>V(\varphi(a_{\rho_2})).$$

因为

$$V(f_{\alpha}(a_{\rho_1})f_{\beta}(a_{\rho_1}))=V(f_{\alpha}(a_{\rho_2})f_{\beta}(a_{\rho_2})),$$

故由 (2)，

$$V(q(a_{\rho_1})\varphi(a_{\rho_1})+f_{\gamma}(a_{\rho_1}))=V(q(a_{\rho_2})\varphi(a_{\rho_2})+f_{\gamma}(a_{\rho_2})). \tag{3}$$

今

$$V(q(a_{\rho_1})\varphi(a_{\rho_1}))>V(q(a_{\rho_2})\varphi(a_{\rho_2})),\ V(f_{\gamma}(a_{\rho_1}))=V(f_{\gamma}(a_{\rho_2}))$$

由此及 (3) 易见

$$V(q(a_{\rho_1})\varphi(a_{\rho_1}))>V(f_{\gamma}(a_{\rho_1})).$$

因之，由 (2)，

$$V(f_{\alpha}(a_{\rho_1})f_{\beta}(a_{\rho_1}))=V(f_{\gamma}(a_{\rho_1})),$$

是即

$$W(\alpha)W(\beta)=W(\gamma).$$

第二个赋值条件容易验证. 所以，W 是 K 的一个赋值而且显然是 V 的开拓. 仿情形甲）可证 θ 为 $\{a_{\rho}\}$ 的似极限. 由定理四及六，K 是 F 的直接扩张.

定理八. 若 $\{a_{\rho}\}$ 似收敛而 $f(x)$ 是一个非常数多项式，则 $\{f(a_{\rho})\}$ 似收敛.

定理九. 若 $\{a_{\rho}\}$ 似收敛，$\dfrac{f(x)}{g(x)}$ 是一个非常数有理式，并且 $\{g(a_{\rho})\}$ 非似收敛于

0，则 $\left\{\frac{f(a_\rho)}{g(a_\rho)}\right\}$ 似收敛.

证. 由定理七，总可以将 F 扩张为一个域 K 使 $\{a_\rho\}$ 在 K 中有一个似极限. 故由定理四五及一立得定理八及九.

参考文献

Schilling, C., F. G. The Theory of Valuations, Math. Surv., No. 4, New York, 1950.

Kaplansky, I., 1942, Maximal Fields with Valuations, Duke 9, p. S03.

Krull, W., 1932, Allgemeine Bewertungstheorie, J. für Math., 167, p. 160.

Ostrowski, A., 1935, Untersuchungen zur arithmetischen Theorie der Körper, Math. Zeit, S9, p. 269.

Chevalley, C., Introduction to the Theory of Algebraic Functions of one variable, Math. Surv, No. 6, New York, 1951.

On Pseudo-Convergence

Abstract

Let F be a valuated field, and K an algebraic extension of F. It is fundamental in valuation theory that the valuation of F can be prolongated to one of K. In his book, *The Theory of Valuations*, Schilling tried to prove this fact by a so to speak, indirect method. But his presentation was based on Kaplansky's theory about maximal fields (1942), and in the proof of Lemma 9, p. 40 in the book, the existence of the abovesaid prolongation was tacitly assumed. So, Schilling's treatment was begging the question.

Of course, the existence of the prolongation can be proved directly. In fact, even the following more general theorem can be proved: For any extension K of F (not necessarily algebraic), the valuation of F can be prolongated to one of K. (For a proof see, for example, Chevalley, algebraic Functions of one variable, Th. 1, p. 6.)

In this note, we proved directly that any pseudo-convergent sequence in F has a pseudo-limit in a suitable immediate extension of F, and deduced the abovementioned Lemma 9 very easily from simple facts about pseudo-convergence and pseudo-limits. Besides, we generalized this lemma to the case of rational functions. In this way, we showed that the theory of pseudo-convergence and maximal fields can be treated without making use of the prolongation theorem and, in turn, the latter theorem can be proved by Schilling's inderect method.

拟赋值环*

引　言

本文只讨论交换环，凡出现的环都是交换的.

设环 R 无零因子，但不是一个域，R 叫作一个拟赋值环，如果其中有一个非零元素 a 具下列性质：R 的任意非零元素整除 a 的若干方. 任意具上述性质的非零元素 a 叫作一个核元素，所有核元素再加上 0 作成 R 的一个理想 P，称为 R 的核. 本文作者（1955）曾证明，任意无非幂零元素的交换环可以表为一组拟赋值环及域的亚直接和. 并且曾讨论到拟赋值环和赋值环的某些类似之点. 这些类似之点启示我们从赋值论的观点来研究拟赋值环，而这就是我们在本文中所要作的.

一个一阶赋值环上面有一个唯一确定的赋值（见§1 中的定义）. 一个拟赋值环上面却可以有不只一个乃至无穷多个赋值（见§3 中的例）. 但是，如果用半赋值的概念代替赋值的概念，却可以证明一个拟赋值环上有一个确定的最强半赋值（§1），而且这个最强半赋值的存在也是拟赋值环的特征属性（§1，定理三）. 在§2 中，我们讨论了拟赋值环上面的赋值，我们证明这些赋值就是拟赋值环上的极大半赋值而且最强半赋值就是所有赋值的下限.

此外，在§3 中，我们通过反例对几个有关问题作了解答，而且提出了本文中没有能够解决的一个问题.

§1. 最强半赋值

设环 R 无零因子，但不是一个域. 设在 R 上规定了一个函数 $v(x)$，其值取非负实数或 ∞. v 叫作 R 上面的一个半赋值，如果

1) $v(a)=\infty$，必要而且只要 $a=0$，

2) $v(ab)\geqslant v(a)+v(b)$，

---

* 本文原载于《东北人民大学自然科学学报》1957 年第 1 期。

3) $v(a^2)=2v(a)$,

4) $v(a+b)\geqslant \min(v(a), v(b))$,

5) R 中有一个非零元素 a 其值 $v(a)>0$.

如果将 2) 改为

2)′ $v(ab)=v(a)+v(b)$,

则 v 说是R 上面的一个赋值. 对于一个域 F,半赋值的定义和上面一样,只有一点不同,$v(x)$ 的值允许为任意实数或 ∞.

这里定义的半赋值和 Mahler (1936) 的准赋值的差别主要只在于这里多一个条件 3).

由 3) 易见对于任意 a, $v(-a)=v(a)$. 3) 可以推广如下:

3)′ $v(a^n)=nv(a)$,

其中 n 是任意正整数. 事实上,若 $a=0$, 则 3)′当然成立. 设 $a\neq 0$, 取 k 使 $2^k>n$. 由 3) 有 $v(a^{2^k})=2^k v(a)$, 于是,由 2),

$$2^k v(a)=v(a^{2^k})\geqslant v(a^n)+v(a^{2^k-n})$$

$$\geqslant v(a^n)+(2^k-n)v(a)\geqslant nv(a)+(2^k-n)v(a)=2^k v(a).$$

因之,

$$v(a^n)+(2^k-n)v(a)=nv(a)+(2^k-n)v(a),$$

即 $v(a^n)=nv(a)$.

设 u, v 是两个半赋值. 若对所有 x, $u(x)\leqslant v(x)$, 则 u 说是小于v, v 说是大于u. 设对任意实数 N, 有实数 M 存在具下列性质:$u(x)>M$ 涵蕴 $v(x)>N$, 这样,我们说 u 强于v, v 弱于u. 显然,如果 u 小于v, 则 u 强于 v. 若 u 强于 v 而 v 也强于 u, 则 u, v 说是等价. 例如,设 v 是一个半赋值,C 是任意正实数. 于是,Cv 也是一个半赋值而且和 v 等价.

若 u 强于 v, 则 $u(x)>0$ 时 $v(x)$ 必然也 >0. 事实上,取 $N=0$, 则由于 u 强于 v, 有 M 存在使 $u(y)>M$ 涵蕴 $v(y)>0$. 今若 $u(x)>0$, 则可取 n 使 $u(x^n)=nu(x)>M$, 因而 $v(x^n)>0$, 故 $v(x)=\dfrac{1}{n}v(x^n)>0$.

现在，设 R 是一个拟赋值环.

命题一. 设 v 是 R 上面的一个半赋值，对于任意核元素 a，有 $v(a)>0$.

证. 由半赋值的定义，有 $b\neq 0$ 存在使 $v(b)>0$. 因为 a 是一个核元素，所以 a 的若干方，比如 a^k，为 b 整除：$a^k=bc$. 从而 $kv(a)\geqslant v(b)+v(c)>0$，故 $v(a)>0$.

命题二. 设 r 是一个核元素，若 r^p 整除 r^q，则 $p\leqslant q$.

证. 假定 $p>q$，设 F 是 R 的商域. 因为 r^p 整除 r^q，故 $r^q=r^pc$，因而 $1=r^{p-q}c$. 此式表示 $1\in R$ 而 r 在 R 中有逆，从而 r 的任意方在 R 中有逆. 但 R 中任意非零元素 a 整除 r 的若干方，故 a 在 R 中有逆. 这表示 R 是一个域，和拟赋环的定义矛盾.

现在，我们用下面的方法在 R 上定义一个半赋值：

取定 R 的一个核元素 r，设 a 是 R 的任意非零元素，看 a 的任意方 a^n. 命 $h(n)=h(a,n)$ 为 a^n 中所包含的 r 的最大方次，换句话说，设 $r^{h(n)}$ 整除 a^n 而 $r^{h(n)+1}$ 不整除 a^n. r 不整除 a^n 时，自然规定 $h(n)=0$.

设 a 整除 r^M，于是，a^n 整除 r^{nM}，因而 $r^{h(n)}$ 整除 r^{nM}，可见 $h(n)\leqslant nM$. 这一方面说明 $h(n)$ 存在，一方面表示 $h(n)/n$，$n=1,2,\cdots$，有上界. 命 $m(a)=m_r(a)$ 代表 $h(n)/n$ 的上确界，今证：

$$m(a)=\lim_{n\to\infty}\frac{h(n)}{n}. \tag{1}$$

设 $\varepsilon>0$，有 M 存在使

$$\frac{h(M)}{M}>m(a)-\frac{\varepsilon}{2}. \tag{2}$$

设 $n>M$，命 $n=qM+p$，$0<p\leqslant M$. 于是 a^{qM} 整除 a^n，因而 $r^{qh(M)}$ 整除 a^n，故 $h(n)\geqslant qh(M)$，而：

$$\frac{h(n)}{n}\geqslant\frac{q}{qM+p}h(M)\geqslant\frac{q}{q+1}\cdot\frac{h(M)}{M}.$$

若 n 很大，则 $\dfrac{q}{q+1}$ 很接近 1，故有 $N\geqslant M$ 存在使 $n>N$ 时

$$\frac{h(n)}{n} > \frac{h(M)}{M} - \frac{\varepsilon}{2}. \tag{3}$$

由（2）（3）可见 $n > N$ 时，

$$\frac{h(n)}{n} > m(a) - \varepsilon.$$

因之，（1）成立.

此外，规定 $m(0) = \infty$. 我们证明 $m(x)$ 是 R 上面的一个半赋值. 条件 1）显然. 今证

$$m(r) = 1,$$

这样就附带证明了条件 5）. 按定义，$r^{h(r,\ n)}$ 整除 r^n，故 $h(r,\ n) \leqslant n$. 但 r^{n-1} 自然整除 r^n，故 $h(r,\ n) \geqslant n-1$. 从而

$$m(r) = \lim_{n\to\infty}\frac{h(r,\ n)}{n} = 1.$$

由定义立知 $h(ab,\ n) \geqslant h(a,\ n) + h(b,\ n)$，两边除以 n 而后取极限得 $m(ab) \geqslant m(a) + m(b)$，这就证明了条件 2）. 今 $h(a^2,\ n) = h(a,\ 2n)$，故

$$\frac{h(a^2,\ n)}{n} = 2\,\frac{h(a,\ 2n)}{2n},$$

取极限得 $m(a^2) = 2m(a)$，因而 3）成立.

现在证明条件 4）. a，b 有一个是 0 或 $m(a)$，$m(b)$ 有一个是 0 时，4）显然. 设 $a \neq 0$，$b \neq 0$，$m(a) \neq 0$，$m(b) \neq 0$. 设 ε 为任意正数 $< \min(m(a),\ m(b))$. 有 M 存在使 $n \geqslant M$ 时，

$$\frac{h(a,\ n)}{n} > m(a) - \varepsilon,$$

$$\frac{h(b,\ n)}{n} > m(b) - \varepsilon.$$

设 $n > 2M$ 而看下列展式：

$$(a+b)^n = a^n + \cdots + \binom{n}{k} a^{n-k} b^k + \cdots + b^n. \tag{4}$$

把此式右边分成三段来看：

i) $0 \leqslant k < M$. 此时 $n - k > n - M > M$，故 a^{n-k} 中因而 $\binom{n}{k} a^{n-k} b^k$ 中所包含

的 r 的最大方次大于 $(n-k)(m(a)-\varepsilon)>(n-M)(m(a)-\varepsilon)$.

ii) $n-M<k\leqslant n$. 此时 b^k 中因而 $\binom{n}{k}a^{n-k}b^k$ 中所包含的 r 的最大方次大于 $k(m(b)-\varepsilon)>(n-M)(m(b)-\varepsilon)$.

iii) $M\leqslant k\leqslant n-M$. 此时 a^{n-k} 中及 b^k 中所包含的 r 的最大方次分别大于 $(n-k)(m(a)-\varepsilon)$ 及 $k(m(b)-\varepsilon)$，故 $\binom{n}{k}a^{n-k}b^k$ 中所包含的 r 的最大方次大于 $n(\min(m(a), m(b))-\varepsilon)$.

总之，(4) 式右边每项所包含的 r 的最大方次都大于 $(n-M)(\min(m(a), m(b))-\varepsilon)$，因之，

$$d\,\frac{h(a+b, n)}{n}>\frac{n-M}{n}(\min(m(a), m(b))-\varepsilon).$$

令 $n\to\infty$而后令 $\varepsilon\to 0$ 即得

$$m(a+b)\geqslant\min(m(a), m(b)).$$

定理一. 对于 R 上面的任意半赋值 v 有

$$m_r(x)\leqslant\frac{v(x)}{v(r)}. \tag{5}$$

因之，$m_r(x)$ 是 R 上面的最强的半赋值.

证. 设 $x^n=r^{h(x, n)}c$，于是

$$nv(x)\geqslant h(x, n)v(r)+v(c),$$

从而

$$\frac{h(x, n)}{n}\leqslant\frac{v(x)}{v(r)}.$$

令 $n\to\infty$即得 (5).

设 s 是另一个核元素. 在 (5) 中取 $v(x)=m_s(x)$，则有 $m_r(x)\leqslant m_s(x)/m_s(r)$，或 $m_s(x)\geqslant m_s(r)m_r(x)$. 因之，

$$m_r(s)m_s(x)\leqslant m_r(x)\leqslant\frac{m_s(x)}{m_s(r)}. \tag{6}$$

在 (6) 中取 $x=r$ 得

$$m_r(s)m_s(r) \leqslant 1.$$

定理二. $a \neq 0$ 是一个核元素，必要而且只要 $m_r(a) > 0$.

证. 由命题一我们知道定理中的条件是必要的. 反之，设 $m_r(a) > 0$，于是，n 充分大时，$h(a, n) > 0$，故 n 充分大时，r 整除 a^n. 因而 r 是一个核元素，由此易见 a 必然也是一个核元素.

定理三. 设 R 无零因子，但不是一个域，若 R 上面有一个极强半赋值 w，则 R 是一个拟赋值环.

证. 设 $w(r) > 0$，$r \neq 0$，和上面一样定义 $h(a, n)$，设 $a^n = r^{h(a, n)}c$ 于是

$$nw(a) \geqslant h(a, n)w(r) + w(c),$$

从而

$$\frac{h(a, n)}{n} \leqslant \frac{w(a)}{w(r)}. \tag{7}$$

(7) 表示 $h(a, n)$ 存在而且 $h(a, n)/n$ 有界，所以我们可以和上面一样可定义 $m_r(a)$ 而且证明 m_r 是一个半赋值. 由 (7)，

$$m_r(a) \leqslant \frac{w(a)}{w(r)},$$

这表示 m_r 强于 w，但题设 w 极强，故 m_r 和 w 等价. 设 $w(a) > 0$，$a \neq 0$，于是 $m_r(a) > 0$，因而 n 充分大时，$h(a, n) > 0$，故 n 充分大时，r 整除 a^n. 但 r 是任意适合 $w(r) > 0$ 的非零元素，而 c 是 R 的任意非零元素时 $w(rc) \geqslant w(r) + w(c) > 0$，可见 rc 也必然整除 a 的若干方，因而 c 整除 a 的若干方，这就证明了 R 是一个拟赋值环.

上面的证明事实上证明了下面的定理：若 R 无零因子但不是一个域，而且若 R 有一个半赋值 w 具下列属性：适合 $w(x) > 0$ 的元素 x 极少，则 R 是一个拟赋值环.

§ 2. 赋值之存在及性质

设 R 是一个拟赋值环. 我们取定一个核元素 r 而考虑在 r 处之值等于 1 的所有半赋值. 这些半赋值实际上代表了 R 上面的所有半赋值，因为任意半赋值 $v(x)$ 等价于 $v(x)/v(r)$ 而后者在 r 处之值为 1. 以下凡不加声明而提到一个半赋值（或赋值）便

指这种标准化了的半赋值（或赋值）而言. 在这样的了解之下，定理一表示 $m(x)=m_r(x)$ 是最小的半赋值.

命题三. 设 F 是 R 的商域. F 中任意元素可表为 c/r^n 的形式，$c\in R$，n 为正整数.

证. 设 b/d 是 F 的任意元素，$b\in R$，$d\in R$. 因为 r 是一个核元素，故 n 充分大时 d 整除 r^n ：$r^n=de$. 因之，$b/d=be/r^n$ 而命题得证.

定理四. 设 a 是 R 的任意非零元素. R 上面有一个赋值 V ，其在 a 处之值 $V(a)=m(a)$.

证. 在 R 的商域 F 中将 1 加到 R 上而得 $R[1]$. $R[1]$ 中的任意元素可以写成 $c+n$ 的形式，其中 $c\in R$，n 是一个整数. r^{-1} 必 $\notin R[1]$ ，否则 $r^{-1}=c+n$，$1=cr+nr$，因而 $1\in R$ 且 r 在 R 中有逆.

现在把 $R[1]$ 再加以扩大. 设 p，q 为正整数，且

$$\frac{q}{p}\geqslant m(a). \tag{1}$$

对于所有这样的 p，q ，将 r^q/a^p 加入 $R[1]$ 而得

$$R^*=R[1,\ r^q/a^p,\ \cdots].$$

我们说 $r^{-1}\notin R^*$. 事实上，假定

$$\frac{1}{r}=\sum b\left(\frac{r^{q_1}}{a^{p_1}}\right)^{k_1}\left(\frac{r^{q_2}}{a^{p_2}}\right)^{k_2}\cdots\left(\frac{r^{q_t}}{a^{p_t}}\right)^{k_t},$$

其中 $b\in R[1]$，k_1，k_2，…，k_t 为正整数. 取 N 为充分大的整数，则有

$$\frac{1}{r}=\frac{1}{a^N}\sum ba^{N-p_1k_1-\cdots-p_tk_t}r^{q_1k_1+\cdots+q_tk_t},$$

或

$$a^N=r\sum ba^{N-p_1k_1-\cdots-p_tk_t}r^{q_1k_1+\cdots+q_tk_t},$$

于是，

$$Nm(a)\geqslant 1+\min\left[(N-p_1k_1-\cdots-p_tk_t)m(a)+q_1k_1+\cdots+q_tk_t\right],$$

因而由（1），

$$Nm(a)\geqslant 1+Nm(a),$$

此为矛盾.

利用 Zorn 引理可以取 F 的一个子环 B 具下性质：$B \supset R^*$，$B \not\ni r^{-1}$，但任意大于 B 的 F 的子环必 $\ni r^{-1}$. 设 F 的子环 $A > B$，于是 $A \ni r^{-1}$，故由命题三，$A = F$. 所以，B 是 F 的一个极大子环. 根据 Krull（1930）的一个定理，B 是一个赋值环而且在 F 中确定一个实值赋值 V. 因为 $r^{-1} \notin B$，故 $V(r) > 0$，因此我们可以将 $V(r)$ 的值取为 1. 因为 $R \subset B$ 所以对于 $x \in R$，$V(x)$ 为非负实数或 ∞，于是 V 在 R 上引起一个赋值，我们仍以 V 来记.

设 $q/p \geqslant m(a)$. 于是 $r^q/a^p \in R^* \subset B$，从而 $V(r^q/a^p) \geqslant 0$，即 $q/p \geqslant V(a)$. 这说明 $V(a) \leqslant m(a)$. 但 m 为最小半赋值，故 $V(a) = m(a)$，而定理得证.

因为 m 是最小半赋值，故由定理四立得

定理五. 令 V 经过 R 上面的所有赋值，有

$$m(a) = \min V(a). \tag{2}$$

利用定理五可以推出下面的公式：

$$m_a(r)^{-1} = \max V(a), \tag{3}$$

但 a 是一个核元素，事实上，$V(x)$ 经过在 r 处之值为 1 的所有赋值时，$V(x)/V(a)$ 便经过在 a 处之值为 1 的所有赋值. 因之，

$$m_a(r) = \min \frac{V(r)}{V(a)} = \min V(a)^{-1},$$

故（3）成立.

由定理二及五有：

定理六. $a \neq 0$ 是一个核元素，必要而且只要对于 R 上面的所有赋值 V，$V(a) > 0$.

定理七. R 上面的半赋值 m 可以开拓为 R 的商域 F 上的一个半赋值.

证. 设 V 是 R 上面的赋值，V 显然可开拓到 F 上. 设 x 是 F 的任意非零元素，由命题三，x 可以表为 c/r^N 的形式，从而

$$V(x) = V(c) - N \geqslant -N.$$

这就是说取定 x，$V(x)$ 有下界，命 $m(x)$ 为 $V(x)$ 的下确界. 由定理五，$m(x)$ 在 R 上和原来的半赋值 $m(x)$ 一致，容易验证 $m(x)$ 是 F 上面的一个半赋值. 比方条

件 2)．对于任意 V 有

$$V(ab)=V(a)+V(b),$$

因之，

$$V(ab)\geqslant m(a)+m(b),$$

从而，

$$m(ab)\geqslant m(a)+m(b),$$

故定理得证.

定理八. 对于 R 上面的任意半赋值 v，有较大的赋值 V 存在.

证. 在 R 的商域 F 中，将 1 加到 R 上得 $R[1]$．设

$$v(a)\geqslant\frac{q}{p}, \tag{4}$$

其中 p，q 为正整数．对于所有这样的 a，p，q，将 a^p/r^q 加入 $R[1]$ 上而得

$$R^*=R\left[1,\ \frac{a^p}{r^q},\ \cdots\right].$$

好象在定理四的证明中一样，可以取 F 的极大子环 B 包含 R^* 而不包含 r^{-1}，B 确定 F 的因而 R 的一个赋值 V 而且 $V(r)$ 可以取为 1．设 $q/p\leqslant v(a)$，于是 $a^p/r^q\in B$，因而 $V(a^p/r^q)\geqslant 0$，即 $q/p\leqslant V(a)$，可见 $V(a)\geqslant v(a)$.

定理九. V 是 R 上面的一个极大半赋值，必要而且只要 V 是一个赋值.

证. 设 V 是一个极大半赋值．由定理八，可以取一个较大的赋值 U．但 V 极大，故 $U=V$．反之，设 V 是一个赋值．假定 V 非极大，则有半赋值 v 较 V 为大而不等于 V．由定理八，有一个赋值 U 较 v 大，因而 U 较 V 为大而不等于 V．U 及 V 可以开拓为 R 的商域 F 的赋值．设 $x\in F$，而 $V(x)>0$．由命题三，x 可以表为 c/r^n 的形式，$c\in R$．于是，$V(c)>n$．既 U 较 V 为大，故 $U(c)>n$，因而 $U(x)>0$，但这表示 U 和 V 在 F 上等价，因此 $U(x)=CV(x)$，$C>0$．取 $x=r$ 得 $1=C$，故 $U=V$，此为矛盾.

§3. 几个有关问题

首先，我们提出下面的作者还没有能够解答的问题：设 U，V 是 R 上面的两个赋

值. 若 U, V 在商域 F 上面的开拓不等价, U, V 在 R 上会不会等价? 在同样假定下, U 在 R 上有没有可能较强于 V?

另一个问题是: 拟赋值环 R 上面会不会有无穷多个不等价的赋值? 下面的例子说明这是可能的.

取任意质数 p. 命 Q 代表有理数域而命 F 代表所有代数数作成的域. 命 v_p 表示 p 在 Q 中所决定的赋值而且取 $v_p(p)=1$. 不难说明 F 中有无穷多个赋值为 v_p 的开拓. 取所有这些赋值在 F 中确定的赋值环的交集 R. R 是 F 子环而 F 是 R 商域. 显然, $p\in R$. 设 $a\in R$ ($a\neq 0$). v_p 在 F 中的所有的开拓在 $Q(a)$ 上引起有限个赋值 V_1, …, V_k. 取 n 大于所有 $V_1(a)$, …$V_k(a)$, 则 $V_i(p^n/a)>0$, 因而 $p^n/a\in R$. 这就是说, R 的任意非零元整除 p 的若干方, 所以 R 是一个拟赋值环. 设 U_1, U_2 是 v_p 在 F 上的两个不同的开拓, 我们证明 U_1, U_2 在 R 上所引起的赋值不等价. U_1, U_2 既不同, 所以必在 Q 的某个有限扩张 Ω 上不同. v_p 在 F 上的所有开拓在 Ω 上引起有限个不同的赋值 U_1, U_2, U_3, …, U_s. 这些赋值在 Ω 上不等价, 故据赋值的逼近定理, 可取 $b\in\Omega$ 使 $U_1(b)>0$, $U_2(b)=0$, $U_3(b)=0$, …, $U_s(b)=0$, 因之, $b\in R$. 因为 $U_1(b)>0$, $U_2(b)=0$, 所以 U_1, U_2 在 R 上不等价.

设 R 是一个拟赋值环. 一个可能的猜想是: R 的核 P 就是 R 的 Jacobson 根理想. 我们知道, R 对于 Köthe 根理想亚直接不可分, 因此, R 不能分为一些域的亚直接和, 由此可见, R 的 Jacobson 根理想必非零. 但 P 是具下列性质的最小的非零理想 N: R/N 没有非零幂零元素, 所以 R 的 Jacobson 根理想必然包含 P. 虽然如此, 下面的例子却说明 Jacobson 根理想确实可以大于 P. 设 K 是一个域, B 是 K 中的一个赋值环, M 是 B 的极大理想. 试看下列所有形式幂级数:

$$a_0+a_1x+\cdots+a_nx^n+\cdots, \tag{1}$$

其中 $a_0\in B$, 而 $n\geqslant 1$ 时, $a_n\in K$. 这些幂级数按照通常的运算作成一个拟赋值环 R. R 的核 P 显然就是常数项 $a_0\neq 0$ 的所有幂级数 (1) 作成的理想. 不难看出, R 的 Jacobson 根理想却是常数项 $a_0\in M$ 的所有幂级数 (1) 作成的理想.

设 R 是一个拟赋值环. 作者在 1955 年的论文中曾以 R 的任意非零理想为 0 的邻域规定了 R 的一个拓扑 T^*. 今 R 的最强半赋值 m 也确定 R 的一个拓扑 T. 现在,

我们来比较 T 和 T^* 的强弱. 设 M 是任意实数. 易见 R 中所有适合 $m(x)>M$ 的元素 x 作成 R 的一个非零理想. 因此, T^* 强于 T. 下面的例子说明 T^* 确实可以真正强于 T.

设 K 是任意域. 试看所有形式幂级数:

$$a_1 r^{\frac{k_1}{2^n}}+\cdots+a_t r^{\frac{k_t}{2^n}}+\cdots, \tag{2}$$

其中 $a_i \in K$, n 及 k_i 都是任意正整数, k_1, k_2, … 作成一个升列, 而且每个 $k_i/2^n$ 都可以化为 $t/2^s$ 的形式, 其中 $t \geqslant s2^s$. 两个这样的级数相加减显然还得到这样的级数. 设在 $t/2^s$ 中 $t \geqslant s2^s$, 在 $q/2^p$ 中 $q \geqslant p2^p$. 设 $q \leqslant s$. 于是,

$$\frac{t}{2^s}+\frac{q}{2^p}=\frac{t+2^{s-p}q}{2^s},$$

右面这个分数的分子 $>t \geqslant s2^s$. 由此可见, 两个形如 (2) 而适合所说的条件的级数相乘仍得到这样的级数. 所以, 所有这些级数作成一个环 R. 今证 R 是一个拟赋值环而 r 是一个核元素. 试看任意形如 (2) 的级数 a, 其中 $a_1 \neq 0$. 取 $M \geqslant (n2^n+k_1)/2^n$, 于是以 a 除 r^M 所得之商的第一项中 r 的方次为

$$\frac{M2^n-k_1}{2^n} \geqslant \frac{n2^n}{2^n},$$

其分子 $\geqslant n2^n$. 以后各项 r 的方次更大, 因而这些方次都可以表为 $t/2^n$ 的形式, 其中 $t \geqslant n2^n$. 所以, a 整除 r^n. 这就证明了 R 是一个拟赋值环而 r 是一个核元素.

现在我们求 R 的最强半赋值 $m(x)=m_r(x)$: 设 a 为 (2) 的形式, 其中 $a_i \neq 0$. a^N 的首项中 r 的方次为 $k_1N/2^n$. 此数减去 $(k_1N/2^n)-n$ 等于 n. 设 $(k_1N/2^n)-n$ 的整数部分为 q. 于是, 以 r^q 除 a^N 所得之商每项中 r 的方次都有大于 n, 因而这些方次都可以表为 $t/2^n$ 的形式, 其中 $t \geqslant n2^n$. 由此可见,

$$h(a_1N) \geqslant q > \frac{k_1N}{2^n}-n-1.$$

另一方面, a^n 中所能提出的 r 的方次显然最多等于 a^N 的首项中 r 的方次, 即 $k_1N/2^n$, 于是

$$\frac{k_1}{2^n}-\frac{n+1}{N}<\frac{h(a_1N)}{N}<\frac{k_1}{2^n},$$

令 $N \to \infty$得：

$$m(a)=\frac{k_1}{2^n},$$

试看 R 的理想 rR．倘 T 强于 T^*，必有 $M>0$ 存在，使 $m(x)>M$ 时，$x \in rR$．取：

$$x=r^{M+\frac{1}{2^m}},$$

则见 $m(x)>M$，但 $x \notin rR$．所以在这个例子中，T^* 真正强于 T．

参考文献

王湘浩，1955，关于 Köthe 半单纯环，人大自然科学学报，143.

Mahler，K.，1936，Über Pseudobewertungen I，Acta Math. 66，79.

Krull，W.，1930，Idealtheorie in unendlichen algebraischen Zahlkörpern II，Math. Zet. 31，527.

On Quasi-Valuation Rings

Abstract

All rings appear will be commutative.

Let R be a ring without zero-divisor, but not a field. R is called a quasi-valuation ring, if it contains a non-zero element a such that any non-zero element of R divides a power of a. The set of all elements a (zero or non-zero) having the above property is an ideal P of R, called the nucleus of R. An element $a \neq 0$ of P is said to be a nuclear element. We proved in a previous note (1955) that any commutative ring without non-zero nilpotents is a subdirect sum of a set of quasi-valuation rings and fields, and noted certain similarities between quasi-valuation rings and ordinary valuation rings. Here, we study the quasi-valuation rings from the viewpoint of valuation theory.

Suppose the ring R has no zero-divisors but is not a field. A function $v(x)$ defined on R taking non-negative real numbers and ∞ as values is said to be a semi-valuation of R, if

1) $v(a)=\infty$, when and only when $a=0$,

2) $v(ab) \geqslant v(a)+v(b)$,

3) $v(a^2)=2v(a)$,

4) $v(a+b) \geqslant \min(v(a), v(b))$,

5) there is an element $a \neq 0$ of R of value $v(a)>0$.

If 2) is replaced by $v(ab)=v(a)+v(b)$, then v is said to be a valuation of R. It can be proved that, for any semi-valuation, $v(-a)=v(a)$, $v(a^n)=nv(a)$.

Semi-valuations of a field are defined in exactly the same way, only that negative values are permitted.

Let u, v be two semi-valuations. If, for all x, $u(x) \geqslant v(x)$, then u is said to be greater than v, and v smaller than u. u is said to be stronger than v, and v weaker than u, if, for any real number N, there is a real number M such that $u(x) > M$ implies $v(x) > N$. If each of u and v is stronger than the other, then they are said to be equivalent.

Now, let R be a quasi-valuation ring. Take a nuclear element r of R. For any $a \neq 0$ in R, denote by $h(a, n)$ the maximum number of times r divides a^n. Let

$$m(a) = \lim_{n \to \infty} \frac{h(a, n)}{n}.$$

It can be shown that the limit on the right hand side actually exists and that $m(x)$ is a semi-valuation of R.

The following theorems can be proved:

Theorem 1. For any semi-valuation v of R,

$$m(x) \leqslant \frac{v(x)}{v(r)}.$$

So, $m(x)$ is the strongest semi-valuation of R.

Theorem 2. $a \neq 0$ is a nuclear element, if and only if $m(a) > 0$.

Theorem 3. Let R be a ring without zero-divisor, but not a field. If R admits a semi-valuation w such that no semi-valuation is stronger than but not equivalent to w, then R is a quasi-valuation ring.

Now, fix the nuclear element r. We normalize the semi-valuations of R by dividing each of them by its value at r. So, every valuation, thus normalized, is of value 1 at r. For valuations of R, we have the following theorems:

Theorem 5. There exist valuations of R, and we have

$$m(a) = \min V(a)$$

where V runs through all valuations of R.

Theorem 6. $a \neq 0$ is a nuclear element, if and only if, for all valuations V of R, $V(a) > 0$.

Theorem 7. The semi-valuation m of R can be prolongated to one of its quotient field F .

Theorem 8. For any semi-valuation of R , there exists a greater valuation.

Theorem 9. A semi-valuation is maximal, if and only if it is a valuation.

Examples can be given showing that R can have infinitely many valuations and that the topology induced in R by the strongest semi-valuation $m(x)$ can be properly weaker than the topology defined by regarding all non-zero ideals as the neighborhoods of 0.

代数方程根的分离的一种方法*

本文给出分离一个实系数方程的实根的一种方法，这个方法是近似求根的连分数法的引申，问题并没有得到最后解决，但在没有重根的情形下我们的方法是永远可用的，而且比用 Sturm 定理的方法要简便得多.

设 $f(x)$ 是一个实系数多项式，次数为 m. 由于 $f(x)$ 的负根就是 $f(-x)$ 的正根，我们只讨论正根.

把 $f(x)$ 变成 $h(x)=f(x+a)$ 我们说是把根减去 a，$h(x)$ 的系数可以用秦九韶程序计算. 把 $f(x)$ 变成 $k(x)=x^m f(\frac{1}{x})$ 我们说是把根取倒数，只要把 $f(x)$ 的系数列倒过来就得到 $k(x)$ 的系数列.

设

$$\alpha=q_0+\frac{1}{q_1}+\frac{1}{q_2}+\cdots\cdots \tag{1}$$

是一个连分数，其中 q_0 是一个非负整数，q_1，q_2，…… 是正整数. 把 $f(x)$ 的根减去 q_0，再取倒数，再减去 q_1，再取倒数，如此继续，我们说是按照连分数（1）变换 $f(x)$.

$f(x)$ 的系数列中的变号数简称为 $f(x)$ 的变号数.

定理. 设 $\alpha>0$ 是 $f(x)$ 的 r 重根，$r\geqslant 0$. 按照连分数（1）变换 $f(x)$，变到若干步以后，所得多项式必将保持恰有 r 个变号.

为了证明定理，我们讨论下面的多项式：

$$\varphi(x)=(x-1)^r(x+t)^s,\ r+s=m,$$

其中 $t>0$. 设 $\varphi(x)$ 的系数列为

$$g_0(t),\ g_1(t),\ \cdots\cdots,\ g_m(t), \tag{2}$$

---

* 本文原载于《吉林大学自然科学学报》1960 年第 1 期。

其中 $g_i(t)$ 是 $x^{(m-i)}$ 的系数，因而 $g_0(t)=1$. 除了一个常数因子，$g_i(t)$ 等于 $\varphi^{m-i}(0)$.

利用 Rolle 定理不难证明

引理 1. 在闭区间〔$-t$，1〕上，$\varphi^{(i)}(x)$ 恰有 $m-i$ 个根，若 0 是 $\varphi^{(i)}(x)$ 的根，则

$$\varphi^{(i-1)}(0)\neq 0，\varphi^{(i+1)}(0)\neq 0,$$

而且 $\varphi^{(i+1)}(x)$ 比 $\varphi^{(i-1)}(x)$ 少一个正根.

由引理 1 易见：

引理 2. 若 $g_i(t)=0$，则 $g_{i-1}(t)$ 与 $g_{i+1}(t)$ 皆非 0 而且符号相反.

现在把 t 看作是变数，试看多项式列

$$g_0(t),\ g_1(t),\ \cdots\cdots,\ g_i(t).$$

设 $g_i(t)$ 的最高次项和最低次项的次数之差为 k，显然 $g_i(t)$ 最多只能有 k 个非 0 根. 容易看出

$$g_0(0),\ g_1(0),\ \cdots\cdots,\ g_i(0)$$

中比

$$g_0(+\infty),\ g_1(+\infty),\ \cdots\cdots,\ g_i(+\infty)$$

中恰多 k 个变号，据此及引理 2，好像证明 Sturm 定理一样，可以证明

引理 3. $g_i(t)$ 恰有 k 个正根而且每个正根都是单根.

同样可以证明

引理 4. 对任意 $t>0$，数列（2）中恰有 r 个变号.

现在证明定理，设

$$\frac{\lambda_n}{\mu_n}=q_0+\frac{1}{q_1}+\cdots+\frac{1}{q_n}$$

是连分数（1）的第 n 个渐近分数，按照连分数（1）变换 $f(x)$，第 $2(n+1)$ 步所得的多项式等于以

$$\frac{\lambda_n x+\lambda_{n-1}}{\mu_n x+\mu_{n-1}}=q_0+\frac{1}{q_1}+\cdots+\frac{1}{q_n}+\frac{1}{x}$$

代 $f(x)$ 中之 x 而后乘以 $(\mu_n x+\mu_{n-1})^m$ 所得的多项式. 设

$$f(x)=(x-\alpha)^r(b_0x^s+b_1x^{s-1}+\cdots+b_s)$$

于是，第 $2(n+1)$ 步所得的多项式等于

$$[(\lambda_n-\alpha\mu_n)x+(\lambda_{n-1}-\alpha\mu_{n-1})]^r[b_0(\lambda_n x+\lambda_{n-1})^s+$$

$$b_1(\lambda_n x+\lambda_{n-1})^{s-1}(\mu_n x+\mu_{n-1})+\cdots+b_s(\mu_n x+\mu_{n-1})^s].$$

若 n 很大，则 $\dfrac{\lambda_{n-1}}{\mu_{n-1}}$，$\dfrac{\lambda_n}{\mu_n}$ 都离 α 很近，据此可以说明上列多项式可以写成常数

$$(\lambda_n-\alpha\mu_n)^r\mu_n^s(b_0\alpha^s+b_1\alpha^{s-1}+\cdots b_s)$$

乘多项式

$$(x-\beta)^r[C_0(1+\varepsilon_0)x^s+C_1(1+\varepsilon_1)\gamma x^{s-1}+\cdots+C_s(1+\varepsilon_s)\gamma^s] \tag{3}$$

的形式，其中

$$\beta=\frac{\alpha\mu_{n-1}-\lambda_{n-1}}{\lambda_n-\alpha\mu_n},$$

$$\gamma=\frac{\mu_{n-1}}{\mu_n},$$

$$C_i=\frac{s!}{i!\ (s-i)!},$$

而 n 足够大时，$\varepsilon_0, \varepsilon_1, \cdots, \varepsilon_s$ 的绝对值可以小于任意预先指定的正数，以 βx 代 x，(3) 变成一个多项式等于 β^m 乘

$$(x-1)^r[C_0(1+\varepsilon_0)x^s+C_1(1+\varepsilon_1)tx^{s-1}+\cdots+C_s(1+\varepsilon_s)t^s], \tag{4}$$

其中 $t=\gamma/\beta$．(4) 的系数的符号与 (3) 的相当系数的符号是一样的。命 (4) 的系数列为

$$g_0^*(t),\ g_1^*(t),\ \cdots,\ g_m^*(t). \tag{5}$$

若 $\varepsilon_0, \varepsilon_1, \cdots, \varepsilon_s$ 绝对值很小，则 $g_i^*(t)$ 的系数和 (2) 中之 $g_i(t)$ 的系数很接近，因而 (5) 中各多项式的正根的相互分割关系和 (2) 中各多项式的正根的相互分割关系一致，由此和引理 2 及 4，不难说明，对任意 $t>0$，(5) 中恰有 r 个变号，因而定理得证，

现在我们看怎样分离 $f(x)$ 的正根。据 Descartes 符号定则，$f(x)$ 的正根数等于 $f(x)$ 的变号数或少一个正偶数．设 q_0 是一个非负整数．据 Budan-Fourier 定则，$f(x)$ 在 q_0 和 q_0+1 之间的根数等于 $f(x+q_0)$ 的变号数减 $f(x+q_0+1)$ 的变号数或

少一个正偶数．用秦九韶程序可以试出，对哪些 q_0，$f(x+q_0+1)$ 的变号数小于 $f(x+q_0)$ 的变号数，因为 $f(x)$ 的变号数只有有限个，这样的 q_0 只有有限个，只对这些 q_0，$f(x)$ 在 q_0 和 q_0+1 之间才可能有根．

对于每个这样的 q_0 把 $f(x)$ 的根减去 q_0 然后取倒数而得 $h(x)$．找出所有具下列性质的整数 $q_1\geqslant 1$：$h(x+q_1+1)$ 的变号数小于 $h(x+q_1)$ 的变号数，对于每个这样的 q_1 把 $h(x)$ 的根减去 q_1 然后取倒数而得 $k(x)$．这样作下去等于按照一些连分数

$$q_0+\frac{1}{q_1}+\frac{1}{q_2}+\cdots$$

交换 $f(x)$．据定理，变到某步以后，所得的各多项式的变号数必皆保持不变，由此可见，用这个方法确有可能把 $f(x)$ 的正根都分离出来而且定出其重数，需要进一步讨论的是：究竟变到哪一步以后多项式的变号数才保持不变．

如果 $f(x)$ 没有重根，这个方法可以肯定地把所有正根分离出来，事实上，我们只要继续作下去直到所得的多项式都只剩一个变号，假定对某个 q_0，某个 q_1，…… 变下去变到第 $2n+1$ 步便只剩一个变号，这样，我们便分离出 $f(x)$ 的一个根在

$$\frac{\lambda_n}{\mu_n},\ \frac{\lambda_n+\lambda_{n-1}}{\mu_n+\mu_{n-1}}.$$

之间．

叙列布尔方程*

引　言

叙列布尔方程对于逻辑网络的综合是很重要的。作者在其 *Circult Synthesis by Solving Sequential Boolean Equations*（Zeitschr. f. math. Logik u. Grundlagen d. Math. 5，1959）一文中曾给出了叙列布尔方程有解的一种判定方法，但没有就方程本身的结构来探讨有解的充要条件。本文的目的就是给出这样的充要条件和解的唯一性定理。此外，我们得到了方程有有限个解的充要条件。这里，我们只讨论所谓“决定性解”。

§1　布尔方程

设 B 是一个布尔代数。若 $F(x_1, \cdots\cdots, x_n)$ 是未知量 x_1，……，x_n 和 B 中的已知量构成的一个布尔式，则

$$F(x_1, \cdots\cdots, x_n)=0$$

说是一个布尔方程。任意布尔方程可以化为

$$\bigcup_{i=0}^{N-1} \alpha_i u_i=0$$

的形式，其中 u_0，……，u_{N-1}，$N=2^n$，是 x_1，……，x_n 构成的 2^n 个极小项。

定义　B 中的一组元素 α_0，……，α_{m-1} 说是饱满的，如果

$$\alpha_0 \cup \cdots\cdots \cup \alpha_{m-1}=1; \tag{2}$$

说是**正交**的，如果（2）成立而且对任意 $i \neq j$，

$$\alpha_i \alpha_j=0 \tag{3}$$

定理　布尔方程（1）有解，必要而且只要系数组

$$\bar{\alpha}_0, \cdots\cdots, \bar{\alpha}_{N-1} \tag{4}$$

---

*　本文原载于国防工业出版社 1965 年出版的《1963 年全国数理逻辑专业学术会议论文选集》。

饱满。(1) 有唯一解，必要而且只要系数组 (4) 正交。

(1) 有唯一解时，其解即由

$$\left.\begin{aligned} &u_0=\bar{\alpha}_0,\\ &\cdots\cdots,\\ &u_{N-1}=\bar{\alpha}_{N-1} \end{aligned}\right\} \tag{5}$$

给出。由 (5) 显然可以计算 x_1，……，x_n 。若 (1) 有解，则其解由

$$\left.\begin{aligned} &u_0\subset\alpha_0,\\ &\cdots\cdots,\\ &u_{N-1}\subset\bar{\alpha}_{N-1} \end{aligned}\right\} \tag{6}$$

给出。由 (6) 可以写出 (1) 的普遍解；B 有限时，还可以求出所有的解。

§2 叙列布尔方程解的存在和唯一性

试看任意“叙列”或“无限长的字”

$$\alpha=\alpha^0\alpha^1\alpha^2\cdots\cdots,$$

其中 $\alpha^i=0$ 或 1。规定

$$\alpha'=\alpha^1\alpha^2\alpha^3\cdots\cdots.$$

若

$$\beta=\beta^0\beta^1\beta^2\cdots\cdots,$$

规定

$$\bar{\alpha}=\bar{\alpha^0}\,\overline{\alpha^1}\,\overline{\alpha^2}\cdots\cdots,$$

$$\alpha\cap\beta=(\alpha^0\cap\beta^0)(\alpha^1\cap\beta^1)\cdots\cdots,$$

$$\alpha\cup\beta=(\alpha^0\cup\beta^0)(\alpha^1\cup\beta^1)\cdots\cdots.$$

设 ξ_0，……，ξ_m 是叙列变量。我们考虑**叙列函数**

$$x=f(\xi_0,\ \cdots\cdots,\ \xi_m)。\tag{7}$$

定义 叙列函数 (7) 说是一个时序函数，如果 x^i 只由 ξ_1^0，……，ξ_m^0，……，ξ_1^i，……，ξ_m^i 决定；说是一个**正则函数**，如果是一个时序函数，而且有一个逻辑网格，以 ξ_1，……，ξ_m 为输入端，以 x 为输出端，实现此函数 (7)，如图 1 所示。

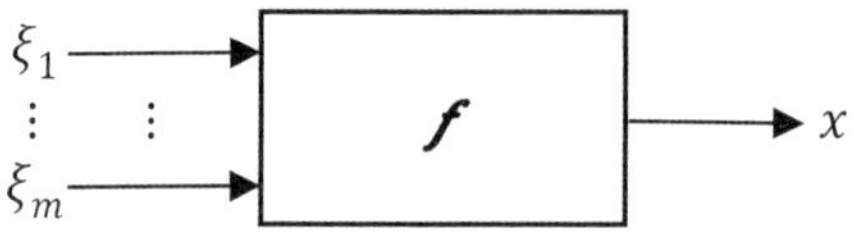

图 1

若 $F(\xi_1,\cdots\cdots,\xi_m,\xi'_1\cdots\cdots,\xi'_m,x_1,\cdots\cdots,x_n,x'_1\cdots\cdots,x'_n)$ 是一个布尔式，其中 $\xi_1,\cdots\cdots,\xi_m$ 是叙列自变量，$x_1,\cdots\cdots,x_n$ 是未知叙列函数，则

$$F(\xi_1,\cdots\cdots,\xi_m,\xi'_1,\cdots\cdots,\xi'_m,x_1,\cdots\cdots,x_n,x'_1,\cdots\cdots,x'_n)=0 \quad (8)$$

说是一个**叙列布尔方程**。任意叙列布尔方程组，其中可含有带多个**撇**的变量，必可化为一个方程（8）的形式。研究（8）的时序解可以给定"起始值"

$$\left.\begin{aligned} &x_1^0=\varphi_1^0(\xi)_1^0,\ \cdots\cdots,\ \xi_m^0),\\ &\cdots\cdots\cdots\cdots\cdots\cdots\cdots\cdots\cdots,\\ &x_n^0=\varphi_n^0(\xi)_1^0,\ \cdots\cdots,\ \xi_m^0),\end{aligned}\right\} \quad (9)$$

而研究以（9）为起始值并适合（8）的函数 $x_1,\cdots\cdots,x_n$ 也可以不给出起始值而要求 $x_1^0,\cdots\cdots,x_n^0$，适合

$$F(0,\cdots\cdots,0,\xi_1^0,\cdots\cdots,\xi_m^0,0,\cdots\cdots,0,x_1^0,\cdots\cdots,x_n^0)=0,$$

这就是说，在开始作输入以前，我们把输入端和输出端的状态都看作是 0。

引进方程

$$\left.\begin{aligned} &y'_1=\xi'_1,\\ &\cdots\cdots,\\ &y'_m=\xi'_m,\end{aligned}\right\} \quad (10)$$

而将（2）改为

$$F(y_1,\cdots\cdots,y_m,y'_1,\cdots\cdots,y'_m,x_1,\cdots\cdots,x_n,x'_1,\cdots\cdots,x'_n)=0, \quad (11)$$

则（10）与（11）合起来作成的方程组和（8）等价，事实上，以（9）为起始值并适合（8）的函数 $x_1,\cdots\cdots,x_n$ 相当于以（9）及

$$\left.\begin{aligned} &y'_1=\xi'_1,\\ &\cdots\cdots,\\ &y'_m=\xi'_m,\end{aligned}\right\} \quad (12)$$

为起始值并适合（10）及（11）的一组函数 x_1，……，x_n，y_1，……，y_m；而不给定起始值时两方面的解也显然是相当的。（10）、（11）可以合并为一个方程。因之，可以限于研究下列形式的方程：

$$F(\xi'_1, \cdots\cdots, \xi'_m, x_1, \cdots\cdots, x_n, x'_1, \cdots\cdots, x'_n)=0. \tag{13}$$

不给定起始值时可以设想把时间推后一步而把 0，……，0 看作是 x_1，……，x_n 的起始值。所以，我们又可以只针对给定了起始值（9）的情形来讨论。

以 u_0，……，u_{N-1} 和 u'_0，……，u'_{N-1} 分别记 x_1，……，x_n 和 x'_1，……，x'_n 作成的极小项。于是，（13）可以化为下面的形式：

$$(u'_0, \cdots\cdots, u'_{N-1})\begin{pmatrix} \alpha_{00} & \alpha_{01} & \cdots & \alpha_{0,N-1} \\ \alpha_{10} & \alpha_{11} & \cdots & \alpha_{1,N-1} \\ \vdots & \vdots & & \vdots \\ \alpha_{N-1,0} & \alpha_{N-1,1} & \cdots & \alpha_{N-1,N-1} \end{pmatrix}\begin{pmatrix} u_0 \\ \vdots \\ \vdots \\ u_{N-1} \end{pmatrix}=0, \tag{14}$$

其中 α_{ij} 是 ξ'_1，……，ξ'_m 作成的布尔式。$A=(\alpha_{ij})$ 称为（14）的系数矩阵。

定义 布尔矩阵 A 说是**饱满**的，如果 A 的每列都是饱满的；说是**正交**的，如果每列都是正交的。

命题 任意正方矩阵 A 必有一个最大的饱满主子矩阵（可以是空矩阵）。

定义 A 的最大饱满主子矩阵 H 称为 A 的**核**。核 H 的一个正交主子矩阵 Z 称为 A 的一个**核子**，如果在 H 中而不在 Z 中且与 Z 中元素同列的所有元素都是 0，如图 2 所示。

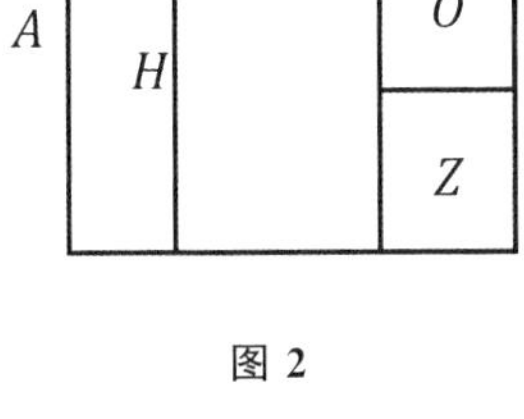

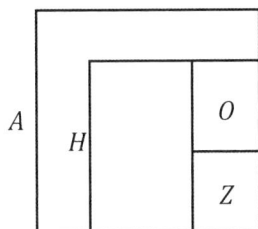

图 2

命题 任意正方矩阵 A 必有一个最大核子（可以是空矩阵）。

定义 A 的最大核子 R 称为 A 的**仁**。

定义 设 A 是（14）的系数矩阵而看 $\overline{A}$ 。设有向量

$$C=(C_0,\ \cdots\cdots,\ C_{N-1}).$$

将 C 转置为一列排在 $\overline{A}$ 的右面。C 称为一个**核向量**，若 C 的非 0 元素只与 $\overline{A}$ 的核 H 的元素同行。类似地定义**仁向量**。

以 ω_0，……，ω_{N-1} 记 φ_1^0，……，φ_n^0 作成的极小项，而命

$$\omega=(\omega_0,\ \cdots\cdots,\ \omega_{N-1})。$$

定理 若（14）有以（9）为起始值的时序解，则 ω 必为核向量；反之，若 ω 为核向量，则（14）必有以（9）为起始值的正则解。

定理 若（14）有以（9）为起始值的唯一正则解，则 ω 必为仁向量；反之，若 ω 为仁向量，则（14）必有以（9）为起始值的唯一时序解。

设 u_0，……，u_{N-1} 为（14）的以（9）为起始值的唯一解。于是，所有向量

$$u'=(u'_0,\ \cdots\cdots,\ u'_{N-1}) \tag{15}$$

都是仁向量。设仁 R 在 $\overline{A}$ 中占 i_0，……，i_{r-1} 各行。这样则对任意 t，

$$\begin{pmatrix} u_{i_0}^{t+1} \\ \vdots \\ u_{i_{r-1}}^{t+1} \end{pmatrix}=R^{t+1}\begin{pmatrix} u_{i_0}^{t} \\ \vdots \\ u_{i_{r-1}}^{t} \end{pmatrix}, \tag{16}$$

其中 R^{t+1} 表示将 R 中所含之 ξ'_1，……，ξ'_m 换成 ξ_1^{t+1}，……，ξ_m^{t+1} 所得之矩阵。

若 u_0，……，u_{N-1} 是（14）的以（9）为起始值的任意时序解，则所有向量（15）都是核向量；而且，若核 H 在 $\overline{A}$ 中占 i_0，……，i_{h-1} 则

$$\begin{pmatrix} u_{i_0}^{t+1} \\ \vdots \\ u_{i_{h-1}}^{t+1} \end{pmatrix}\subset H^{t+1}\begin{pmatrix} u_{i_0}^{t} \\ \vdots \\ u_{i_{h-1}}^{t} \end{pmatrix}。 \tag{17}$$

§3　有限个或无穷多个解的问题

定理 若（14）以（9）为起始值有无穷多个时序解，则必有无穷多个正则解。

若只有有限个正则解，则这些正则解便是所有时序解。

求出适合下列条件的所有正交矩阵 J ：

$$J \subset H。$$

这样的矩阵只有有限个。把这些矩阵 J 中的非 0 元素都换成 1，我们得到有限个矩阵

$$J_1，\cdots\cdots，J_i。\tag{18}$$

设 H 在 $\overline{A}$ 中占 i_0，……，i_{h-1} 各行。试看所有 2^h 个向量

$$(\alpha_0，\cdots\cdots，\alpha_{h-1})，\tag{19}$$

其中 $\alpha_i=0$ 或 1。以所有向量（19）为状态，所有矩阵（18）为输入，造驱动机 $\mathfrak{A}$，其驱动方程规定为

$$\begin{pmatrix}\alpha_0\\ \vdots\\ \alpha_{h-1}\end{pmatrix}'=J_i\begin{pmatrix}\alpha_0\\ \vdots\\ \alpha_{h-1}\end{pmatrix}。$$

若 $\omega=(\omega_0，\cdots\cdots，\omega_{N-1})$ 不是核向量，则方程无解。设 ω 是核向量。将足码不是 i_0，……，i_{h-1} 的分量删掉而将其余分量中的非 0 元素都换成 1，我们得到一个形如（19）的向量 ω^* 。试看 ω^* 在 $\mathfrak{A}$ 中生成的子驱动机。取 ω^* 为起始状态，而以此子驱动机中之非仁向量为终止状态，我们得到一个 Myhill 型自动机 $\mathfrak{A}^*$ 。

定理 方程（14）以（9）为起始值只有有限个解，必要而且只要 $\mathfrak{A}^*$ 所认记的事件中之字长有上界。此事件中之字长有上界必要而且只要 $\mathfrak{A}^*$ 之图象中没有仅含终止状态之顺向圈。

方程只有有限个解时，这些解可以能行地得到。

有限集合上缺值及不缺值函数的结构理论*

引　言

设 $E=E^k$ 是含有 K 个元素的一个集合，$k\geqslant 2$. E^k 中的元素用 0，1，…，k－1 表示. 命 P_k 表定义在 E^k 上并在 E^k 中取值的所有一元或多元函数之集合. 若扩大函数的范围使包括所有缺值函数，即在自变数的某些值组上无定义的函数，则其集合记为 P_k^* .

本文的目的是研究 P_k^* 的结构. 关于 P_k 的结构，函数集 $\{x\vee y,\bar{x}\}$ 及 $\{x\wedge y,\bar{x}\}$ 在 P_2 中完备（定义见§1）早为逻辑学家熟知，而这一事实，近代在电路设计中得到广泛的应用. Sheffer 指出用一个函数 $x\mid y=\overline{x\wedge y}$ 即可叠合出 P_2 中所有函数. Post（1921）证明：E^k 上的函数 $\max(x,y)$，$\min(x,y)$，$x+1(\mathrm{mod}\,k)$ 作成 P_k 的一个完备集. Post（1941）并定出了 P_2 中的所有封闭集（定义见§1）. Яблонский（1952）重新论证了 P_2 中有五个极大封闭集（定义见§1），并且（1954）定出了 P_3 中所有的 18 个极大封闭集. Кузнецов，Яблонский（1958）和另一些苏联作者进一步对 P_k 的结构作了讨论，定出了几类极大封闭集. 罗铸楷（1963，Ⅰ及Ⅱ）扩大了极大封闭集类中的线性函数集类和 T 型集类，并得到范围很广的一类极大封闭集. 庞云阶（1962）补充论证了有一个能行的方法求出 P_k 的所有极大封闭集. 但是，关于 P_k 的极大封闭集之全面、系统的理论尚有待于建立.

本文作者在自动机理论的研究中看到应对 P_k^* 的结构进行探讨. P_k^* 似乎比 P_k 要复杂，但我们所得到的结果说明并非如此. 我们具体刻划了 P_k^* 的极大封闭集，得到了对于给定的 k 求出 P_k^* 的所有极大封闭集的实际可行的方法，这些在 P_k 的情形都还是没有得到解决的问题. 我们的方法和结果，对于研究 P_k 的极大封闭集也是有用的.

---

* 本文原载于《吉林大学自然科学学报》1963 年第 2 期。1962 年 12 月 15 日，王湘浩先生在吉林大学数学系庆祝建系十周年科学报告会上宣讲。

本文所引用的关于 P_k 的一些基本概念和事实均见 Яблонский (1958) 的论文.

§1. 函数集之完备性

在一般地提出函数结构问题时，可以不限制 E 为有限集合，但假定 E 至少有两个元素. 命 P 表示所有定义在 E 上并在 E 中取值的一元或多元函数之集合，而命 P^* 表 P 添上所有缺值函数作成的集合.

函数 $f(x_1, \cdots, x_n)$ 可以简记为 $f(\tilde{x})$，其中，$\tilde{x}$ 表自变数列 $(x_1, \cdots, x_n)$. 同样，值组 $(a_1, \cdots, a_n)$ 可以简记为 $\tilde{a}$. 若 $f(\tilde{x})$ 对值组 $\tilde{a}$ 无定义，则我们说

$$f(\tilde{a}) = *.$$

设 $f(t_1, \cdots, t_m)$, $g_1(\tilde{x})$, $\cdots$, $g_m(\tilde{x}) \in P^*$. 我们规定复合函数

$$h(\tilde{x}) = f(g_1(\tilde{x}), \cdots, g_m(\tilde{x})) \tag{1}$$

如下：若 $g_1(\tilde{a})$, $\cdots$, $g_m(\tilde{a})$ 都有定义，则规定

$$h(\tilde{a}) = f(g_1(\tilde{a}), \cdots, g_m(\tilde{a}));$$

若 $g_1(\tilde{a})$, $\cdots$, $g_m(\tilde{a})$ 有的无定义，则规定

$$h(\tilde{a}) = *.$$

设 $A \subset P^*$. 若 $f(t_1, \cdots, t_m) \in A$，而 $g_i(x_1, \cdots, x_n)$, $i = 1, 2, \cdots, m$，或为 A 中的函数或为自变数 x_1, $\cdots$, x_n 之一，则说复合函数 (1) 是由 A 中的函数复合而得到的一个函数. 由 A 中的函数复合而得到的所有函数之集合记为 A'. 自然，$A' \supset A$. 若 $A' = A$，则 A 说是一个封闭集. 易见，任意一些封闭集的交集仍是封闭集. 命

$$\overline{A} = A' \cup A'' \cup A''' \cup \cdots.$$

$\overline{A}$ 中的函数说是由 A 中函数叠合而得到的函数，或说是由 A 中函数生出的函数. 不难说明，$\overline{A}$ 等于包含 A 的所有封闭集的交集，而 A 为封闭集必要而且只要 $\overline{A} = A$. 我们有

1) $A \subset \overline{A}$；

2) $\overline{\overline{A}} = \overline{A}$；

3) $\overline{\phi}=\phi$;

4) 若 $A\subset B$ ，则 $\overline{A}\subset\overline{B}$.

由 3) 及 4)，P^* 成为一个邻域空间（Alexandroff & Hopf，1935，S. 42. aufgabe）. 但 $\overline{A+B}=\overline{A}+\overline{B}$ 不成立，所以 P^* 不是一个拓扑空间.

若 $\overline{A}=P^*$ ，即若 A 在 P^* 中处处稠密，则 A 说是（在 P^* 中）完备. 函数结构理论的中心问题是如何判断一个函数集 A 是否完备.

以下我们限制 $E=E^k=\{0, 1, \cdots, k-1\}$.

处处无定义的函数称为空函数，记为 $*$.

定理 1 若所有不缺值函数可由 A 中函数生出，而 A 中有一个非空缺值函数，则 A 完备.

证 先证由 A 中函数可以叠合出下列函数

$$\tau(x)=\begin{cases}k-1, & x=k-1;\\ *, & x\neq k-1.\end{cases}$$

由题设，A 中有一个非空缺值函数 $\varphi(t_1, \cdots, t_m)$. 因为 $\varphi(\tilde{t})$ 非空缺值，故有值组 $\tilde{\alpha}$, $\tilde{\beta}$ ，使 $\varphi(\tilde{\alpha})=$ 某 $c\in E^k$，$\varphi(\tilde{\beta})=*$. 命

$$g(x)=\begin{cases}k-1, & x=c;\\ 0, & x\neq c,\end{cases}$$

$$h_i(x)=\begin{cases}\alpha_i, & x=k-1;\\ \beta_i, & x\neq k-1,\end{cases}$$

$i=1, 2, \cdots, m$. 于是，

$$\tau(x)=g(\varphi(h_1(x), \cdots, h_m(x))).$$

但由题设，$g(x)$，$h_i(x)\in\overline{A}$，故 $\tau(x)\in\overline{A}$

今设 $f(\tilde{x})$ 为任意函数 $\in P_k^*$ ，其定义域为 M. 将 $f(\tilde{x})$ 扩充为一个不缺值函数 $F(\tilde{x})$ 如下：

$$F(\tilde{x})=\begin{cases}f(\tilde{x}), & \tilde{x}\in M;\\ 0, & \tilde{x}\notin M,\end{cases}$$

并命

$$G(\tilde{x})=\begin{cases}k-1, & \tilde{x}\in M;\\ 0, & \tilde{x}\notin M.\end{cases}$$

我们有

$$f(\tilde{x})=\min(F(\tilde{x}),\ \tau(G(\tilde{x}))).$$

但 $F(\tilde{x})$，$G(\tilde{x})$，$\min(x, y)\in\overline{A}$，故 $f(\tilde{x})\in\overline{A}$.

由定理 1，若 $A\subset P_k$ 在 P_k 中完备，则 A 添上任意一个非空缺值函数后，便在 P_k^* 中完备.

命题 1 函数

$$m(x, y, z)=\begin{cases}\max(x, y)+1, & y=z;\\ *, & y\neq z,\end{cases}$$

在 P_k^* 中完备.

证 我们知道 $\max(x, y)+1=m(x, y, y)$ 在 P_k 中完备. 但 $m(x, y, z)$ 非空缺值，故由定理 1，命题得证.

§2. 极大封闭集

若 M 是一个封闭集 $<P_k^*$，而 M 与 P_k^* 之间没有另外的封闭集，则 M 称为一个极大封闭集. 极大封闭集 M 说是属于第一种，如果有的不缺值一元函数 $\notin M$；属于第二种，如果所有不缺值一元函数 $\in M$，但有的不缺值二元函数 $\notin M$；属于第三种，如果所有不缺值二元函数 $\in M$.

命题 2 设 M 是一个封闭集，$<P_k^*$. M 极大封闭，必要而且只要 M 添上不属于 M 的任意函数即成为完备集.

证 设 M 极大封闭，$f\notin M$. 因为 $\overline{M\cup\{f\}}>M$，所以 $\overline{M\cup\{f\}}=P_k^*$，即 $M\cup\{f\}$ 完备. 反之，设 M 添上不属于 M 的任意函数即成为完备集. 于是，封闭集 $A>M$ 时，A 即为完备集，因而 $A=P_k^*$，所以 M 极大封闭.

定理 2 第三种极大封闭集只有一个，即 $P_k\cup\{*\}$.

证 易见 $P_k\cup\{*\}$ 封闭且 $<P_k^*$. 由定理 1，$P_k\cup\{*\}$ 添上任意一个不属于它

的函数即成为完备集，所以 $P_k \cup \{*\}$ 极大封闭. 今设极大封闭集 M 包含所有不缺值二元函数. 于是，由 M 中的函数可以叠合出所有不缺值函数，因而 $M \supset P_k$. 但 $M < P_k^*$ ，所以由定理 1, M 不含有非空缺值函数，可见 $M \subset P_k \cup \{*\}$. 但 M 极大封闭, $M < P_k \cup \{*\}$ 是不可能的. 因之, $M = P_k \cup \{*\}$.

命题 3 设 A 封闭而且 $< P_k^*$. 于是，将自变数加入 A 所得的集合 B 仍封闭而且 $< P_k^*$.

证 设 $f(t_1, \cdots, t_m) \in B$, $g_i(\tilde{x})$ 或 $\in B$ 或为自变数, $i=1, 2, \cdots, m$. 试看复合函数

$$h(\tilde{x}) = f(g_1(\tilde{x}), \cdots, g_m(\tilde{x})).$$

若 $f(\tilde{t}) \in A$, 则 $h(\tilde{x}) \in A' = A \subset B$. 若 $f(\tilde{t})$ 是自变数，比方等于 t_i ，则 $h(\tilde{x}) = g_i(\tilde{x}) \in B$. 因之, $h(\tilde{x}) \in B$. 所以 B 封闭.

另外，因为 $A < P_k^*$ ，所以 A 不包含命题 1 中之函数 $m(x, y, z)$. 因之, $B \not\ni m(x, y, z)$ ，故 $B < P_k^*$.

由命题 3 立得

命题 4 任意极大封闭集包含自变数.

证 设 N 是 $x_1, \cdots, x_n$ 的一个不缺值函数集. $f(t_1, \cdots, t_m)$ 说是保 N ，如果对任意 $g_1(\tilde{x}), \cdots, g_m(\tilde{x}) \in N$ ，复合函数 $f(g_1(\tilde{x}), \cdots, g_m(\tilde{x}))$ 或缺值或仍 $\in N$. 所有保 N 函数之集合记为 $T(N)$. 若对任意 $f(\tilde{x}), g_1(\tilde{x}), \cdots, g_n(\tilde{x}) \in N$ ，复合函数 $f(g_1(\tilde{x}), \cdots, g_n(\tilde{x}))$ 仍 $\in N$ ，则 N 说是一个 n 元封闭集.

设 N 是一个一元封闭集，包含自变数 x 但非包含 x 的所有不缺值函数. 这样一个函数集 N 称为一个一元正则集. 不缺值函数 $f(x, y)$ 说是一个假二元函数，若 $f(x, y)$ 不依赖于 x 或不依赖于 y .

设 N 是一个二元封闭集，包含自变数 x, y 的所有假二元函数但非包含 x , y 的所有不缺值函数. 这样一个函数集 N 称为一个二元正则集.

命题 5 一个封闭集 A 所包含的 $x_1, \cdots, x_n$ 的所有函数作成一个 n 元封闭集.

证 设 $f(\tilde{x}), g_1(\tilde{x}), \cdots, g_n(\tilde{x}) \in A$. 因为 A 封闭，故 $f(g_1(\tilde{x}),$

$g_2(\tilde{x})$，…，$g_n(\tilde{x})$）仍 $\in A$．

命题 6 设 N 是一个 n 元封闭集．于是，$T(N)$ 是一个封闭集．

证 若 $f(t_1, \cdots, t_m)$ 保 N，而 $g_1(\tilde{y})$，…，$g_m(\tilde{y})$ 或保 N 或为自变数，则 $f(g_1(\tilde{y}), \cdots, g_m(\tilde{y}))$ 仍保 N.

定理 3 若 M 是一个第一种极大封闭集，则 M 所包含的 x 的所有函数作成一个一元正则集 $N(M)$，而 $M=T(N(M))$．若 M 是一个第二种极大封闭集，则 M 所包含的 x，y 的所有函数作成一个二元正则集 $N(M)$ 而 $M=T(N(M))$．

证 只证第一句话，第二句话可类似地证明．因为 M 是第一种极大封闭集，所以有的 x 的函数不在 $N(M)$ 内．由命题 4，$N(M)\ni x$．由命题 5，$N(M)$ 是一元封闭集．因之，$N(M)$ 是一元正则集.

试看 $T=T(N(M))$．因为 M 封闭，所以 M 中的函数保 $N(M)$，因而 $M\subset T$．我们知道有的 x 的不缺值函数不在 $N(M)$ 内．设 $\varphi(x)\notin N(M)$．因为 $x\in N(M)$，所以 $\varphi(x)$ 不保 $N(M)$，因而 $\varphi(x)\notin T$，故 $T<P_k^*$．因之，$M\subset T<P_k^*$．但由命题 6，T 封闭，而 M 极大封闭，故 $T=M$．

由定理 3，第一种极大封闭集的个数不超过 x 的函数集的个数 2^{k^k}，第二种极大封闭集的个数不超过 x，y 的函数集的个数 $2^{k^{k^2}}$．因之，

定理 4 P_k^* 只有有限个极大封闭集.

证 N 是一个一元正则集时，$T(N)$ 称为一个第一种 T 型集．N 是一个二元正则集时，$T(N)$ 称为一个第二种 T 型集．$P_k\cup\{*\}$ 称为第三种 T 型集．凡 T 型集必 $<P_k^*$．一个 T 型集说是极大，如果不包含在另外的 T 型集之内.

命题 7 任意封闭集 $A<P_k^*$ 可以扩充为一个极大 T 型集.

证 由命题 3，将自变数加入 A 所得到的集合 B 仍封闭而且 $<P_k^*$．假定有的 x 的函数不在 B 内．由命题 5，B 所包含的 x 的所有函数作成一个一元正则集 N．因为 B 封闭，所以 $B\subset T(N)$．这就是说，B 可以扩充为一个第一种 T 型集．同理，若 x 的函数都在 B 内而有的 x，y 的函数不在 B 内，则 B 可以扩充为一个第二种 T 型集．若 x，y 的函数都在 B 内，则 B 可以扩充为 $P_k\cup\{*\}$．总之，B 可以扩充为一个 T 型集．因为 T 型集的个数有限，由这一 T 型集出发，总可以找到包含它的一个极大

T 型集.

定理 5 M 是一个极大封闭集必要而且只要 M 是一个极大 T 型集.

证 设 M 极大封闭，由定理 2 及 3，M 是一个 T 型集. 由于 M 极大封闭，不可能有另外的 T 型集包含 M，所以 M 是一个极大 T 型集.

反之，设 M 是一个极大 T 型集. 若 M 非极大封闭，则有封闭集 A 适合 $M < A < P_k^*$. 由命题 7，A 可以扩充为一个极大 T 型集 T，因而 $M < T$，与 M 为极大 T 型集矛盾. 所以 M 极大封闭.

由命题 7 及定理 5 有

定理 6 任意封闭集 $A < P_k^*$ 可以扩充为一个极大封闭集.

定理 7 A 是一个完备集必要而且只要 A 不包含在任何极大封闭集之内.

证 设 A 完备，M 极大封闭. 若 $A \subset M$，则 $\overline{A} \subset M < P_k^*$，与 $\overline{A} = P_k^*$ 矛盾. 所以 $A \not\subset M$. 反之，设 A 不包含在任何极大封闭集之内. 于是，$\overline{A}$ 更不包含在任何极大封闭集之内，故由定理 6，$\overline{A} = P_k^*$.

定理 7 将完备集之刻画问题转化为极大封闭集之刻画问题.

命题 8 一个第一种 T 型集 $T(N)$ 为极大 T 型集必要而且只要 $T(N)$ 不包含在另外的第一种或第二种 T 型集之内. 一个第二种 T 型集 $T(N)$ 为极大 T 型集必要而且只要 $T(N)$ 不包含在另外的第二种 T 型集之内.

证 设 $T(N)$ 为第一种或第二种 T 型集. 我们证明 $T(N) \not\subset P_k \cup \{*\}$. 试看函数

$$\varphi(x) = \begin{cases} 0, & x = 0; \\ *, & x \neq 0. \end{cases}$$

对任意 $g \in N$ 若 $g \equiv 0$ 则 $\varphi(g) \equiv 0 \equiv g \in N$；若 $g \neq 0$，则 $\varphi(g)$ 缺值. 因之，$\varphi(x) \in T(N)$. 但 $\varphi(x) \notin P_k \cup \{*\}$. 故 $T(N) \not\subset P_k \cup \{*\}$. 今若 N 是一个二元正则集，则 $T(N)$ 包含所有一元函数. 但第一种 T 型集不能包含所有一元函数，所以 $T(N)$ 不能包含在任何第一种 T 型集之内. 这样，我们便得到命题中的结论.

§3. 第二种极大封闭集

现在，我们定出所有第二种极大封闭集. 首先，我们定出所有二元正则集，然后证明，对任意二元正则集 N，$T(N)$ 必是一个极大封闭集.

命题 9 若二元正则集 N 含有一个取 $r(\geqslant 3)$ 个值的真二元函数 $f(x, y)$，则 N 包含 x，y 的所有至多取 r 个值的函数.

证 因为 N 含所有假二元函数，可设 $f(x, y)$ 所取的 r 个值为 $0, 1, \cdots, r-1$. 根据 Яблонский（1958）的论文 §11 中的引理，又可设 $f(x, y)$ 在 $E^{r-1}=\{0, 1, \cdots, r-2\}$ 上取此 r 个值. 于是，$f(x, y)$ 在 $E^r=\{0, 1, \cdots, r-1\}$ 上是取 r 个值 $0, 1, \cdots, r-1$ 的真二元函数. 因之，利用假二元函数和 $f(x, y)$ 可以迭合出两个函数 $\varphi(x, y)$，$\psi(x, y)$，其在 E^r 上之限制函数分别为 $\max(x, y)$，$\min(x, y)$.

今证 N 包含 x，y 的任意至多取 r 个值的函数 $g(x, y)$. 因为 N 包含所有假二元函数，不失一般性可假定 $g(x, y)$ 所取的值都在 E^r 内. 命

$$j'_i(x)=\begin{cases}r-1, & x=i;\\ 0, & x\neq i,\end{cases}$$

$i=0, 1, \cdots, k-1$，且命

$$\varphi(x_1, \cdots, x_n)=\varphi(x_1, \varphi(x_2, \cdots, \varphi(x_{n-1}, x_n)\cdots)).$$

我们有

$$g(x, y)=\varphi[\psi(j'_0(y), g(x, 0)), \cdots, \psi(j'_{k-1}(y), g(x, k-1))],$$

从而命题得证.

函数 $l(x, y)$ 说是一个局部线性函数，若对任意 $\alpha_1, \alpha_2, \beta_1, \beta_2\in E^k$，$l(\alpha_1, \beta_1)$，$l(\alpha_1, \beta_2)$，$l(\alpha_2, \beta_1)$，$l(\alpha_2, \beta_2)$ 或全相等，或分成两对，每对相等.

命题 10 假二元函数是局部线性函数.

证 设 $f(x, y)$ 只依赖于 x. 于是，$f(\alpha_1, \beta_1)=f(\alpha_1, \beta_2)$，$f(\alpha_2, \beta_1)=f(\alpha_2, \beta_2)$. 因之，$f(x, y)$ 是局部线性函数.

命题 11 若局部线性函数 $l(x, y)$ 是真二元函数，则 $l(x, y)$ 恰取两个值.

证 因为 $l(x,y)$ 是真二元函数，所以不只取一个值．若 $l(x,y)$ 取三个或更多的值，则有 α_1，α_2，β_1，β_2 使 $l(x,y)$ 在 $\{\alpha_1,\alpha_2\}\times\{\beta_1,\beta_2\}$ 上至少取三个值，与 $l(x,y)$ 为局部线性函数矛盾．

命题 12 若局部线性函数 $l(x,y)$ 是真二元函数，取值 c_1 及 c_2，则 $E^k=A_1\cup A_2=B_1\cup B_2$，而 $l(x,y)$ 在 $A_1\times B_1$ 及 $A_2\times B_2$ 上取值 c_1，在 $A_1\times B_2$ 及 $A_2\times B_2$ 上取值 c_2．

证 设

$$l(\alpha_1,\beta_1)=c_1. \tag{1}$$

命 A_1 为使 $l(x,\beta_1)=c_1$ 的所有元素 x 之集合，$A_2=E^k-A_1$．命 B_1 为使 $l(\alpha_1,y)=c_1$ 的所有元素 y 之集合，$B_2=E^k-B_1$．

设 $\alpha'_1\in A_1$，$\beta'_1\in B_1$．于是，

$$l(\alpha'_1,\beta_1)=c_1,\ l(\alpha_1,\beta'_1)=c_1.$$

由此及（1），根据局部线性函数的定义，有 $l(\alpha'_1,\beta'_1)=c_1$．因之，

$$l(A_1,B_1)=c_1.$$

设 $\alpha'_2\in A_2$，$\beta'_2\in B_2$．于是，

$$l(\alpha_1,\beta'_2)=c_2,\ l(\alpha'_2,\beta_1)=c_2$$

由此及（1）有 $l(\alpha'_2,\beta'_2)=c_1$．因之，

$$l(A_2,B_2)=c_1.$$

设 $\alpha'_1\in A_1$，$\beta'_2\in B_2$．于是，

$$l(\alpha'_1,\beta_1)=c_1,\ l(\alpha_1,\beta'_2)=c_2.$$

由此及（1）有 $l(\alpha'_1,\beta'_2)=c_2$，．因之，

$$l(A_1,B_2)=c_2$$

同理可证

$$l(A_2,B_1)=c_2.$$

命题 13 所有局部线性函数作成一个二元正则集 N_1．

证 N_1 包含所有假二元函数而且有的二元函数不在 N_1 内．今证 N_1 为二元封闭集．设 f，g，$h\in N_1$．命 $\varphi=f(g,h)$ 而证 $\varphi\in N_1$．试看任意 α_1，α_2，β_1，$\beta_2\in$

E^k，因为 $g \in N_1$，g 在 $\{\alpha_1, \alpha_2\} \times \{\beta_1, \beta_2\}$ 上所取的值有三种可能情形如下表所示：

x	α_1	α_1	α_2	α_2
y	β_1	β_2	β_1	β_2
$g(x, y)$	c_1	c_1	c_2	c_2
	c_1	c_2	c_1	c_2
	c_1	c_2	c_2	c_1

表中 c_1 和 c_2 可以相等．同样，对于 h 有下面的表：

x	α_1	α_1	α_2	α_2
y	β_1	β_2	β_1	β_2
$h(x, y)$	d_1	d_1	d_2	d_2
	d_1	d_2	d_1	d_2
	d_1	d_2	d_2	d_1

表中 d_1 和 d_2 可以相等．关于 g 的三种情形和关于 h 的三种情形相配有九种可能情形．这九种情形分为两种类型以下列两表为代表：

x	α_1	α_1	α_2	α_2
y	β_1	β_2	β_1	β_2
g	c_1	c_1	c_2	c_2
h	d_1	d_1	d_2	d_2

（2）

x	α_1	α_1	α_2	α_2
y	β_1	β_2	β_1	β_2
g	c_1	c_1	c_2	c_2
h	d_1	d_2	d_1	d_2

（3）

第一种类型的特点是 c_1 只与 d_1 配，c_2 只与 d_2 配．第二种类型的特点是四个值组 (c_1, d_1)，(c_1, d_2)，(c_2, d_1)，(c_2, d_2) 都出现．不论是哪种情形，由于 $f \in N_1$，φ 在 $\{\alpha_1, \alpha_2\} \times \{\beta_1, \beta_2\}$ 上所取的四个值或全相等或分成两对，每对相等．因之．$\varphi \in N_1$．

命题 14 若二元正则集 N 含有一个真二元局部线性函数 l_0，则 N 包含所有局部线性函数．

证 设 l 为任意真二元局部线性函数. 因为 N 包含所有假二元函数，可以假定 l_0 所取的值为 0，1，而且不失普遍性，也可以假定 l 所取的值为 0，1. 设 $E^k=A_1\cup A_2=B_1\cup B_2$，$l(A_1,B_1)=l(A_2,B_2)=0$，$l(A_1,B_2)=l(A_2,B_1)=1$. 设 $l_0(\alpha_1,\beta_1)=l_0(\alpha_2,\beta_2)=0$，$l_0(\alpha_1,\beta_2)=l_0(\alpha_2,\beta_1)=1$. 规定

$$\varphi(x)=\begin{cases}\alpha_1, & x\in A_1;\\ \alpha_2, & x\in A_2,\end{cases}\qquad \psi(y)=\begin{cases}\beta_1, & y\in B_1;\\ \beta_2, & y\in B_2.\end{cases}$$

我们有

$$l(x,y)=l_0(\varphi(x),\psi(y)).$$

因而 $l\in N$.

命题 15 若二元正则集 N 含有一个取两个值的非局部线性函数 φ，则 N 包含 x，y 的所有取两个值的函数.

证 因为 φ 是取两个值的非局部线性函数，有 α_1，α_2，β_1，$\beta_2\in E^k$ 使 φ 在 $\{\alpha_1,\alpha_2\}\times\{\beta_1,\beta_2\}$ 上所取的四个值中有三个相等而其余一个不同. 由于 N 包含所有假二元函数，不失普遍性，可设 α_1，α_2，β_1，β_2，为 0，1，而设 $\varphi(0,0)=0$，$\varphi(0,1)=\varphi(1,0)=\varphi(1,1)=1$. 这就是说，$\varphi(x,y)$ 在 E^2 上之限制函数为 $x\vee y=\max(x,y)$. 因之，用 $\varphi(x,y)$ 和假二元函数可以叠合出一个函数 $\psi(x,y)$，其在 E^2 上之限制函数为 $x\wedge y=\min(x,y)$. 于是，用命题 9 证明中第二段的方法可以证明 N 包含取两个值的任意函数 $g(x,y)$.

所有假二元函数作成一个二元正则集 N_0. 容易说明，N_0 添上 x，y 的所有至多取 $r(2\leqslant r\leqslant k-1)$ 个值的函数作成一个二元正则集 N_r. 这样，由命题 9，14，15，我们便有

定理 8 二元正则集恰有 k 个，即

$$N_0,\ N_1,\ N_2,\ \cdots,\ N_{k-1}.$$

并且我们有 $N_r<N_{r+1}$.

我们用 $N(g,h)$ 代表所有函数 $f(g,h)$ 之集合，其中，f 经过 N 中之所有函数.

命题 16 设 g，$h\in N_r$. 若 $N_{r+1}(g,h)\subset N_r$，则对任意函数 $\varphi(x,y)$，$\varphi(g,h)\in N_r$.

证 $i)r=0$. 若 g，h 都只依赖于 x，则对任意 φ，$\varphi(g, h)$ 只依赖于 x，因而 $\in N_0$，故命题成立. 同样，g，h 都只依赖于 y 时，命题亦成立. 今设 $g=g(x)$，$h=h(y)$，而且都不是常数. 于是，有 α_1，α_2，使 $g(\alpha_1)\neq g(\alpha_2)$；有 β_1，β_2，使 $h(\beta_1)\neq h(\beta_2)$. 将 E^k 分为 A_1，A_2，使 $A_1\ni g(\alpha_1)$，$A_2\ni g(\alpha_2)$；又将 E^k 分为 B_1，B_2，使 $B_1\ni h(\beta_1)$，$B_2\ni h(\beta_2)$. 规定函数 $l(x, y)$ 如下：

$$l(A_1, B_1)=l(A_2, B_2)=0,$$

$$l(A_1, B_2)=l(A_2, B_1)=1,$$

则 $l\in N_1$. 因为

$$l(g(\alpha_1), h(\beta_1))=l(g(\alpha_2), h(\beta_2))=0,$$

$$l(g(\alpha_1), h(\beta_2))=l(g(\alpha_2), h(\beta_1))=1,$$

故 $l(g, h)\notin N_0$. 所以在这种情形下，不可能 $N_1(g, h)\subset N_0$.

$ii)r=1$. 试对任意 $\{\alpha_1, \alpha_2\}\times\{\beta_1, \beta_2\}$ 看 g，h 之四个值组. 在命题13的证明中，我们看到，这四个值组有两种类型，分别以表（2）和表（3）为代表. 在表（3）中，我们假定 $c_1\neq c_2$，$d_1\neq d_2$，盖否则仍归于第一种类型. 在表（2）的情形下，对任意 $\varphi(x, y)$，$\varphi(g, h)$ 的四个值或全相等，或分成两对，每对相等. 因之，若对任意 $\{\alpha_1, \alpha_2\}\times\{\beta_1, \beta_2\}$，$g$，$h$ 的四个值组互为第一种类型，则对任意 $\varphi(x, y)$，$\varphi(g, h)\in N_1$. 因而命题成立. 今设对某 $\{\alpha_1, \alpha_2\}\times\{\beta_1, \beta_2\}$，$g$，$h$ 的四个值组为第二种类型，例如表（3）所示. 规定 $f(x, y)$ 如下：

$$f(x, y)=\begin{cases}0, & (x, y)=(c_1, d_1);\\ 1, & (x, y)\neq(c_1, d_1).\end{cases}$$

在 $\{\alpha_1, \alpha_2\}\times\{\beta_1, \beta_2\}$ 上，$f(g, h)$ 的四个值中有三个相等而其余一个不同. 因之，$f\in N_2$，而 $f(g, h)\notin N_1$，所以，不可能 $N_2(g, h)\subset N_1$.

$iii)\ 2\leqslant r\leqslant k-2$. 用 A^r，B^r 代表 E^k 的 r 元子集. 假定对所有 $A^r\times B^r$，g，h 最多有 r 个不同的值组. 设 $\varphi(x, y)$ 任意. 若 $\varphi(g, h)$ 为假二元函数，则 $\varphi(g, h)\in M_r$. 若 $\varphi(g, h)$ 为真二元函数，则 $\varphi(g, h)$ 最多取 r 个值. 事实上，若 $\varphi(g, h)$ 不只取 r 个值，则对某 $A^r\times B^r$，$\varphi(g, h)$ 不只取 r 个值，与所作假定矛盾. 因之，不论 $\varphi(g, h)$ 是否假二元函数，都有 $\varphi(g, h)\in N_r$，故命题成立. 今设对某 $A^r\times$

B^r，g，h 不只有 r 个值组，例如，值组

$$(c_0, d_0), (c_1, d_1), \cdots, (c_r, d_r) \tag{4}$$

皆不同. 规定 $f(x, y)$ 如下：

$$f(x, y)=\begin{cases} i, & (x, y)=(c_i, d_i); \\ 0, & (x, y)\neq \text{任何}(c_i, d_i). \end{cases}$$

在 $A^r \times B^r$ 上，$f(g, h)$ 取 $r+1$ 个值，因而 $f(g, h)$ 是真二元函数，所以 $f(g, h)\notin N_r$. 但 $f\in N_{r+1}$，故不可能 $N_{r+1}(g, h)\subset N_r$.

定理 9 第二种极大封闭集恰有 k 个，即 $T(N_0)$，$T(N_1)$，$T(N_2)$，…，$T(N_{k-1})$.

证 根据命题 8，只要证明这些 T 型集互不包含. 若 $T(N_r)\subset T(N_s)$，则 $N_r\subset T(N_r)\subset T(N_s)$，但 $T(N_s)$ 中包含的 x，y 的函数恰是 N_s 中的函数，故 $N_r\subset N_s$. 由此可见，$T(N_{k-1})$ 是极大封闭集，而且要证 $T(N_r)$，$0\leqslant r\leqslant k-2$，是极大封闭集，只要证明对任意 $s>r$，$T(N_r)\not\subset T(N_s)$.

设 $0\leqslant r\leqslant k-2$，$s>r$. 设 N_{r+1} 中的函数是

$$\psi_1(x, y), \psi_2(x, y), \cdots, \psi_m(x, y).$$

取一个取 k 个值的真二元函数 $\varphi(x, y)$，规定一个函数 $F(x, y, t_1, \cdots, t_m)$ 如下：对任意 α，β，在 $x=\alpha$，$y=\beta$，$t_1=\psi_1(\alpha, \beta)$，$t_2=\psi_2(\alpha, \beta)$，…，$t_m=\psi_m(\alpha, \beta)$ 时，$F=\varphi(\alpha, \beta)$；而在其余情形下，$F=*$. 因为 $N_{r+1}\subset N_s$ 而

$$F(x, y, \psi_1(x, y), \cdots, \psi_m(x, y))=\varphi(x, y)\notin N_s$$

故 $F\notin T(N_s)$. 今证 $F\in T(N_r)$. 试以 N_r 中的函数 g，h，g_1，g_2，…，g_m 代 x，y，t_1，…，t_m. 只有当 $g_1=\psi_1(g, h)$，$g_2=\psi_2(g, h)$，…，$g_m=\psi_m(g, h)$ 时，$F(g, h, g_1, \cdots, g_m)$ 才不是缺值函数. 但这时 $N_{r+1}(g, h)\subset N_r$，而 $F(g, h, g_1, \cdots, g_m)=\varphi(g, h)$. 由命题 16，$\varphi(g, h)\in N_r$. 这就证明了 $F\subset T(N_r)$. 因之，$T(N_r)\not\subset T(N_s)$.

§4. 第一种极大封闭集

为了刻划第一种极大封闭集，我们先讨论两个第一种 T 型集在什么条件下相互

包含.

命 Q 表 x 的所有不缺值函数作成的集合.

这样一个函数可以看作是 E^k 的一个变换，而两个函数作复合等于两个变换相乘. 因此，在函数的复合运算之下，Q 可以看作是 E^k 的所有变换作成的变换半群. 函数 x 是半群 Q 的单位元素 e. 一个（一元）正则集就是 Q 的含 e 的一个真子半群.

设 A，B，C 是 Q 的三个子集. 用

$$(A : B)_C$$

表 C 中所有适合 $B\varphi \subset A$ 的元素 φ 作成的集合. 用

$$_C(A : B)$$

表 C 中所有适合 $\varphi B \subset A$ 的元素 φ 作成的集合.

设 G，H 是两个正则集. 我们说 G 低于 H，H 高于 G，如果

$$_Q(G : (G : H)_G) = H. \tag{1}$$

G 低于 H 记为 $G \to H$. 由 $(G : H)_G$ 的定义有

$$H(G : H)_G = G,$$

因而（1）的左边永远包含（1）的右边. 所以 $G \to H$ 必要而且只要

$$_Q(G : (G : H)_G) \subset H. \tag{2}$$

定理 10 设 G，H 是两个正则集. $T(G) \subset T(H)$ 必要而且只要 $G \to H$.

证 设 $G \to H$，证明 $T(G) \subset T(H)$. 假定 $T(G) \not\subset T(H)$，则有 $f(t_1, \cdots, t_n) \in T(G)$，$\notin T(H)$. 因之，有 $h_1(x), \cdots, h_n(x) \in H$ 使 $\psi(x) = f(h_1(x), \cdots, h_n(x))$ 不缺值而且 $\notin H$. 若 $g \in (G : H)_G$，则 $h_1(g), \cdots, h_n(g) \in G$，因而 $\psi(g) = f(h_1(g), \cdots, h_n(g)) \in G$. 可见，$\psi(G : H)_G \subset G$，故 $\psi \in {}_Q(G : (G : H)_G) \subset H$，此为矛盾.

反之，设 $T(G) \subset T(H)$，证 $G \to H$. 假定 G 不低于 H，则有 $\varphi \notin H$ 但

$$\varphi(G : H)_G \subset G \tag{3}$$

设 H 之元素为 $h_1, \cdots, h_m$. 规定一个函数 $F(x, t_1 \cdots, t_m)$ 如下：对任意 α，在 $x = \alpha$，$t_1 = h_1(\alpha), \cdots, t_m = h_m(\alpha)$ 时，$F = \varphi(\alpha)$；而在其余情形，$F = *$，以任意 g，$g_1, \cdots, g_m \in G$ 代入 F. 只有在 $g_1 = h_1 g, \cdots, g_m = h_m g$ 时，

$$F(g, g_1, \cdots, g_m)$$

才不缺值，而这时 $g \in (G:H)_G$，故由 (3) $F(g, g_1, \cdots, g_m)=\varphi g \in G$，因之，$F \in T(G)$. 但 $F(x, h_1(x), \cdots, h_m(x))=\varphi(x) \notin H$，故 $F \notin T(H)$，此为矛盾.

易见若 $T(G) \subset T(H)$，则 $G \subset H$，故若 $G \to H$ 则 $G \subset H$. 此外，作为定理 10 的简单推论又有

1) $G \to G$；

2) 若 $G \to H$，$H \to G$，则 $G = H$；

3) 若 $G \to H$，$H \to K$，则 $G \to K$.

这表示低于关系确定正则集之间的一种半序. 这些事实也可以由定义直接推证.

设 G 是任意半群. G 的子集 J 说是 G 的理想，如果 $GJ \subset J$，$JG \subset J$. 若 $GJ \subset J$，则 J 说是 G 的左理想；若 $JG \subset J$，则 J 说是 G 的右理想. 由定义，空集必是 G 的理想.

E^k 的一对一变换就是 E^k 的置换，非一对一的变换我们称之为奇异变换. 对于置换，我们采用通常的轮换写法. 写一个奇异变换，我们把 E^k 的元素按其影象分组写下来而把影象写在各组的上面. 例如

$$\begin{matrix} 1 & 2 \\ (023, & 14) \end{matrix}$$

代表变 0，2，3 为 1，变 1，4 为 2 的变换. 半群，理想等前面加“奇异”二字表示其元素都是奇异变换.

正则集 G 说是极高，如果没有另外的正则集高于 G. 下面的两个命题是显然的：

命题 17 $T(G)$ 不包含在另外的第一种 T 型集之内必要而且只要 G 极高.

命题 18 G 极高，必要而且只要对所有正则集 $H > G$ 有 $\varphi \notin H$ 使

$$\varphi(G:H)_G \subset G. \tag{4}$$

命题 19 设 $G \subset H$. $J=(G:H)_G$ 是 G 的右理想，H 的左理想.

证 $HJG \subset GG \subset G$，故 $JG \subset (G:H)_G = J$. $HHJ \supset HJ \subset G$，故 $HJ \subset (G:H)_G \subset J$

命题 20 G 极高的充要条件是，对 G 的所有非空奇异理想 J，若

$$G <_Q (J : J) < Q \tag{5}$$

则有 $\varphi \notin_Q (J : J)$ 使

$$\varphi J \subset G. \tag{6}$$

证 设 G 极高，设对 G 的理想 J，（5）成立．命 $H =_Q (J : J)$．$HHJ = H(HJ) \subset HJ \subset J$，故 $HH \subset H$．由此及（5）知 H 是一个正则集而且 $H > G$．因之，有 $\varphi \notin H$ 使（4）成立．今 $HJ \subset J \subset G$，故 $J \subset (G : H)_G$，因而（6）成立．

反之，设条件成立．设正则集 $H > G$，求证有 $\varphi \notin H$ 使（4）成立．命 $J = (G : H)_G$．

由命题 19，J 是 G 的理想．若 J 为空集，取任意 $\varphi \notin H$（4）即成立．所以可以假定 J 非空．若 J 的元素有的是置换，比方 σ，则 $H\sigma \subset G$，从而 $H \subset G\sigma^{-1} \subset G$，与前设 $H > G$ 矛盾．因之，J 是 G 的非空奇异理想．命 $K =_Q (J : J)$．因为 J 是 H 的左理想，故 $H \subset K$，今 $KJ \subset J$，故若 $H < K$，则有 $\varphi \notin H$ 使 $\varphi J \subset J \subset G$，即（4）成立．若 $H = K$，则（5）成立，因而有 $\varphi \notin K = H$ 使（6），即（4），成立．

设 G 是 Q 的任意子半群．于是，$G = \widetilde{G} \cup G^0$，其中 $\widetilde{G}$ 含 G 中所有置换，G^0 含 G 中所有奇异变换．$\widetilde{G}$ 称为 G 的非奇异部分，G^0 称为 G 的奇异部分．若非空，$\widetilde{G}$ 是一个置换群，G^0 是一个奇异半群．

命题 21 若正则集 G 的奇异部分 G^0 是 Q 的左理想，则 G 极高．

证 设正则集 $H > G$．于是，$(G : H)_G$ 是 G 的奇异理想，故 $\subset G^0$．取任意 $\varphi \notin H$，则 $\varphi (G : H)_G \subset \varphi G^0 \subset G^0 \subset G$，即（4）成立．所以 G 极高．

利用命题 21，不难定出奇异部分是 Q 的左理想的所有极高正则集．设有奇异变换 g，g 所生出之主左理想 Qg 的元素很容易写出来．例如 $g = \begin{pmatrix} 0 & 1 & 3 \\ (01, & 2, & 34) \end{pmatrix}$．于是，$Qg$ 中的元素其“分组”必为 01，2，34 或为这些组的进一步合并，即（012，34），（0134，2），（01，234）；（01234）．对于这些分组，以所有可能方法写上影象元素，就得出 Qg 的所有元素．我们就用 g 的分组来代表 Qg．例如对于上面的 g，$Qg = (01, 2, 34)$．

Q 的任意左理想是一些 Qg 的并集，因而不过是把一些“分组”合起来得到的．这样，不难定出所有可能的奇异左理想 G^0．若 $G=\widetilde{G}\cup G^0$ 是一个正则集，则 $G^0\widetilde{G}\subset G^0$．命 H 为适合 $G^0\sigma\subset G^0$ 的所有置换 σ 作成的置换群．取 $\widetilde{G}$ 为 H 的任意子群，则 $(\widetilde{G}\cup G^0)(\widetilde{G}\cup G^0)=\widetilde{G}\widetilde{G}\cup G^0\widetilde{G}\cup\widetilde{G}G^0\cup G^0G^0\subset\widetilde{G}\cup G^0\cup G^0\cup G^0=\widetilde{G}\cup G^0$，所以 $G=\widetilde{G}\cup G^0$ 是一个子半群．只要在 $G^0=Q^0$ 时不取 $\widetilde{G}=\widetilde{Q}$，则 G 是一个正则集．由命题 21，这样就得到了奇异部分为 Q 的左理想的所有极高正则集．

命题 22　设极高正则集 G 的奇异部分 G^0 不是 Q 的左理想，$g\in G^0$，$Qg\not\subset G$．命题 $G(g)=Qg\cap G$，则

$$G={}_Q(G(g):G(g)).$$

证　命 $H={}_Q(G(g):G(g))$．于是，$GG(g)\subset GG\subset G$，$GG(g)\subset GQg\subset Qg$，故 $GG(g)\subset G(g)$，因而 $G\subset H$．因为 $Qg\not\subset G$，故更有 $QG(g)\not\subset G(g)$，因而 $H<Q$．今 $HHG(g)\subset HG(g)\subset G(g)$，所以 $HH\subset H$．因之，H 是一个正则集．假定 $H>G$．因为 G 极高，应有 $\varphi\notin H$ 使（4）成立．因为 $HG(g)\subset G(g)\subset G$，故 $G(g)\subset(G:H)_G$，所以 $\varphi G(g)\subset G$，但 $\varphi G(g)\subset\varphi Qg\subset Qg$，故 $\varphi G(g)\subset G(g)$，即 $\varphi\in H$，此为矛盾．

利用命题 22 可以如下定出奇异部分不是 Q 的左理想的所有极高正则集：取所有可能的奇异左主理想 Qg，按照前面的讨论，这样一个左理想不过是一个“分组”，定出 Qg 的所有子半群，并看作是 $G(g)$ 计算 $G={}_Q(G(g):G(g))$．在这些 G 中，利用命题 18 或命题 20 便可以选出奇异部分不定 Q 的左理想的所有极高正则集．

现在我们讨论一个第一种 T 型集在什么条件下包含在一个第二种 T 型集之内．

设 N 是一个二元正则集．对于函数 $g(x)$，$h(x)$ 用 $N(g,h)$ 表示所有函数 $f(g(x),h(x))$ 之集合，其中 $f\in N$．再设 G 是一个一元正则集．用 $(G:N)$ 表示所有函数偶 (g,h) 之集合，其中 $g\in G$，$h\in G$，而 $N(g,h)\subset G$．对于函数 $\varphi(x,y)$，用 $\varphi(G:N)$ 代表所有 $\varphi(g,h)$ 之集合，其中 $(g,h)\in(G:N)$．

相当于定理 10，我们有

定理 11　设 G 为一元正则集，N 为二元正则集．$T(G)\not\subset T(N)$ 必要而且只要有不缺值函数 $\varphi(x,y)\notin N$ 使

$$\varphi(G:N)\subset G \tag{1}$$

证 设 $T(G)\not\subset T(N)$. 于是，有 $f(t_1, \cdots, t_r)\in T(G)$, $\notin T(N)$. 因之，有 $n_1(x, y)$, $\cdots n_r(x, y)\in N$ 使 $\varphi(x, y)=f(n_1(x, y), \cdots, n_r(x, y))$ 不缺值而且 $\notin N$. 若 $(g, h)\in(G:N)$，则 $n_1(g, h), \cdots, n_r(g, h)\in G$，因而 $\varphi(g, h)=f(n_1(g, h), \cdots, n_r(g, h))\in G$，故（1）成立.

反之，设对不缺值函数 $\varphi(x, y)\notin N$，（1）成立.

设 N 中的函数为 $n_1(x, y), \cdots, n_s(x, y)$. 规定一个函数 $F(x, y, t_1, \cdots, t_s)$ 如下：对任意 α，β 在 $x=\alpha$，$y=\beta$，$t_1=n_1(\alpha, \beta)$，$\cdots$，$t_s=n_s(\alpha, \beta)$ 时，$F=\varphi(\alpha, \beta)$；在其余情形下，$F=*$. 以任意 $g, h, g_1, \cdots, g_s\in G$ 代入 F. 只有在 $g_1=n_1(g, h), \cdots, g_s=n_s(g, h)$ 时，$F(g, h, g_1, \cdots, g_s)$ 才不缺值，而这时 $(g, h)\in(G:N)$，故由（1），$F(g, h, g_1, \cdots, g_s)=\varphi(g, h)\in G$. 因之，$F\in T(G)$. 但 $F(x, y, n_1(x, y), \cdots, n_s(x, y))=\varphi(x, y)\notin N$，故 $F\notin T(N)$. 因之，$T(G)\not\subset T(N)$.

由定理 10 及 11，我们有

定理 12 第一种 T 型集 $T(G)$ 是一个极大封闭集，必要而且只要

1）对所有一元正则集 $H>G$ 有不缺值函数 $\varphi(x)\notin H$ 使

$$\varphi(G:H)_G\subset G,$$

2）对所有二元正则集 N 有不缺值函数 $\psi(x, y)\notin N$ 使

$$\psi(G:H)\subset G$$

前面我们说明了怎样定出所有极高正则集. 只要再利用定理 11 对这些极高正则集进行检查就可以定出所有第一种极大封闭集. 下面的简单命题是很有用的：

命题 23 若 $(g, h)\in(G:N)$，则 $Qg\subset G$，$Qh\subset G$.

证 因为 N 包含所有假二元函数，故若 $(g, h)\in(G:N)$，则对任意 $\varphi\in Q$，将 (g, h) 代入 $\varphi(x)$ 有 $\varphi(g)\in G$；将 (g, h) 代入 $\varphi(y)$ 有 $\varphi(h)\in G$.

现在，我们讨论一下 P_k^* 的极大封闭集和 P_k 的极大封闭集的关系. 设 N 是 $x_1, \cdots, x_n$ 的一个不缺值函数集. 用 $\widetilde{T}(N)$ 代表保 N 而且不缺值的所有函数作成的集合. 换句话说，$\widetilde{T}(N)=T(N)\cap P_k$. 可以证明（见庞云阶（1962）），$P_k$ 的极大封

闭集或是 $\widetilde{T}(N_{k-1})$ 或是 $\widetilde{T}(G)$ 的形式，其中 G 是一个一元正则集.

命题 24 若 $\widetilde{T}(G)$ 是 P_k 的极大封闭集，则 $T(G)$ 是 P_k^* 的极大封闭集.

证 只要证明将任意 $f(t_1,\cdots,t_m)\notin T(G)$ 加入 $T(G)$ 则成为 P_k^* 的完备集. 由对 f 的假定，有 $g_1(x),\cdots,g_m(x)\in G\subset T(G)$ 使 $\varphi(x)=f(g_1(x),\cdots,g_m(x))$ 不缺值而且 $\notin G$. 今 $\widetilde{T}(G)$ 所包含的 x 的函数只有 G 中的函数，故 $\varphi(x)\notin\widetilde{T}(G)$. 因为 $\widetilde{T}(G)$ 在 P_k 中极大封闭，所以由 $\varphi(x)$ 与 $\widetilde{T}(G)$ 中的函数可以生出 P_k 中的所有函数. 因之，$\overline{T(G)\cup\{\varphi\}}\supset P_k$. 但我们知道 $T(G)$ 中有非空缺值函数，故 $\overline{T(G)\cup\{\varphi\}}=P_k^*$.

由命题 24 可见，如果我们定出了 P_k^* 的所有第一种极大封闭集 $T(G_1)$, $T(G_2)$, …，则 P_k 的极大封闭集，除 $\widetilde{T}(N_{k-1})$ 外，都在 $\widetilde{T}(G_1)$, $\widetilde{T}(G_2)$, …，之中. 这就大大缩小了求 P_k 的极大封闭集时所要考虑的函数集的范围.

利用极大封闭集可以判定一个函数集是否完备. 设对于给定的 k 求出了所有极大封闭集. 第三种和第二种极大封闭集作为已经知道，所谓求出了所有极大封闭集就是说定出了所有对应极大封闭集的一元正则集. 假定给出了一个有限函数集 A. 所谓给出了一个函数 $f(x_1,\cdots,x_n)$ 意思就是说，对于 $x_1,\cdots,x_n$ 的任意值组，有确定的方法在有限步内求出 f 对应的值或断定 f 在此值组处缺值. 不难说明，这样便有法判断 A 是否包含在某个极大封闭集之内，因而判定 A 是否完备.

虽然如此，判定 A 是否完备却不一定要先求出所有极大封闭集，而且即使求出了所有极大封闭集，用上面说的方法判定 A 的完备性也不一定是最简单的. 极大封闭集在理论上有重要意义. 例如，通过对极大封闭集的刻画我们便从本质上全面地掌握了完备集，但判定一个具体的函数集是否完备而拿所有的极大封闭集来试却未必是最好的方法.

除了完备性的判定问题，我们提出下面的相关问题：设 A 是完备集，怎样用 A 中的函数叠合出给定的函数 f？现在我们叙述一个方法判断 A 是否完备，并在 A 完备时能行地叠合出给定的函数 f. 这个方法在 P_k 的情形也是适用的，我们只就 P_k^* 的情形来说明.

首先，看 A 中的函数是否都保 x，换句话说，看 A 是否 $\subset T(x)$，若 $A\subset T(x)$，则 A 不完备. 若 $A\not\subset T(x)$，则在 A 中找到一个函数 $h(t_1, t_2, \cdots)$ 使 $\varphi_0(x)=h(x, x, \cdots)$ 不缺值而且 $\neq x$. 求出 x 和 $\varphi_0(x)$ 所生成的一元封闭集 G_0. 若 $G_0<Q$，看 A 中的函数是否都保 G_0. 若 $A\subset T(G_0)$，则 A 不完备. 若 $A\not\subset T(G_0)$，则在 A 中找到一个函数 $h_1(t_1, t_2, \cdots)$ 而且在 G_0 中找到函数 $g_1^0(x)$，$g_2^0(x)$，$\cdots$ 使 $\varphi_1(x)=h_1(g_1^0(x), g_2^0(x), \cdots)$ 不缺值而且 $\notin G_0$. 如此类推，得 G_0，G_1，G_2，$\cdots$. 若到某个 $G_i<Q$，发现 $A\subset T(G_i)$，则 A 不完备. 若没有这样的 G_i，则因为 Q 中函数有限，必达到一个 $G_i=Q$.

然后，看 A 是否 $\subset T(N_0)$. 若 $A\not\subset T(N_0)$，则可以合出一个不缺值函数 $\psi_1(x, y)\notin N_0$. 设 $N_{r-1}\not\ni\psi_1\in N_r$. 于是，$N_0$ 与 ψ_1 生出 N_r. 再看 A 是否 $\subset T(N_r)$. 如此类推，若 $A\not\subset$ 任意 $T(N_i)$，最后必达到所有不缺值二元函数之集合 N_k. 由 N_k 可以迭合出 P_k. 最后，看 A 中是否有非空缺值函数而判断 A 是否 $\subset P_k\cup\{*\}$.

根据本文中和 Яблонский (1958) 论文中有关的论证，N_r，$\cdots$，N_k，P_k，P_k^* 中的函数都可以逐步能行地叠合出来. 因此，若 A 完备，则对给定的函数 f，可以按照上述判定 A 是否完备的步骤，特别是最后利用定理 1 的证明，逐步由 A 叠合出 f.

§5. 二值三值情形

现在把上面的理论应用于 $k=2$ 和 $k=3$ 的特殊情形来定出 P_2^* 和 P_3^* 的所有极大封闭集.

先看 P_2^*. Q 的元素是单位元素 e，交换 (01)，两个常数 0 和 1. Q 的左理想有两个：空集及 $\{0, 1\}$. 以空集为奇异部分的极高正则集有两个：$\{e\}$，$\{e, (01)\}$. 这两个极高正则集中不包含任何 Qg，故由命题 23，$T(e)$，$T(e, (01))$ 都是极大封闭集. e 就是函数 x，(0, 1) 就是函数 $\bar{x}$，因此 $T(e)$ 也写作 $T(x)$，$T(e, (01))$ 也写作 $T(x, \bar{x})$.

以 $\{0, 1\}$ 为奇异部分的极高正则集为 $\{0, 1, e\}$. 设 N 为二元正则集. 由命题 23，若 $(g, h)\in(\{0, 1, e\}:N)$，则 g 和 h 都是常数，但这样则对任意 $\varphi(x, y)$

$\notin N$，$\varphi(g, h)$ 也是常数，因而 $\varphi(g, h) \in \{0, 1, e\}$. 可见 $T(0, 1, e) = T(0, 1, x)$ 是极大封闭集.

现在定出奇异部分不是 Q 的左理想的所有极高正则集. 因为 Q 只有两个奇异元素 0 和 1，可能的 $G(g)$ 只有 $\{0\}$ 和 $\{1\}$. 计算 ${}_Q(G(g) : G(g))$ 得 $\{0, e\}$ 和 $\{1, e\}$；由命题 20 知是极高正则集. $\{0, e\}$ 中不包含任何 Qg，所以由命题 23，$T(0, e) = T(0, x)$ 是极大封闭集. 同理，$T(1, e) = T(1, x)$ 是极大封闭集.

这样，我们就得到了 P_2^* 的所有极大封闭集：$T(x)$，$T(x, \bar{x})$，$T(0, 1, x)$，$T(0, x)$，$T(1, x)$，$T(N_0)$，$T(N_1)$，$P_2 \cup \{*\}$，共八个. 八个极大封闭集中有五个和 P_2 的极大封闭集相当. 事实上，P_2 的五个极大封闭集是：自对偶函数集 $\widetilde{T}(x, \bar{x})$，单调函数集 $\widetilde{T}(0, 1, x)$，保 0 函数集 $\widetilde{T}(0, x)$，保 1 函数集 $\widetilde{T}(1, x)$，线性函数集 $\widetilde{T}(N_1)$.

现在我们看 $k = 3$ 的情形. 先证明下面的两个命题：

命题 25 若一元正则集 G 所包含的 Q 的主左理想中有一个最大的，则对任意二元正则集 N，$T(G) \not\subset T(N)$.

证 设这个最大的主左理想为 Qg_0. 取任意 $\varphi(x, y) \notin N$. 若 $(g, h) \in (G : N)$，则 Qg，$Qh \subset G$，因而 Qg，$Qh \subset Qg_0$. 所以 g 和 h 的"分组"与 g_0 的分组一样或等于 g_0 的分组加以合并. 由此不难说明 $\varphi(g, h)$ 的分组与 g_0 的分组一样或等于 g_0 的分组加以合并，故 $\varphi(g, h) \in Qg_0 \subset G$. 因之，$\varphi(G : N) \subset G$. 所以，$T(G) \not\subset T(N)$.

命题 26 设 $k = 3$，G 是一个一元正则集. 若 G^0 不是 Q 的左理想而对 G^0 中适合 $Qg \not\subset G$ 的所有 g 恒有 $G = {}_Q(G(g) : G(g))$，则 $T(G)$ 是极大封闭集.

证 命 S 为 G 中所包含的所有常数之集合. 先设 S 非空而且 $\neq E^3$. 设 $\alpha \in S$. 于是，$\alpha \in G^0$，$Q\alpha^0 = E^3 \not\subset G$，$G(\alpha) = S$. 因之，$G = {}_Q(S : S)$. 设 N 是 G 的任意非空奇异理想. 我们有 $S = SN \subset GN \subset N$. 若 $h \in {}_Q(N : N)$，则 $hS \subset hN \subset N \subset G$. 但 hS 中都是常数，故 $hS \subset S$，因而 $h \in {}_Q(S : S) = G$. 这表示 ${}_Q(N : N) = G$，从而命题 20 中之 (5) 不可能成立. 因之，G 极高.

次设 S 为空集. 设 J 是 G 的任意非空奇异理想. 取任意 $g_0 \in J$. 因为 G 中没有

常数，故 $Qg_0\not\subset G$，因而 $G={}_Q(G(g_0):G(g_0))$．不失去普遍性，假定 $g_0=(\overset{\alpha}{0}\overset{\beta}{1},2)$．$\alpha$，$\beta$ 不可能一个是0，一个是1，否则 g_0^2 将为常数，与 G 中没有常数矛盾．若 $(\alpha,\beta)=(2,0)$ 或 $(2,1)$，则 $g_0^2=(\overset{0}{0}\overset{2}{1},2)$ 或 $(\overset{1}{0}\overset{2}{1},2)$．所以，不失去普遍性，可设 $(\alpha,\beta)=(0,2)$ 或 $(1,2)$．但这样则易见 $G(g_0)g_0=G(g_0)$，从而 $G(g_0)\subset G(g_0)g_0\subset GJ\subset J$．若 $h\in{}_Q(J:J)$，则 $hG(g_0)\subset hJ\subset J\subset G$．但 $hG(g_0)\subset hQg_0\subset Qg_0$，故 $hG(g_0)\subset G(g_0)$．因之，$h\in{}_Q(G(g_0):G(g_0))=G$．这表示 ${}_Q(J:J)=G$，从而命题 20 中的（5）不可能成立．因此，G 极高．

在以上两种情形下，都有 $E^3\not\subset G$．但任意 $Qg\subset E^3$，故 G 中不包含任何 Qg，因而由命题 23，$T(G)$ 为极大封闭集．

今设 $S=E^3$．Q 只有四个主左理想，即 E^3，$(01,2)$，$(02,1)$，$(12,0)$．我们证明 G 最多包含后三个主左理想之一．假定 G 包含这三个主左理想中的两个，比方 $(01,2)$，$(02,1)$．因为 G^0 不是 Q 的左理想，必有 $g_0=(\overset{\alpha}{1}\overset{\beta}{2},0)\in G$ 而 $Qg_0\not\subset G$．若 $(\alpha,\beta)=(0,2)$，$(2,0)$，$(1,2)$ 或 $(2,1)$，易见 $(01,2)g_0=Qg_0$；若 $(\alpha,\beta)=(0,1)$ 或 $(1,0)$，则 $(02,1)g_0=Qg_0$．此与 $Qg_0\not\subset G$ 矛盾．

既然 G 最多包含 $(01,2)$，$(02,1)$，$(12,0)$ 之一，所以由命题25，对任意二元正则集 N，$T(G)\not\subset T(N)$．

今设 J 为 G 的任意非空奇异理想．我们证明，若命题 20 中之（5）成立，则有 $\varphi\notin{}_Q(J:J)$ 使（6）成立．这样就证明了 $T(G)$ 是极大封闭集．分三种情形来看：

i) J 是 Q 的左理想．这时，${}_Q(J:J)=Q$，故命题 20 中的（5）不成立．

ii) J 不是 Q 的左理想，但其元素都在 G 所包含的 Q 的一个左理想之内．这时，G 包含 $(01,2)$，$(02,1)$，$(12,0)$ 之一．由以上所证，G 最多也只能包含这三个中的一个．由对称，可假定 $G\supset(01,2)$，因而 $J<(01,2)$．设 $j=(\overset{\alpha}{1}\overset{\beta-1}{0},2)\in J$，$(\alpha,\beta)$ 不可能等于 $(0,2)$，$(2,0)$，$(1,2)$，$(2,1)$．盖否则 $(01,2)j=(01,2)$

而 $J \supset (01,2)$，与 $J < (01,2)$ 矛盾. 所以，$j=\overset{0\ 1}{(01,2)}$ 或 $\overset{1\ 0}{(01,2)}$，从而 $J \supset j(01,2)=\{0,1,\overset{0\ 1}{(01,2)},\overset{1\ 0}{(01,2)}\}$. 因之，$J=\{0,1,2,\overset{0\ 1}{(01,2)},\overset{1\ 0}{(01,2)}\}$ 取 $\varphi=\overset{0\ 2}{(02,1)}$，则 $\varphi\notin_Q(J:J)$，而 $\varphi J\subset(01,2)\subset G$，即命题 20 中的（6）成立.

iii) J 中有一个元素 j_0 适合 $Qj_0\not\subset G$. 这时，$G=_Q(G(j_0):G(j_0))$. 由对称，可设 $j_0=\overset{\alpha\ \beta}{(01,2)}$，若 $(\alpha,\beta)=(0,2),(2,0),(1,2)$ 或 $(2,1)$，则和 S 为空集的情形一样可以说明命题 20 中的（5）不成立. 设 $(\alpha,\beta)=(0,1)$ 或 $(1,0)$. 由对称，可假定 $(\alpha,\beta)=(0,1)$. 可以证明 $h_0=\overset{1\ 0}{(12,0)}\in_Q(G(j_0):G(j_0))$. 事实上.

$$\overset{1\ 0}{(12,0)}\overset{0\ 1}{(01,2)}=\overset{0\ 1}{(01,2)},\quad \overset{1\ 0}{(12,0)}\overset{1\ 0}{(01,2)}=\overset{1\ 0}{(01,2)},$$

$$\overset{1\ 0}{(12,0)}\overset{1\ 2}{(01,2)}=\overset{1\ 0}{(12,0)}\overset{2\ 1}{(01,2)}=1\quad \overset{1\ 0}{(12,0)}\overset{0\ 2}{(01,2)}=\overset{0\ 1}{(01,2)}.$$

若 $\overset{2\ 0}{(01,2)}\in G(j_0)$. 则 $\overset{1\ 0}{(01,2)}=\overset{0\ 1}{(01,2)}\overset{2\ 0}{(01,2)}\in G(j_0)$.

而 $\overset{1\ 0}{(12,0)}\overset{2\ 0}{(01,2)}=\overset{1\ 0}{(01,2)}\in G(j_0)$.

所以. $h_0=\overset{1\ 0}{(12,0)}\in_Q(G(j_0):G(j_0))=G$. G 中必包含一个 $h_1=\overset{\gamma\ \delta}{(12,0)}$. 其中 $(\gamma,\delta)=(0,2),(2,0),(1,2)$ 或 $(2,1)$. 否则 $G=_Q(G(h_0):G(h_0))$ 将包含 $(01,2)$，与 $Qg_0\not\subset G$ 矛盾. 但如此则 $j_1=\overset{\delta\ \gamma}{(01,2)}=h_1j_0\in J$，其中 $(\delta,\gamma)=(2,0),(0,2),(2,1)$ 或 $(1,2)$，从而可以象 S 为空集的情形一样说明命题 20 中

之（5）不成立.

现在我们定出奇异部分不是 Q 的左理想的所有极高正则集．子半群 K 和 $\widetilde{K}$ 说是共轭，如果把 E^3 中的元素颠倒一下，K 就变成 $\widetilde{K}$．我们先求出互不共轭的那些极高正则集，则其余的可以将 E^3 的元素加以颠倒而得到.

首先，假定命题 26 证明中之 S 非空且 $\neq E^3$．由命题 22，若 G 极高，必有 $G = {}_Q(S:S)$；而由命题 26 之证明，这时 $T(G)$ 必为极大封闭集．除去共轭情形，S 有两种可能：$S=\{0\}$，$S=\{0,1\}$．利用 $G={}_Q(S:S)$ 计算得

0　1　　0　2　　0　1　　0　2　　1　0　　2　0

G_1：0，(01，2)，(01，2)，(02，1)，(02，1)，(12，0)，(12，0)，e，(12).

0　1　　1　0　　0　2　　1　2　　0　1　　1　0　　1　0

G_2：0，1，(01，2)，(01，2)，(01，2)，(01，2)，(02，1)，(02，1)，(12，0)，

0　1

(12，0)，e，(01).

其次，假定 S 为空集，由对称，可设 G 包含 (01，2) 的一个元素 g_0．求 (01，2) 的不包含常数的所有子半群，以之为 $G(g_0)$ 而定 G．除去共轭情形，我们得到下面的可能的

0　2　　　0　2　　1　2　　　0　2　　2　0

$G(g_0)$：{(01，2)}，{(01，2)，(01，2)}，{(01，2)，(01，2)}

0　2　　2　0　　1　2　　2　1

{(01，2)，(01，2)，(01，2)，(01，2)}

从而由计算得

0　2　　2　0

G_3：(01，2)，(12，0，) e.

0　2　　1　2

G_4：(01，2)，(01，2)，e，(01).

0　2　　2　0　　0　2　　2　0

G_5：(01，2)，(01，2)，(12，0)，(12，0)，e，(02).

G_6：$\overset{0\ 2}{(01,2)}$，$\overset{2\ 0}{(01,2)}$，$\overset{1\ 2}{(01,2)}$，$\overset{2\ 1}{(01,2)}$，e，(01).

今设 $S=E^3$. 由对称，可假定 G 包含(01，2)的一个元素 g_0 而 $Qg_0 \not\subset G$. 求(01，2)的 $>E^3$ 的所有真子半群，以之为 $G(g_0)$ 而定 G. 这样一个 $G(g_0)$ 必包含一个 $\overset{\alpha\ \beta}{(01,2)}$，其中 $(\alpha, \beta)=(0, 2)$，(2，0)，(1，2)或(2，1)，盖否则 $G={}_Q(G(g_0):G(g_0))$ 将包含(01，2). 除去共轭情形，我们得到下面的可能的 $G(g_0)$：

0，1，2，$\overset{0\ 1}{(01,2)}$，$\overset{0\ 2}{(01,2)}$.

0，1，2，$\overset{0\ 1}{(01,2)}$，$\overset{1\ 2}{(01,2)}$.

0，1，2，$\overset{0\ 1}{(01,2)}$，$\overset{1\ 0}{(01,2)}$，$\overset{2\ 0}{(01,2)}$，$\overset{0\ 2}{(01,2)}$

0，1，2，$\overset{0\ 2}{(01,2)}$.

0，1，2，$\overset{0\ 2}{(01,2)}$，$\overset{2\ 0}{(01,2)}$

0，1，2，$\overset{0\ 1}{(01,2)}$，$\overset{0\ 2}{(01,2)}$，$\overset{1\ 2}{(01,2)}$.

0，1，2，$\overset{0\ 2}{(01,2)}$，$\overset{1\ 2}{(01,2)}$.

0，1，2，$\overset{0\ 2}{(01,2)}$，$\overset{2\ 0}{(01,2)}$，$\overset{1\ 2}{(01,2)}$，$\overset{2\ 1}{(01,2)}$.

0，1，2，$\overset{0\ 1}{(01,2)}$，$\overset{1\ 0}{(01,2)}$，$\overset{0\ 2}{(01,2)}$.

0，1，2，$\overset{0\ 1}{(01,2)}$，$\overset{1\ 0}{(01,2)}$，$\overset{0\ 2}{(01,2)}$，$\overset{1\ 2}{(01,2)}$.

从而由计算得

0 1　0 2　0 1　0 2　1 0　2 0

G_7：0，1，2，(01，2)，(01，2)，(02，1)，(02，1)，(12，0)，(12，0)，e，(12)

0 1　1 2　1 0　2 1

G_8：0，1，2，(01，2)，(01，2)，(12，0)，(12，0)，e.

0 1　1 0　2 0　0 2　0 1　1 0

G_9：0，1，2，(01，2)，(01，2)，(01，2)，(01，2)，(02，1)，(02，1)，

0 2　2 0　0 1　1 0　0 2　2 0

(02，1)，(02，1)，(12，0)，(12，0)，(12，0)，(12，0)，e，(12).

0 2　2 0

G_{10}：(01，2)，(02，1)，(12，0)，e.

0 2　2 0　0 2　2 0

G_{11}：(01，2)，(01，2)，(02，1)，(12，0)，(12，0)，e，(02).

0 1　0 2　1 2　1 0　2 0　2 1

G_{12}：0，1，2，(01，2)，(01，2)，(01，2)，(12，0)，(12，0)，(12，0)，e.

0 2　1 2　2 1　2 0　2 1　2 0

H_1：0，1，2，(01，2)，(01，2)，(02，1)，(02，1)，(12，0)，(12，0)，e，(01).

0 2　2 0　1 2　2 1　0 2　2 0

H_2：0，1，2，(01，2)，(01，2)，(01，2)，(01，2)，(02，1)，(02，1)，

1 2　2 1　0 2　2 0　1 2　2 1

(02，1)，(02，1)，(12，0)，(12，0)，(12，0)，(12，0)，e，(01).

0 1　1 0　0 2　0 1　1 0　1 0

H_3：0，1，2，(01，2)，(01，2)，(01，2)，(02，1)，(02，1)，(12，0)，

0 1

(12，0)，e.

0 1　1 0　0 2　1 2　0 1　1 0

H_4：0，1，2，(01，2)，(01，2)，(01，2)，(01，2)，(02，1)，(02，1)，

0　1　　1　0

(12，0)，(12，0)，e，(01).

在这些正则集中，H_1，H_2 分别与 G_7，G_9 共轭，故应删去 H_1，H_2. 在 H_3，H_4 中，

　　0　1

取　　　　，我们看到 ${}_Q(G(g_0):G(g_0))$ 大于 H_3 和 H_4，因而 H_3 和 H_4 应删去.

(02，1)

对于 G_7，…，G_{12}，通过检查，我们看到它们适合命题 26 的题设，因而 $T(G_7)$，…，$T(G_{12})$ 都是极大封闭集.

为了求出和一个子半群 K 共轭的子半群的个数，我们用下面的方法. 用一个置换 σ 变 K 中的元素看是否使 K 不变. 所有使 K 不变的 σ 作成一个置换群，以此群的元数除对称群 $\widetilde{Q}$ 的元数 6 即得与 K 共轭的子半群的个数. 例如使 G_3 不变的置换为 e，(02)，故 G_3 有 3 个共轭正则集. 用此法进行计算，我们看到 G_1，…，G_{12} 每个有 3 个共轭正则集. 因此，这里我们共得到 36 个极大封闭集.

现在定出奇异部分是 Q 的左理想的那些极高正则集. 除去共轭情形，Q 共有五个奇异左理想：G_{13}^0 = 空集，$G_{14}^0=E^3$，$G_{15}^0=(01，2)$，$H_5^0=(01，2)\cup(02，1)$，$H_6^0=Q^0$. 根据命题 21 及在这一命题下面所说的话，G_{13}^0 确定六个极高正则集，即对称群 $\widetilde{Q}$ 的六个子群，G_{14}^0 确定六个极高正则集，即 G_{14}^0 添上 $\widetilde{Q}$ 的六个子群所得到的六个正则集. G_{15}^0 确定两个极高正则集，即 $G_{15}^0\cup\{e，(01)\}$，$G_{15}^0\cup\{e\}$. H_5^0 确定两个极高正则集，即

H_5^1：(01，2)，(02，1)，e，(12)；

H_5^2：(01，2)，(02，1)，e.

H_5^0 确定五个极高正则集，即 H_5^0 添上 $\widetilde{Q}$ 的五个真子群所得的五个正则集.

根据命题 23 及 25，对于 G_{13}^0，G_{14}^0，G_{15}^0 所确定的任意极高正则集 G，$T(G)$ 必是极大封闭集. 因为和 G_{15}^0 共轭的左理想共有 3 个，所以这里共有 18 个极大封闭集.

现在我们证明 H_5^0，H_6^0 不对应极大封闭集. 先看 H_6^0，设 $H=\overline{H}\cup H_6^0$ 是 H_6^0 所确定的五个极高正则集之一，我们证明 $T(H)\subset T(N_2)$. 事实上，$(H:N_2)$ = 所有 $(g，h)$ 之集合，其中 $g\in H_6^0$，$h\in H_6^0$. 设不缺值函数 $\varphi(x，y)\notin N_2$. 于是，$\varphi(x，$

y）是取所有三个值 0，1，2 的真二元函数．因之，在某 $\{\alpha_1, \alpha_2\} \times \{\beta_1, \beta_2\}$ 上，$\varphi(x, y)$ 取三个值 0，1，2．不失去普遍性，假定 $\varphi(\alpha_1, \beta_1)=0$，$\varphi(\alpha_1, \beta_2)=1$，$\varphi(\alpha_2, \beta_1)=2$．设 a_0，a_1，a_2 是 0，1，2 的任意排列．规定

$$g(x)=\begin{cases}\alpha_1, & x=a_0;\\ \alpha_1, & x=a_1;\\ \alpha_2, & x=a_2,\end{cases}$$

$$h(x)=\begin{cases}\beta_1, & x=a_0;\\ \beta_2, & x=a_1;\\ \beta_1, & x=a_2.\end{cases}$$

于是，$g, h \in H_6^0$，而函数 $\varphi(g(x), h(x))$ 相当置换 $\begin{pmatrix} a_0 & a_1 & a_2 \\ 0 & 1 & 2 \end{pmatrix}$．可见，$\varphi(H: N_2)$ 包含对称群 $\widetilde{Q}$．但 $H \neq \widetilde{Q}$，故 $\varphi(H: N_2) \not\subset H$．这就是说，任何不缺值函数 $\varphi(x, y) \notin N_2$ 都不能适合 $\varphi(H: N_2) \subset H$，故 $T(H) \subset T(N_2)$．

对于 H_5^0，我们证明 $T(H_5^1) \subset T(N_0)$，$T(H_5^2) \subset T(N_0)$．命 H 表 H_5^1，H_5^2 之一．我们有 $(H: N_0)$＝所有 (g, h) 之集合，其中 $g \in H_5^0$，$h \in H_5^0$．设不缺值函数 $\varphi(x, y) \notin N_0$．于是，$\varphi(x, y)$ 是一个真二元函数．可取 α_1，α_2，β_1，β_2 使 $\varphi(\alpha_1, \beta_1) \neq \varphi(\alpha_1, \beta_2)$，$\varphi(\alpha_1, \beta_1) \neq \varphi(\alpha_2, \beta_1)$．将 $\varphi(\alpha_1, \beta_1)$，$\varphi(\alpha_1, \beta_2)$，$\varphi(\alpha_2, \beta_1)$ 分别记为 c_0，c_1，c_2．于是，$c_0 \neq c_2$，$c_0 \neq c_1$．

先假定 $c_1=c_2$．规定

$$g(x)=\begin{cases}\alpha_1, & x=0;\\ \alpha_1, & x=1;\\ \alpha_2, & x=2.\end{cases}\quad h(x)=\begin{cases}\beta_1, & x=0;\\ \beta_2, & x=1;\\ \beta_1, & x=2.\end{cases}\tag{1}$$

于是，$g\in(01,\ 2)$，$h\in(02,\ 1)$，而 $\varphi(g,\ h)=(\overset{c_1}{12},\ \overset{c_0}{0})\notin H$.

今设 $c_1\neq c_2$. 因为 c_0，c_1，c_2 都不同，$c_3=\varphi(\alpha_2,\ \beta_2)$ 必为 c_0，c_1，c_2 之一. 分为两种情形讨论：

i）$c_3=c_0$. 规定

$$g_1(x)=\begin{cases}\alpha_1, & x=0;\\ \alpha_1, & x=1;\\ \alpha_2, & x=2.\end{cases}\quad h_1(x)=\begin{cases}\beta_2, & x=0;\\ \beta_1, & x=1;\\ \beta_2, & x=2.\end{cases}\tag{2}$$

于是，$g_1\in(01,\ 2)$，$h_1\in(02,\ 1)$，而 $\varphi(g_1,\ h_1)=(\overset{c_0}{12},\ \overset{c_1}{0})\notin H$.

ii）$c_3=c_1$ 或 c_2. 规定 g，h 如（1）. 假定 $\varphi(g,\ h)\in H$. 于是，$(c_0,\ c_1,\ c_2)=(0,\ 1,\ 2)$ 或 $(0,\ 2,\ 1)$. $c_3=c_1$ 时，规定

$$g_2(x)=\begin{cases}\alpha_2, & x=0;\\ \alpha_2, & x=1;\\ \alpha_1, & x=2.\end{cases}\quad h_2(x)=\begin{cases}\beta_1, & x=0;\\ \beta_2, & x=1;\\ \beta_1, & x=2.\end{cases}$$

于是，$g_2 \in (01, 2)$，$h_2 \in (02, 1)$，而 $\varphi(g_2, h_2) = (02)$ 或 (012)，都 $\notin H$. $c_3 = c_2$ 时，规定 g_1，h_1 如 (2). 于是，$g_1 \in (01, 2)$，$h_1 \in (02, 1)$ 而 $\varphi(g_1, h_1) = (01)$ 或 (021)，都 $\notin H$. 这就证明了 $T(H) \subset T(N_0)$.

因此，P_3^* 共有 54 个第一种极大封闭集. 我们知道，P_3^* 有三个第二种极大封闭集 $T(N_0)$，$T(N_1)$，$T(N_2)$，有一个第三种极大封闭集 $P_3 \cup \{*\}$. 所以，P_3^* 共有 58 个极大封闭集.

参考文献

Sheffer, Trans. Am. Math. Soc, Vol. Xiv. PP. 481—488.

Post E., 1921, Introduction to a General Theory of Elementary Propositions. Amer. J. Math., 43, 163—185.

Post E., 1941. Two-Valued Iterative Systems, 1941.

Alexandroff and Hopf, 1935, Topologie I.

Яблонский С. В., 1952, О Суперпозициях Функций алгебры Логики, Матем. сб., 30 (72): 2, 329—348.

Яблонский С. В., 1954, О Функциональной Полноте В Трехзначном исчислении, ДАН СССР, 95, No 6, 1153—1156.

Яблонский С. В., 1958, Функциональные построения в к-значной логике, Труды Матем. Ин-та АН СССР 51.

Мартинов В. В. 1960, Исследование некоторых классов функций в многозначных логиках, проблемы кибернетики 3.

庞云阶，1962，在多值逻辑中求全部极大封闭集的一个能行方法，吉林大学自然科学学报.

罗铸楷，1963，Ⅰ，线性函数集和环的极大封闭性，吉林大学自然科学学报.

罗铸楷，1963，Ⅱ，关于保分划之函数集的极大封闭性（I），吉林大学自然科学学报.

Structure Theory of Total and Partial Functions Defined in a Finite Set

1. Let $E = E^k$ be a set with $k \geqslant 2$ elements. Denote by $P = P_k$ the set of all functions of any finite number of variables defined on E. If the totality of functions is enlarged to include the partial functions defined in E, then the set is denoted by $P^* = P_k^*$. Heretofore, several authors have studied the structure of P_k, especially the cases $k = 2$, 3. The purpose of this paper is to investigate the structure of P_k^*. Our results obtained and the method used are also helpful for studying P_k.

2. The closure $\overline{A}$ of a set $A \subset P^*$ is the set of all functions generated by those is A and by independent variables through compositions. A is said to be closed if $\overline{A} = A$; to be complete if $\overline{A} = P^*$. A closed set $M < P^*$ is said to be maximal, if there exists no closed set M_1 such that $M < M_1 < P^*$. It can be proved that in P^* there are only finitely many maximal closed sets and that A is complete if and only if $A \not\subset$ any maximal closed set. So, an important problem is to determine all the maximal closed sets in P^*.

3. A maximal closed set M is said to be of the first kind, if not all total functions of a single variable $\in M$; of the second kind, if all total functions of a single variable $\in M$, but not all total functions of two variables $\in M$; of the third kind, if all total functions of two variables $\in M$.

Theorem. There is only a single maximal closed set of the third kind, namely $P_k \cup \{*\}$, $*$ being the function nowhere defined.

4. Let A be a set of total functions of r variables $x_1, \cdots, x_r$. $f(t_1, \cdots, t_n)$ is

said to preserve A, if each function obtained by substituting into f any n functions $\in A$ is either a partial function or a function in A. Denote by $T(A)$ the set of all functions preserving A. A total function $\varphi(x, y)$ is said to be locally linear, if, for any α_1, α_2, β_1, $\beta_2 \in E$, $\varphi(\alpha_1, \beta_1)$, $\varphi(\alpha_1, \beta_2)$, $\varphi(\alpha_2, \beta_1)$, $\varphi(\alpha_2, \beta_2)$ are either all equal or equal in pairs. Denote by N_1 the set of all locally linear functions; by N_0 the set of all total functions $\psi(x, y)$ depending actually on only a single variable x or y; by N_i the set obtained by adjoining to N_0 all total functions $\rho(x, y)$ taking at most i values, $2 \leqslant i \leqslant k-1$. We have $N_0 < N_1 < N_2 < \cdots < N_{k-1}$.

Theorem. There are exactly k maximal closed sets of the second kind, namely,

$$T(N_0),\ T(N_1),\ T(N_2),\ \cdots,\ T(N_{k-1}).$$

5. Let Q be the set of all total functions of x. Q is a semigroup under the operation Af of composition. Any maximal closed set of the first kind is of the form $T(G)$, where G is a proper sub-semigroup of Q, containing the function $f(x)=x$. a set of the form $T(G)$ is a maximal closed set, if and only if, for any proper sub-semigroup $H > G$, $T(G) \not\subset T(H)$, and, for each N_i, $T(G) \not\subset T(N_i)$.

Theorem Let $H > G$. $T(G) \not\subset T(H)$, If and only if there exists a total function $\varphi(x) \notin H$ such that $Hg \subset G$ implies $\varphi g \in G$.

Theorem $T(G) \not\subset T(N_i)$, if and only if there exists a total function $\psi(x, y) \notin N_i$ such that $N_i(g, h) \subset G$ implies $\psi(g, h) \in G$.

Theorem If $T(G) \cap P$ is a maximal closed set in P, then $T(G)$ is a maximal closed set in P^*.

6. By the general theory sketched above, it is not difficult to determine all the maximal closed sets in P_2^* and P_3^*. There are 8 maximal closed sets in P_2^*: $T(x)$, $T(0)$, $T(1)$, $T(0, 1, x)$, $T(x, \bar{x})$, $T(N_0)$, $T(N_1)$, $P_2 \cup \{*\}$. The number of maximal closed sets in P_3^* is 58.

广义归结*

摘　要

本文对 1965 年 J. A. Robinson 提出的归结方法（Resolution Principle）做了推广，提出了广义归结方法．由于广义归结方法允许对所要证明的定理做比较自然的描述，因此，使用广义归结方法去证明定理会简单，自然．本文证明了广义归结方法的完备性和一种广义锁归结方法、广义线性归结方法、广义语义归结方法的完备性．

一、引　言

用 1965 年 Robinson 提出的归结方法证明定理，要将定理的否定写成一阶逻辑中的公式，再将公式化成 Skolem 标准型，并将标准型的母式化成合取范式．合取范式中的每一项（是一个析取式）称为一个子句，于是合取范式对应着一个子句集．用归结方法证明这个子句集是不可满足的，就相当于证明了该定理．

但是，将 Skolem 标准型的母式化成合取范式，然后，对子句集使用 Robinson 的归结方法，未必永远是最好的．

例如，文献［2］中（p. 193），关于群论的一个定理，其对应的子句集为：

A：$P(i(x),x,e)$，

B：$P(e,x,x)$，

C：$\sim P(x,y,u)\vee\sim P(y,z,v)\vee\sim P(u,z,w)\vee P(x,v,w)$，

D：$\sim P(x,y,u)\vee\sim P(y,z,v)\vee\sim P(x,v,w)\vee P(u,z,w)$，

E：$\sim P(a,e,a)$．

其中 C，D 两式是由母式

$$F:\ \mathrm{P}(x,y,u)\wedge P(y,z,v)\rightarrow(P(u,z,w)\leftrightarrow P(x,v,w))$$

拆开来的．母式 F 描述了群论中的结合律．显然，直接使用母式 F 就更自然，更简

---

* 本文由王湘浩、刘叙华共同创作，原载于《计算机学报》1982 年第 2 期。

便些.

我们可以将母式 F 看做是一个广义子句. 为此，我们于 1979 年 7 月提出了广义归结方法，并得到了一些结果（见文献［1］，对得到的结果没有给出证明）.

用自然推导法去证明数学定理（所谓自然推导法是指模拟人在证明定理时的推导过程，具有 Heuristic 因素的方法）比归结方法更具有智能的特征，因此，自然推导法是人工智能中的一个重要研究课题. 我们认为，自然推导法和归结方法应当冶于一炉，取长补短. 在自然推导系统中，在推导过程的某些验证中使用归结方法是可取的.

由于广义归结方法对问题的描述更为自然，因此，在自然推导法中使用广义归结方法，比使用 Robinson 的归结方法会更好些.

本文证明了文献［1］中的论断，并得到新的结果. 亦即本文将证明广义归结、一种广义锁归结、广义线性归结、广义语义归结都是完备的.

二、广义归结方法

本文用 1，0 代表真与假.

定义 设 A_1，…，A_n 是一些原子.

$$\Phi(A_1, A_2, \cdots, A_n)$$

是以命题逻辑符号

$$\sim, \quad \vee, \quad \wedge, \quad \rightarrow, \quad \leftrightarrow$$

连结这些原子和一些 0，1 作成的公式. 称这样的一个公式为一个广义子句.

定义 如果一个广义子句只由 0，1，逻辑符号和括号组成，则称此广义子句为常子句. 特别，其值为 0 的常子句称为零子句，其值为 1 的常子句称为壹子句.

设 $\Phi(A_1, \cdots, A_n)$ 是一个广义子句. 在我们特别注意 Φ 中某些原子

$$A_{i_1}, \cdots, A_{i_r}$$

时，我们可以把 Φ 写成 $\Phi(A_{i_1}, \cdots, A_{i_r})$.

例如，若特别注意 A_1 时，我们可以将 $\Phi(A_1, \cdots, A_n)$ 写成 $\Phi(A_1)$. $\Phi(A_1=0)$ 表示以 0 代 Φ 中 A_1 的所有出现得到的广义子句.

以替换 σ 变 Φ，可能合一 A_1 与某些原子，$\Phi^\sigma(A_1^\sigma=0)$ 表示在 Φ^σ 中以 0 代这些合一后的所有出现得到的广义子句.

定义 设 $\Phi(A_1, \cdots, A_n)$ 是一个广义子句，$A_{i_1}, \cdots, A_{i_r}$ 是 Φ 中一些原子，其中 $r \geqslant 1$. 若 $A_{i_1}, \cdots, A_{i_r}$ 有一个 mgu（最一般合一替换）σ，则说 Φ^σ 是合一 $A_{i_1}, \cdots, A_{i_r}$ 所得到的 Φ 的因子.

定义 设 Φ，Ψ 是两个广义子句，换名使无公共变量. 设 Φ^σ 是合一

$$A_{i_1}, \cdots, A_{i_r}$$

所得到的 Φ 的因子. Ψ^τ 是合一

$$B_{j_1}, \cdots, B_{j_s}$$

所得到的 Ψ 的因子. 设 ρ 是 $A_{i_1}^\sigma$ 和 $B_{j_1}^\tau$ 的 mgu. 于是

$$\Phi^{\sigma\rho}(A_{i_1}^{\sigma\rho}=0) \vee \Psi^{\tau\rho}(B_{j_1}^{\tau\rho}=1),$$

$$\Phi^{\sigma\rho}(A_{i_1}^{\sigma\rho}=1) \vee \Psi^{\tau\rho}(B_{j_1}^{\tau\rho}=0)$$

都称为 Φ 与 Ψ 的广义归结式. 其中 A_{i_1}，B_{j_1} 分别称为 Φ，Ψ 的归结原子.

定义 设 S 是一个广义子句集，Φ 是一个广义子句. 从 S 推出 Φ 的一个演绎是一个有限广义子句序列：

$$\Phi_1, \cdots, \Phi_n.$$

其中 (1) Φ_i 或者是 S 中广义子句，或者是 Φ_k，Φ_l 的广义归结式，$k<i$，$l<i$. (2) $\Phi_n=\Phi$.

不难看出，Robinson 的归结方法是我们引进的广义归结方法的一个特例.

命题 1 设 Φ，Ψ 是广义子句，于是 Φ 与 Ψ 的广义归结式是 Φ 和 Ψ 的逻辑推论.

定理 1 广义基子句集 S 不可满足，当且仅当由 S 可逐步归结出零子句.

证 必要性：设 M 是 S 的原子集，对 M 的元素数用归纳法.

当 M 由一个元素组成时，设为 A. 若 S 中有零子句，则结论显然. 若 S 中无零子句，因为 S 不可满足，令解释 I_1 指定 A 为 0，则 I_1 弄假 S，不妨设 I_1 弄假 S 中子句 $\Phi(A)$. 于是，$\Phi(A=0)$ 是零子句. 令解释 I_2 指定 A 为 1，则 I_2 弄假 S，不妨设 I_2 弄假 S 中子句 $\Psi(A)$. 于是，$\Psi(A)$ 是零子句. 因此，S 中子句 Φ 与 Ψ 的广义归结式

$$\Phi(A=0) \vee \Psi(A=1)$$

是零子句. 亦即，由 S 可归结出零子句.

假定 M 的元素数小于 n 时，结论成立. 证明 M 的元素数为 n 时，结论也成立（$n>1$）.

设 A 是 M 中一个原子. 令

$$S'=\{\Phi' \mid \Phi\in S \wedge \Phi'=\Phi(A=0)\}.$$

显然 S' 不可满足，且 S' 中原子数小于 n. 由归纳假设，存在从 S' 出发推出零子句的演绎 D'_1. 将 D'_1 中初始节点上的 S' 中子句，原来是 A 的位置上再恢复 A，于是得从 S 出发的一个演绎 D_1. 而 D_1 演绎出的广义子句，或是零子句；或为只含原子 A 的广义子句 Φ，且 $\Phi(A=0)$ 是零子句.

若是前者，则归纳法完成，若是后者，则令

$$S''=\{\Phi'' \mid \Phi\in S \wedge \Phi(A=1)=\Phi''\}.$$

显然 S'' 不可满足，且 S'' 中原子数小于 n. 与上同理，存在由 S 出发的一个演绎 D_2，而 D_2 演绎出的广义子句，或为零子句；或为只含原子 A 的广义子句 Ψ，且 $\Psi(A=1)$ 是零子句.

若是前者，则归纳法完成. 若是后者，则由 Φ，Ψ 出发可得一个推出零子句的演绎 D_3. 因为 Φ 与 Ψ 的广义归结式

$$\Phi(A=0)\vee\Psi(A=1)$$

是一个零子句.

将 D_1，D_2，D_3 连结起来，得从 S 出发推出零子句的演绎 D. 归纳法完成.

充分性：若存在从 S 出发推出零子句的演绎 D，则 S 不可满足.

若不然，设解释 I 满足 S，令演绎 D 为如下广义子句序列：

$$G_1,\ \cdots,\ G_k.$$

由演绎的定义及命题 1 知，I 满足每一个 $G_i(i=1,\ \cdots,\ k)$. 特别是 I 满足 G_k. 但是，G_k 是零子句，故矛盾.

由文献［2］可以看出，Herbrand 定理和提升引理的证明并不依赖于子句是关于文字的析取式这一假定. 因此这两个定理对于广义子句集也是正确的. 但是，文献［2］关于提升引理的证明有漏洞，故在下面证明提升引理.

提升引理（Ⅰ）[①] 设 Φ' 和 Ψ' 分别是广义子句 Φ 和 Ψ 的例，P' 是 Φ' 和 Ψ' 的广义归结式. 于是，可取 Φ 和 Ψ 的广义归结式 P，使 P' 是 P 的例.

证 对 Φ 和 Ψ 中变量适当改名，使得 Φ 和 Ψ 无公共变量. 设 Φ 中所有变量符号为 x_1，…，x_n；Ψ 中所有变量符号为 y_1，…，y_m. 令

$$P'=\Phi'^{\sigma'\rho'}(A'^{\sigma'\rho'}_{i_1}=0)\vee\Psi'^{\tau'\rho'}(B'^{\tau'\rho'}_{j_1}=1).$$

其中 σ' 是 Φ' 中原子 A'_{i_1}，…，A'_{i_r} 的 mgu，τ' 是 Ψ' 中原子 B'_{j_1}，…，B'_{j_s} 的 mgu，ρ' 是 A'_{i_1}，B'_{j_1} 的 mgu.

A'_{i_1}，B'_{j_1} 分别是 Φ'，Ψ' 中的归结原子. 因为 Φ'，Ψ' 分别是 Φ，Ψ 的例，令

$$\Phi'=\Phi^{\theta_1},\quad\Psi'=\Psi^{\theta_2}$$

（下面我们将替换 $\{t_1/x_1,\cdots,t_n/x_n\}$ 中的 x_1，…，x_n 称为替换的分母；t_1，…，t_n 称为替换的分子）. 不妨设 θ_1 的分母只有 x_i，θ_2 的分母只有 $y_j(1\leqslant i\leqslant n,1\leqslant j\leqslant m)$. σ' 的分母只有 x_i，τ' 的分母只有 y_j. 于是，可做出如下替换：

$$\theta=\theta_1\cdot\sigma'\cup\theta_2\cdot\tau'.$$

因此，$\Phi'^{\sigma'}=\Phi^{\theta_1\cdot\sigma'}=\Phi^{\theta}$，$\Psi'^{\tau'}=\Psi^{\theta_2\cdot\tau'}=\Psi^{\theta}$.

故

$$P'=\Phi^{\theta\rho'}(A^{\theta\rho'}_{i_1}=0)\vee\Psi^{\theta\rho'}(B^{\theta\rho'}_{j_1}=1).$$

其中 A_{i_1}，B_{j_1} 分别是 Φ，Ψ 中原子，且 $A^{\theta_1}_{i_1}=A'_{i_1}$，$B^{\theta_2}_{j_1}=B'_{j_1}$.

令 A_{i_2}，…，A_{i_r} 是 Φ 中原子，且 $A^{\theta_1}_{i_2}=A'_{i_2}$，…，$A^{\theta_1}_{i_r}=A'_{i_r}$.

B_{j_2}，…，B_{j_s} 是 Ψ 中原子，且 $B^{\theta_2}_{j_2}=B'_{j_2}$，…，$B^{\theta_2}_{j_s}=B'_{j_s}$.

A_1，…，A_{r_1} 是 Φ 中所有在 $\theta\cdot\rho'$ 下可与 A_{i_1} 合一的原子.

B_1，…，B_{r_2} 是 Ψ 中所有在 $\theta\cdot\rho'$ 下可与 B_{j_1} 合一的原子.

显然，$\{A_{i_1},\cdots,A_{i_r}\}\subseteq\{A_1,\cdots,A_{r_1}\}$，$\{B_{j_1},\cdots,B_{j_s}\}\subseteq\{B_1,\cdots,B_{r_2}\}$.

设

λ_1 是 $\{A_1,\cdots,A_{r_1}\}$ 的 mgu.

λ_2 是 $\{B_1,\cdots,B_{r_2}\}$ 的 mgu.

---

① 研究生王玉书、黄秉超参加了此引理的证明.

不妨设 $\lambda_1=\{t_1/x_1, \cdots, t_n/x_n\}$，$\lambda_2=\{u_1/y_1, \cdots, u_m/y_m\}$ 且 $t_1, \cdots, t_n$ 中变量符号只有 x，$u_1, \cdots, u_m$ 中变量符号只有 y.

设 ρ 是 $A_1^{\lambda_1}$ 和 $B_1^{\lambda_2}$ 的 mgu（关于 $A_1^{\lambda_1}$ 和 $B_1^{\lambda_2}$ 是可合一的将在下面证明）.

于是，可以证明：$\lambda \cdot \rho$ 是 $\{A_1, \cdots, A_{r_1}, B_1, \cdots, B_{r_2}\}$ 的 mgu，其中 $\lambda=\lambda_1 \cup \lambda_2$.

事实上，设 η 是 $\{A_1, \cdots, A_{r_1}, B_1, \cdots, B_{r_2}\}$ 的一个合一. 于是，η 既是 $\{A_1, \cdots, A_n\}$ 的合一，也是 $\{B_1, \cdots, B_{r_2}\}$ 的合一.

不妨设 η 的分母只有 x_i 和 $y_j(1 \leqslant i \leqslant n, 1 \leqslant j \leqslant m)$，将 η 中分母为 x 的元素组成的替换记为 η_1，η 中分母为 y 的元素组成的替换为 η_2. 显然，η_1 是 $\{A_1, \cdots, A_{r_1}\}$ 的合一，η_2 是 $\{B_1, \cdots, B_{r_2}\}$ 的合一. 所以存在替换 δ_1，δ_2，使得

$$\eta_1=\lambda_1 \cdot \delta_1,$$

$$\eta_2=\lambda_2 \cdot \delta_2.$$

由于 η_1 的分母中只有 x，所以 δ_1 的分母中也只有 x；同理，δ_2 的分母中只有 y. 于是

$$\begin{aligned}
&(\lambda_1 \cup \lambda_2) \cdot (\delta_1 \cup \delta_2) \\
&=\{t_1^{(\delta_1 \cup \delta_2)}/x_1, \cdots, t_n^{(\delta_1 \cup \delta_2)}/x_n, u_1^{(\delta_1 \cup \delta_2)}/y_1, \cdots, u_m^{(\delta_1 \cup \delta_2)}/y_m, \delta_1, \delta_2\} \\
&=\{t_1^{\delta_1}/x_1, \cdots, t_n^{\delta_1}/x_n, u_1^{\delta_2}/y_1, \cdots, u_m^{\delta_2}/y_m, \delta_1, \delta_2\} \\
&=\{t_1^{\delta_1}/x_1, \cdots, t_n^{\delta_1}/x_n, \delta_1, u_1^{\delta_2}/y_1, \cdots, u_m^{\delta_2}/y_m, \delta_2\} \\
&=\lambda_1 \cdot \delta_1 \cup \lambda_2 \cdot \delta_2.
\end{aligned}$$

令 $\delta=\delta_1 \cup \delta_2$，于是

$$\begin{aligned}
\eta=\eta_1 \cup \eta_2 &=\lambda_1 \cdot \delta_1 \cup \lambda_2 \cdot \delta_2 \\
&=(\lambda_1 \cup \lambda_2) \cdot (\delta_1 \cup \delta_2)=\lambda \cdot \delta.
\end{aligned}$$

因为 $A_1^{\eta}=B_1^{\eta}$，所以 $A_1^{\lambda \cdot \delta}=B_1^{\lambda \cdot \delta}$

亦即

$$(A_1^{(\lambda_1 \cup \lambda_2)})^{\delta}=(B_1^{(\lambda_1 \cup \lambda_2)})^{\delta},$$

$$(A_1^{\lambda_1})^{\delta}=(B_1^{\lambda_2})^{\delta}.$$

亦即 δ 是 $A_1^{\lambda_1}$ 和 $B_1^{\lambda_2}$ 的合一.

(注:至此证明了:若 η 是 $\{A_1, \cdots, A_{r_1}, B_1, \cdots, B_{r_2}\}$ 的合一,则可找到 $A_1^{\lambda_1}$, $B_1^{\lambda_2}$ 的合一 δ. 而 $\theta \cdot \rho'$ 确是 $\{A_1, \cdots, A_{r_1}, B_1, \cdots, B_{r_2}\}$ 的合一,所以 $A_1^{\lambda_1}$, $B_1^{\lambda_2}$ 是可合一的,于是前面令 ρ 是 $A_1^{\lambda_1}$ 和 $B_1^{\lambda_2}$ 的 mgu 是合理的.)

因此,$\delta=\rho \cdot \zeta$. 故

$$\eta=\lambda \cdot \delta=(\lambda \cdot \rho) \cdot \zeta.$$

而 $\lambda \cdot \rho$ 是 $\{A_1, \cdots, A_{r_1}, B_1, \cdots, B_{r_2}\}$ 的合一是明显的,所以我们证明了:$\lambda \cdot \rho$ 是

$$\{A_1, \cdots, A_{r_1}, B_1, \cdots, B_{r_2}\}$$

的 mgu.

令

$$P=\Phi^{\lambda_1 \cdot \rho}(A_{i_1}^{\lambda_1 \cdot \rho}=0) \vee \Psi^{\lambda_2 \cdot \rho}(B_{j_1}^{\lambda_2 \cdot \rho}=1),$$

显然,P 是 Φ 与 Ψ 的广义归结式. 设

$$\theta \cdot \rho'=(\lambda \cdot \rho) \cdot \sigma,$$

于是有

$$\begin{aligned} P' &= \Phi^{\theta \cdot \rho'}(A_{i_1}^{\theta \cdot \rho'}=0) \vee \Psi^{\theta \cdot \rho'}(B_{j_1}^{\theta \cdot \rho'}=1) \\ &= \Phi^{\lambda \cdot \rho \cdot \sigma}(A_{i_1}^{\lambda\rho\sigma}=0) \vee \Psi^{\lambda\rho\sigma}(B_{j_1}^{\lambda\rho\sigma}=1). \end{aligned}$$

因为 Φ 中在 $\lambda \cdot \rho \cdot \sigma$(即 $\theta \cdot \rho'$)下与 A_{i_1} 可合一的所有原子为 $A_1, \cdots, A_{r_1}$,且 $\lambda \cdot \rho$ 可合一 $A_1, \cdots, A_r$,故 Φ 中在 $\lambda \cdot \rho$ 下可与 A_{i_1} 合一的原子就只有 $A_1, \cdots, A_{r_1}$. 同理,Ψ 中在 $\lambda \cdot \rho$ 下可与 B_{j_1} 合一的原子也只有 $B_1, \cdots, B_{r_2}$.

因此,

$$\begin{aligned} P' &= \Phi^{\lambda \cdot \rho \cdot \sigma}(A_{i_1}^{\lambda \cdot \rho \cdot \sigma}=0) \vee \Psi^{\lambda \cdot \rho \cdot \sigma}(B_{j_1}^{\lambda \cdot \rho \cdot \sigma}=1) \\ &= (\Phi^{\lambda \cdot \rho}(A_{i_1}^{\lambda \cdot \rho}=0) \vee \Psi^{\lambda \cdot \rho}(B_{j_1}^{\lambda \cdot \rho}=1))^{\sigma} \\ &= (\Phi^{\lambda_1 \cdot \rho}(A_{i_1}^{\lambda_1 \cdot \rho}=0) \vee \Psi^{\lambda_2 \cdot \rho}(B_{j_1}^{\lambda_2 \cdot \rho}=1))^{\sigma} \\ &= P^{\sigma}. \end{aligned}$$

定理 2(广义归结的完备性) 广义子句集 S 不可满足,当且仅当由 S 可逐步归结出零子句.

证 必要性. 若 S 不可满足，由 Herbrand 定理，则存在有限的不可满足的 S 的基例集 S' .

由定理 1 知，存在从 S' 出发推出零子句的演绎 D' .

将演绎树 D' 中所有初始节点上的子句，都换成 S 中子句，且使旧子句是新子句的例. 将每个非初始节点上的子句，也换成新子句，使新子句是其前任节点上新子句的广义归结式，且这些非初始节点上的旧子句也是新子句的例（由提升引理，这是能做到的）. 因为只有常子句的例才能是常子句，特别是，只有零子句的例，才能是零子句. 故演绎树 D' 的终点上的零子句，仍以零子句代替. 于是，D' 做如上修改后，得由 S 推出零子句的演绎 D .

充分性的证明与定理 1 相同，略去.

三、广义锁归结方法

定义 将广义子句 Φ 中出现的所有原子，都标以一个整数（相同原子的不同出现可标以不同整数），则称广义子句 Φ 已配锁.

我们规定：广义子句中的 0，1 不配锁，且已配锁广义子句 Φ 中原子 A 代以 0，1 时，原子 A 的锁也自动消失.

定义 设 Φ，Ψ 是已配锁广义子句，$R(\Phi,\ \Psi)$ 是 Φ 与 Ψ 的广义归结式. 如果 Φ，Ψ 的归结原子分别是 Φ，Ψ 中有最小锁原子，则称 $R(\Phi,\ \Psi)$ 为 Φ 与 Ψ 的广义锁归结式.

类似地，可定义出广义锁演绎的概念.

定理 3 设 S 是不可满足广义基子句集. 若将 S 中的相同原子配以相同锁，则存在从 S 出发推出零子句的广义锁演绎.

完全模仿定理 1 的必要性证法即可证得此定理. 故略去.

提升引理（Ⅱ） 设 Φ'，Ψ' 分别是已配锁广义子句 Φ，Ψ 的例，P' 是 Φ' 和 Ψ' 的广义锁归结式. 于是，可取 Φ 和 Ψ 的广义锁归结式 P ，使 P' 是 P 的例.

定理 4 设 S 是不可满足广义子句集. 若将 S 中相同的谓词符号配以相同锁，则存在从 S 出发推出零子句的锁演绎.

使用 Herbrand 定理，定理 3 和提升引理（II）不难证得此定理.

定理 3，4 说明，一种特殊的广义锁归结是完备的. 但是，一般的广义锁归结甚至对于基子句集也不完备.

例 1 $S=\{P\leftrightarrow Q,\ \sim(P\leftrightarrow Q)\}$.

$$\left.\begin{array}{l}(1)\ {}_1P\leftrightarrow{}_2Q\\(2)\ \sim({}_4P\leftrightarrow{}_3Q)\end{array}\right\}S_0,$$

则相应的广义锁演绎为：

$$\left.\begin{array}{l}(3)\ (0\leftrightarrow{}_2Q)\vee(1\leftrightarrow{}_2Q)\\(4)\ \sim({}_4P\leftrightarrow 0)\vee\sim({}_4P\leftrightarrow 1)\end{array}\right\}S_1\qquad\begin{array}{l}(1),(1)\\(2),(2)\end{array}$$

$$\left.\begin{array}{l}(5)\ (0\leftrightarrow{}_2Q)\vee\sim(1\leftrightarrow 0)\vee\sim(1\leftrightarrow 1)\\(6)\ (1\leftrightarrow{}_2Q)\vee\sim(0\leftrightarrow 0)\vee\sim(0\leftrightarrow 1)\\(7)\ \sim({}_4P\leftrightarrow 0)\vee(0\leftrightarrow 1)\vee\sim(1\leftrightarrow 1)\\(8)\ \sim({}_4P\leftrightarrow 1)\vee(0\leftrightarrow 0)\vee\sim(1\leftrightarrow 0)\\(9)\ \text{壹子句}\\(10)\ \text{壹子句}\end{array}\right\}S_2\qquad\begin{array}{l}(1),(4)\\(1),(4)\\(2),(3)\\(2),(3)\\(3),(3)\\(4),(4)\end{array}$$

不难看出，令

$$S_3=\{R(\Phi,\Psi)\mid(\Phi\in S_0\cup S_1\cup S_2)\wedge(\Psi\in S_2)\},$$

其中 $R(\Phi,\Psi)$ 表示 Φ 与 Ψ 的广义归结式. S_3 中的广义子句，或者是壹子句，或者是等价（命题逻辑中通常意义下的等价）于壹子句的广义子句. 因此，永远也归结不出零子句.

例 2 令

$$S=\{{}_1P(x)\leftrightarrow{}_2Q(x),\ \sim({}_4P(f(y))\leftrightarrow{}_3Q(f(y)))\}.$$

与例 1 同理，没有从 S 推出零子句的广义锁演绎，而 S 是不可满足的，是明显的.

四、广义线性归结方法

定义 设 S 是一个广义子句集，Φ_0 是 S 中广义子句. 以 Φ_0 为顶，从 S 到 Φ_n 的一个线性演绎，是如下图的一个演绎.

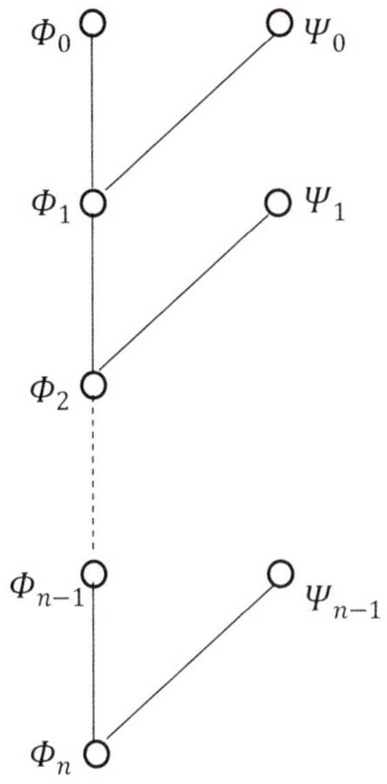

(1) 对于 $i=0$，…，$n-1$，Φ_{i+1} 是 Φ_i（广义中心子句）和 Ψ_i（广义边子句）的广义归结式.

(2) 每个 Ψ_i 或者属于 S，或者是一个 Φ_j，其中 $j \leqslant i$.

下面，我们称两个广义基子句 Φ，Ψ 是等价的，并且记以 $\Phi \equiv \Psi$，是指 Φ，Ψ 作为命题逻辑中的公式是等价的，并且 Φ 中所含原子与 Ψ 中所含原子相同.

命题 2 设 Φ_1，Φ_2，Ψ_1，Ψ_2 是广义基子句，且

$$\Phi_1 \equiv \Phi_2,\ \Psi_1 \equiv \Psi_2.$$

如果 Φ_1 和 Ψ_1 可归结，则 Φ_2 和 Ψ_2 也可归结，且以相同归结原子做出的相应广义归结式等价. 亦即

$$(\Phi_1(A=0) \vee \Psi_1(A=1)) \equiv (\Phi_2(A=0) \vee \Psi_2(A=1)),$$

$$(\Phi_1(A=1) \vee \Psi_1(A=0)) \equiv (\Phi_2(A=1) \vee \Psi_2(A=0)).$$

定义 设 $S=\{\Phi_1,\ \cdots,\ \Phi_n\}$，$S^*=\{\Phi_1^*,\ \cdots,\ \Phi_n^*\}$ 是两个广义基子句集，并且

$$\Phi_i \equiv \Phi_i^*, \ i=1, \ \cdots, \ n.$$

若 G_1，…，G_k 是从 S 推出 G_k 的一个演绎 D，则广义子句序列：

$$G_1^*, \ \cdots, \ G_k^*$$

称为从 S^* 推出 G_k^* 的一个与 D 等价的演绎 D^*，其中

（1）若 $G_i = \Phi_j$，则 $G_i^* = \Phi_j^*$，$1 \leqslant i \leqslant k$，$1 \leqslant j \leqslant n$.

（2）若 G_i 是 G_l 与 G_j 的广义归结式，则 G_i^* 是 G_l^* 与 G_j^* 的广义归结式，

$$l < i, \ j < i.$$

命题 3 若 $S = \{\Phi_1, \ \cdots, \ \Phi_n\}$，$S^* = \{\Phi_1^*, \ \cdots, \ \Phi_n^*\}$ 是两个广义基子句集，且

$$\Phi_i \equiv \Phi_i^*, \ i=1, \ \cdots, \ n,$$

则从 S，S^* 出发的两个等价演绎所推出的两个广义子句仍然等价.

命题 4 设广义基子句 $\Phi = \Phi_1 \wedge \cdots \wedge \Phi_m$，其中 Φ_i 是普通子句，$i=1$，…，m. 若其中某个 Φ_i 是零子句，则存在从 Φ 出发推出零子句的演绎.

如果我们将空子句集（即子句集是空集）看做是可满足的，则有下面定理.

定理 5 设 S 是不可满足广义基子句集，$\Phi_0 \in S$. 如果 $S - \{\Phi_0\}$ 是可满足的，则存在以 Φ_0 为顶，从 S 出发推出零子句的广义线性演绎.

证 设 $S = \{\Phi_0, \ \Phi_1, \ \cdots, \ \Phi_n\}$，将每个广义子句 Φ_i 化为合取范式：

$$\begin{cases} \Phi_0 \equiv \Phi_{01} \wedge \cdots \wedge \Phi_{0m_0}, \\ \qquad \vdots \\ \Phi_n \equiv \Phi_{n1} \wedge \cdots \wedge \Phi_{nm_n}. \end{cases}$$

将上面等价式组的右端分别记以 Φ_0^*，…，Φ_n^*，令

$$S^* = \{\Phi_0^*, \ \cdots, \ \Phi_n^*\},$$

$$S' = \{\Phi_{01}, \ \cdots, \ \Phi_{0m_0}, \ \cdots, \ \Phi_{n1}, \ \cdots, \ \Phi_{nm_n}\}.$$

显然，S' 是普通子句集，不可满足，且 $S' - \{\Phi_{01}, \ \cdots, \ \Phi_{0m_0}\}$ 可满足. 不妨设

$$S'' = \{S' - \{\Phi_{01}, \ \cdots, \ \Phi_{0m_0}\}\} \cup \{\Phi_{01}, \ \cdots, \ \Phi_{0i}\}$$

不可满足（$1 \leqslant i \leqslant m_0$），而 $S'' - \{\Phi_{0i}\}$ 是可满足的.

于是，存在从 S'' 出发，以 Φ_{0i} 为顶，使用普通归结方法推出零子句的线性演绎

D'（如图 1）：

$$G'_1, \cdots, G'_k.$$

其中

(1) $G'_1 = \Phi_{0i}$，$G'_k =$ 零子句.

(2) G'_{2i-1} 是 G'_{2i-3} 和 G'_{2i-2} 的广义归结式，$i = 2, 3, \cdots, k+1/2$.

(3) 每个 G'_{2i}，或者属于 S''，或者是一个

$$G'_{2j-1}(2j-1 \leqslant 2i-1, i = 1, \cdots, k-1/2).$$

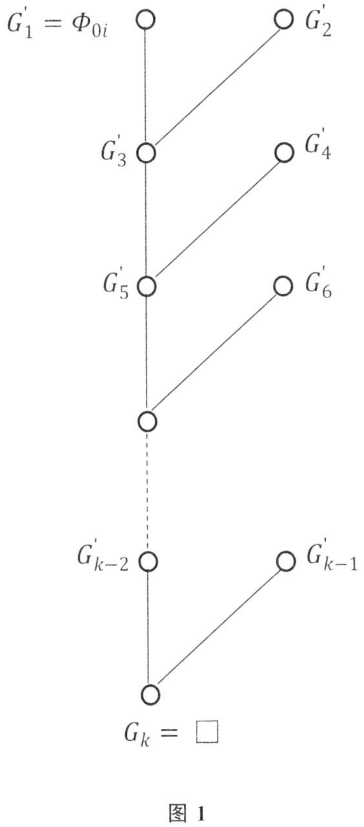

图 1

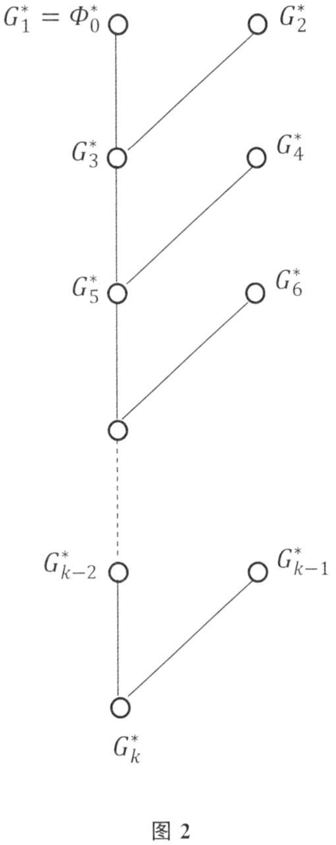

图 2

下面我们归纳定义对应于 D' 的广义子句序列 D^*_1（如图 2）：

$$G^*_1, \cdots, G^*_k.$$

(1) $G^*_1 = \Phi^*_0$.

(2) 假设 G^*_i $(1 \leqslant i \leqslant h, h \leqslant k-1)$ 已经定义.

若 $G'_{h+1} = \Phi_{pq}$，则令 $G^*_{h+1} = \Phi^*_p$，$0 \leqslant p \leqslant n$.

若

$$G'_{h+1}=R(G'_j, G'_l)=G'_j(A=0)\vee G'_l(A=1),$$

其中 $1\leqslant j, l\leqslant h$，则令

$$G^*_{h+1}=R(G^*_j, G^*_l)=G^*_j(A=0)\vee G^*_l(A=1).$$

显然，D^*_1 是以 Φ^*_0 为顶的从 S^* 出发推出 G^*_k 的一个线性演绎.

下面用归纳法证明：对任意 i，$1\leqslant i\leqslant k$，有

$$G^*_i\equiv G'_i\wedge H_i.$$

其中 H_i 是一个合取范式.

当 $i=1$ 时，$G^*_1=\Phi^*_0=\Phi_{01}\wedge\cdots\wedge\Phi_{0m_0}=G'_1\wedge H_1$.

如果 $i\leqslant h$ 时，命题已成立.

设 $i=h+1$. 若 $G^*_{h+1}=\Phi^*_p$，则 $G'_{h+1}=\Phi_{pq}$，故 $G^*_{h+1}=G'_{h+1}\wedge H_{h+1}$.

若 $G^*_{h+1}=R(G^*_j, G^*_l)=G^*_j(A=0)\vee G^*_l(A=1)$，$1\leqslant j, l\leqslant h$.

由归纳假设知，$G^*_j\equiv G'_j\wedge H_j$，$G^*_l\equiv G'_l\wedge H_l$，，所以

$$\begin{aligned}G^*_{h+1}&\equiv(G'_j\wedge H_j)(A=0)\vee(G'_l\wedge H_l)(A=1)\\&\equiv(G'_j(A=0)\wedge H_j(A=0))\vee(G'_l(A-1)\wedge H_l(A=1))\\&\equiv(G'_j(A=0)\vee G'_l(A=1))\wedge H_{h+1}\\&\equiv G'_{h+1}\wedge H_{h+1}.\end{aligned}$$

归纳法完成.

因为 $G^*_k\equiv G'_k\wedge H_k$，而 G'_k 是零子句，由命题 3，4 知，存在以 G^*_k 为顶，从 G^*_k 出发推出零子句的线性演绎 D^*_2.

将 D^*_1，D^*_2 连结起来，得以 Φ^*_0 为顶，从 S^* 出发推出零子句 G^* 的线性演绎 D^*.

对于广义子句集 S^* 与 S，与 D^* 等价的一个演绎 D，是一个以 Φ_0 为顶，从 S 出发推出广义子句 G 的一个线性演绎.

由命题 3 知，$G\equiv G^*$，故 G 是零子句. 定理证毕.

定理 6 设 S 是不可满足广义子句集，Φ_0 是 S 中一个广义子句. 如果 $S-\{\Phi_0\}$ 是可满足的，则存在以 Φ_0 为顶，从 S 推出零子句的线性演绎.

证 因为 S 不可满足，由 Herbrand 定理知，存在不可满足的 S 的有限基例集 S'. 显然，至少有一个 Φ_0 的基例 Φ'_0 在 S' 中.

不妨设 $S'-\{\Phi'_0\}$ 是可满足的. 由定理 5 知，存在以 Φ'_0 为顶，从 S' 推出零子句的线性演绎 D'. 应用提升引理，由 D' 可得以 Φ_0 为顶，从 S 出发推出零子句的线性演绎 D.

五、广义语义归结方法

设 S 是广义基子句集，I 是 S 的一个解释. 由命题 4 知，零子句不能由 S 中在 I 下为真的子句，通过广义归结方法得到. 所以，引入广义语义归结方法是有效的.

定义 设 E_1，…，E_q，N，$q\geqslant 1$，是一个广义子句序列，I 是这些子句的一个解释，P 是这些子句中谓词符号的一个顺序. 称（E_1，…，E_q，N）为关于 P 和 I 的语义互撞（简称 PI -互撞），当且仅当 E_1，…，E_q，N 满足下面条件：

(1) E_i 在 I 下其值为 0，$i=1$，…，q .

(2) 令 $R_1=N$ ，对每个 $i=1$，…，q ，存在 R_i 和 E_i 的广义归结式 R_{i+1}.

(3) E_i 中的归结原子是 E_i 中最大谓词符号.

(4) R_{q+1} 在 I 下其值为 0.

其中 E_1，…，E_q 称为此互撞的电子，N 称为此互撞的核，R_{q+1} 称为此互撞的 PI -归结式.

定义 设 I 是广义子句集 S 的一个解释，P 是出现在 S 中谓词符号的一个顺序. 从 S 出发的一个演绎称为 PI -演绎，当且仅当演绎中的每一个广义子句，或者是 S 中子句，或者是一个 PI -归结式.

定理 7 设 S 是广义基子句集，I 是 S 的一个解释，P 是出现在 S 中的谓词符号的一个顺序. 如果 S 是不可满足的，则存在从 S 推出零子句的 PI -演绎.

证 设 M 是 S 的原子集，对 M 的元素数用归纳法.

设 M 的元素数为 1. 令 $M=\{A\}$.

取 $I_1=\{\sim A\}$ ，因 I_1 弄假 S ，所以 S 中必有广义子句 Φ ，使 $\Phi(A=0)$ 为零子句.

取 $I_2=\{A\}$，因 I_2 弄假 S，所以 S 中必有广义子句 Ψ，使 $\Psi(A=1)$ 为零子句.

对 S 的解释 I（无非是 I_1 或（I_2），Φ，Ψ 中必有一个在 I 下其值为 0，不妨设为 Φ，于是

$$R(\Phi, \Psi)=\Phi(A=0) \vee \Psi(A=1)=\text{零子句}.$$

显然（Φ，Ψ）是 PI -互撞，其 PI -归结式是零子句.

如果 M 的元数 $<n$ 时，定理成立.

设 M 的元数 $=n$.

（1）若 S 中含有在 I 下其值为 0 的单元广义子句（即，只含有一个原子的广义子句），令为 $\Phi_u(A)$，不妨设 I 指定 A 为 0. 令

$$S'=\{\Phi' \mid \Phi \in S \wedge \Phi'=\Phi(A=1)\}.$$

显然，S' 不可满足，且 S' 的原子集元数 $<n$，由归纳假设，存在从 S' 推出零子句的 PI' -演绎 D'，其中 $I' \cup \{\sim A\}=I$.

对演绎树 D' 中每一个互撞（E'_1，…，E'_q，N'），其中 E'_1，…，E'_q，N' 是附着在 D' 中初始节点上的广义子句：

如果 $E'_i=E_i(A=1)$，其中 $E_i \in S$，则在 E'_i 的节点上面附着一个 PI -互撞

$$(\Phi_u, E_i)=\Phi_u(A=0) \vee E_i(A=1).$$

如果 $N'=N(A=1)$，其中 $N \in S$，则用 PI -互撞（Φ_u，E'_1，…，E'_q，N）代替 PI' -互撞（E'_1，…，E'_q，N'）.

于是，演绎树 D' 经上面修改后，成为从 S 出发推出零子句的 PI -演绎.

（2）若 S 中不含有在 I 下其值为 0 的单元广义子句，则取原子集 M 中元素 A，使 A 是 M 中的最小谓词符号.

不妨设 I 指定 A 为 0，令

$$S'=\{\Phi' \mid \Phi \in S \wedge \Phi'=\Phi(A=0)\}.$$

显然，S' 不可满足，且 S' 的原子集元数 $<n$，由归纳假设，存在从 S' 推出零子句的 PI' -演绎 D'，其中 $I' \cup \{\sim A\}=I$.

将演绎树 D' 的初始节点上的所有子句 $\Phi'=\Phi(A=0)$ 换为子句 Φ，因为 A 是 S 中的最小谓词符号，所以，原来 D' 中的 PI' -互撞，经此修改后，仍是一个 PI -互撞，

故经此修改后的演绎树 D'，变成一个从 S 出发推出零子句或推出含原子 A 的单元子句 Φ_u 的 PI -演绎 D_1.

若 D_1 是前者，则归纳法完成.

若 D_1 是后者，则考虑广义子句集 $S \cup \{\Phi_u\}$，因为 $\Phi_u(A=0)$ 是零子句，所以，由（1）知存在从 $S \cup \{\Phi_u\}$ 推出零子句的 PI -演绎 D_2.

将 D_1，D_2 连结起来，得从 S 推出零子句的 PI -演绎 D. 定理证毕.

定理 8 设 S 是广义子句集，I 是 S 的一个解释，P 是出现在 S 中的谓词符号的一个顺序. 如果 S 是不可满足的，则存在从 S 推出零子句的 PI -演绎.

类似于文献［2］中关于此定理的习惯证法，使用 Herbrand 定理和提升引理，不难证明此定理. 故略去.

参考文献

［1］王湘浩，广义归结，计算机科学暑期讨论会资料，1979，7.

［2］C. L. Chang，R. C. T. Lee，Symbolic Logic and Mechanical Theorem Proving，Academic Press，New York，1973.

Generalized Resolution

Abstract

In this paper, we will generalize J. A. Robinson's resolution principle so that resolvents can be obtained not only from two disjunctive clauses, but also from any two matrices. We will also lock resolution, linear resolution and semantic resolution. These resolutions have all been proved complete. As it is more natural to retain the matrix forms in the statement of a theorem than to transform them into several disjunctive clauses, it is the authors' hope that generalized resolutions shall be useful in theorem proving by the natural deductive method.

关于归结原理中的取因子问题*

摘　要

本文提出了约化式的概念. 证明了：在归结原理中，使用约化式和二元归结式的概念，而不使用因子的概念，归结原理仍然是完备的.

一、引　言

1965 年由 Robinson 提出的关于定理机器证明的归结原理，可以应用到人工智能的很多领域中. 后来，人们又对归结原理做了各种改进（见文献［1］）. 不论是原始的还是各种改进的归结原理，两个子句进行归结时，不仅要考虑这两个子句的自身，还要考虑这两个子句的所有因子.

本文引进了约化的概念，并证明了在归结原理中，取因子这件事可以由约化过程代替，而一个子句的一次约化等价于该子句与前面演绎出的某子句的某个例进行一次不取因子的二元归结. 而约化的机器实现要比取因子的机器实现简单易行.

二、合一替换

定义 1　一个替换 σ 是形如

$$\{t_1/x_1, \cdots, t_n/x_n\}$$

的一个有限集合，其中 $x_1, \cdots, x_n$ 是变量符号，t_i 是不同于 x_i 的项（$i=1, \cdots, n$），并且 $x_1, \cdots, x_n$ 互不相同. 特别，没有元素的替换称为空替换，记以 ε。以 $t_1, \cdots, t_n$ 为 σ 的分子，$x_1, \cdots, x_n$ 为 σ 的分母.

定义 2　设替换

$$\sigma=\{t_1/x_1, \cdots, t_n/x_n\},$$

如果 $t_1, \cdots, t_n$ 是不同的变量符号，则称 σ 为一个改名替换，简称为改名.

---

*　本文由王湘浩、刘叙华共同创作，原载于《中国科学（A 辑）》1982 年第 11 期。

关于替换的乘法，合一替换，最一般合一替换等概念均见文献［1］. 本文将最一般合一替换简记为 mgu.

命题 1 若一组表达式 E_1，…，E_n 是可合一的，则它们的 mgu 除了相差一个改名外，是唯一确定的.

证 设 σ，λ 是 E_1，…，E_n 的两个 mgu，于是有替换 φ_1，φ_2，使得

$$\sigma=\lambda\cdot\varphi_1，\lambda=\sigma\cdot\varphi_2.$$

将 φ_2 中所有如下元素删除：若 $t/x\in\varphi_2$，并且 x 出现在 σ 的分母中，但不出现在 σ 的分子中. 如此得到替换 φ_2^*. 显然，仍有

$$\lambda=\sigma\cdot\varphi_2^*.$$

同理，可将 φ_1 处理成 φ_1^*，并且

$$\sigma=\lambda\cdot\varphi_1^*.$$

任取 $t/x\in\varphi_2^*$.

1）若 x 出现在 σ 的分子中，因为

$$\sigma=\sigma\cdot\varphi_2^*\cdot\varphi_1^*.$$

不妨设 σ 中元素 u/y 的分子中有 x，将 u 记以 $u(x)$，于是有：

$$u(x)=(u(x))^{\varphi_2^*\cdot\varphi_1^*}.$$

因此，t 必须是一个变量符号.

2）若 x 不出现在 σ 的分母中，则由 $\lambda=\sigma\cdot\varphi_2^*$ 知，$t/x\in\lambda$. 再由 $\sigma=\lambda\cdot\varphi_1^*$ 知，t 必是一个变量符号.

综合 1)，2)，φ_2^* 是如下一个替换：

$$\varphi_2^*=\{u_1/y_1，\cdots，u_k/y_k，v_1/z_1，\cdots，v_l/z_l\},$$

其中 u_1，…，u_k，v_1，…，v_l 是变量符号，z_1，…，z_l 不出现在 σ 的分母中，y_i，…，y_k 出现在 σ 的分母中，从而也就出现在 σ 的分子中.

由于 $\sigma=\sigma\cdot(\varphi_2^*\cdot\varphi_1^*)$，所以替换 $(\varphi_2^*\cdot\varphi_1^*)$ 的分母必然是既出现在 σ 的分母中，又不出现在 σ 的分子中. 因此，$(\varphi_2^*\cdot\varphi_1^*)$ 的分母中既没有 y_1，…，y_k，也没有 z_1，…，z_l. 所以，只有 u_1，…，u_k，v_1，…，v_l 互不相同，并且都出现在 σ 的分母中，才能找到 φ_1^*，使得 $\sigma=\sigma\cdot(\varphi_2^*\cdot\varphi_1^*)$.

故 φ_2^* 类是一个改名，并且 φ_2^* 的分子都出现在 σ 的分母中.

定义 3 设 σ 是表达式 E_1，…，E_n 的 mgu，E_1，…，E_n 中出现的所有变量为 x_1，…，x_m. 若 σ 的分母中没有除 x_1，…，x_m 以外的变量符号，则称 σ 为一个正规 mgu.

推论 1 设表达式 E_1，…，E_n 中出现的所有变量符号为 x_1，…，x_m. 若 σ 是 E_1，…，E_n 的正规 mgu，则 σ 的分子中没有除 x_1，…，x_m 以外的变量符号.

证 由文献［1］的 Unification 算法知，有 E_1，…，E_n 的一个正规 mgu λ，且 λ 的分子中也没有除 x_1，…，x_m 以外的变量符号.

由命题 1 知，$\lambda=\sigma\cdot\varphi$，并且 φ 是一个改名. 若 σ 的分子中有变量符号 $y\notin\{x_1, \cdots, x_m\}$，则显然 φ 是什么样的改名替换，也不能使 $\lambda=\sigma\cdot\varphi$ 成立.

命题 2 设

$$E_1, \cdots, E_m, E_{m+1}, \cdots, E_n \tag{1}$$

是一组表达式. 假如 E_1，…，E_m 有一个 mgu σ，于是，

$$E_1^\sigma, E_{m+1}^\sigma, \cdots, E_n^\sigma \tag{2}$$

是可合一的，当且仅当（1）是可合一的. 若 τ 是（2）式的 mgu，则 $\sigma\cdot\tau$ 是（1）的 mgu.

证 1）若（2）式是可合一的，设 θ 是（2）式的合一替换，于是有

$$E_1^{\sigma\cdot\theta}=E_{m+1}^{\sigma\cdot\theta}=\cdots=E_n^{\sigma\cdot\theta},$$

而 σ 是 E_1，…，E_m 的 mgu，所以

$$E_1^\sigma=\cdots=E_m^\sigma,$$

故 $E_1^{\sigma\cdot\theta}=\cdots=E_m^{\sigma\cdot\theta}$. 显然，$\sigma\cdot\theta$ 是（1）式的合一替换，即（1）式是可合一的.

2）若（1）式是可合一的，设 θ 是（1）式的合一替换，于是有

$$E_1^\theta=\cdots=E_m^\theta=E_{m+1}^\theta=\cdots=E_n^\theta$$

因此，θ 也是 E_1，…，E_m 的合一替换，故

$$\theta=\sigma\cdot\lambda.$$

于是，$E_1^{\sigma\cdot\lambda}=E_{m+1}^{\sigma\cdot\lambda}=\cdots=E_n^{\sigma\cdot\lambda}$，即，$\lambda$ 是（2）式的合一替换，亦即（2）式是可合一的.

3）若 τ 是（2）式的 mgu，则

$$E_1^{\sigma\cdot\tau}=E_{m+1}^{\sigma\cdot\tau}=\cdots=E_n^{\sigma\cdot\tau}.$$

而 $E_1^{\sigma}=\cdots=E_m^{\sigma}$，所以 $\sigma\cdot\tau$ 是（1）的合一替换.

设 θ 是（1）式的任意一个合一替换，当然 θ 也是 E_1，…，E_m 的合一替换，于是有

$$\theta=\sigma\cdot\varphi.$$

因为 $E_1^{\theta}=\cdots=E_{m+1}^{\theta}=\cdots=E_n^{\theta}$，所以，$\varphi$ 是（2）式的合一替换，故 $\varphi=\tau\cdot\psi$. 因此 $\theta=\sigma\cdot\tau\cdot\psi$，亦即，$\sigma\cdot\tau$ 是（1）式的 mgu.

命题 3 考虑下面两个表达式集合：

$$E_1,\ \cdots,\ E_m, \tag{3}$$

$$F_1,\ \cdots,\ F_n, \tag{4}$$

且（3）和（4）式无公共变量. 若 σ，τ 分别是（3），（4）式的正规 mgu，λ 是 E_1^{σ} 和 F_1^{τ} 的 mgu，则 $(\sigma\cup\tau)\cdot\lambda$ 是（3）和（4）式并在一起组成的表达式集合的 mgu.

证 设（3）式中所有变量符号为 x_1，…，x_n，（4）式中所有变量符号为 y_1，…，y_m. 再设

$$\sigma=\{t_1/x_1,\ \cdots,\ t_n/x_n\},$$

$$\tau=\{u_1/y_1,\ \cdots,\ u_m/y_m\},$$

其中项 $t_i(i=1,\ \cdots,\ n)$ 中只含 $x_j(j=1,\ \cdots,\ n)$，项 $u_i(i=1,\ \cdots,\ m)$ 中只含 $y_j(j=1,\ \cdots,\ m)$.

设 η 是 $\{(3),(4)\}$ 的一个合一替换，于是 η 既是（3）式的合一替换，也是（4）式的合一替换.

不妨设 η 的分母中只有 x_i 和 $y_j(1\leqslant i\leqslant n,\ 1\leqslant j\leqslant m)$（在整个证明完成后，可以看出，若 η 的分母中还有不同于 x_i 和 y_j 的变量符号 z，则将证明中的某些地方稍做修改即可）. 将 η 中分母为 x（意指 x_1，…，x_n）的元素组成的替换记为 η_1，分母为 y（意指 y_1，…，y_m）的元素组成的替换记为 η_2. 显然，η_1 是（3）式的合一替换，η_2 是（4）式的合一替换. 于是，存在替换 δ_1，δ_2，使得

$$\eta_1=\sigma\cdot\delta_1,\quad \eta_2=\tau\cdot\delta_2.$$

由于 η_1 的分母中只有 x，所以 δ_1 的分母中也只有 x；同理，δ_2 的分母中只有 y.

于是，

$(\sigma\cup\tau)\cdot(\delta_1\cup\delta_2)$

$=\{t_1^{(\delta_1\cup\delta_2)}/x_1,\ \cdots,\ t_n^{(\delta_1\cup\delta_2)}/x_n,\ u_1^{(\delta_1\cup\delta_2)}/y_1,\ \cdots,\ u_m^{(\delta_1\cup\delta_2)}/y_m\mid\delta_1,\ \delta_2\}^*$

$=\{t_1^{\delta_1}/x_1,\ \cdots,\ t_n^{\delta_1}/x_n,\ u_1^{\delta_2}/y_1,\ \cdots,\ u_m^{\delta_2}/y_m\mid\delta_1,\ \delta_2\}^*$

$=\{t_1^{\delta_1}/x_1,\ \cdots,\ t_n^{\delta_1}/x_n\mid\delta_1\}^*\cup\{u_1^{\delta_2}/y_1,\ \cdots,\ u_m^{\delta_2}/y_m\mid\delta_2\}^*$

$=\sigma\cdot\delta_1\cup\tau\cdot\delta_2$

其中 $\{S\mid\delta\}^*$ 表示从 $S\cup\delta$ 中删除如下元素所得的集合：首先删除 δ 中那样元素：其分母等于 S 中某元素的分母；其次删除 S 中那样元素：分子与分母相同.

令 $\delta=\delta_1\cup\delta_2$，于是，

$$\eta=\eta_1\cup\eta_2=\sigma\cdot\delta_1\cup\tau\cdot\delta_2$$
$$=(\sigma\cup\tau)\cdot(\delta_1\cup\delta_2)=(\sigma\cup\tau)\cdot\delta.$$

因为 η 是 $\{(3),(4)\}$ 的合一替换，所以 $E_1^{\eta}=F_1^{\eta}$，所以 $E_1^{(\sigma\cup\tau)\cdot\delta}=F_1^{(\sigma\cup\tau)\cdot\delta}$，故 $E_1^{\sigma\cdot\delta}=F_1^{\tau\cdot\delta}$. 即 δ 是 E_1^{σ} 和 F_1^{τ} 的合一替换，故有替换 ζ，使得

$$\delta=\lambda\cdot\zeta,$$

故得

$$\eta=(\sigma\cup\tau)\cdot\lambda\cdot\zeta,$$

而 $(\sigma\cup\tau)\cdot\lambda$ 是 $\{(3),(4)\}$ 的合一替换是显然的. 综合上述，$(\sigma\cup\tau)\cdot\lambda$ 是 $\{(3),(4)\}$ 的 mgu.

命题 4 设 $\{E_1,\ \cdots,\ E_n\}$ 是表达式集合. 若 $\{E_1,\ \cdots,\ E_n\}$ 是可合一的，则

1）令 $\sigma_1=\varepsilon$，对于 $i=1,\ \cdots,\ n-1$，$E_i^{\sigma_1\cdots\sigma_i}$ 和 $E_{i+1}^{\sigma_1\cdots\sigma_i}$ 是可合一的，设其 mgu 为 σ_{i+1}，

2）$\sigma_1,\ \cdots,\ \sigma_n$ 是 $\{E_1,\ \cdots,\ E_n\}$ 的 mgu.

证 对 n 用归纳法.

当 $n=2$ 时，命题显然成立.

如果对于 $(n-1)$ 命题成立，下面证明对于 n 命题也成立 $(n\geqslant 3)$.

由归纳假设知，对于 $i=1,\ \cdots,\ n-2$，$E_i^{\sigma_1\cdots\sigma_i}$ 和 $E_{i+1}^{\sigma_1\cdots\sigma_i}$ 是可合一的，设其 mgu 为

σ_{i+1} 于是，$\sigma_1 \cdot \cdots \cdot \sigma_{n-1}$ 是 $\{E_1, \cdots, E_{n-1}\}$ 的 mgu.

1）因为 $\{E_1, \cdots, E_{n-1}, E_n\}$ 是可合一的，所以由命题 2 知，$E_{n-1}^{\sigma_1 \cdots \sigma_{n-1}}$ 和 $E_n^{\sigma_1 \cdots \sigma_{n-1}}$ 是可合一的，设其 mgu 为 σ_n.

2）因为 $\sigma_1 \cdot \cdots \cdot \sigma_{n-1}$ 是 $E_1, \cdots, E_{n-1}$ 的 mgu，σ_n 是 $E_{n-1}^{\sigma_1 \cdots \sigma_{n-1}}$ 和 $E_n^{\sigma_1 \cdots \sigma_{n-1}}$ 的 mgu，由命题 2 知，$\sigma_1 \cdot \cdots \cdot \sigma_{n-1} \cdot \sigma_n$ 是 $E_1, \cdots, E_n$ 的 mgu.

命题 5 设 $E_1, \cdots, E_n$ 是一组表达式，E_i 和 E_j 无公共变量（$i \neq j$，$i, j = 1, \cdots, n$），$\theta_1, \cdots, \theta_n$ 是一组替换．若 $E_1^{\theta_1}, \cdots, E_n^{\theta_n}$ 是可合一的，则 $E_1, \cdots, E_n$ 是可合一的.

证明略.

三、用约化代替取因子

定义 4 设 C_1，C_2 是无公共变量的子句．L_1，L_2 分别是 C_1，C_2 中文字．如果 L_1 与 $\sim L_2$ 有 mgu σ，则子句

$$\boxed{L_1^\sigma} \cup (C_1^\sigma - L_1^\sigma) \cup \boxed{L_2^\sigma} \cup (C_2^\sigma - L_2^\sigma)$$

称为 C_1 和 C_2 的二元归结式，记为 $R(C_1, C_2)$．其中若 $(C_1^\sigma - L_1^\sigma)$ 和 $(C_2^\sigma - L_2^\sigma)$ 中有框文字，则一律删除.

定义 5 两个子句称为相等，当且仅当其中非框文字组成的子句相等.

定义 6 设 $C = \boxed{L'} \cup C' \cup \boxed{L''} \cup C''$ 是一个子句.

1）若 C' 中有文字 L 与 L' 可合一，且其 mgu 为 σ_1，则

$$\boxed{L'^{\sigma_1}} \cup (C'^{\sigma_1} - L^{\sigma_1}) \cup \boxed{L''^{\sigma_1}} \cup C''^{\sigma_1}$$

称为子句 C 的约化式，记以 $R(C)$.

2）若 C'' 中有文字 L 与 L' 可合一，且其 mgu 为 σ_2，则

$$\boxed{L'^{\sigma_2}} \cup C'^{\sigma_2} \cup \boxed{L''^{\sigma_2}} \cup (C''^{\sigma_2} - L^{\sigma_2})$$

也称为子句 C 的约化式，也记以 $R(C)$.

定义 7 设 S 是一个子句集，$C_1, \cdots, C_n$ 是一个子句序列．如果对每个 i（$i =$

1，…，n），C_i 或者属于 S，或者是 $C_j(j<i)$ 的约化式，或者是 C_l 和 $C_k(l<i, k<i)$ 的二元归结式（记以 $R(C_l, C_k)$），则称此序列为从 S 推出 C_n 的一个演绎.

命题 6 设 D 是一个演绎，C 是 D 中一个子句. 于是，C 的约化式必是 C 与前面演绎出的某个子句的例的二元归结式.

证 由于 C 能约化，所以 C 必是含有框文字的子句. 设

$$C=\boxed{L'}\cup C'\cup\boxed{L''}\cup C'',$$

并且 C'' 中文字 L 与 L'' 有 mgu λ，于是，

$$R(C)=\boxed{L'^{\lambda}}\cup C'^{\lambda}\cup\boxed{L''^{\lambda}}\cup(C''^{\lambda}-L^{\lambda})$$

由于约化并不增加也不减少子句的框文字，所以在演绎中，子句 C 的前面必有两个子句

$$C_p=L_p\vee C'_p,$$

$$C_q=L_q\vee C'_q,$$

使得 C_p 与 C_q 的二元归结式 $R(C_p, C_q)$ 经若干次（可以是零次）约化得到 C. 不妨设

$$R(C_p, C_q)=\boxed{L_p^{\sigma}}\cup(C_p^{\sigma}-L_p^{\sigma})\cup\boxed{L_q^{\sigma}}\cup(C_q^{\sigma}-L_q^{\sigma}).$$

设 $R(C_p, C_q)$ 经 n 次约化（对应的替换乘积为 η）得 C，于是经 $(n+1)$ 次约化（显然，对应的替换乘积应是 $\eta\cdot\lambda$）得 $R(C)$. 即

$$\begin{aligned}R^{n+1}(R(C_p, C_q))&=\boxed{L_p^{\sigma\cdot\eta\cdot\lambda}}\cup(C_p^{\sigma\cdot\eta\cdot\lambda}-S_{L_p})\cup\boxed{L_q^{\sigma\cdot\eta\cdot\lambda}}\cup(C_q^{\sigma\cdot\eta\cdot\lambda}-S_{L_q}).\\&=R(C)\end{aligned}$$

其中 1) S_{L_p} 是 C_p 中某些文字的集合，且 S_{L_p} 中任意文字 A，有 $A^{\sigma\cdot\eta\cdot\lambda}=L_p^{\sigma\cdot\eta\cdot\lambda}$.

2) S_{L_q} 是 C_q 中某些文字的集合，且 S_{L_q} 中任意文字 B，有 $B^{\sigma\cdot\eta\cdot\lambda}=L_q^{\sigma\cdot\eta\cdot\lambda}$.

3) $L_p^{\sigma\cdot\eta}=L'$，$L_q^{\sigma\cdot\eta}=L''$.

4) $C'^{\lambda}=(C_p^{\sigma\cdot n\cdot\lambda}-S_{Lp})$.

5) $(C''^{\lambda}-L^{\lambda})=(C_p^{\sigma\cdot n\cdot\lambda}-S_{Lq})$.

考虑子句 $C_p^{\sigma\cdot\eta}$ 与子句 C 的二元归结式，其中

$$C_p^{\sigma\cdot\eta}=L_p^{\sigma\cdot\eta}\vee C'^{\sigma\cdot\eta}_p,$$

$$C=\boxed{L'}\vee C'\vee\boxed{L''}\vee C''.$$

因为 $\sim L_P^{\sigma\cdot\eta}=\sim L'=L''$，而 L'' 与 C'' 中文字 L 有 mgu λ，所以

$$R(C_P^{\sigma\cdot\eta},\ C)=\boxed{L_P^{\sigma\cdot\eta\cdot\lambda}}\cup(C_p^{\sigma\cdot\eta\cdot\lambda}-L_P^{\sigma\cdot\eta\cdot\lambda})\cup\boxed{L^{\lambda}}\cup(C'^{\lambda}\cup C''^{\lambda}-L^{\lambda}).$$

因为 $(C_p^{\sigma\cdot\eta\cdot\lambda}-L_P^{\sigma\cdot\eta\cdot\lambda})$ 是将子句 C_p 经替换 $\sigma\cdot\eta\cdot\lambda$ 后，删除所有与 $L_P^{\sigma\cdot\eta\cdot\lambda}$ 相同的文字所得之子句，所以

$$(C_p^{\sigma\cdot\eta\cdot\lambda}-L_P^{\sigma\cdot\eta\cdot\lambda})\subseteq(C_p^{\sigma\cdot\eta\cdot\lambda}-S_{L_p})=C'^{\lambda}$$

而 $L^{\lambda}=\sim L_p^{\sigma\cdot\eta\cdot\lambda}\neq L_p^{\sigma\cdot\eta\cdot\lambda}$，所以 C_p 中所有经替换 $\sigma\cdot\eta\cdot\lambda$ 后，与 L^{λ} 相同的文字仍在子句 $(C_p^{\sigma\cdot\eta\cdot\lambda}-L_P^{\sigma\cdot\eta\cdot\lambda})$ 中，故

$$(C_p^{\sigma\cdot\eta\cdot\lambda}-L_P^{\sigma\cdot\eta\cdot\lambda})\cup(C'^{\lambda}-L^{\lambda})=C'^{\lambda},$$

故

$$\begin{aligned}R(C_P^{\sigma\cdot\eta},\ C)&=(C_p^{\sigma\cdot\eta\cdot\lambda}-L_P^{\sigma\cdot\eta\cdot\lambda})\cup(C'^{\lambda}\cup C''^{\lambda}-L^{\lambda})\\&=(C_p^{\sigma\cdot\eta\cdot\lambda}-L_P^{\sigma\cdot\eta\cdot\lambda})\cup(C'^{\lambda}-L^{\lambda})\cup(C''^{\lambda}-L^{\lambda})\\&=C'^{\lambda}\cup(C''^{\lambda}-L^{\lambda})=R(C),\end{aligned}$$

即子句 C 的约化式是子句 C_p 的例 $C_p^{\sigma\cdot\eta}$ 与子句 C 的二元归结式.

定义 7. 设子句集 $S=\{C_1,\ C_2\}$，子句序列 $C_1,\ C_2,\ C_3,\ C_4\cdots,\ C_n(n\geqslant 3)$ 称为一个从 S 出发的简单演绎，如果

1) $C_3=R(C_1,\ C_2)$

2) $C_{i+1}=R(C_i)$，当 $3\leqslant i\leqslant n-1$.

定理 1（提升引理） 若子句 C'_1，C'_2 分别是子句 C_1，C_2 的例，C' 是 C'_1 和 C'_2 的二元归结式，则存在子句 C，使 C' 是 C 的例，并且 C 可以从 $\{C_1,\ C_2\}$ 出发简单演绎出来.

证 1) 由文献 [1, 2] 中的提升引理知，存在子句 C，

$$C=(C_1^{\lambda_1\cdot\sigma}-\{L_1^1,\ \cdots,\ L_1^{r_1}\}^{\lambda_1\cdot\sigma})\cup(C_2^{\lambda_2\cdot\sigma}-\{L_2^1,\ \cdots,\ L_2^{r_2}\}^{\lambda_2\cdot\sigma})$$

（来说明的各符号如 L_1^1，…，$L_1^{r_1}$ 等含义见文献 [1, 2]），使得 C' 是 C 的例，亦即 $C'=C^{\eta}$. 其中，λ_1 是 $\{L_1^1,\ \cdots,\ L_1^{r_1}\}$ 的正规 mgu；λ_2 是 $\{L_2^1,\ \cdots,\ L_2^{r_2}\}$ 的正规 mgu；σ 是 $\{L_1^{1\lambda_1},\ \sim L_2^{1\lambda_2}\}$ 的 mgu.

2）因为$\{L_1^1, \ldots, L_1^{r_1}, \sim L_2^1, \cdots, \sim L_2^{r_2}\}$可合一，所以$\{L_1^1, \sim L_2^1, \cdots, \sim L_2^{r_2}\}$可合一.

由命题4知，可设$(L_1^1)^{\tau_0 \cdot \tau_1 \cdots \tau_{i-1}}$和$(\sim L_2^i)^{\tau_0 \tau_1 \cdots \tau_{i-1}}$的mgu为$\tau_i$，其中$\tau_0 = \varepsilon$，$i = 1, \cdots, r_2$.

令$\tau = \tau_0 \cdot \tau_1 \cdots \tau_{r_2}$，于是$\tau$是$\{L_1^1, \sim L_2^1, \cdots, \sim L_2^{r_2}\}$的mgu.

因为$\{\sim L_2^1, \cdots, \sim L_2^{r_2}, L_1^1, \cdots, L_1^{r_1}\}$可合一，而$\{\sim L_2^1, \cdots, \sim L_2^{r_2}, L_1^1\}$的mgu为$\tau$，由命题2知，$\{\sim L_2^{1^\tau}, L_1^{1^\tau}, \cdots, L_1^{r_1^\tau}\}$可合一.

由命题4知，可设$(\sim L_2^1)^{\tau \cdot \delta_0 \cdots \delta_{i-1}}$与$(L_1^i)^{\tau \cdot \delta_0 \cdots \delta_{i-1}}$的mgu为$\delta_i$，其中$\delta_0 = \varepsilon$，$i = 1, \cdots, r_1$. 显然，$\delta_1 = \varepsilon$.

令$\delta = \delta_0 \cdot \delta_1 \cdots \delta_{r_1}$，于是$\delta$是$\{\sim L_2^{1^\tau}, L_1^{1^\tau}, \cdots, L_1^{r_1^\tau}\}$的mgu.

由命题2知，$\tau \cdot \delta$是$\{\sim L_2^1, \cdots, \sim L_2^{r_2}, L_1^1, \cdots, L_1^{r_1}\}$的mgu.

3）因为λ_1是$\{L_1^1, \cdots, L_1^{r_1}\}$的正规mgu，$\lambda_2$是$\{L_2^1, \cdots, L_2^{r_2}\}$的正规mgu，$\sigma$是$\{L_1^{1^{\lambda_1}}, \sim L_2^{1^{\lambda_2}}\}$的mgu. 由命题3知，$(\lambda_1 \cup \lambda_2) \cdot \sigma$是$\{L_1^1, \cdots, L_1^{r_1}, \sim L_2^1, \cdots, \sim L_2^{r_2}\}$的mgu. 由命题1知，

$$(\lambda_1 \cup \lambda_2) \cdot \sigma = \tau \cdot \delta \cdot \varphi,$$

其中φ是一个改名.

由文献[1，2]知，$L_2^1, \cdots, L_2^{r_2}$是C_2中在$\lambda_2 \cdot \sigma$下与L_2^1可合一的所有文字，故C_2中在$\tau \cdot \delta$下和L_2^1可合一的所有文字也是$L_2^1, \cdots, L_2^{r_2}$，因此，C_2中在τ下和L_2^1可合一的所有文字也是$L_2^1, \cdots, L_2^{r_2}$. 同理，C_1中在$\tau \cdot \delta$下和L_1^1可合一的所有文字是$L_1^1, \cdots, L_1^{r_1}$.

4）因为L_1^1与$\sim L_2^1$的mgu为τ_1，所以，可做子句C_1和C_2的二元归结如下：

$$R(C_1, C_2) = \boxed{L_1^{1^{\tau_1}}} \cup (C_1^{\tau_1} - L_1^{1^{\tau_1}}) \cup \boxed{L_2^{1^{\tau_1}}} \cup (C_2^{\tau_1} - L_2^{1^{\tau_1}}).$$

由2）知，首先可对$R(C_1, C_2)$进行m次（$m \leqslant r_2$）约化，以求在子句C_2中删去文字$L_2^1, \cdots, L_2^{r_2}$，即

$$R^m(R(C_1, C_2)) = \boxed{L_1^{1^\tau}} \cup (C_1^\tau - L_1^{1^\tau}) \cup \boxed{L_2^{1^\tau}} \cup (C_2^\tau - \{L_2^1, \cdots, L_2^{r_2}\}^\tau).$$

其次，可再对 $R^m(R(C_1, C_2))$ 进行 n 次（$n \leqslant r_1$）约化，以求在子句 C_1 中删去文字 L_1^1，…，$L_1^{r_1}$，即

$$R^n(R^m(R(C_1, C_2))) = \boxed{L_1^{1^{\tau\cdot\delta}}} \cup (C_1^{\tau\cdot\delta} - \{L_1^1, \cdots, L_1^{r_1}\}^{\tau\cdot\delta}) \cup$$

$$\boxed{L_2^{1^{\tau\cdot\delta}}} \cup (C_2^{\tau\cdot\delta} - \{L_2^1, \cdots, L_2^{r_2}\}^{\tau\cdot\delta}).$$

由于 φ 是一个改名，所以，

$$((C_1^{\tau\cdot\delta} - \{L_1^1, \cdots, L_1^{r_1}\}^{\tau\cdot\delta}) \cup (C_2^{\tau\cdot\delta} - \{L_2^1, \cdots, L_2^{r_2}\}^{\tau\cdot\delta}))^\varphi$$

$$= (C_1^{\tau\cdot\delta\cdot\varphi} - \{L_1^1, \cdots, L_1^{r_1}\}^{\tau\cdot\delta\cdot\varphi}) \cup (C_2^{\tau\cdot\delta\cdot\varphi} - \{L_2^1, \cdots, L_2^{r_2}\}^{\tau\cdot\delta\cdot\varphi})$$

$$= (C_1^{\lambda_1\cdot\sigma} - \{L_1^1, \cdots, L_1^{r_1}\}^{\lambda_1\cdot\sigma}) \cup (C_2^{\lambda_2\cdot\sigma} - \{L_2^1, \cdots, L_2^{r_2}\}^{\lambda_2\cdot\sigma})$$

$$= C,$$

所以，

$$C = (R^n(R^m(R(C_1, C_2))))^\varphi,$$

故

$$C' = C^\eta = (R^n(R^m(R(C_1, C_2))))^{\varphi\cdot\eta},$$

亦即 C' 是 $R^n(R^m(R(C_1, C_2)))$ 的例. 而 $R^n(R^m(R(C_1, C_2)))$ 显然是从 $\{C_1, C_2\}$ 出发的一个简单演绎结果.

定理 2（使用约化的归结的完备性） 若 S 是不可满足子句集，则存在从 S 推出空子句□的一个演绎.

证 设 S 是不可满足的，由 Herbrand 定理，有 S 的基例集 S' 不可满足. 由文献[1] 中归结原理的完备性知，存在从 S' 推出□的归结演绎 D'. 并且 D' 中的归结式都是二元归结式，D' 中无约化式.

由定理 1（提升引理）知，可将 D' 提升为一个从 S 出发推出□的，其中只含有二元归结式和约化式的演绎 D.

由于我们定义的简单演绎是线性的，所以不难证明，只使用约化而不使用因子的线性归结是完备的.

在锁归结和语义归结中，使用约化而不使用因子的方法也不破坏它们的完备性.

参考文献

[1] Chang，C. L. & Lee，R. C. T.，Symbolic Logic and Mechanical Theorem Proving，Academic Press，New York，1973.

[2] 王湘浩、刘叙华，计算机学报，5（1982）2：81—92.

关于人工智能*

一、人工智能的性质

人类在生产劳动中，创造了各种各样的生产工具。所有这些工具本来都只是为了代替人类的某些体力劳动。人类的这个创造过程，到 18 世纪瓦特发明蒸汽机时，取得了重大成就，从而使第一次工业革命得以实现。

远在十七世纪，人们就提出过这样一个问题：能不能用工具或机器来代替人的某些脑力劳动？直到廿世纪，电子计算机诞生了，这一问题才有正面回答的实际可能，而人工智能这一学科才得以建立。

作为计算机科学的一个重要分支，人工智能的研究目的是提高计算机应用的灵巧性，也就是说，使计算机具有更多的智能因素。因此，人工智能是计算机应用研究的最前沿，人工智能的研究应以在计算机上实现各种应用系统为目标。人工智能也有它自身的理论，但它的理论研究，也是紧密围绕着这一目标而进行的。人工智能的一切成果都要经受实验的考验。只有用机器实现了的智能，才是人工智能。所以人工智能是一个实验性学科。

人工智能的研究领域极其广泛，它几乎涉及到人类创造的所有重要学科，诸如数学、物理、计算机科学、心理学、生理学、医学、语言学、逻辑学、经济、法律、哲学等等，因此，人工智能又是一门综合性的边缘学科因此，人工智能又是一门综合性的边缘学科。

二、人工智能的研究课题

人工智能研究的课题虽然很多，但从根本上来说，无非是如何处理知识的问题。

---

* 本文由王湘浩、李家治、管纪文、刘叙华、何志钧、石纯一、马希文共同创作，系《计算机科学》周年纪念征文，原载于《计算机科学》1983 年第 2 期。

即是：知识的获取，知识的表示和知识的利用这三个基本问题。

关于知识的获取，有如下课题：

1. 学习

2. 概念的形成

3. 自然语言理解

4. 模式识别与景物分析

5. 计算机视觉系统与听觉系统

关于知识的表示，有如下方法：

6. 用一阶逻辑描述知识

7. 语义网络法

8. 框架知识表示法

9. 产生式系统

关于知识的利用，有如下课题：

10. 定理证明和公式推演

11. 问题求解系统

12. 归纳推理

13. 程序正确性证明及自动程序设计

14. 专家咨询系统

15. 博奕

16. 作曲、绘画

17. 智能机器人

三、人工智能的研究现状

（一）一些重要的工作

1956 年到 1961 年，这是人工智能的形成时期。这个时期，重要的工作有以下几个。

1. 从 1956 年开始，Newell-Simon-Shaw 建立了一般问题求解程序 GPS。这个程序使用心理学方法，模拟数学家的思维过程来求解问题。其所能解的问题达十数种之多。这开创了定理机器证明中的启发式方法。

2. 1960 年 Mccarthy 提出了人工智能的程序语言 LISP。这种语言对以后人工智能的发展，起到了重要的作用。

3. 1955 年 Samuel 设计了一个下棋程序，这个程序具有自改善、自适应的学习能力。四年后，这个程序战赢了设计者本人。又三年后，这个程序战胜了美国一个保持了八年之久的常胜不败的世界冠军。这项工作，对机器模仿人的学习过程，进行了卓有成效的探索。

4. 1958 年 Rosenblatt 研制了一种具有学习能力的感知器，想制造出“类大脑”的计算机来。这是从结构上模拟人脑的一种尝试。

近年来，我国在非经典逻辑及知识表示方面开展了研究。北京大学马希文等研究了有关“知道”的模态逻辑，并研究了 LISP 语言。

（二）数学定理的机器证明

A. 问题的意义

人工智能研究中，碰到的第一问题是：什么是人的智能？人们对智能有着各种各样的定义，日本科学家渡边慧说：人的智能分为两种，一种是演绎能力，一种是归纳能力。

演绎是依靠某种逻辑进行的，而当前计算机是能进行逻辑运算的，因此可以想象，人类的演绎能力，至少在某种程度上是可以教给计算机的。但是，要想将依赖于经验（可能是很难说清楚的经验）的归纳能力教给计算机，则就是一个比较困难的问题了，因为使用归纳能力证明一个定理时，首先需要制定假说，然后对假说的置信度进行评价（亦即进行归纳）。但是，人是怎样制定假说的，直到目前，还拿不出像样的理论。可以想象，科学家在制定假说时的思维过程是相当复杂的。

无论如何，持各种不同观点的人都不否认，演绎能力是人类智能中很重要的一部分，而数学定理的证明是人类演绎能力的最集中表现。因此，使数学定理的证明实现机械化，应该是人工智能中的重要研究课题。况且，定理机器证明的重大进展是离不开将人的归纳思维机械化的。因此，定理机器证明的研究将关系着人工智能是否能够成功。

B. 方法与现状

数学定理的机器证明的研究，主要沿着下面几种路线进行着，并已取得一些成果。

1. 自然推导法：分析数学家证明数学定理时的思维过程，从而将这一过程编成程序，赋给机器，这是定理机器证明的研究中，关键而又困难的一条道路。

例如，给出一个公理集合 A，推理规则集合 I。在这个公理系统中要证明如下一个定理：在前提集合 P 下，有结论 C。

于是，计算机可以从 $A \cup P$ 出发，做各种组合，按照规则 I 得出一些结论。如果这些结论中有 C，则得到要证定理的一个证明；否则，将这些新得到的定理，做为新的公理或前提，再做各种组合，按照 I，再得出一些定理，……，重复此过程，直到得出结论 C 为止。

按照上面的试探过程，一级一级得出的新定理的数目，是按指数函数的速度增长的。因此得到一个定理的证明所需的时间，可能会是惊人的长。但是，如果人类证明，弄得巧，可能只需很短时间。

问题就在这里，因此，模拟人的那种抄近路的试样方法是必要的。

一种抄近路的方法是很有趣的。为了证明一个定理，先求出所谓只要它真就好办的中间定理，一旦找到这个中间定理，那么证明这个中间定理的道路就比证明原定理的道路近了很多。

但是，寻找中间定理的推理，是从特定的定理出发，反过来得出以这个特定定理为结论的前面级别的定理，这已经不是演绎，而是归纳了。因此，我们必须将数学家的一些发明手段，亦即启发式方法（Heuristic Method）编进计算机程序。

例如 Gelrnter 的平面几何定理的证明程序就是采用这种方法。

Newell 等人在 1956 年研制出的启发式程序，就把人在证明定理时所用的分解法，代入法，替换法等编入程序，根据机器内存贮的公理系统，让机器试探着去证明定理。这个程序证明了罗素的“数学原理”第二章中的定理。

1960 年，Slagle 编制了求解不定积分的启发式程序。这个程序具有人在求解不定积分时的思维技巧。该程序解决这类问题的能力达到了大学一，二年级的优秀学生水平。

1960 年 Newell 等人在以前工作的基础上，通过心理学实验，总结了人在解题过程中的共同思维活动。这种共同思维活动可归结为三个阶段：第一，想出大致的解题计划；第二，根据记忆中的理论和推理规则组织解题；第三，进行方法和目的分析。Newell 等人，根据这个规律编制了“通用解题程序”GPS，这个程序可解十几种性质不同的课题，使启发式程序有了较大普遍性。

吉林大学王湘浩、管纪文等编制出了解算术问题，求解高次方程问题，求三角方程的根，证明三角恒等式，进行代数式因式分解等初等数学的启发式程序，这些程序的解题能力都达到了优秀中学生水平。

2. 对一类问题找出统一的计算机上可实现的算法。对一类问题是否存在统一的算法解，这是数理逻辑长期研究的课题之一，即所谓判定问题。1936 年 Church 和 Turing 证明了一阶逻辑中判定问题是不可解的。但是，一阶逻辑中很多很有意义的子类，其判定问题却是可解的。

著名的如 Tarski 和 Mckinsey 证明的初等代数和初等几何判定法，Szmielew 证明的 Abel 群判定法等等。

这些算法虽然在理论上是能行的，但在实际上，即使电子计算机也难以实现。比如 Tarski 算法就有很多人做过研究，但仍未改进到足以在计算机上实现的程度。我国数学家吴文俊教授于 1977 年给出了一个在计算机上可实现的算法，对初等几何的一个很大子类（其中包括很多很困难的几何题目）能进行判定。

3. 定理证明器。众所周知，许多数学理论都可在一阶逻辑中得到表示，虽然一阶逻辑被证明其判定问题是不可解的。但是，1930 年 Herbrand 给出了一个一阶逻辑的半可判定算法。人们曾试图在计算机上实现 Herbrand 算法，结果效率低得使人无法忍受。

1965 年 Robinson 改进了 Herbrand 算法，提出了归结原理（Resolution）。这是一种有实用价值的算法。之后，人们对归结原理进行了各种改进，以提高其效率。著名的有 Slagle 的语义归结，Boyer 的锁归结和 Loveland，Luckham 的线性归结，吉林大学王湘浩、刘叙华等，也对归结原理做了很多改进，提出了广义归结，不取因子的归结和锁语义归结，并指出了国外提出的OL-归结的错误，并提出修改的线性有序归结（MOL-归结）。

最近R. S. Boyer和J. S. Moore两人已将归结法引入了定理证明器。使得定理证明器有了更高的效能。该系统在计算机上证明了一些较难的数学定理和程序的正确性，例如优化表达式编译程序的正确性证明，快速串搜索算法的正确性证明，唯一质因分解定理的证明。

4. 计算机辅助证明

直到目前为止，所有上述三种方法还都不能证明数学上较难和较大的定理。定理机器证明还没有取得令人信服的成果。因此，人机联合进行数学定理证明就是一条不容忽视的方向。由人来提出一些有创造意义的猜测或假设，由机器来处理大量的信息，这也许就能收到只用人，或者只用机器都做不出的很好的结果。这方面最令人信服的例子是，1976年美国Illinois大学的K. Appel，W. Haken，J. Koch使用机器，证明了一百多来年未能解决的难题——四色问题。

（三）“专家系统”

A. 什么是“专家系统”

在海上，一个巨大的石油钻机正在工作。突然，钻机自动关闭了。原来，这个复杂的电子机械系统中某处出了故障。那么，问题是否严重？采取什么相应对策？这些都必须迅速回答，以便采取紧急措施，对付危急的局面。在地质勘探方面，都是由地质学家审查来自矿床的技术数据，判断矿床是否有开采价值。还有许多类似的问题，一般都是由人类专家来处理。因为这些问题中包含大量只有领域专家才能“做决定”的问题。但是现在，这类“决定”可成功地用称为“专家系统”的新型计算机程序系统完成。

什么是一个“专家系统”？它是人工智能研究的分支之一，是具有大量专门知识的计算机程序系统。“专家系统”模拟人类专家“做决定”的过程来解决复杂的问题。在某些例子中，已超过人类专家的水平。

专家系统能从存贮的专门任务知识得出结论，因此，这样的系统又称为基于知识的专家系统。这样的系统包含大量与人类专家在他们专长领域中做决定时使用的相同种类的规则。系统中的知识编码成符号形式，由系统加以处理。系统通过逻辑的或可能性的推理得出结论。

专家系统的一个重要特性是“透明性”（Transporency）。——它能按用户要求用人类语言形式显示推理过程，即不仅提供准确的回答而且告诉你如何得到这一回答。这一点是很有意义的。

专家系统在探求结论时，不被限于某一特殊路径，而且在可选路径中进行探索。系统可对事实和假设加权，以便对给定的问题，在某种意义上，给出最恰当的结论。

B. 如何生成一个“专家系统”?

在人工智能的初期探索中，人们企图寻求一个通用的技术来征服问题解答。后来，这种做法在很大程度上停止了。当前探索集中在直接为特定任务研制高性能的问题解答程序。

建立一个专家系统的前提条件是存在领域专家，这一点是首先要考虑的。一个专家系统是否成功不仅依赖于是否“抓住”了书本知识而且依赖于人类专家的试探法——直觉，非正式规则，和经验知识。所有这些信息组成了专家系统的知识库。因此，我们可以说，一个专家系统的效能实际上取决于知识库的质量——它的完善性，可用性和确实性。

在研制一个专家系统时，首先要考虑的就是要建立一个高质量的知识库问题。编码专家知识的任务称为“知识工程”。知识工程师（计算机科学家）和研制者必须在一定程度上了解专家的领域，询问专家，把他们非正式规则转换成 IF-THEN 形式，最后把这些规则和书本知识编码到计算机程序中。很显然，知识工程要求与专家进行大量合作。而这一过程对专家也大为有益，使他们的专长和判断力得到加强，使他们更好地理解自己的课题，成为更好的专家。John G. Guschnig 说“研制一个专家系统，就是实行一个学科的编码。”

获得知识是建立专家系统的关键问题，因此，任何的知识获得工具都是有意义的。R. Reboh 为 Prospector 系统研制了一个程序，称为 KAS（Knowledge-Acquistion System），是一个知识获得工具，利用它可为 Prospector 扩充老模型生成新模型。

为了使专家系统能模拟人类专家的做决定的过程，系统必须是灵活的和可扩充的。这些特性使系统通过增加新规则和更正知识库而加以扩充。在这个意义下，系统不是处于“静止”状态，而是能“学习”。在构造专家系统时，我们也要考虑“学习”问题。

C. 专家系统有什么用途

专家系统首先设计用于帮助人们解决复杂的问题。现已广泛用于勘探与地质学，化学结构，医疗诊断，遗传工程，空中交通控制和商业等领域。

第一专家系统 Dendral，1965 在斯坦福大学开始研制。其目标是帮助有机化学家确定分子的键结构。对这一问题，传统的计算机方法无法应对，以致人们不得不寻求新的途径来解决这种艰巨的问题，这就是利用“基于知识的系统”——专家方法。这个系统获得很大成功。今天已在世界广泛应用。同时，又产生许多新的系统，例如，Meta-Dendral 等，对化学家们产生更大的吸引力。

七十年代中期，斯坦福大学又研制另一专家系统 MYCIN，其目的是诊断脑膜炎和其它细菌传染病并给出处置建议。脑膜炎特别适于用专家诊断系统，因为快速诊断和处置可使病人免于大脑被破坏或死亡。

第一个地质勘探专家系统——Propector 于 1976 年在斯坦福大学开始研制，其目标是为地质学家在勘探不同矿床时，如斑岩铜矿，镍硫化矿和铀矿等，提供某些建议。据文献报导，该系统预报了哥伦比亚地区铜矿床的位置。钻井后发现的矿床的实际轮廓线与预报相符。现在勘探专家系统已广泛用于矿床定位和石油井位等。例如，最近 Prospector 成功地预报了华盛顿州 Tolman 山区铜矿床位置。系统预报准确度达地质学家水平。

总之，专家系统以它的准确性和经济性吸引了人们的注意。可期望出现在下面领域显示威力：

1. 癌症化学治疗。以 Shortliffe 领导的斯坦福医学院的一个小组已研制了一个专家系统 Oncocin，用于辅助医生处置癌病人，指导复杂的药物治疗安排，使疗效最高，毒化作用最小。

2. 目标识别。斯坦福的 Feigenbaum 和 H. Penny Nii 等为美国联邦的 ARPA 完成一个分类目标的课题，其中包括生成一个专家系统 SU/X，它根据粗略的传感器信息来识别和跟踪某些移动的目标。

3. 空中交通控制。飞机调度控制是一个十分复杂的问题。现在伊利诺伊大学(Illinois)正在研制一个空中交通控制专家模型。

4. 化学研究。化学家常常需要快速确定和分析化学结构。通过专家系统，他们可利用最好的化学家的创造性。因此，专家系统已成为化学家构造和分析物质的强有力的工具。人工智能的应用研究，最接近实用的，是专家系统。各种专家系统正在不断涌现，犹如雨后春笋一般。我国在专家系统方面也已有一些工作，如吉林大学管纪文等和中医研究院朱仁康等合作的中医诊疗专家系统，浙江大学何志钧等的规则基专家咨询系统设计方法研究，清华大学林尧瑞、石纯一等关于课表管理和交通运输管理专家系统的研究，等等。

我们热切希望，我国能出现更多更好的实用系统，使人工智能为我国四化做出直接的贡献。

(四) 自然语言理解

在五十年代，人们曾热衷于机器翻译，认为不同的两种语言，其间 80%的词可以逐字逐词一对一地进行翻译，剩下的 20%可以用特殊的语法加以解决。结果遭到了失败。事实证明，这 20%很难解决，而且常出笑话。

但是，要普及计算机，让计算机走出少数专家才能使用的小圈子，就不可避免地要让计算机与人类日常使用的语言打交道，就离不开自然语言的处理。

这样，经过了六十年代初的一段沉默之后，这方面的工作又活跃起来了。人们认识到，没有语言的理解，语言的翻译是不会成功的。因此，从 60 年代开始，人们转向自然语言理解的研究。

近年来关于自然语言理解的一个相当成功的系统，是 Winograd 设计的 *W* 系统。这系统使用机器人来搬运各种形状、大小和颜色的积木块。人向机器人发命令和提问题，机器人接受命令处理积木块并作简单回答。人与机器之间用自然语言谈话。诚然，谈话的范围是限制了的，但机器所能理解的句子可以非常复杂。机器像人一样，用其全部智力和知识来理解语言，语义、语法和推理三者交叉使用。

目前，处理自然语言的研究，已从实验室走向工程实用。不少工程实用系统，都具有一定的处理自然语言的能力。此外，结合语音识别的成果，自然语言处理还应用于飞机订票系统以及家庭自动电话秘书等等方面。

世界上对英文的理解研究最早，欧洲各国及日本也都取得了一些成果。我国对机器理解汉语方面，也有所研究，并取得了一定的成果。

（五）机器人

美国麻省理工学院的机器人由一台 PDP-6 型计算机、一部电视摄像机和一只人工手臂组成。这只手能抓住和捡起各种尺寸的积木块，在一个给定的区域内组装成规定的形状。

日本日立中心研究所的机器人有两只眼，一只眼看图纸，另一只眼与机械手进行装配作业。依靠两只眼的协调配合，完成按图纸装配的工作。日立公司的一种具有视觉和触觉的机器人，已用来制造水泥杆，将螺钉拧到水泥杆的模型上。当机器人走近螺钉时，电视摄像机搜索目标，识别其形状与位置，再由带触角的手进行确认后，用扳手紧固螺钉。识别与紧固一个螺钉，用 2.5 秒时间。

美国斯坦福大学计算机科学系有一个人工智能实验室，主要从事景物分析研究，具体是搞智能机器人和人造手臂。这种机器人能对周围环境作简单判断，能自动确定本身的行动方向，人造手臂使用了神经元元件能自动选择控件位置，能自动装卸机件和螺径，还会写字。

最引人注目的是斯坦福大学的模拟猴子摘香蕉的智能机器人。1963 年 Mccarthy 教授，利用通用解题程序 GPS，在计算机上模拟了猴子摘香蕉问题。猴子从位置 1 走到位置 2，把箱子推到位置 3，爬上箱子，摘下挂在天花板上的香蕉。1969 年，他让机器人当猴子，进行了实验。房间里有一个高出地面的平台，平台上面有一只箱子。机器人走向平台。围绕平台转了 20 分钟，怎么也爬不上平台去。后来发现房角处有一块斜面板。于是走向斜面板，把它推到平台边，沿斜面爬上了平台，推下箱子。这表现了机器人利用简单工具解决问题的智能。

在人不宜去工作的地方，用机器人代替人去工作是有意义的。例如管道的探查和检修工作，或宇宙空间的工作等等。苏联的“月球20号”，是无人驾驶的。它降落在月球上，由机器人钻削岩石，并把岩石样品放入一个球形囊器，携回地球。

（六）自动程序设计

早期的计算机没有高级语言。人们必须通过机器代码来与计算机打交道。后来出现了算法语言，人们通过算法语言来打交道。机器代码枯燥无味，难学难懂。算法语言好多了，但仍很不方便。人们自然提出这样一个问题：如何根据原始问题的要求，自动地产生出直接与计算机打交道的解题程序来？人们希望，把当前的必须告诉计算机“如何解题”的状况，变为只要告诉计算机“解什么题”的状况。计算机内的智能系统将自动地安排出解决这个问题的具体细节，并最终生成可执行的程序。这当然是很困难的问题。但也已获得不少成果。在我国，有北京航空学院孙怀民等人利用子目标演绎和结构归纳法的自动编程技术。

（七）其它

用计算机进行作曲、绘画方面，国内也已经取得研究成果。山东翻译出古琴谱，浙江大学何志钧等用计算机设计图案应用于花布设计，吉林大学庞云阶等用计算机作国画的工作在国外受到很大重视。

四、人工智能研究动向

1977年以前，人工智能研究的重点是在定理证明和一般问题求解方面。从近三届（1977，1979，1981）人工智能国际会议提供论文的比例分析，研究重点有转向自然语言、机器视觉、学习与专家系统等方面。

专家系统的成就将更为可观。人工智能系统向实用性、综合性发展。

从人工智能这门学科本身的发展需要而论，应更多地注重智能机器人、学习机和自动程序设计的研究，以便从中发展人工智能的一般原理，使这门学科由经验阶段向理论阶段发展。

Factoring Problem in Resolution*

Abstract

This paper presents the concept of reductant and proves that if we use the concept of reductant rather than factor, then resolution is still complete.

Ⅰ. Introduction

The resolution principle in theorem proved by Robinson in 1965 can be applied to various areas in artificial intelligence. Later, many important refinements of resolution are presented[1]. In the original unrestrieted form of the principle and in these refinements, each time resolution is applied to two clauses, one must consider not only the clauses themselves, but also their factors. The main purpose of this paper is to show that factoring can be replaced by a reduction process which is equivalent to taking binary resolution of a clause and a clearly defined instance of another. Implementation of the reduction is simpler than that of factoring in machine.

Ⅱ. Unifiers

Definition 1. A substitution σ is a finite set of the form

$$\{t_1/x_1, \cdots, t_n/x_n\},$$

where every x_i is a variable, every t_i is a term different from x_i and no two elements in the set $\{x_1, \cdots, x_n\}$ are the same. Especially, the substitution that consists of no elements is called the empty substitution and is denoted by ε. The $\{t_1, \cdots, t_n\}$ is called numerator of σ, the $\{x_1, \cdots, x_n\}$ is called denominator of σ.

---

* 本文由王湘浩、刘叙华共同创作，原载于《中国科学（A 辑）》（*Science in China*, *Ser. A*）1983 年第 7 期。

Definition 2. Let $\sigma = \{t_1/x_1, \cdots, t_n/x_n\}$ be a substitution. If $t_1, \cdots, t_n$ are distinct variables, σ is said to be a renomination.

As to concepts of composition of two substitutions, unifier and the most general unifier, see [1].

In this paper, the most general unifier is denoted as mgu.

Proposition 1. *If a number of expressions* $E_1, \cdots, E_n$ *is unifiable, their mgu is uniquely determined except renomination.*

Proof. Let σ, λ be the mgu of $E_1, \cdots, E_n$. Thus, there are substitutions φ_1, φ_2, such that

$$\sigma = \lambda \cdot \varphi_1, \quad \lambda = \sigma \cdot \varphi_2.$$

These elements in φ_2 are deleted such that $t/x \in \varphi_2$, and x occurs in denominator of σ but does not occur in numerator of σ. Then we obtain substitution φ_2^*. Clearly,

$$\lambda = \sigma \cdot \varphi_2^*.$$

Simiarly, we can obtain φ_1^* from φ_1, and

$$\sigma = \lambda \cdot \varphi_1^*.$$

Let t/x be an arbitrary element in φ_2^*.

1) If x occurs in the numerator of σ, without loss of generality, we may assume that x occurs in the numerator of the element u/y, where $u/y \in \sigma$.

Since

$$\sigma = \sigma \cdot \varphi_2^* \cdot \varphi_1^*,$$

therefore

$$u(x) = (u(x))^{\varphi_2^* \cdot \varphi_1^*}.$$

Clearly, t must be a variable.

2) If x does not occur in the denominator of σ, since $\lambda = \sigma \cdot \varphi_2^*$, $t/x \in \lambda$. Because $\sigma = \lambda \cdot \varphi_1^*$, t must be a variable.

According to 1) and 2), $\varphi_2^* = \{u_1/y_1, \cdots, u_k/y_k, v_1/z_1, \cdots, v_l/z_l\}$ where $u_1, \cdots, u_k, v_1, \cdots, v_l$ are variables, $z_1, \cdots, z_l$ do not occur in the denominator of

σ, and y_1, …, y_k occur in the denominator of σ. Thus, y_1, …, y_k, occur in the numerator of σ.

Since $\sigma = \sigma \cdot (\varphi_2^* \cdot \varphi_1^*)$, the denominator of $(\varphi_2^* \cdot \varphi_1^*)$ must be in the denominator of σ, but not in the numerator of σ. Hence, there are neither y_1, …, y_k nor z_1, …, z_l in the denominator of $(\varphi_2^* \cdot \varphi_1^*)$. Therefore, only if u_1, …, u_k, v_1, …, v_l are different from one another and occur in the denominator of σ, we can find φ_1^* such that

$$\sigma = \sigma \cdot (\varphi_2^* \cdot \varphi_1^*).$$

Therefore, φ_2^* is a renomination, and all the numerators of φ_2^* occur in the denominator of σ.

Definition 3. Let σ be an mgu of expressions E_1, …, E_n, and let all variables occurring in E_1, …, E_n be x_1, …, x_m. If there is no variables other than x_1, …, x_m in the denominator of σ, then σ is called a normal mgu.

Corollary 1. *Let all variables occurring in E_1, …, E_n, be x_1, …, x_m. If σ is a normal mgu of E_1, …, E_n, there is no variables other than x_1, …, x_m in the numerator of σ.*

Proof. From the unification algorithm in [1], we know that there is a normal mgu λ of E_1, …, E_n, there are no variables other than x_1, …, x_m, in the numerator of λ.

According to the Proposition 1, $\lambda = \sigma \cdot \varphi$, and φ is a renomination. If there is variable $y \notin \{x_1, \cdots, x_m\}$ in the numerator of σ, clearly, no matter what renomination φ is, the expression $\lambda = \sigma \cdot \varphi$ could not hold.

Proposition 2. *Let*

$$E_1, \cdots, E_m, E_{m+1}, \cdots, E_n \tag{1}$$

be a set of expressions. Suppose E_1, …, E_m have an mgu σ. Then,

$$E_1^\sigma, E_{m+1}^\sigma, \cdots, E_n^\sigma \tag{2}$$

are unifiable if and only if (1) is unifiable. If τ is the mgu of (2), then $\sigma\tau$ is that *of* (1).

Proof. 1) If (2) is unifiable, let θ be a unifier of (2), then

$$E_1^{\sigma\cdot\theta}=E_{m+1}^{\sigma\cdot\theta}=\cdots=E_n^{\sigma\cdot\theta}.$$

Since σ is an mgu of E_1, …, E_m.

$$E_1^{\sigma}=\cdots=E_m^{\sigma},$$

therefore, $E_1^{\sigma\cdot\theta}=\cdots=E_n^{\sigma\cdot\theta}$. Clearly, $\sigma\cdot\theta$ is a unifier of (1). Namely, (1) is unifiable.

2) If (1) is unifiable, let θ be a unifier of (1), then

$$E_1^{\theta}=\cdots=E_m^{\theta}=E_{m+1}^{\theta}=\cdots=E_n^{\theta}.$$

Hence, θ is a unifier of E_1, …, E_m. Therefore

$$\theta=\sigma\cdot\lambda.$$

Clearly, $E_1^{\sigma\cdot\lambda}=E_{m+1}^{\sigma\cdot\lambda}=\cdots=E_n^{\sigma\cdot\lambda}$, namely, λ is a unifier of (2), i. e. (2) is unifiable

3) If τ is an mgu of (2),

$$E_1^{\sigma\cdot\tau}=E_{m+1}^{\sigma\cdot\tau}=\cdots=E_n^{\sigma\cdot\tau}$$

Since $E_1^{\sigma}=\cdots=E_m^{\sigma}$, therefore $\sigma\cdot\tau$ is a unifier of (1).

Let θ be an arbitrary unifier of (1), of course, θ is a unifier of E_1, …, E_m. Hence,

$$\theta=\sigma\cdot\varphi.$$

Since $E_1^{\theta}=\cdots=E_{m+1}^{\theta}=\cdots=E_n^{\theta}$, φ is a unifier of (2), namely, $\varphi=\tau\cdot\psi$. Hence, $\theta=\sigma\cdot\tau\cdot\psi$, i. e. $\sigma\cdot\tau$ is an mgu of (1).

Proposition 3. Consider the following two sets of expressions:

$$E_1,\ \cdots,\ E_m, \tag{3}$$

$$F_1,\ \cdots,\ F_n, \tag{4}$$

Suppose (3) *and* (4) *have no common variables. Let* $\sigma\cdot\tau$ *be normal mgu of* (3) *and* (4) *respectively, and* λ *be an mgu of* E_1^{σ}, F_1^{σ}. *Then*, $(\sigma\cup\tau)\lambda$ *is an mgu of expression set formed by the union of* (3) *and* (4).

Proof. Let all variables in (3) be x_1, …, x_n and all variables in (4) be

y_1, …, y_m. Let

$$\sigma = \{t_1/x_1, \cdots, t_n/x_n\},$$

$$\tau = \{u_1/y_1, \cdots, u_m/y_m\},$$

where term $t_i(i=1, \cdots, n)$ contains only $x_j(j=1, \cdots, n)$, and $u_i(i=1, \cdots, m)$ contains only $y_j(j=1, \cdots, m)$.

Let η be a unifier of { (3), (4) }, then η is a unifier of (3), and also of (4).

Without loss of generality, we may assume that there are only x_i and $y_j(1 \leqslant i \leqslant n, 1 \leqslant j \leqslant m)$ in the denominator of η, we denote the substitution $\eta_1(\eta_2)$, in which the substitution is composed of the elements of η whose denominator is $x_i(i=1, \cdots, n)(y_j(j=1, \cdots, m))$. Clearly, η_1 is a unifier of (3), and η_2 is a unifier of (4). Therefore, there are and δ_1 and δ_2 such that

$$\eta_1 = \sigma \cdot \delta_1, \quad \eta_2 = \tau \cdot \delta_2.$$

Since there are only x_1, …, x_n in the denominator of η_1, there are only x_1, …, x_n in the denominator of δ_1. Similarly, there are only y_1, …, y_m in the denominator of δ_2. Thus,

$$\begin{aligned}
&(\sigma \cup \tau) \cdot (\delta_1 \cup \delta_2) \\
&= \{t_1^{(\delta_1 \cup \delta_2)}/x_1, \cdots, t_n^{(\delta_1 \cup \delta_2)}/x_n, u_1^{(\delta_1 \cup \delta_2)}/y_1, \cdots, u_m^{(\delta_1 \cup \delta_2)}/y_m \mid \delta_1, \delta_2\}^* \\
&= \{t_1^{\delta_1}/x_1, \cdots, t_n^{\delta_1}/x_n, u_1^{\delta_2}/y_1, \cdots, u_m^{\delta_2}/y_m \mid \delta_1, \delta_2\}^* \\
&= \{t_1^{\delta_1}/x_1, \cdots, t_n^{\delta_1}/x_n \mid \delta_1\}^* \cup \{u_1^{\delta_2}/y_1, \cdots, u_m^{\delta_2}/y_m \mid \delta_2\}^* \\
&= \sigma \cdot \delta_1 \cup \tau \cdot \delta_2,
\end{aligned}$$

where $\{S \mid \delta\}^*$ is a set obtained from $(S \cup \delta)$ by deleting the following elements: First, those elements in δ whose denominator is equal to the denominator of some elements in S, then those in S whose numerator and denominator are the same.

Let $\delta = \delta_1 \cup \delta_2$, then

$$\eta = \eta_1 \cup \eta_2 = \sigma \cdot \delta_1 \cup \tau \cdot \delta_2 = (\sigma \cup \tau) \cdot (\delta_1 \cup \delta_2) = (\sigma \cup \tau) \cdot \delta.$$

Since η is a unifier of { (3), (4) }, $E_1^{\eta} = F_1^{\eta}$, $E_1^{(\sigma \cup \tau) \cdot \delta} = F_1^{(\sigma \cup \tau) \cdot \delta}$, $E_1^{\sigma \cdot \delta} = F_1^{\tau \cdot \delta}$. Hence,

δ is a unifier of E_1^{σ} and F_1^{τ}. Therefore there is a substitution ξ such that

$$\delta = \lambda \cdot \zeta.$$

Hence,

$$\eta = (\sigma \cup \tau) \cdot \lambda \cdot \zeta.$$

Clearly, $(\sigma \cup \tau) \cdot \lambda$ is a unifier of { (3), (4) }, therefore $(\sigma \cup \tau) \cdot \lambda$ is an mgu of { (3), (4) }.

Proposition 4. *Let* $\{E_1, \cdots, E_n\}$ *be a set of expressions. If* $\{E_1, \cdots, E_n\}$ *is unifiable, then*

1) *Let* $\sigma_1 = \varepsilon$, *then for* $i = 1, \cdots, n-1$, $E_i^{\sigma_1 \cdots \sigma_i}$, *and* $E_{i+1}^{\sigma_1 \cdots \sigma_i}$ *are unifiable, and assume their mgu to be* σ_{i+1}.

2) $\sigma_1, \cdots, \sigma_n$ *is an mgu of* $\{E_1, \cdots, E_n\}$.

Proof. We prove this proposition by induction on n.

If $n = 2$. Proposition 4 is obvious. Assume that proposition 4 holds for $(n-1)$. To complete the induction, we consider the case for $n (n \geqslant 3)$.

By the induction hypothesis, $E_i^{\sigma_1 \cdots \sigma_i}$ and $E_{i+1}^{\sigma_1 \cdot \cdots \cdot \sigma_i}$ are unifiable for $i = 1, \cdots, n-2$. Suppose their mgu is σ_{i+1}, then $\sigma_1, \cdots, \sigma_{n-1}$ is an mgu of $\{E_1, \cdots, E_{n-1}\}$.

1) Since $\{E_1, \cdots, E_{n-1}, E_n\}$ is unifiable, from Proposition 2, $E_{n-1}^{\sigma_1 \cdots \sigma_{n-1}}$ and $E_n^{\sigma_1 \cdots \sigma_{n-1}}$ are unifiable. Here suppose their mgu is σ_n.

2) Since $\sigma_1, \cdots, \sigma_{n-1}$ is an mgu of $E_1, \cdots, E_{n-1}$ and σ_n is an mgu of $E_{n-1}^{\sigma_1 \cdots \sigma_{n-1}}$ and $E_n^{\sigma_1 \cdot \cdots \cdot \sigma_{n-1}}$, from Proposition 2, $\sigma_1, \cdots, \sigma_{n-1}, \sigma_n$ is an mgu of $E_1, \cdots, E_n$.

Proposition 5. *Let* $E_1, \cdots, E_n$ *be a set of expressions,* E_i *and* E_j *have no common variables* $(i \neq j, i, j = 1, \cdots, n)$, *and* $\theta_1, \cdots, \theta_n$ *be a set of substitution. If* $E_1^{\theta_1}, \cdots, E_n^{\theta_n}$ *are unifiable,* $E_1, \cdots, E_n$ *are unifiable.*

The proof of Proposition 5 is omitted.

Ⅲ. Resolution by Reduction

Definition 4. Let C_1 and C_2 be clauses without common variables, L_1 and L_2 be literals in C_1 and C_2 respectively. If L_1 and $\sim L_2$ have an mgu σ, then clause

$$\boxed{L_1^\sigma} \cup (C_1^\sigma - L_1^\sigma) \cup \boxed{L_2^\sigma} \cup (C_2^\sigma - L_2^\sigma)$$

is called a binary resolvent of C_1 and C_2, denoted by $R(C_1, C_2)$, in which if there is any framed literal in $(C_1^\sigma - L_1^\sigma)$ and $(C_2^\sigma - L_2^\sigma)$, then delete them out.

Definition 5. Two clauses are taken to be equal, if and only if the clauses that consist of their unframed literals are equal.

Definition 6. Let $C = \boxed{L'} \cup C' \cup \boxed{L''} \cup C''$ be a clauses.

1) If a literal L in C' and L' have an mgu σ_1, then

$$\boxed{L'^{\sigma_1}} \cup (C'^{\sigma_1} - L^{\sigma_1}) \cup \boxed{L''^{\sigma_1}} \cup C''^{\sigma_1}$$

is a reductant of C, denoted by $R(C)$.

2) If a literal L in C'' and L'' have an mgu σ_2, then

$$\boxed{L'^{\sigma_2}} \cup C'^{\sigma_2} \cup \boxed{L'^{\sigma_2}} \cup (C''^{\sigma_2} - L^{\sigma_2})$$

is a reductant of C, denoted by $R(C)$.

Definition 7. Let S be set of clauses and C_1, …, C_n be a sequence of clauses. For each i $(i=1, \cdots, n)$, if C_i is a clause in S or a reductant of a deduction of C_j $(j < i)$, or a binary resolvent of C_l and C_k $(l < i, k < i)$, the sequence is a deduction of C_n from S.

Proposition 6. *Let D be deduction, and C be a clause in D. Then a reductant of C is a binary resolvent of C and an instance of some clause deduced before.*

Proof. Without loss of generality, we may assume that

$$C = \boxed{L'} \cup C' \cup \boxed{L''} \cup C'',$$

and the literal L in C'' and L'' has an mgu λ. Then,

$$R(C) = \boxed{L'^{\lambda}} \cup C'^{\lambda} \cup \boxed{L''^{\lambda}} \cup (C''^{\lambda} - L^{\lambda}).$$

Because the reduction would neither increase nor decrease the numbers of framed literal in clauses, in the deduction there must be two clauses before C:

$$C_p = L_p \vee C'_p,$$

$$C_q = L_q \vee C'_q,$$

such that C is obtained from $R(C_p, C_q)$ by reducing the clause some times (including zero time). Let

$$R(C_p, C_q) = \boxed{L_p^\sigma} \cup (C_p^\sigma - L_p^\sigma) \cup \boxed{L_q^\sigma} \cup (C_q^\sigma - L_q^\sigma)$$

Suppose C is obtained by reducing $R(C_p, C_q)$ n times, with the corresponding substitution being η. Through reduction $(n+1)$ times for $R(C_p, C_q)$, (with the corresponding substitution being $\eta \cdot \lambda$), we obtain $R(C)$, namely,

$$R^{n+1}(R(C_p, C_q)) = \boxed{L_p^{\sigma\cdot\eta\cdot\lambda}} \cup (C_p^{\sigma\cdot\eta\cdot\lambda} - S_{Lp}) \cup \boxed{L_q^{\sigma\cdot\eta\cdot\lambda}} \cup (C_q^{\sigma\cdot\eta\cdot\lambda} - S_{Lq}) = R(C)$$

where

1) S_{Lp} is a set of some literals in C_p and $A^{\sigma\cdot\eta\cdot\lambda} = L_p^{\sigma\cdot\eta\cdot\lambda}$ holds for any literal A in S_{Lp};

2) S_{Lq} is a set of some literals in C_q and $B^{\sigma\cdot\eta\cdot\lambda} = L_q^{\sigma\cdot\eta\cdot\lambda}$ holds for any literal B in S_{Lq};

3) $L_p^{\sigma\cdot\eta} = L'$, $L_q^{\sigma\cdot\eta} = L''$;

4) $C'^{\lambda} = (C_p^{\sigma\cdot\eta\cdot\lambda} - S_{Lp})$;

5) $(C''^{\lambda} - L^{\lambda}) = (C_q^{\sigma\cdot\eta\cdot\lambda} - S_{Lq})$;

Let us consider a binary resolvent of $C_p^{\sigma\cdot\eta}$ and C, in which

$$C_p^{\sigma\cdot\eta} = L_p^{\sigma\cdot\eta} \vee C'^{\sigma\cdot\eta}_p,$$

$$C = \boxed{L'} \vee C' \vee \boxed{L''} \vee C''.$$

Since $\sim L_p^{\sigma\cdot\eta} = \sim L' = L''$, L'' and L in C'' have an mgu λ,

$$R(C_p^{\sigma\cdot\eta}, C) = \boxed{L_p^{\sigma\cdot\eta\cdot\lambda}} \cup (C_p^{\sigma\cdot\eta\cdot\lambda} - L_p^{\sigma\cdot\eta\cdot\lambda}) \cup \boxed{L^{\lambda}} \cup (C'^{\lambda} \cup C''^{\lambda} - L^{\lambda}),$$

$(C_p^{\sigma\cdot\eta\cdot\lambda} - L_p^{\sigma\cdot\eta\cdot\lambda})$ is obtained from $C_p^{\sigma\cdot\eta\cdot\lambda}$ by deleting all literals that are the same as $L_p^{\sigma\cdot\eta\cdot\lambda}$, therefore,

$$(C_p^{\sigma\cdot\eta\cdot\lambda} - L_p^{\sigma\cdot\eta\cdot\lambda}) \subseteq (C_p^{\sigma\cdot\eta\cdot\lambda} - S_{Lp}) = C'^{\lambda}$$

Since $L^{\lambda} = \sim L_p^{\sigma\cdot\eta\cdot\lambda} \neq L_p^{\sigma\cdot\eta\cdot\lambda}$, if the literal L^{λ} occur in C'^{λ}, the literal L^{λ} must occur in

$(C_p^{\sigma\cdot\eta\cdot\lambda} - L_p^{\sigma\cdot\eta\cdot\lambda})$. Hence.

$$(C_p^{\sigma\cdot\eta\cdot\lambda} - L_p^{\sigma\cdot\eta\cdot\lambda}) \cup (C'^{\lambda} - L^{\lambda}) = C'^{\lambda}.$$

Therefore,

$$\begin{aligned} R(C_p^{\sigma\cdot\lambda}, C) &= (C_p^{\sigma\cdot\eta\cdot\lambda} - L_p^{\sigma\cdot\eta\cdot\lambda}) \cup (C'^{\lambda} \cup C''^{\lambda} - L^{\lambda}) \\ &= (C_p^{\sigma\cdot\eta\cdot\lambda} - L_p^{\sigma\cdot\eta\cdot\lambda}) \cup (C'^{\lambda} - L^{\lambda}) \cup (C''^{\lambda} - L^{\lambda}) \\ &= C'^{\lambda} \cup (C''^{\lambda} - L^{\lambda}) \\ &= R(C). \end{aligned}$$

Hence, the reductant of C is a binary resolvent of C and an instance $C_p^{\sigma\cdot\eta}$ of C_p.

Definition 7. Let $S = \{C_1, C_2\}$ be a set of clauses. A sequence of clauses C_1, C_2, C_3, C_4, …, $C_n (n \geqslant 3)$ is called a simple deduction from S, if and only if

1) $C_3 = R(C_1, C_2)$;

2) $C_{i+1} = R(C_i)$, when $3 \leqslant i \leqslant n-1$.

Theorem 1. *(Lifting Lemma). If C'_1, C'_2 are instances of C_1, C_2 respectively, and C' is a binary resolvent of C'_1 and C'_2, there is a simple deduction of C from $\{C_1, C_2\}$ and C' is an instance of C.*

Proof. 1) From Lifting Lemma in [1] and [2], we know that there is clause C,

$$C = (C_1^{\lambda_1\cdot\sigma} - \{L_1^1, \cdots, L_1^{r_1}\}^{\lambda_1\cdot\sigma}) \cup (C_2^{\lambda_2\cdot\sigma} - \{L_2^1, \cdots, L_2^{r_2}\}^{\lambda_2\cdot\sigma}),$$

such that C' is an instance of C, namely, $C' = C^{\eta}$, where λ_1 is a normal mgu of $\{L_1^1, \cdots, L_1^{r_1}\}$, λ_2 is a normal mgu of $\{L_2^1, \cdots, L_2^{r_2}\}$, σ is an mgu of $\{L_1^{1\lambda_1}, \sim L_2^{1\lambda_2}\}$.

2) Because $\{L_1^1, \cdots, L_1^{r_1}, \sim L_2^1, \cdots, \sim L_2^{r_2}\}$ are unifiable, $\{L_1^1, \sim L_2^1, \cdots, \sim L_2^{r_2}\}$ are unifiable.

From Proposition 4, Let mgu of $(L_1^1)^{\tau_0\cdot\tau_1\cdots\tau_{i-1}}$ and $(\sim L_2^i)^{\tau_0\cdot\tau_1\cdots\tau_{i-1}}$ be τ_i, where $\tau_0 = \varepsilon$, $i = 1, \cdots, r_2$, and also let $\tau = \tau_0 \cdot \tau_1 \cdot \cdots \cdot \tau_{r_2}$, then τ is an mgu of $\{L_1^1, \sim L_2^1, \cdots, \sim L_2^{r_2}\}$.

Since $\{\sim L_2^1, \cdots, \sim L_2^{r_2}, L_1^1, \cdots, L_1^{r_1}\}$ is unifiable, τ is an mgu of

$\{\sim L_2^1, \cdots, \sim L_2^r, L_1^1\}$, from Proposition 2, $\{\sim L_2^{1\tau}, L_1^{1\tau}, \cdots, L_1^{r_1\tau}\}$ are unifiable. From Proposition 4, let mgu of $(\sim L_2^1)^{\tau\cdot\delta_0\cdots\delta_{i-1}}$ and $(L_1^r)^{\tau\cdot\delta_0\cdots\delta_{i-1}}$ be δ_i, where $\delta_0=\varepsilon$, $i=1, \cdots, r_1$, then clearly, $\delta_1=\varepsilon$.

Let $\delta=\delta_0\cdot\delta_1\cdots\delta_{r_1}$, then δ is an mgu of $\{\sim L_2^{1\tau}, L_1^{1\tau}, \cdots, L_1^{r_1\tau}\}$. From Proposition 2, $\tau\cdot\delta$ is an mgu of $\{\sim L_2^1, \cdots, \sim L_2^{r_2}, L_1^1, \cdots, L_1^{r_1}\}$.

3) Since λ_1 is a normal mgu of $\{L_1^1, \cdots, L_1^{r_1}\}$, λ_2 is a normal mgu of $\{L_2^1, \cdots, L_2^{r_2}\}$, σ is an mgu of $\{L_1^{1\lambda_1}, \sim L_2^{1\lambda_2}\}$. From Proposition 3, $(\lambda_1\cup\lambda_2)\cdot\sigma$ is an mgu of $\{L_1^1, \cdots, L_1^{r_1}, \sim L_2^1, \cdots, \sim L_2^{r_2}\}$, and from Proposition 1,

$$(\lambda_1\cup\lambda_2)\cdot\sigma=\tau\cdot\delta\cdot\varphi,$$

where φ is a renomination.

From [1] and [2], $L_2^1, \cdots, L_2^{r_2}$ are all literals in C_2 and each of these literals and L_2^1 have a unifier $\lambda_2\cdot\sigma$. Therefore, all literals in C_2 with each of these literals and L_2^1 having a unifier $\tau\cdot\sigma$ are $L_2^1, \cdots, L_2^r$, too. Hence, all such literals in C_2 with each of these literals and L_2^1 having a unifier τ are $L_2^1, \cdots, L_2^r$. Similarly, all literals in C_1 with each of these literals and L_1^1 having a unifier $\tau\cdot\delta$ are $L_1^1, \cdots, L_1^{r_1}$.

4) Since τ_1 is an mgu of $\{L_1^1, \sim L_2^1\}$,

$$R(C_1, C_2)=\boxed{L_1^{1\tau_1}}\cup(C_1^{\tau_1}-L_1^{1\tau_1})\cup\boxed{L_2^{1\tau_1}}\cup(C_2^{\tau_1}-L_2^{1\tau_1}).$$

From 2) to delete literals $L_2^1, \cdots, L_2^{r_2}$ in C_2, we should first carry out $m(m\leqslant r_2)$ times reduction for $R(C_1, C_2)$, namely,

$$R^m(R(C_1, C_2))=\boxed{L_1^{1\tau}}\cup(C_1^\tau-L_1^{1\tau})\cup\boxed{L_2^{1\tau}}\cup(C_2^\tau-\{L_2^1, \cdots, L_2^{r_2}\}^\tau)$$

Then, we carry out $n(n\leqslant r_1)$ times reduction for $R^m(R(C_1, C_2))$ to delete literals $L_1^1, \cdots, L_1^{r_1}$ in C_1, namely,

$$R^n(R^m(R(C_1, C_2)))=\boxed{L_1^{1\tau\cdot\delta}}\cup(C_1^{\tau\delta}-\{L_1^1, \cdots, L_1^{r_1}\}^{\tau\delta})$$

$$\cup\boxed{L_2^{1\tau\delta}}\cup(C_2^{\tau\delta}-\{L_2^1, \cdots, L_2^{r_2}\}^{\tau\delta}).$$

As φ is a renomination, so

$$((C_1^{\tau\cdot\delta}-\{L_1^1, \cdots, L_1^{r_1}\}^{\tau\cdot\delta})\cup(C_2^{\tau\cdot\delta}-\{L_2^1, \cdots, L_2^{r_2}\}^{\tau\cdot\delta}))^{\varphi}$$

$$=(C_1^{\tau\cdot\delta\cdot\varphi}-\{L_1^1, \cdots, L_1^{r_1}\}^{\tau\cdot\delta\cdot\varphi})\cup(C_2^{\tau\cdot\delta\cdot\varphi}-\{L_2^1, \cdots, L_2^{r_2}\}^{\tau\cdot\delta\cdot\varphi})$$

$$=(C_1^{\lambda_1\cdot\sigma}-\{L_1^1, \cdots, L_1^{r_1}\}^{\lambda_1\cdot\sigma})\cup(C_2^{\lambda_2\cdot\sigma}-\{L_2^1, \cdots, L_2^{r_2}\}^{\lambda_2\cdot\sigma})=C.$$

Therefore,

$$C=(R^n(R^m(R(C_1, C_2))))^{\varphi}.$$

Namely,

$$C'=C_\eta=(R^n(R^m)R(C_1, C_2))))^{\varphi\cdot\eta}$$

We have proved that C' is an instance of $R^n(R^m(R(C_1, C_2)))$ which is deducted from $\{C_1, C_2\}$ by simple deduction.

Theorem 2 (*Completeness of Resolution by Reduction*) *If S is an unsatisfiable set of clauses, there is a deduction of the empty clause* $\square$ *from S.*

Proof. Since S is unsatisfiable, by Herbrand's theorem there is a finite unsatisfiable set S' of ground instances of clauses in S. From completeness of resolution in [1], there is a resolution deduction D' of $\square$ from S', and every resolvent in D' is a binary resolvent without reductants in D'.

By Theorem 1, we can lift D' to a deduction D of $\square$ from S in which only binary resolvent and reductant are contained. This completes the proof.

Because the simple deduction in this paper is linear, it is not difficult to prove that linear resolution is complete in which only reduction, not factor is used.

Using reduction instead of factoring in semantic resolution and lock resolution, they are still complete.

References

[1] Chang, C. L. & Lee, R. C. T., *Symbolic Logic and Mechanical Theorem Proving*, Academic Press, New York, 1973.

[2] 王湘浩，刘叙华，计算机学报，5 (1982)，2：81—92.

欧氏环中唯一分解定理的直接证明*

本文在三种不同定义的欧氏环中，用数学归纳法和最小数原理直接证明了唯一分解定理，以使近世代数教学中因子分解问题的处理有利于使学生加深对初等数学中有关问题的理解。这里所用的方法是 Zermelo 在整数环情形下著名证法的引申和变通。

一、问题的提出

数论中证明有理整数的唯一分解定理（即所谓算术的基本定理），通常是先证明

1. 任意二整数 a 和 b 有一个最大公约数 d 。

2. d 可以表为 $sa+tb$ 的形式。

3. 若素数 p 整除 ab ，则 p 整除 a 与 b 之一。

然后利用了证明分解的唯一性。但是，我们在小学算术中却是先学因数分解，然后学最大公约数及其简单求法。如果学得多一点，那就再学用辗转相除法求最大公约数。上述数论中的讲法不符合我们在初等数学中的学习过程，使我们感到是把事情完全弄颠倒了。由于人们对此不满意，就想了一些办法利用数学归纳法（或用最小数原理）直接证明唯一分解定理[1]，其中以 Zermelo 的证法[2] 最为著名。有了唯一分解定理，上面所说的 1 和 3 就成为十分明显的事实，至于辗转相除法不妨靠后一些再讲，那时再证明 2，并说明怎样简便地计算 s 与 t ，而这就涉及秦九韶的求一术了。按这样的顺序来讲就显得比较自然。

Hasse[3] 把 Zermelo 的方法用于域上的一元多项式环，因而在此情形下也可以在讲过唯一分解定理后再证明对应的 1 与 3。有了这两个例子，人们自然会提出下面的问题：这种方法能否推广到一般的欧氏环？

近世代数书讲因子分解时，通常是在主理想环上证明唯一分解定理，然后证明欧

* 本文原载于《辽宁师范大学学报（自然科学版）》1986 第 2 期。

氏环是主理想环，从而推出唯一分解定理对欧氏环成立。Van der Waerden[4] 为了避免选择公理，宁愿不讨论主理想环中元素分解为素元的可能性；但证明欧氏环中分解的唯一性时却仍以主理想环为桥梁。本文作者在辽师参加《近世代数》编写中感到，主理想环似乎和初等数学距离较远，以它为基础来讲唯一分解定理不易使学生加深对初等数学的理解。虽然主理想环较广一些[5]，但重要的还是欧氏环，学时不多的近世代数课似乎可以不讲主理想环而以欧氏环为本讲因子分解。如果认为学生也应该知道主理想环，不妨安排几个习题让他们自己来处理。若在欧氏环中可用数学归纳法或最小数原理直接证明唯一分解定理，则近世代数课中因子分解的讲法就与初等数学有更多的联系了。

二、欧氏环的不同定义

各近世代数书中欧氏环的定义是不很一致的，本文作者看到的有三种[4,6,7]。为了区别这三种欧氏环，我们引进下面的定义：

定义 1 设 R 是一个整环，δ 是 R 到非负整数集内的映射。δ 说是 R 中的一个欧氏范函数，如果

1）$\delta(a)=0$ 当且仅当 $a=0$

2）对 R 的任意元素 a，b，

$$\delta(ab)=\delta(a)\delta(b) \tag{1}$$

3）对 R 的任意非零元素 a 和任意元素 b 有 R 的元素 q 和 r 存在使

$$b=qa+r，\delta(r)<\delta(a) \tag{2}$$

若在 R 中取定了一个欧氏范函数，则 R 说是一个第一种欧氏环。

定义 2 设 R 是一个整环，ψ 是 R 的非零元素集到非负整数集内的映射。ψ 说是 R 中的一个欧氏阶函数，如果

i）若 b 是（$a\neq 0$）的因子，则

$$\psi(b)\leqslant\psi(a) \tag{3}$$

ii）对 R 的任意非零元素 a 和任意元素 b 有 R 的元素 q 和 r 存在使

$$b=qa+r，r=0 \text{ 或 } \psi(r)<\psi(a) \tag{4}$$

若在 R 中取定了一个欧氏阶函数，则 R 说是一个**第二种欧氏环**。

定义 3 设 R 是一个整环，φ 是 R 的非零元素集到非负整数集内的映射。φ 说是 R 中的一个欧氏标函数，如果对 R 的任意非零元素 a 和任意元素 b 有 R 的元素 q 和 r 存在使

$$b = qa + r,\ r = 0 \text{ 或 } \varphi(r) < \varphi(a) \tag{5}$$

若在 R 中取定了一个欧氏标函数，则 R 说是一个**第三种欧氏环**。

定理 1 若 ψ 是 R 中的欧氏阶函数，则 ψ 也是 R 中的一个欧氏标函数。若 δ 是 R 中的欧氏范函数，则 δ 在 R 的非零元素集上的限制 ψ 是 R 中的一个欧氏阶函数。

证 定理的第一句话自明。设 δ 是 R 中的欧氏范函数，ψ 是 δ 在 R 的非零元素集上的限制。由 3)，显然 ii) 成立。设 b 是 $a(\neq 0)$ 的因子。于是，有 c 使 $a = bc$。$\psi(a) = \delta(bc) = \delta(b)\delta(c) = \psi(b)\delta(c) \geqslant \psi(b)$，因为由 1)，$\delta(c) \geqslant 1$。这就证明了 i) 成立。

以下，我们分别对三种欧氏环用数学归纳法及最小数原理证明唯一分解定理。

三、第一种欧氏环

设 R 是第一种欧氏环，δ 是取定的欧氏范函数。先证明几个简单的定理：

定理 2 $\delta(e) = 1$ 当且仅当 e 为单位。

证 因为 $1 \cdot 1 = 1$，所以 $\delta(1)\delta(1) = \delta(1)$，因而 $\delta(1) = 1$。若 e 是单位，则 $ee^{-1} = 1$，因而 $\delta(e)\delta(e^{-1}) = \delta(1) = 1$，故 $\delta(e) = 1$。反之，设 $\delta(e) = 1$。有 q 及 r 使

$$1 = qe + r,\ \delta(r) < \delta(e) \tag{6}$$

既 $\delta(e) = 1$，$\delta(r)$ 必等于 0，故 $r = 0$，从而 $1 = qe$，这表示 e 是单位。

定理 3 若 b 是 $a(\neq 0)$ 的真因子，则

$$\delta(b) < \delta(a) \tag{7}$$

证 b 是 a 的真因子时，有非单位 c 使

$$a = bc \tag{8}$$

于是，$\delta(a) = \delta(b)\delta(c)$，而由定理 2，$\delta(c) > 1$，故 (7) 成立。

定理 4 若 $\delta(a) < \delta(b)$，$c \neq 0$，则 $\delta(ac) < \delta(bc)$

证 $c \neq 0$ 时 $\delta(c) > 0$。故若 $\delta(a) < \delta(b)$，两边乘以 $\delta(c)$ 得 $\delta(a)\delta(c) <$

$\delta(b)\delta(c)$ 所以 $\delta(ac)<\delta(bc)$ 。

设有等式

$$p_1p_2\cdots p_m=q_1q_2\cdots q_n \tag{9}$$

若 $m=n$ ，且可颠倒各因子的次序使两边对应位置的元素相伴，则（9）的两边说是本质相同，否则说是本质相异。

下面用 Hasse[3] 的方法证明唯一分解定理：

定理 5 在第一种欧氏环中，任意非零元素 x 可以唯一地分解为素元的乘积。

证 作下列归纳法假定：设 $a\neq 0$，设对所有满足

$$\delta(x)<\delta(a) \tag{10}$$

的非零元素 x ，定理成立。在此假定下，试证 $x=a$ 时定理也成立。若 a 是单位或是素元，定理当然对 a 成立。设 a 非单位非素元。于是，a 可分为真因子 b，c 的乘积：

$$a=bc \tag{11}$$

由定理 3，$\delta(b)<\delta(a)$ ，$\delta(c)<\delta(a)$ 。故由归纳法假定，b 与 c 可分为素元的乘积。把这两个乘积接起来就得到 a 的一个素元分解式。

今证 a 的素元分解式唯一。假若不然，则 a 有两个本质相异的素元分解式如下：

$$a=p_1p_2\cdots p_m=q_1q_2\cdots q_n \tag{12}$$

不妨设

$$\delta(p_1)\leqslant\delta(q_1) \tag{13}$$

p_1 和 q_1 必非相伴。因若 $q_1=p_1e$ ，e 是单位，则由（12）得

$$p_2\cdots p_m=(eq_2)\cdots q_n \tag{14}$$

命

$$b=p_2\cdots p_m \tag{15}$$

b 是 a 的真因子，故 $\delta(b)<\delta(a)$ 。但（14）两边是 b 的两个本质相异的素元分解式，此与归纳法假定矛盾。因之，p_1 和 q_1 非相伴。同理，p_1 和任意 q_1 非相伴。有 h 与 r 存在使

$$q_1=hp_1+r \tag{16}$$

$$\delta(r)<\delta(p_1) \tag{17}$$

因 p_1 不能整除 q_1，故由（16），

$$r \neq 0 \tag{18}$$

命

$$d = rq_2 \cdots rq_n \tag{19}$$

由（13），（17），$\delta(r) < \delta(q_1)$，故由定理 4，

$$\delta(d) < \delta(a) \tag{20}$$

以 $q_2 \cdots q_n$ 乘（16）的两边可见 p_1 整除 d，因而 d 有一个含 p_1 的素元分解式。但由（18），（17），p_1 不能整除 r，故 r 的分解式中不含与 p_1 相伴的素元，因而由（19），d 又有一个不含 p_1 的素元分解式。这就是说，d 有两个本质相异的素元分解式。由（20），此与归纳法假定矛盾。这就完成了归纳法。

四、第二种欧氏环

现在，设 R 是第二种欧氏环，ψ 是取定的欧氏阶函数。

定理 6 若 b 是 $a(\neq 0)$ 的真因子，则

$$\psi(b) < \psi(a) \tag{21}$$

证 设 b 是 $a(\neq 0)$ 的真因子。有 q 及 r 使

$$b = qa + r \tag{22}$$

$$r = 0 \text{ 或 } \psi(r) < \psi(a) \tag{23}$$

因 b 是 a 的真因子，故 a 不整除 b，因而 $r \neq 0$。

由（23）有

$$\psi(r) < \psi(a) \tag{24}$$

b 既是 a 的因子，由（22）知 b 是 r 的因子，故由 i）有

$$\psi(b) \leqslant \psi(r) \tag{25}$$

由（24）及（25）即得（21）。

现在证明唯一分解定理。分解的可能性可象第一种欧氏环中一样地证明，只不过在下面的证明中，我们用最小数原理代替数学归纳法。证明分解的唯一性却要用另外的方法，因为 Zermelo 及 Hasse 的证法在第二种欧氏环的情形下不适用。事实上，我

们推出（20）要用定理 4，而下例说明定理 4 对欧氏阶函数不成立：

例 1　设 R 是整数环。规定 ψ 如下：$a \neq 0$ 及 ± 21 时，$\psi(a)=|a|$；$\psi(\pm 21)=11$。不难说明，ψ 是 R 中的一个欧式阶函数。但是，

$$\psi(5)<\psi(7),\ \psi(5\times 3)>\psi(7\times 3)$$

定理 7　在第二种欧氏环 R 中，任意非零元素可以唯一地分解为素元的积。

证　先证分解的可能性。假定有的非零元素 x 不能分解为素元的乘积。命 a 为此种 x 中之 ψ 值最小者。a 当然非单位非素元，因为，单位及素元都算是已经分解为素元的乘积。因之，a 可分为真因子 b，c 的乘积：

$$a=bc \tag{26}$$

由定理 6，$\psi(b)<\psi(a)$，$\psi(c)<\psi(a)$。根据 $\psi(a)$ 的最小性知 b 与 c 都可分解为素元的乘积。这两个乘积相接乃是 a 的一个素元分解式，此为矛盾。

今证分解的唯一性。假定有的非零元素有本质相异的素元分解式。于是，可取本质相异但相等的两个素元乘积：

$$p_1p_2\cdots p_m=q_1q_2\cdots q_n \tag{27}$$

若在（27）中有的 p_i 与 q_j 相伴，则可以从两边消去 p_i。这样消下去，可使两边无相伴素元，因而不妨设本来（27）中任意 p_i 与 q_j 即非相伴。称这样一个等式（27）为一个反常等式，并称出现在反常等式中的素元为反常素元。设 p 为 ψ 值最小的反常素元，并设（27）为含 p 的一个反常等式，比方说

$$p=p_1 \tag{28}$$

在左边含 p 的所有反常等式中，不妨设（27）右边素元的个数最少。有 d 与 r 使

$$q_1=dp_1+r \tag{29}$$

$$r=0\ \text{或}\ \psi(r)<\psi(p_1) \tag{30}$$

因 p_1 与 q_1 非相伴，故 $r\neq 0$，因而由（30），

$$\psi(r)<\psi(p_1) \tag{31}$$

命

$$f=rq_2\cdots q_n \tag{32}$$

将 r 分解为素元的乘积：

$$r = r_1 r_2 \cdots r_k \tag{33}$$

以 $q_2 \cdots q_n$ 乘（29）的两边可看出 p_1 整除 f 。因之，f 有两个素元分解式：

$$p_1 s_1 s_2 \cdots s_k = r_1 r_2 \cdots r_k q_2 \cdots q_n \tag{34}$$

因 p_1 与 q_1 非相伴，故由（29），p_1 不整除 r ，因而 p_1 与（34）右边任意素元非相伴。所以（34）两边是本质相异的两个素元乘积。由（34）两边消去相伴素元可得一反常等式

$$p_1 s_1' s_2' \cdots = r_1' r_2' \cdots q_1' q_2' \cdots \tag{35}$$

其中 $s_1' s_2' \cdots$ ，$r_1' r_2' \cdots$ ，$q_1' q_2' \cdots$ 分别为由（34）作上述消去后 $s_1 s_2 \cdots s_k$ ，$r_1 r_2 \cdots r_k$ ，$q_2 \cdots q_n$ 之所剩者。(35）右边不能只剩下 $q_1' q_2' \cdots$ 而无 $r_1' r_2' \cdots$ ，因为，这与（27）右边素元个数最少矛盾。但这样，则（35）表示 r_1' 是一个反常素元。易见 $\psi(r_1') < \psi(p_1)$ ，此与 p_1 为 φ 值最小的反常素元矛盾。

五、第三种欧氏环

设 R 是第三种欧氏环，φ 是取定的欧氏标函数。

定理 8 若 b 是 $a(\neq 0)$ 的真因子，则有 b 的相伴元 b' 满足

$$\varphi(b') < \varphi(a) \tag{36}$$

证 作下列的归纳法假定：设 $a \neq 0$，设对所有适合

$$\varphi(x) < \varphi(a) \tag{37}$$

的 x 和 x 的真因子 y ，有 y 的相伴元 y' 满足

$$\varphi(y') < \varphi(x) \tag{38}$$

试证 b 是 a 的真因子时，也有 b 的相伴元 b' 满足（36）。b 既是 a 的真因子，故有 q 及非零元素 r 存在使

$$b = qa + r \tag{39}$$

$$\varphi(r) < \varphi(a) \tag{40}$$

由（39）知 b 整除 r 。若 b 与 r 相伴，则取 $b' = r$ ，（36）便成立。若 b 是 r 的真因子，则由（40）及归纳法假定，有 b 的相伴元 b' 满足 $\varphi(b') < \varphi(r)$ ，因而满足（36）。这就完成了归纳法。

下面的例子说明定理 6 对欧氏标函数不成立：

例 2 设 R 是整数环。规定

$$\varphi(a)=\begin{cases}a, & a>0\text{ 时}\\ |a|+100, & a<0\text{ 时}\end{cases}$$

易见 φ 是 R 中的一个欧氏标函数。今 -3 是 15 的真因子，但 $\varphi(15)=15$，$\varphi(-3)=103$。

定理 7 的证明只要稍作变动便适用于第三种欧氏环。

定理 9 在第三种欧氏环中，任意非零元素可以唯一地分解为素元的乘积。

证 我们来修改定理 7 的证明。首先，凡 ψ 当然都改为 φ。在（26）式后，可由定理 8 推出：有 b 的相伴元 b' 和 c 的相伴元 c' 满足 $\varphi(b')<\varphi(a)$，$\varphi(c')<\varphi(a)$。由此可见，b' 与 c' 都可分解为素元的乘积，因而 b 与 c 亦可分解为素元的乘积。这样修改后，证明的第一部分即可用于定理 9。证明的第二部分只须修改最后那两句话如下：易见有 r_1' 的相伴元 r_1'' 满足 $\varphi(r_1'')<\varphi(p_1)$，设 $r_1''=er_1'$，e 为单位。由（35）有

$$(ep_1)s_1's_2'\cdots=r_1''r_2'\cdots q_1'q_2'\cdots$$

因之，r_1'' 是一个反常素元，此与 p_1 为 φ 值最小反常素元矛盾。

六、比较

现在我们比较欧氏环的三种定义，看看它们的优缺点。

第一种欧氏环中，分解的可能性和唯一性可以用一次数学归纳法同时证明，毕其功于一役。在其余两种欧氏环中却要分开证明。但是 δ 的定义中要求过多，这对于唯一分解定理是很不必要的。欧氏环所概括的最重要的环是整数环和域 F 上的一元多项式环 $F[x]$，在 $F[x]$ 的情形下，不能直接规定 $f(x)$ 的次数为 $\delta(f(x))$，却要规定

$$\delta(f(x))=c^{\text{次}f(x)} \tag{41}$$

其中 c 是大于 1 的一个整数。这显得很不自然。

第二种欧氏环中，分解的可能性和唯一性要分开证明，这比第一种欧氏环中的证明复杂一些。但 ψ 的定义中要求比 δ 要少。

第二三两种欧氏环相比较，定理 8 不如定理 6 干净自然，从而定理 9 的证明也显

得有些绕弯子。欧氏标函数的定义要求最少，这是它的优点，但我们有下面的定理：

定理 10 设 φ 是 R 中的欧氏标函数。$a \neq 0$ 时规定

$$\psi(a) = \min_{x\text{与}a\text{相伴}} \varphi(x) \tag{42}$$

则 ψ 是一个欧氏阶函数。

证 先证 i)，设 b 是 $a(\neq 0)$ 的因子，要证

$$\psi(b) \leqslant \psi(a) \tag{43}$$

若 b 与 a 相伴，则据 ψ 的定义，$\psi(b)=\psi(a)$，故（43）成立。设 b 是 a 的真因子。由（42），有 a 的相伴元 a' 使 $\psi(a)=\varphi(a')$。b 也是 a' 的真因子，由定理 8，有 b 的相伴元 b' 使 $\varphi(b')<\varphi(a')$。因之，$\psi(b)=\psi(b')\leqslant\varphi(b')<\varphi(a')=\psi(a)$。

次证 ii)，设 $a\neq 0$，b 任意。据 ψ 的定义，有 a 的相伴元 ea 使 $\psi(a)=\varphi(ea)$。可取 q' 和 r 使

$$b = q'(ea) + r,\quad r = 0 \text{ 或 } \varphi(r) < \varphi(ea) \tag{44}$$

命 $q = q'e$，$b = qa + r$，$r = 0$ 或 $\psi(r) \leqslant \varphi(r) < \varphi(ea) = \psi(a)$，故（5）成立。

定理 10 表明，R 是第三种欧氏环时，只要用（42）把 φ“规范化”R 便在 ψ 之下成为第二种欧氏环。

参考文献

[1] G. H. Hardy and E. M. Wright, an Introduction to the Theory of Numbers 21—22, Clarendon Press, Oxford, 1954

[2] Zermelo, Gottinger Nachrichten, i (1934), 43—44

[3] H. Hasse, Zahlentheorie, 1949, 6—7

[4] B. L. Van der Waerden, Moderne Algebra, Leipzig, 62—66

[5] Motzkin, The Eucliolean Algorithm, Bull. Amer. Math. Soc. 55, 1142—1146, 1949

[6] 张禾瑞《近世代数基础》高等教育出版社 1978，138—139

[7] N. Jacobson, Lectures in Abstract Algebra, 1951, Vol. 1, p. 122

On Vincent's Theorem *

Vincent's theorem is the theoretical foundation on which the computer algorithms so far considered best to isolate polynomial real roots are based. In this paper, we present an improvement and an extension of Vincent's theorem. Several theorems are given about the number of coefficient sign variations of polynomials under continued fraction transformation. Using these theorems, a new, fast computer algorithm to isolate polynomial real roots has been designed and implemented.

Theorem 1. (Improved Vincent-Uspensky Theorem)

Let $p(x)=0$ be a polynomial equation of rational coefficients with degree $n>1$ and $p(x)$ square-free; let $\Delta>0$ be the minimum root separation of $p(x)$; let F_k be the Kth number of the Fibonacci sequence; let m be the smallest positive integer such that

$$F_{m-1}F_m\Delta>1+\frac{1}{\varepsilon_n} \tag{1}$$

where $\varepsilon_n=(1+1/n)^{\frac{1}{n-1}}-1$; let a_1, a_2, $\cdots$, a_m be arbitrary positive integers; then the transformation

$$x=a_1+\frac{1}{a_2}+\cdots+\frac{1}{a_m}+\frac{1}{y} \tag{2}$$

will eventually transform $p(x)=0$ into $\overline{p}(y)=0$ with at most one sigh variation in the coefficient sequence.

In the Vincent-Uspensky theorem [1], the m specified as in theorem 1 must also satisfy the condition " $F_{m-1}\Delta>2$" which is deleted in our theorem.

---

* 本文由王湘浩、陈建华共同创作，原载于《数学进展》1988 年第 1 期。

Now we extend the Vincent theorem to the case of $p(x)$ having multiple roots.

Theorem 2.

Let $p(x)=0$ be rational polynomial equation of degree $n>1$ and $\Delta>0$ be the minimum root separation of $p(x)$; let t be the smallest index such that

$$F_t F_{t-1}\Delta>1$$

and m' be the smallest index such that

$$m'>1+\mathrm{Log}\phi n/2,$$

Where $\phi=1.618\cdots$, F_k is the Kth number of the Fibonacci sequence; Iet $m=t+m'$. The transformation (2) turns $p(x)$ into $\bar{p}(y)=0$ with coefficient sign variation of $\bar{p}(y)$ as r. If $r=0$, $p(x)$ has no root in the open interval with endpoints p_m/q_m, p_{m-1}/q_{m-1}, and if $r>0$, $p(x)$ has a unique real root in the open interval and r is just the multiplicity of that root.

In the following, we denote the number of coefficient sign variations of real polynomial $p(x)$ as $V(p(x))$.

Theorem 3.

Given an integral polynomial $p(x)$ of degree n, and $g(x)=x^n p(1/x)$. Suppose $x_0=1$ be a root of $p(x)$ of multiplicity $r\geqslant 0$. Then

$$V(g(x+1))=V(p(x))-V(p(x+1))-r-2e,$$

where e is a non-negative integer.

Theorem 4.

Given a square-free, integral polynomial $p(x)$ of degree n and a continued fraction

$$a_1+\frac{1}{a_2}+\cdots+\frac{1}{a_m}+\cdots, \tag{3}$$

where a_1 is non—negative, a_2, ⋯, a_m positive integers. Transforming $p(x)=0$ by (3) gives out

$$p_i(x)=x^m p_{i-1}(a_i+1/x),\ p_0(x)=p(x),\ i=1,\ 2,\ \cdots,\ m,\ \cdots.$$

Then we have

$$V(p_0(x))\geqslant V(p_1(x))\geqslant\cdots\geqslant V(p_m(x))\geqslant\cdots, \tag{4}$$

and when m is large enough, $V(p_m(x))\leqslant 1$.

Theorem 5[2].

Given an integral polynomial $p(x)$. Suppose continued fraction (3) be a root of $p(x)$ of multiplicity $r\geqslant 0$. Transforming $p(x)=0$ by (3), we get $Pi(x)$ as described in theorem 4. Then (4) also holds, and when m is large enough, we have

$$V(p_m(x))=r$$

Theorem 6.

Let $p(x)$ be a real polynomial of degree n, then

$$V(p(x))+V(p(-x))\leqslant n.$$

Theorem 2 can be used in designing a new, fast computer algorithm for isolating real roots of integral polynomial equations. We have devised such algorithm [3] and implemented it on computer algebra system.

Reference

[1] Akritas, A. G., A Correction on a Theorem by Uspensky, Bull. of Greck Math. Soc., 19 (1978), 278—285.

[2] Wang Xianghao, A Method to Isolate Roots of Algebraic Equations, Jilin Univ. Academic Press, 1 (1960), 99—101.

[3] Chen Jianhua, A New Algorithm for the Isolation of Real Roots of Polynomial Equations, Proc. of 2nd Int. Conf. on computer and applications, Beijing, 1987.

A Factorization Problem *

Abstract

The factorization problem discussed in [1] is investigated further here. As a tool, a cycle algebra A is introduced, and the structure of its sub-algebras A, is determined. A method for finding all decompositions of an element into irreducible factors is given.

Key Words Automaton; Factorization; Cycle Algebra

Guan Chi-Wen analyzed in [1] the structure of the diagrams of non-singular linear autonomous automata. A factorization problem was proposed and discussed. The aim of the present paper is to investigate this factorization problem further, and to find a solution.

§ 1. Diagram-Type of a Non-Singular Autonomous Automaton

Let M be a non-empty finite set, whose elements are called states. Let σ be a transformation of M. The pair (M, σ) is called an **autonomous automaton.** The automaton (M, σ) is said to be **non-singular**, if σ is one-one, i. e.. if σ is a permutation.

Let (M, σ) be a non-singular autonomous automaton. Since σ can be expressed as a product of disjoint cyclic permutations, the states of the automaton (M, σ) fall into several disjoint cycles. Suppose the number of cycles having h states is c_h, $c_h \geqslant 0$. $h = 1, 2, \cdots, m$. The expression

$$c_1 z_1 + c_2 z_2 + \cdots + c_m z_m, \tag{1}$$

where z_h is a symbol denoting a typical cycle of h states, is called the **diagram-**

---

* 本文原载于《东北数学》1990 年第 6 卷第 1 期。

type of the automaton (M, σ).

Let (M, σ) and (N, τ) be non-singular autonomous automata. Form the Cartesion product $M \times N$. Define the permutation $\sigma \times \tau$ as follows: for $u \in M$, $\varepsilon \in N$,

$$(\sigma \times \tau)(u, \varepsilon) = (\sigma u, \tau\varepsilon).$$

The non-singular autonomous automaton $(M \times N, \sigma \times \tau)$ is called the product of the automata (M, σ) and (N, τ).

Suppose the diagram-types of (M, σ) and (N, τ) are respectively (1) and

$$d_1 z_1 + d_2 z_2 + \cdots + d_m z_m. \tag{2}$$

Let $(u, \tau) \in M \times N$. Suppose u is in a cycle z_i and τ is in a cycle z_j, Let $[i, j]$ and (i, j) denote respectively the l. c. m. and h. c. d. of i and j. Then, under $\sigma \times \tau$, (u, τ) is in a cycle $z_{[i, j]}$, and $z_i \times z_j$ falls into

$$ij/[i, j] = (i, j)$$

cycles $z_{[i, j]}$. It follows that the diagram-type of the product of (M, σ) and (N, τ) is the product of the diagram-types (1) and (2) of these automata, the latter product being formed by multiplying (1) and (2) out as in ordinary algebra, replacing $z_i z_j$, by $(i, j) z_{[i, j]}$, and collecting terms.

§ 2. The Cycle Semi-Ring and the Cycle Algebra

Consider all expressions of the form (1). We add these expressions as ordinary linear forms and multiply them as explained in § 1. The following laws hold:

$$\alpha + \beta = \beta + \alpha, \tag{3}$$

$$\alpha + (\beta + \gamma) = (\alpha + \beta) + \gamma, \tag{4}$$

$$\alpha\beta = \beta\alpha, \tag{5}$$

$$\alpha(\beta\gamma) = (\alpha\beta)\gamma, \tag{6}$$

$$\alpha(\beta + \gamma) = \alpha\beta + \alpha\gamma. \tag{7}$$

(3), (4), (5), (7) are obvious. Now,

$$z_i(z_j z_k)=(j, k)(i, [j, k])z_{[i, j, k]},$$

$$(z_i z_j)z_k=(i, j)([i, j], k)z_{[i, j, k]},$$

$$(j, k)(i, [j, k])=\frac{jk}{[j, k]}\frac{i, [j, k]}{[i, j, k]}=\frac{ijk}{[i, j, k]},$$

$$(i, j)([i, j], k)=\frac{ij}{[i, j]}\frac{[i, j], k}{[i, j, k]}=\frac{ijk}{[i, j, k]}.$$

Hence,

$$z_i(z)_j z_k)=(z)_i z_j)z_k, \tag{8}$$

and (6) follows from (8). We call the algebraic system S formed by the expressions (1) the cycle semi-ring.

Since it is assumed that in (1) not all the coefficients c_1, c_2, $\cdots$, c_m are zero, the semi-ring S contains no zero-element, It has the identity z_i, for $z_i z_j=(1, i)z_{[1, i]}=z_i$. So, (1) may be written as

$$c_1+c_2 z_2+\cdots+c_m z_m. \tag{9}$$

We call c_1 the constant term of (9).

If rational coefficients are allowed in the expression (9), and additions and multiplications are performed as above, then the resulted algebraic system A is a commutative associative algebra over the rational field Q, called the cycle algebra. A contains S as a sub-system.

A has no non-zero nilpotents. For, let

$$\alpha=c_i z_i+c_{i+1}z_{i+1}+\cdots+c_m z_m, \quad c_i\neq 0$$

Then, $\alpha^k=c_i^k i^{k-1}z_i+$ terms of higher z-subscrlpts. So, $\alpha^k\neq 0$.

Let k be any positive integer, The elements of the form

$$\alpha=\sum_{k\mid n}c_k z_k \tag{10}$$

in A form a sub algebre A_n of A of degrce $d(n)$ over Q, $d(n)$ being the number of divisors of n. Since there are no non-zero nilpotent elements, A_n is semi-simple, and is the direct sum of a number of fields. For our purpose, we shall find the explicit decomposition.

It is easily seen that

$$\frac{1}{k}z_k,\ 1-\frac{1}{k}z_k$$

are idempotent elements. So, the elements

$$x_k=\frac{1}{k}z_k\prod_{\substack{d\mid n\\ d>k}}\left(1-\frac{1}{d}z_d\right),\ k\mid n \tag{11}$$

are also idempotent elements. Expanding of the right hand side of (11) shows that $x_k=\frac{1}{k}z_k$ +terms of higher z-subscripts, and hence $x_k\neq 0$. Let $k\mid R$, $k\mid R$, $h<k$. The right hand side of (11) contains $1-\frac{1}{k}z_k$ as a factor. Since

$$z_k\left(1-\frac{1}{k}z_k\right)=0 \tag{12}$$

We have $z_k x_k=0$. Denote the elements $\frac{1}{k}z_k$ and the clements x_k, in increasing order of their subscripts by

$$y^{(1)},\ y^{(2)},\ \cdots,\ y^{(d(n))};$$
$$x^{(1)},\ x^{(2)},\ \cdots,\ x^{(d(n))}.$$

We have

$$\begin{aligned}x^{(1)}+x^{(2)}&=(1-y^{(2)})(1-y^{(3)})\cdots(1-y^{(d(n))})\\&\quad+y^{(2)}(1-y^{(2)})\cdots(1-y^{(d(n))})\\&=(1-y^{(3)})\cdots(1-y^{(d(n))}),\\x_1+x_2+x_3&=(1-y^{(3)})(1-y^{(4)})\cdots(1-y^{(d(n))})\\&\quad+y^{(3)}(1-y^{(4)})\cdots(1-y^{(d(n))})\\&=(1-y^{(4)})\cdots(1-y^{(d(n))})\end{aligned}$$

etc., and have finally

$$x^{(1)}+x^{(2)}+\cdots+x^{(d(n))}=(1-y^{(d(n))})+y^{(d(n))}=1.$$

So, we have proved

Theorem 1 The elements (11) form a set of orthogonal idempotent elements whose sum is 1:

$$x_k^2 = x_k$$

$$x_k x_k = 0, \quad \text{for } k \neq k,$$

$$\sum_{k \mid x} x_k = 1. \tag{13}$$

Theorem 1 implies that the $d(n)$ elements x_k are linearly independent over Q and hence form a basis of A_n over Q. So, any element $a \in A_n$, can be expressed uniquely as a linear combination of the x_k's.

Suppose

$$\alpha = \sum_{d \mid n} c_d z_d. \tag{14}$$

We have by (13)

$$\alpha = \sum_{k \mid n} \alpha x_k.$$

Now,

$$\alpha x_k = \frac{1}{h} \sum_{d \mid n} c_d z_d c_k \prod_{k < k \mid n} \left(1 - \frac{1}{k} z_k\right).$$

When $d \nmid k$, $[d, k] > k$, one of $1 - \frac{1}{k} z_k$ is$] - \frac{1}{[d, h]}$ So, by (12),

$$z_d z_k \prod_k \left(1 - \frac{1}{k} z_k\right) = (d, h) z_d z_k \prod_k \left(1 - \frac{1}{k} z_k\right) = 0.$$

When $d \mid k$, $z_d z_k = d z_k$,

$$z_d z_k \prod_k \left(1 - \frac{1}{k} z_k\right) = d z_k \prod_k \left(1 - \frac{1}{k} z_k\right).$$

Hence,

$$\alpha x_k = \frac{1}{h} \sum_{d \mid n} d z_d c_k \prod_k \left(1 - \frac{1}{k} z_k\right) = \left(\sum_{d \mid h} d c_d\right) x_k.$$

Writing

$$\alpha_k = \sum_{d \mid h} d c_d. \tag{15}$$

We have

$$\alpha=\sum_{d\mid h}\alpha_h x_h. \tag{16}$$

By Möbius inversion formula,

$$c_k=\frac{1}{h}\sum_{d\mid h}\mu\left(\frac{h}{d}\right)\alpha_d. \tag{17}$$

where μ is the Mobius function.

Theorem 2 A_n is the direct sum of the $d(n)$ fields Q_{xh}. each of which is a replica of Q. If α is given by (10) and a_h given by (15), then (16) holds; if α is given by (16) and c_k given by (l7), then (10) holds.

§ 3. Factorization in the Cycle Semi-Ring

Decomposition of an element in S into factors corresponds to that of a non-singular autonomous automaton into a product of automata. If not stated otherwise, all divisibility properties in this section will be understood to mean properties with respect to S. For example, for α, $\beta\in S$, $\alpha\mid\beta$ means that there is an element $\gamma\in S$ such that $\beta=\alpha\gamma$.

For given α, β, it is not difficult to decide whether $\alpha\mid\beta$, and when $\alpha\mid\beta$, to find γ such that $\beta=\alpha\gamma$. Let n be the l. c. m. of the k's for which c_k appears in the expressions of α, β. Then, α, $\beta\in A_n$. Express α, β in terms of the x_k's as follows:

$$\alpha=\sum_{k\mid n}\alpha_k x_k, \tag{18}$$

$$\beta=\sum_{k\mid n}b_k x_k \tag{19}$$

Since the z-coefficients of α, β are non-negative integers, so also are all the a_k's and b_k's. If there is a γ such that

$$\beta=\alpha\gamma \tag{20}$$

γ cannot contain any z_k, with $k\nmid n$, and with positive coefficient. For otherwise, β would contain such a z_k by (20), contrary to the fact that $\beta\in A_n$. Hence, $\gamma\in A_n$. Suppose

$$\gamma = \sum_{k \mid n} c_k x_k, \tag{21}$$

We have

$$b_k = a_k c_k. \tag{22}$$

Now, we consider the following three possible cases:

i) For some k, $a_k = 0$ but $a_k \neq 0$, For such an k, (22) is not solvable for c_k. So, in this case, $\alpha \nmid \beta$. 0

ii) $a_k \neq 0$ for all $k \mid n$. This is the case when a has a non-zero constant term. In this case, for all $k \mid n$, (22) has the unique solution $c_k = b_k / a_k$, and the y in (21) is uniquely determined. If for some k, b_k / a_k is not integral, then $\alpha \nmid \beta$. If all b_k / a_k are integral, express γ in the $z_k{}'s$, say

$$\gamma = \sum_{k \mid n} g_k x_k. \tag{23}$$

If one of the $g_k{}'s$ is fraclional or negative, then $\alpha \nmid \beta$. If all the $g_k{}'s$ are non-negative integers, then $\alpha \mid \beta$, and the γ obtained is the unique quotient of β divided by α.

iii) Suppose one of the $a_k{}'s$ is zero, but, whenever $a_k = 0$, we have $b_k = 0$ also. Since one of the $a_k{}'s$ is zero, a_1 must be zero, and this is the case when a has a zero constant term. If, for one of the non-zero $a_k{}'s$, b_k / a_k, is fractional, then again $\alpha \nmid \beta$. Suppose, for all $a_k \neq 0$, b_k / a_k are integral. For $a_k = b_k = 0$, (22) has infinitely many solutions. But c_k must be chosen as an integer satisfying

$$0 \leqslant c_k \leqslant \frac{b_k}{a_k}, \tag{24}$$

and the number of these $c_k{}'s$ is finite. Besides, the tws must be chosen so that, when y is expressed in the $z_k{}'s$ as shown in (23), all $g_k{}'s$ should be non-negative integers. If this requirement cannot be fulfilled, then $\alpha \nmid \beta$. If fulfilled, then $\alpha \mid \beta$. The "quotient" may not be unique, but they can only be finitely many, and all of them can be found by appropriate choices of the $c_k{}'s$.

Example 1 Let $a = z_3 + 3z_6$, $\beta = 5z_3 + 22z_6$. In A_6, express α, β in the $x_k{}'s$:

$$a = 3x_3 + 21x_6, \quad \beta = 15x_3 + 147x_6.$$

We have

$$\gamma = c_1 x_1 + c_2 x_2 + 5x_3 + 7x_6 .$$

$5 - c_1$, must be a non-negative multiple of 3. Hence, $c_1 = 2$ or 5. Take $c_1 = 2$, then $c_2 - 2$ must be a non-negative multiple of 2, and $7 + 2 - 5 - c_2$ that of 6. Hence, $c_2 = 4$. We have

$$\gamma = 2x_1 + 4x_2 + 5x_3 + 7x_6 = 2 + z_2 + z_3 .$$

Now, take $c_1 = 5$. Reasoning as above, we obtain $c_2 = 7$ and

$$\gamma = 5x_1 + 7x_2 + 5x_3 + 7x_6 = 5 + z_2 .$$

So, $\alpha \mid \beta$ and there are two "quotienls".

Theorem 3 If $\alpha \mid \beta$ and $\beta \mid \alpha$, then $\alpha = \beta$. In other words, the only associate of an element is itself.

Proof Take n such that $\alpha \in A_n$, $\beta \in A_n$. Express α, β in the x_k's as shown in (18), (19). Since α, β divide each other, we must have $a_k \mid b_k$, $b_k \mid a_k$ for all $k \mid n$. So, $a_k = b_k$ for all $k \mid n$, and hence $\alpha = \beta$.

Theorem 4 1 the only unit in S.

Proof Suppose ε is a unit in S. Then, $\varepsilon \mid 1$, $1 \mid \varepsilon$. By Theorem 3, $\varepsilon = 1$.

An element $a \in S$ is said to be cancelable, if $\alpha\gamma = \alpha\delta$ implies $\gamma = \delta$ for any γ, $\delta \in S$.

Theorem 5 a is cancelable, if and only if its constant term is not zero.

Proof Suppose the constant term of a is not zero. Let $\beta = \alpha\gamma = \alpha\delta$. We have shown in ii) above that in this case the quotient β/α is uniquely determined. Hence, $\gamma = \delta$. Now, suppose the constant term of a is zero. Let n be taken such that $\alpha \in A_n$. Express α in the x_k's as shown in (18). we have $a_1 = 0$. Let $\beta = x_1$. Obviously, obviously, $\alpha\beta = 0$. Express β in the z_k's. We see from the inversion formula (17) that all the z-coefficients of $\alpha\beta$ are integral. Let $\alpha\beta = \gamma - \delta$, γ being the sum of the terms in $\alpha\beta$ with positive coefficients and $-\delta$ that of the terms with negative coefficients. We have $\alpha\gamma = \alpha\delta$, γ and $\delta \in A$, and $\gamma \neq \delta$ because $\beta \neq 0$. Hence, α is not cancelable.

Example 2 Let $\alpha = az_2 + bz_3 + cz_6$, $\alpha \in A_6$. Let $\beta = x_1$. In the z-form, $6\beta = 6 - 3z_2 - 2z_3 + z_6$

So

$$(az_2 + bz_3 + cz_6)(6 + z_6) = (az_2 + bz_3 + cz_6)(3z_2' + 2z_3).$$

An element α is said to be irreducible [2], if α is not a unit and its only divisors are units and its associates. An element a is said to be prlme [2], if α is not a unit and if $\alpha \mid \beta\gamma$ implies $\alpha \mid \beta$ or $\alpha \mid \gamma$.

Example 3 Ordinary prime numbers are irreducible in S, because ordinary integers can only be factored in S into ordinary integers. Now, let us see, for what valves of k, z_k is irreducible. Since $z_1 = 1$, z_1 is by definitlon not irreduclble. If k is not a prime power, k can be written as the product of two proper factors f and g such that $(f, g) = 1$. We have $z_f z_g = z_k$, which shows z_k is not irreducible. Let k be a prime power p^r, $r \geqslant 1$, and $z_k = \alpha\beta$. Take any term $a_f z_f$, of α and any term $b_g z_g$, of β, $a_f \neq 0$, $b_g \neq 0$. Their product $a_f b_g (f, g) z_{[f, g]}$ must be z_k, because otherwise $\alpha\beta$ would contain terms with subscripts different from k, or contain a term $c_k z_k$ with $c_k > 1$. Thus, $a_f = 1$, $b_g = 1$, $(f, g) = 1$, $[f, g] = k = p^r$. R follows that one of f and g is l and the other is p^r itself, say $f = 1$, $g = p^r$. So, α and β contain the terms l and z_k respectlvely. In order that $\alpha\beta$ does not contain more things than z_k, a must contain only the term 1, and β only the term z_k. This shows that z_k is irreducible when $k = p^r$, $r \geqslant 1$.

Theorem 6 There exist no prime elements in S.

Proof 1 is by definition not prime, let $a (\neq 1) \in S$, Take n such that $a \in A_n$, Suppose

$$a = \sum_{k \mid n} a_k z_k,$$

and let

$$c=\sum_{k\mid n} ka_k$$

Since $a \neq l$, $c>1$. Consider the element z. We say

$$a \nmid z$$

In fact, if

$$\alpha\beta = z,$$

let $b_k z_k$, be a term of β, $b_k \neq 0$. The product of this term with any non-zero term $a_k z_k$ of a is $a_k b_k (k, k) z_{[k, k]}$. $[k, k]$ must be $n^2 c$. Since $k \mid a$, if a prime number p goes $t>0$ times in $n^2 c$, r times in k and s times in k, then $t>r$. But $t=\max(r, s)$. So, $\max(r, s)>r$. Therefore, $\max(r, s)=s$, $\min(r, s)=r$. It follows $k=n^2 c$, $(k, k)=h$, and

$$a(b_k z_k)=bz\sum_{k\mid n} ka_k = bcz$$

Since $c>1$, this contradicts (26). Now,

$$zz=n^2 cz \quad az=cz$$

Hence, $\alpha \mid zz$. This and (25) show that a is not a prime element.

Theorem 7 An element $a \in S$ has only finitely many factors.

Proof Take A_n such that $a \in A_n$. Suppose β is a factor of α, β must $\in A_n$. Let

$$\alpha=\sum_{k\mid a} a_k z_k, \quad \beta=\sum_{k\mid a} b_k x_k$$

Since $\beta \mid \alpha$, $b_n \mid a_n$, b_n is the largest among the $b_n{}'s$. So, for any $k \mid n$,

$$0 \leqslant b_n \leqslant a_n$$

which shows that the number of possible β's is finite.

From this theorem, the following theorem follows immediately:

Theorem 8 Any element in S can be written as a product of irreducible factors.

The decomposition is of course not necessarily unique. We describe the factorization method through the following example:

Example 4 Let us factorize $\alpha=3+16z_1+13z_3+221z_6$. Any factor or a must be in A_n. Express a in the $x's$:

$$\alpha = 3x_1 + 35x_2 + 42x_3 + 1400x_5.$$

Suppose $\alpha = \beta\gamma$, and

$$\beta = b_1x_1 + b_2x_2 + b_3x_3 + b_5x_5,$$

$$\gamma = c_1x_1 + c_2x_2 + c_3x_3 + c_6x_6$$

We have $b_1c_1 = 3$. It may be assumed that $b_1 = 1$, $c_1 = 3$. b_2 and c_2 must fulfil $b_2c_2 = 35$, and $b_2 - b_1$ and $c_2 - c_1$ must be non-negative multiples of 2. Hence, three combinations are possible:

$$\binom{b_2}{c_2} = \binom{1}{35}, \binom{5}{7}, \binom{7}{5}.$$

b_2 and c_3 must fulfil $b_2c_3 = 42$, and $b_2 - 1$ and $c_3 - 3$ must be non-negative multiples of 3. Two combinations are possible:

$$\binom{b_2}{c_3} = \binom{1}{42}, \binom{7}{6}.$$

First, take the combination

$$\begin{pmatrix} b_2 & b_2 \\ c_2 & c_3 \end{pmatrix} = \begin{pmatrix} 1 & 1 \\ 35 & 42 \end{pmatrix}.$$

b_2 and c_3 must fulfil $b_2c_3 = 1400$, and $c_3 + 3 - 35 - 42$ and $b_2 + 1 - 1 - 1$ must be non-negadve multiples of 6. Only the combination

$$b_2 = 7, \ c_3 = 200$$

is possible. Thus,

$$\beta = x_1 + x_2 + x_3 + 7x_5, \ \gamma = 3x_1 + 35x_2 + 42x_3 + 200x_5$$

By similar arguments, we find the following decomposition of y:

$$\gamma = (x)_1 + 7x_2 + 7x_3 + 25z_5)(3x_1 + 5x_2 + 6x_3 + 8x_6),$$

and prove that β and both of the above factors of y are irreducible. Proceeding in this way, we see that α can be written in four essentially different ways as products of irreducibie factors, which, expressed in the z's, are

$$\alpha = (1 + z_6)(1 + 3z_2 + 2z_3 + 2z_6)(3 + z_2 + z_3)$$

$$= (1 + 2z_3)(1 + 2z_2)(3 + 2z_2 + z_3 + 5z_6)$$

$$=(1+2z_3)(1+3z_2+3z_6)(3+z_2+z_3)$$
$$=(1+2z_3+3z_6)(1+3z_2)(3+z_2+z_3).$$

Guan proved in [1] that there exist in two elements having no h. c. d.. In fact. concrete examples can be given. We prove first the following general theorem:

Theorem 9 Let G be a commutative semi-group with an identity 1. Let π, α, β be elements of G, π being irreducible and a cancelable. If $\pi \nmid \alpha$, $\pi \nmid \beta$, but $\pi | \alpha\beta$, then $\alpha\beta$, $\pi\alpha$ have no h. c. d., and π, α have no l. c. m..

Proof If δ is an h. c. d. of $\alpha\beta$ and $\pi\alpha$, or if δ is an l. c. m. of π and α, then δ divides $\alpha\beta$, $\pi\alpha$. and is divisible by π, α. We shall prove that such a δ cannot exist. Suppose on the contrary that there is such a δ, and that

$$\delta=\lambda\alpha, \tag{27}$$

$$\pi\alpha=\mu\delta.$$

Thus, $\pi\alpha=\mu\lambda\alpha$, which implies

$$\pi=\mu\lambda \tag{28}$$

If λ were a unit, it would follow from (27) that $\delta | \alpha$. Since $\pi | \delta$, we would have $\pi | \alpha$. If λ were an associate of π, we would have by (27) that $\pi\alpha | \delta$. But $\delta | \alpha\beta$. So, $\pi\alpha | \alpha\beta$. which would imply $\pi | \beta$. Hence, λ is neither a unit nor an associate of π. This contradicts (28), which claims that λ is a divisor of the irreducible element π.

Jacobson proved in [2] that if every pair of elements has an h. c. d., then every irreducible element is prime. Theorem 9 is more concrete.

Example 5 Using the decompositions in Example 4, two elements having no h. c. d., can be eastly exhibited, We give a more interesting example, in which only one of the elements involved is cancelable. Let $\pi=z_2$, $\alpha=2+2z_2$, $\beta=z_4$. We have $\alpha\beta=6z_4=(3z_4)\pi$, $\pi\alpha=6z_2$. All the assumptions in Theorem 9 are fulfilled. So, $6z_4$, and $6z_2$ have no h. c. d..

References

[1] Guan. Chiwen, On linear Autonomous Machines (1) (in Chinese, with a

Detailed English Abstract), Acta Sa. Nat. Jilin Unw. (2) (1963), 117 146.

［2］ Jacobson, N., Lectures in Abstract Algebra, Vol. 1, Chapter IV, Van Nostrand, New York, 1951, 114－127.

代数学（节选）*

引 言

代数学是数学的一门古老的分支，也是重要的基础数学分支之一．一百多年来，尤其是最近几十年来，由于数学本身的发展以及应用的需要，代数学的研究对象及其研究方法发生了巨大的变革．一系列的新的代数领域被建立起来，大大地扩充了代数学的研究范围，形成了所谓近世代数学．它与以代数方程的根的计算与分布为研究中心的古典代数学有所不同，近世代数学以研究各种代数结构——群、环、域、格、结合及非结合代数等——的性质为其中心问题，目前仍在蓬勃发展之中．它的方法和结果渗透到那些与它相接近的各个不同的学科中，成为一些有着新面貌和新内容的学科——代数数论、代数几何、拓扑代数、李（Lie）群和李（Lie）代数、代数拓扑、泛函分析等．这样，近世代数学就对于全部现代数学的发展显著地产生了很大影响，而且对于理论物理学等其它学科也有很多的应用．最近十几年来，由于快速电子计算机的发展，提出了代数计算机械化的重要课题，以及有如线路的代数理论等代数学的新的问题．

我国古代在代数学方面有着悠久的历史和光辉的成就，象秦九韶（1247 年）的高次代数方程的数值解法、杨辉（1261 年）的由二项式系数所组成的三角形、以及朱世杰（1303 年）的级数论等．但是，近代我国代数学研究工作的发展则是比较晚的，除了个别的早期代数工作者［如曾炯之（1933—1934 年）对函数域上可除代数的工作］之外，可以说自 1938 年华罗庚所主持的群论讨论会才开始得到初步发展．通过这个讨论会，得到了一些有限群论方面的研究成果，这是以后在群论方面研究工作的开端．解放以前，在矩阵几何及典型群等方面，也有了一系列的工作．此外，在李群和

* 本文由华罗庚、段学复、王湘浩共同创作，原载于科学出版社 1959 年出版的《十年来的中国科学 数学 1949－1959》。

李代数以及代数数论等方面的工作则初步开始.[①]

§3. 代数数论及赋值论

1. 代数数论的中心理论是所谓类域论，而类域论的主要论题是：对于任意有限次代数数域 F，讨论 F 有哪些阿培尔扩张并研究这些阿培尔扩张之间的关系. 我们可以提出下面的普遍问题：对于任意域 E，讨论 E 有哪些代数扩张，并研究这些扩张之间的关系. 这一普遍问题自然是十分困难的. 上述类域论的论题所以能够很好地解决，主要在于对于代数数域，我们有一套算术理论，就是说，理想理论和赋值论.

试看 F 的所有赋值 $\mathfrak{P}$[阿基米德（Archimedes）的和非阿基米德的]，和完满化域 $F_{\mathfrak{P}}$. 在每个 $F_{\mathfrak{P}}$ 中取一个非零元素 $\mathfrak{U}_{\mathfrak{P}}$ 作“向量”$(\cdots, \mathfrak{U}_{\mathfrak{P}}, \cdots)$，我们要求除有限个 $\mathfrak{P}$ 外，$\mathfrak{U}_{\mathfrak{P}}$ 都是 $\mathfrak{P}$ 进单位（就是说，除有限个 $\mathfrak{P}$ 外，$\mathfrak{U}_{\mathfrak{P}}$ 的赋值都是 1）. 每个这样的向量 $\mathfrak{P}$ 叫作一个伊德尔（idile），$\mathfrak{U}_{\mathfrak{P}}$ 叫作它的 $\mathfrak{P}$ 分量. 用分量分别相乘的办法规定两个伊德尔的积，所有伊德尔作成一个群 J_F，叫 F 的基本群. 显而易见，F 的非零元素作成的群 P_F 可以看作是 J_F 的子群.

设 K 是代数数域 F 的一个有限次阿培尔扩张，对于 F 的任意赋值 $\mathfrak{P}$，$\mathfrak{P}$ 在 K 中的所有开拓确定 K 的同一个完满化域 $K^{\mathfrak{P}}/F_{\mathfrak{P}}$. 十分明显，$J_F$ 可以看作是 J_K 的子群，而且不难看出怎样定义 K 的伊德尔 $\mathfrak{U}$ 的模 $N_{KF}\mathfrak{U}$ 使得这个定义用在 K 的元素 A 上时和 $N_{KF}A$ 一致. 用 $N_{KF}J_K$ 代表 K 的所有伊德尔的模作成的 J_F 的子群，而把 $P_K N_{KF} J_K$ 称为 K/F 的模群. 类域论证明了下面的定理：K/F 的伽罗瓦（Galois）群 G 和 J_F 对于 K/F 的模群的商群同构. 此外，可以用一种统一的方法定义 J_F 到 G 的一个同态映象，使其核恰为上述模群，而这就是所谓互反律的内容. 伊德尔 $\mathfrak{U}$ 在这个同态映象下的影象记为

$$\left(\frac{K/F}{\mathfrak{A}}\right).$$

类域论证明，J_F 的子群 N 是 F 的某个有限阿培尔扩张的模群必要而且只要：1) $P_F \subset N$，2) 有 n 存在使每个伊德尔 $\mathfrak{A}$ 的 n 方在 N 内，3) 有 $\mathfrak{P}_1$，…，$\mathfrak{P}_{\mathfrak{M}}$ 存在使适

---

① 参见段学复，近代中国数学家在代数方面的贡献 [J]. 数学进展，1955 (03)：609—614.

合下列条件的 $\mathfrak{A}$ 都在 N 内：$\mathfrak{A}$ 在 $\mathfrak{P}_1$，…，$\mathfrak{P}_m$ 处的分量为 1，$\mathfrak{A}$ 在其余 $\mathfrak{P}$ 处的分量为 $\mathfrak{P}$ 进单位．此外，可以证明，F 的所有有限阿培尔扩张所作成的“格”和 J_K 的所有模子群（即适合上述三条件的子群）作成的格反相似．类域论就这样解决了它的主要问题．

2．类域论是研究有关代数数域的问题的有力工具．例如，在代数构造理论中一个著名的问题是：是否任意单纯代数都是巡回代数？这个问题所以能在代数数域上面的单纯代数的情形下得到解决在于一个类域论的问题——葛伦瓦尔特（Grunwald）扩张的存在问题——能够得到解决．设 F 是任意有限次代数数域，S 是 F 的一组有限个赋值，G 是一个有限阿培尔群．设对 S 中任意 $\mathfrak{P}$，给定阿培尔扩张 $K^{\mathfrak{P}}/F_{\mathfrak{P}}$，其伽罗瓦群同构于 G 的一个子群．F 的一个阿培尔扩张 K，其伽罗瓦群同构于 G，称为 F 的一个葛伦瓦尔特扩张，若对 S 中的任意 $\mathfrak{P}$，$\mathfrak{P}$ 在 K 中的开拓所确定的完满化域恰为 $K^{\mathfrak{P}}$，葛伦瓦尔特定理就是说葛伦瓦尔特扩张永远存在．

王湘浩在 1948 年曾设法证明下面的定理：代数数域上一个单纯代数的所有正则元素作成的乘法群的换位子群等于代数中所有缩减模为 1 的元素作成的群．为了证明这一定理，需要把葛伦瓦尔特定理加以推广，但他发现葛伦瓦尔特扩张并不是永远存在的．后来，王湘浩找到了在 G 为巡回群而元数为一个质数方 l^s 时葛伦瓦尔特扩张存在的充要条件．

F 中的赋值 $\mathfrak{P}$，若使有理质数 2 的赋值小于 1，称为偶赋值，偶赋值 $\mathfrak{P}$ 称为奇偶赋值，如果对于任意正整数 x，

$$(F_{\mathfrak{P}}(e^{2\pi i/2^x}):F_{\mathfrak{P}})=(F(e^{2\pi i/2^x}):F).$$

命 $t=t(F)$ 为使 $\cos 2\pi/2^x$ 在 F 内之最大正整数 x，试看上述赋值相 S 和质数方 l^s．下面的情况称为“特殊情况”：1）$l=2$，2）$F(e^{2\pi i/2^s})/F$ 非巡回，3）S 外无奇偶赋值．王湘浩证明了下列定理：设 G 为巡回群，元数为质数方 l^s．不在特殊情况时，葛伦瓦尔特扩张必然存在；而在特殊情况时，葛伦瓦尔特扩张存在必要而且只要使 $(\sec 2\pi/2^{t+1})2^s$ 非 $K^{\mathfrak{P}}$ 中某元素对于 $F_{\mathfrak{P}}$ 之模之奇偶赋值 $\mathfrak{P}$ 的个数为偶数．利用这个定理同样可以证明代数数域上面的任意单纯代数必是巡回代数．王湘浩在文献中实际上证明了一个更广的定理，从而在文献中证明了上述关于单纯代数的换位子群的定理．

以后，王湘浩在文献中解决了 G 任意时葛伦瓦尔特扩张何时存在的问题.

王湘浩曾提出判断上述关于代数数域上单纯代数的换位子群的定理对于任意域上的单纯代数 A 是否成立的问题. 他只在 A 的指数无平方因子时正面回答了这个问题. 儿玉哲夫在 1956 年曾尝试对于示性数为 0 的任意域证明定理成立，但他的证明中包含着一个带根本性的错误，因而这个问题仍旧停留在原来的地方.

由于葛伦瓦尔特定理中所包含的错误，关于单纯代数的马斯（Maass）模定理的原证不适用. 王湘浩曾利用一个特殊设计的葛伦瓦尔特定理证明这个模定理. 爱许勒尔（M. Eichler）在 1938 年曾证明这个定理，他不用葛伦瓦尔特定理而利用关于单纯代数中的理想的某些较复杂的讨论. 后来，王湘浩得到了一个极为简易的证明，这个定理乃成为一个显然的结论.

聂灵沼采用一种直接方法来建立有限代数数域 F 上面的无限类域论. 设 K 是 F 的最大阿培尔扩张，其对于 F 的伽罗瓦群为 G_F. 如所周知，以 K 对于所有有限扩张 E/F 的伽罗瓦群 G_E 为 1 的邻域系，G_F 成为一个紧致拓扑群. 以 J_F 的所有模子群为 1 的邻域系，J_F 也成为一个拓扑群（不适合分离公理 T_1）. 我们自然希望象在有限情形那样能用 J_F 刻划 G_F ，而由于 K 包含 F 的所有阿培尔扩张，只要这事能作到，无限类域论的主要问题也便解决了. C. 薛佛倡曾在 1940 年用群指标的理论来建立无限类域论，但他并没有作到直接用 J_F 刻划 G_F . 聂灵沼的论文在 1956 年发表，但他的这一工作是在 1951 年完成的，聂灵沼直接定义了 J_F 到 G_F 的一个连续而且开的同态映象

$$\mathfrak{A} \to \left(\frac{K/F}{\mathfrak{A}}\right).$$

他证明这一同态映象的影象集合在 G_F 中稠密而且证明 J_F 是一个紧致群，因而断定映射是到 G_K 全部的. 设映射的核为 D ，则 $J_F/D \simeq G_F$. 此外，他又基本上确定了核 D ，这样就达到了用 J_F 刻划 G_F 的要求.

3. 在赋值论方面，聂灵沼改进了关于离散赋值完满域的构造问题的处理方法. 哈塞（H. Hasse）、施密特（F. K. Schmidt）、魏脱（E. Witt）、梯许米勒（O. Teichmüller）等曾证明这样一个域 K 的构造基本上由剩余类域决定（在示性数不等情形至少在绝对分歧指数为 1 时是这样）. 以前，$\mathfrak{R}$ 非完备时，要用一种间接方法来

处理，这样就并不能清楚地显示出 K 的构造. 聂灵沼就 $\mathfrak{R}$ 的示性数为 p 的情形，给了一个一致的证明，他取定一个“p-基底”作出了 $\mathfrak{R}$ 的一个正规代表系，在完备情形，这个正规代表系就是通常的保乘代表系. 这使得他不但能用统一的方法证明定理而且能够清楚地抽绘出 K 的构造.

管纪文解决了离散赋值完满域的乘法群在剩余类域无限的情形下的构造问题. 在剩余类域有限的情形，这一问题是亨斯尔 K. Hensel 和 H. 哈赛解决的，剩余类域无限时，沙伐略维奇（И. Р. Шафаревич）曾得到部分的结果. 管纪文除得到普遍结果还就在局部类域论中有基本意义的广义正则域和正则域的情形得到了比较整齐的结果，他利用这一结果来构造反例以说明在广义局部类域论中正则域 K 在一定条件下必包含指数有限但非模群之子群，这样他便简化并且扩充了 M. Moriya 的论证和结果.

在广义赋值论中，似收敛的概念和完满域的概念是有基本意义的. 设 F 是一个赋值域，其赋值记为 V. 设 $\{a_\rho\}$ 是 F 中一个无最后项的整序叙列，我们说 $\{a_\rho\}$ 以 a 为似极限，如果从 ρ 的某个值开始，$V(a-a_\rho)$ 严格上升. 若对充分大的 $\tau>\sigma>\rho$ 有 $V(a_\tau-a_\sigma)>V(a_\sigma-a_\rho)$，则 $\{a_\rho\}$ 说是似收敛. 希林（O. Schilling）在其“赋值论”一书中利用完满域的存在来论证赋值开拓的可能性，但他的论证有循环的缺点，因为他在证明关于多项式的似收敛的一个定理时，用到了赋值开拓的存在. 王湘浩用直接证明任意似收敛数列在 F 的一个适当的直接扩张中有一个似极限的方法克服了上述缺点并且澄清了关于似收敛的某些基本性质. 以后，何伯和发现可以用直接计算的方法来证明上述关于多项式的定理.

戴执中叙述并且论证了下面的定理：赋值域 K 完满的充要条件是“幅”为质理想的似收敛数列在 K 中有似极限，并且指出，对于可数无限阶赋值，完满性与极大性等价的充要条件是赋值为离散的.

戴执中推广了奥斯特洛夫斯基（A. Ostrowski）关于一阶赋值的完满域的下列几个定理：1）完满域 K 的任意有限扩张也是完满域；2）若 K 的代数扩张 L 中含有对 K 的次数任意高的元素，则 L 非完满，3）K 的代数封闭域 L 为完满域的充要条件是 L 对 K 有限. 戴执中证明，这几个定理对于可数无限阶赋值也是对的，只有在推广第一个定理时对于扩张加上了可离的限制.

在广义赋值论中，完满性的定义是多种多样的．黄明游定义了一种代数极大赋值域和一种最小相对完满化域并且进行了初步研究，对这两种域的关系还没有得到肯定的结果．

§4．环论及代数论

1．环论的中心问题是环的构造问题．研究这一问题的一个最重要的方法是象下面这样：我们把最常 R 最重要的环，例如整区、体、体上的全阵环等，看作是一个极端；把和这些最常见最重要的环差别最大而具有另一方面的简单性的环看作是另一个极端．这第二类的环我们称之为根环．若环 R 的两边理想 N 是一个根环，则 N 称为 R 的一个根理想．在选取哪些环为根环时，我们要求任意环 R 必须有最大的根理想 P，称为 R 的根，而且要求 R/P 的根为（0）．根为（0）的环叫作半单纯环，这些半单纯环自然就包括那些最常见最重要的环．为了研究一般环的构造，我们应当研究三方面的问题：1）半单纯环的构造，2）根环，根理想和根的性质，3）任意环 R 是怎样由它的根 P 和剩余环 R/P 结合起来的？

例如，我们可以把幂零元素环算作是根环．这样确定的根叫作蔻脱（Kothe）根．一个环叫作一个拜尔（Baer）根环，如果它的任意非零同态映象必然包含非零幂零理想．由拜尔根环确定的根叫拜尔根．除了这两种根，人们还定义了另外一些根．对于适合所谓降链条件的环，这些根都是一样的，而这就是所谓古典根．

对于适合降链条件的环，问题 1）已经完全解决：任意半单纯环恰有一法分解为单纯环的直和，而单纯环等于体上的全阵环．对于域上的有限阶代数，问题 3）也已经较好地解决：若代数 A 对于根 P 的剩余代数 A/P 可离，则 A 可以分解为 P 和一个子代数 B 的群直和，而且对于任意可离子代数 C 必有 A 的自同构（基本上是一个内自同构）将 C 变入 B 内．

2．为了对于一般环研究问题 1），人们用亚直和的概念代替直和的概念．对于拜尔根和另外一些根，以前已经得到了半单纯环的亚直和分解并且对于分出来的那些较简单的环进行了一些研究．王湘浩得到了蔻脱半单纯环的分解定理，他并且详细地讨论了交换环的情形．对于交换环，以前已知的结果是任意无幂零元素的交换环可以表示为一些无零因子的环的亚直和. 王湘浩证明任意无幂零元素的交换环可以表为一组在保持无幂零元素的条件下的亚直不可分环的亚直和，而这种亚直不可分环或者是一个域或者是一种所谓拟赋值环. 一个无零因子但不是一个域的交换环 R 叫做一个拟赋值环，如果其中有一个非零元素 a 具有下列性质：R 中任意非零元素整除 a 的若干方. 拟赋值环是赋值环的推广. 王湘浩引进了半赋值的概念：设交换环 R 无零因子，但不是一个域，规定在 R 上面的一个函数 $v(x)$，其值取非负实数或 ∞，叫做 R 上面的一个半赋值，如果 1) $v(a)=\infty$，必要而且只要 $a=0$，2) $v(ab)\geqslant v(a)+v(b)$，3) $v(a^2)=2v(a)$，4) $v(a+b)\geqslant \min(v(a), v(b))$，5) R 中有一个非零元素 a 其值 $v(a)>0$. 王湘浩证明，R 是一个拟赋值环，必要而且只要 R 上面有一个最强半赋值. 此外，他还讨论了拟赋值环上面的赋值和半赋值特别和最强半赋值的关系. 以后，谢邦杰把王湘浩关于蔻脱半单纯环的构造定理推广到非结合环.

3. 谢邦杰在环论方面的工作主要是关于幂零元素理想和幂零理想的，这些工作属于上面所说的问题 2) 方面. 谢邦杰得到了下面的定理："任意环 R 的上指数为 $n(>2)$ 的幂零元素左（右）理想恒含有 R 的上指数为 2 的幂零元素左（右）理想；而 R 的上指数为 2 的幂零元素左（右）理想恒为 R 的若干个幂零指数为 2 的幂零左（右）理想的并集". 所谓一个幂零元素理想 N 的上指数就是使得任意 $a\in N$ 的 n 方为 0 的最小正整数 n. 作为这个结果的应用，他讨论了拜尔上下根何时相等的问题，库罗什-贾考勃生（Курош-Jacobson）问题，蔻脱问题，得到某些进一步的结果或将前人的论证加以简化.

谢邦杰在环论的工作，充分利用了零化子这个工具. 他并且引进了所谓零化子的升链、降链和双链条件. 用这些代替通常的链条件，他改进了前人关于幂零元素环的某些结果. 对于通常的升链和降链条件，谢邦杰作了一些分析性工作. 他证明："环 R 的右理想适合升（降）链条件必要而且只要对于两边理想 M_1，…，M_n，$R/M_i(i=1, \cdots, n)$ 之右理想适合升（降）链条件而且 R 的含于 $M=\bigcap\limits_{i=1}^{n} M_i$ 的右理想适合升（降）链条件". 根据这个定理，在 $n=1$ 时的特殊情形可以看出过去在环论的某些工作中虽然在形式上似乎减弱了前人的定理中的条件，但实质上却并没有减弱. 谢邦杰还用同样的方法处理了子环的链条件得到了相应的结果，并因而简化了霍布金

(C. Hopkins)一个定理的证明而且推广了定理本身.

4. 在问题3)方面，在库罗什(А.Г.Курош)的指导下，刘绍学系统地讨论了将上述关于有限阶代数的定理和关于有限阶交错代数、李代数、约当代数的相应定理推广到无穷阶代数的问题. 他的工作概括并扩充了克梯斯(C.W.Curtis)、欣克曼(E. Schenkmann)、库洛奇金(В.М.Курочкин)等人在这一方面的工作. 一个代数是局部有限(可离)的，如果任意有限个元素包含在一个有限阶(有限阶可离)子代数之内，类似地可以定义局部幂零代数等等. 以局部幂零代数为根环确定的根叫局部幂零根. 刘绍学主要就局部幂零根和局部了解根来推广上述四方面的定理. 关于剩余代数 A/P ，则分别不同情况，假定其为可数阶局部可离的，可数阶局部半单纯的，或这样一些代数的直和. 他首先用公理方法统一地讨论了将上述四方面的定理推广到局部有限代数的问题，然后通过对上述四种代数的分别研究把这些定理再加以推广.

5. 在代数论方面，谢邦杰修正了亚尔倍脱(A.A.Albert)代数构造一书中的一个错误结论. 设 A 是域 F 上一个有1的有限阶代数，S 是 A 在 F 上的一个自同构，子代数 B 说是一个 S -不变子代数，如果 $B^S=B$ ，A 就是 S -不可约的，如果 A 不能表为两个非零 S -不变子代数的直和，设 A 是 S -不可约的，于是，$A=D\times J$ ，其中 D 是一个唯一确定的 n 阶 S -不变对角子代数，J 是一个在等阶意义下唯一确定的不可约子代数. 可以选 J 使之为 S^n -不变. 设 J_0 是这样一个 J ，问题是，是否任意 J 都是 S^n -不变的？A.A.亚尔倍脱忽略了一般的 J 和 J_0 的区别，谢邦杰证明了下面的定理：任取 J 必为 S^n -不变的充要条件为 S^n 在 J_0 上和 J_0 在 F 上的所有同构可交换. 在这个定理的基础上，所谓循环系的相似概念才能正确地引进.

另外，在将有限维结合代数、交错代数、李代数和约当代数糅合在一类较广的代数之中，以便这四种代数中平行的概念和定理可以统一处理的问题方面，刘绍学综合了乌兹可夫(А.УзКОБ)和A.A.亚尔倍脱的方法而定义了一种包括上述四种代数的 λ -对称代数，并对于这种代数定义了根、半单纯和单纯等概念，并且证明了半单纯代数就是单纯代数的直和.

6. 在交换环的研究方面，周伯壎定义了一种素性环：设 R 是一个整区，$x\in R$ 说是素于 R 的理想 A ，如果只有 $a\in A$ 时才能使 $ax\in A$. 若不素于 A 所有元素 x 也作

成一个理想 A'，则根据福克斯（L. Fuchs，1950），A 称为素性理想，如果除 R 本身外，R 的所有理想都是素性的，则 R 说是一个素性环. 周伯壎证明，R 是素性环的充要条件是其所有素理想组成的集全有序. 他并且得到了素性环为赋值环的充要条件. 在文献，他进一步讨论了关于某些素性环上的特殊化问题，关于相应的特殊化环得到某些结果. 此外，周伯壎并将素性环的概念以及上述关于素性环的充要条件推广到非交换环.

7. 此外，在环论方面，还应当提到下面的一些工作，谢邦杰，谢邦杰与诸勤华，谢邦杰与李学淑，邵震豪，吴宏国在根的方面进行了工作，他们引进了一些根并对若干旧有的根的性质作了讨论. 王世强对于实向量作成的有序环进行了研究. 刘荫南和张功安修正了库罗什的一个错误，得到了抽象根能在具有降链条件的环类上引起古典根的充要条件. 宫德荣证明了关于将一种矩阵环中的极大布耳（Boole）环化为对角形式的一个定理，李希民得到了关于蔻脱问题的一个较简单的解法. 作为克立福特（Clifford）代数的推广，周怀生定义了一种反交换重对称代数，他研究了这种代数的构造及表示.

科普文章

复数的定义*

虚数在过去一般中学教科书里面是名副其实的『虚』数. 编者告诉学生说，以前我说过，负数没有平方根，现在请大家承认 i 的平方等于－1. 于是学生仿照例题计算复数的加减乘除，计算含有虚数的复杂的算式，求方程式的虚根，却完全不知道计算的是些什么东西，也不知道学了这种计算会有什么用处.

另一方面，大学教科书形式地界说复数为所谓“实数偶”. 两个实数偶怎样相加呢？定义是 (a，b) ＋ (c，d) ＝ (a＋c，b＋d). 这似乎有些道理，比方，a 斤米 b 斤豆加 c 斤米 d 斤豆等于 a＋c 斤米 b＋d 斤豆. 但两个实数偶怎样相乘呢？定义是 (a，b) (c，d) ＝ (ac－bd，ad＋bc). 这样界说乘法，理由是什么呢？

在这篇文字里，我们给了一个复数的定义，希望这个定义能对中学的同学们有一些帮助，因为这个定义至少使复数有了一种实际的意义.

一、向　量

量有多种，例如长度，面积，体积，质量，时间，速度，加速度，力. 有的量只有单纯的大小，这样的量叫纯量，例如质量. 但有的量，单纯的大小不足以完满地表达它们的意义，还需要一种方向的概念，这样的量叫向量，例如力. 纯量中以长度最为基本，度量时最为直接. 一个长度可以用一条线段代表，因此我们可以说向量中以定向线段最为基本. 一条线段 AB 有两个方向，从 A 到 B 和从 B 到 A；定向线段就是取定了一个固定的方向的线段. 定向线段可以看作是其他各种向量的数学代表，比方向南的千斤的力可以用向南的千尺的定向线段代表. 由于这个缘故，以后我们只讨论定向线段，凡说向量即指定向线段而言.

设 u 是一个向量，从 A 到 B. A 叫 u 的起点，B 叫 u 的终点. 为了表明 u 的方

* 本文原载于《中国数学杂志》1951 年第 1 期。

向，作图时我们在 AB 线段上作一个箭头指向 B 点，而书写时，u 可以写作 $\overrightarrow{AB}$：$u=\overrightarrow{AB}$.

向量本来可以有空间的各种方向，但本文只讨论平面向量，以后假定所有向量在一个固定平面上.

两个向量如果长度相等方向相同说是相等. 如果长度相等而方向相反，便说是互为负向量. 例如 $\overrightarrow{AB}=\overrightarrow{CD}$ 时，则 $\overrightarrow{BA}=-\overrightarrow{CD}$，又如 $\overrightarrow{AB}=-\overrightarrow{BA}$. $u=\overrightarrow{AB}$ 时，$|u|$ 表示线段 AB，因而 $|\overrightarrow{AB}|=|\overrightarrow{BA}|$.

设 u 是任意向量. 若是 a 一个正实数，取向量 v 使 v 和 u 的方向相同而 $|v|$ 和 $|u|$ 的比等于 a：$\frac{|v|}{|u|}=a$；若 a 是一个负实数，比方 $a=-b$，取 v 和 u 的方向相反而 $\frac{|v|}{|u|}=b$. 我们说 v 是 u 的 a 倍：

$$v=au,$$

或说 v 和 u 的比等于 a：

$$\frac{v}{u}=a.$$

长度等于 0 的向量叫零向量，记为 $\vec{0}$. 例如一个物体不受外力时，则加在这个物体上的力便是一个零向量. $\overrightarrow{AA}$ 代表 A 处的零向量，所有零向量作为相等. 我们规定 $0u=\vec{0}$，$a\vec{0}=\vec{0}$，$-\vec{0}=\vec{0}$，$u\neq\vec{0}$ 时，规定 $\frac{\vec{0}}{u}=0$.

不难说明 u 的 b 倍的 a 倍等于 u 的 ab 倍：

$$a(bu)=(ab)u,$$

因此，u，v，w 互相平行时，

$$\frac{u}{v}\frac{v}{w}=\frac{u}{w}. \tag{1}$$

现在我们规定向量合并的方法. 设 u，v 是两个向量. 取 $\overrightarrow{AB}=u$，$\overrightarrow{BC}=v$，则 $\overrightarrow{AC}$ 叫做 u，v 的和，记为 $u+v$. 我们这样界说加法，是因为物理上各种向量合并的方法都是这样. 例如移动. 从 A 到 B 的移动这个物理观念只和起点终点有关系，和中间的路线没有关系，所以从 A 走到 B 再从 B 走到 C 等于从 A 走到 C.

由定义易见

(2) $$u+\vec{0}=u,$$

(3) $$u+(-u)=\vec{0},$$

而且不难说明 $au+bu=(a+b)u$，换句话说，若 u，v，w 互相平行，则

(4) $$\frac{u}{w}+\frac{v}{w}=\frac{u+v}{w}.$$

图一中，$\overrightarrow{AB}+\overrightarrow{BC}=\overrightarrow{AC}=\overrightarrow{AD}+\overrightarrow{DC}$，所以，

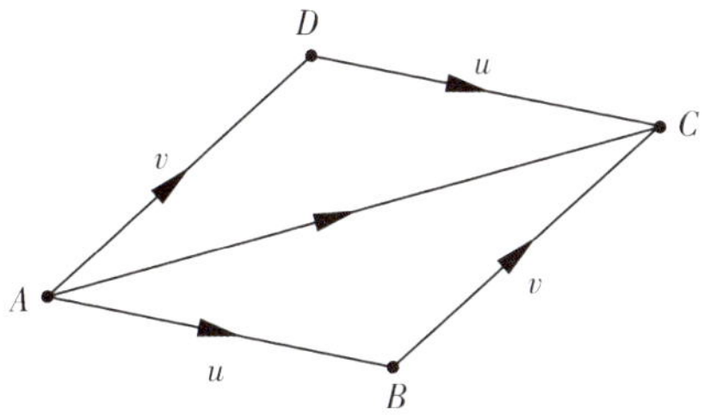

图一

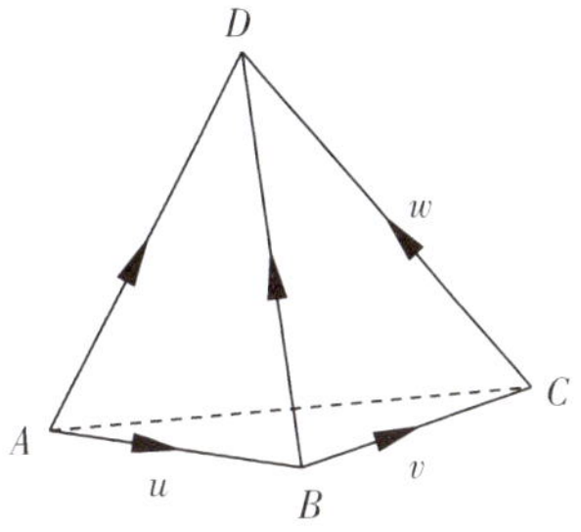

图二

(5) $$u+v=v+u.$$

又图二中，$\overrightarrow{AC}+\overrightarrow{CD}=\overrightarrow{AD}=\overrightarrow{AB}+\overrightarrow{BD}$，所以，

(6) $$(u+v)+w=u+(v+w).$$

设 u，v 是两个不等于 0 的向量，取 $\overrightarrow{AB}=u$，$\overrightarrow{AC}=v$，从 $\overrightarrow{AC}$ 到 $\overrightarrow{AB}$ 的角度 θ 叫做 u 和 v 的方向差，记为$\angle(u,v)$. θ 是一个广义角，可以是正若干度或负若干度或零度. $u=\vec{0}$，$v\neq\vec{0}$ 时，任意度数都算是$\angle(u,v)$. 根据广义角的意义可知

$$\angle(u,v)=-\angle(v,u),$$

$$\angle(u,v)+\angle(v,w)=-\angle(u,w)$$

二、复 数

上面说过，长度是最基本的量. 因此，我们最好利用线段界说实数，然后应用到其他各种量的量法上面. 正实数可以界说为线段的比，正负实数可以界说为一条直线

上向量的比，或平行向量的比．这样界说了实数，可以利用比例线段的理论说明实数的基本算律，因而建立起实数的理论．但这不在本文的范围以内．这里，我们假定实数的理论，而在第一节里面，说明了两个平行向量的比是一个实数．

复数是什么呢？

定义．复数是平面上向量的比．

但向量的比是什么呢？

向量的概念是长度和方向两个概念的结合．因此，比较两个向量时，应该比较它们的长度和方向，换句话说，求它们的长度比和方向差．只有长度比或只有方向差不能完满地说明两个向量的关系，但是长度比和方向差联合起来便把两个向量的关系说明白了．所以，向量比这个概念就是长度比和方向差两个概念的结合，上面的定义说，复数不是别的，就是这种向量比．

设 u，v 是平面上的两个向量，$v\neq\vec{0}$，u 和 v 的比记为 $\frac{u}{v}$，根据复数的定义，$\frac{u}{v}$ 也就是一个复数．若 u 和 v 平行，则以前已经说过了，$\frac{u}{v}$ 是一个实数．

设 $x=\frac{u}{v}$，$y=\frac{w}{t}$ 是两个复数．向量比既然是长度比和方向差的结合，所以若 $\frac{|u|}{|v|}=\frac{|w|}{|t|}$ 而且 $\angle(u,v)=\angle(w,t)$，则 $\frac{u}{v}$ 和 $\frac{w}{t}$ 相等：$\frac{u}{v}=\frac{w}{t}$，而且只有这样才相等．

定理一　若 $\frac{u}{v}=\frac{w}{t}$，则 $\frac{u}{w}=\frac{v}{t}$.

证　由假定 $\frac{|u|}{|v|}=\frac{|w|}{|t|}$，故 $\frac{|u|}{|w|}=\frac{|v|}{|t|}$．令 $\angle(u,w)=\angle(u,v)+\angle(v,w)$，$\angle(v,t)=\angle(v,w)+\angle(w,t)$．但由假定，$\angle(u,v)=\angle(w,t)$，故 $\angle(u,w)=\angle(v,t)$．因之，$\frac{u}{w}=\frac{v}{t}$.

定理二　$\frac{u_1}{v_1}=\frac{u_2}{v_2}$，$\frac{v_1}{w_1}=\frac{v_2}{w_2}$，则 $\frac{u_1}{w_1}=\frac{u_2}{w_2}$.

证　由假定，$\frac{u_1}{u_2}=\frac{v_1}{v_2}$，$\frac{v_1}{v_2}=\frac{w_1}{w_2}$，故 $\frac{u_1}{u_2}=\frac{w_1}{w_2}$，因而 $\frac{u_1}{w_1}=\frac{u_2}{w_2}$.

定理三 $\dfrac{u}{v}=\dfrac{w}{t}$，则 $\dfrac{u+v}{v}=\dfrac{w+t}{t}$.

证 由假定知在图三的两个三角形中，

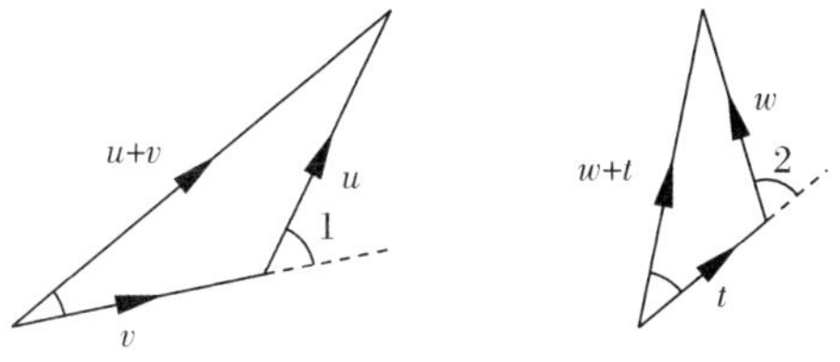

图三

$$\frac{|u|}{|v|}=\frac{|w|}{|t|}，\angle 1=\angle 2，$$

故两个三角形相似，因而 $\dfrac{|u+v|}{|v|}=\dfrac{|w+t|}{|t|}$，$\angle(u+v，v)=\angle(w+t，t)$. 所以$\dfrac{u+v}{v}=\dfrac{w+t}{t}$.

定理四 $\dfrac{u_1}{w_1}=\dfrac{u_2}{w_2}$，$\dfrac{v_1}{w_1}=\dfrac{v_2}{w_2}$，则 $\dfrac{u_1+v_1}{w_1}=\dfrac{u_2+v_2}{w_2}$.

证 $\dfrac{u_1}{u_2}=\dfrac{w_1}{w_2}$，$\dfrac{v_1}{v_2}=\dfrac{w_1}{w_2}$，故 $\dfrac{u_1}{u_2}=\dfrac{v_1}{v_2}$，因而 $\dfrac{u_1}{v_1}=\dfrac{u_2}{v_2}$. 由定理三，$\dfrac{u_1+v_1}{v_1}=\dfrac{u_2+v_2}{v_2}$，故由定理二，$\dfrac{u_1+v_1}{w_1}=\dfrac{u_2+v_2}{w_2}$.

以上各定理的证明中，没有讨论特别情形，读者可以自己补充.

现在我们界说复数的运算.

设 $x=\dfrac{u}{w}$，$y=\dfrac{v}{t}$ 是两个复数. 取任意非零向量 w_1，取 u_1，v_1 使 $\dfrac{u_1}{w_1}=\dfrac{u}{w}=x$，$\dfrac{v_1}{w_1}=\dfrac{v}{t}=y$. 复数$\dfrac{u_1+v_1}{w_1}$称为 x，y 的和，记为 $x+y$. 若 w_2 是另一非零向量，而 $\dfrac{u_2}{w_2}=x$，$\dfrac{v_2}{w_2}=y$，则由定理四，$\dfrac{u_1+v_1}{w_1}=\dfrac{u_2+v_2}{w_2}$. 因之，$x+y$ 由 x，y 确定，和所取的向量 w_1 没有关系，加法的定义就是说，若 u_1 是 w_1 的 x 倍，v_1 是 w_1 的 y 倍，则 u_1+v_1 是 w_1 的 $x+y$ 倍. 由（4），这个定义用在实数上，和实数原有的加法结合.

设 $x=\dfrac{u}{v}$，$y=\dfrac{w}{t}$，$y\neq 0$. 取任意非零向量 w_1，取 u_1，v_1 使 $\dfrac{v_1}{w_1}=\dfrac{w}{t}=y$，$\dfrac{u_1}{v_1}=\dfrac{u}{v}=x$. 复数 $\dfrac{u_1}{w_1}$ 称为 x，y 的积，记为 xy. 若 w_2 是另一非零向量，而 $\dfrac{v_2}{w_2}=y$，$\dfrac{u_2}{v_2}=x$，则由定理二，$\dfrac{u_1}{w_1}=\dfrac{u_2}{w_2}$. 因之，$xy$ 由 x，y 确定，和所取的向量 w_1 没有关系，乘法的定义就是说，若 v_1 是 w_1 的 y 倍，u_1 是 v_1 的 x 倍，则 u_1 是 w_1 的 xy 倍. 若 $w=\vec{0}$，则按第一节，$\dfrac{w}{t}$ 是实数 0，换句话说，$y=0$. $y=0$ 时，界说 $xy=0$. 由（1），乘法的定义用在实数上，和实数原有的乘法结合.

我们有下面的算律：

（一）$x+y=y+x$，

（二）$(x+y)+z=x+(y+z)$，

（三）$x+0=x$，

（四）对于任意复数 x，有一个确定的复数 y，使 $x+y=0$，

（五）$xy=yx$，

（六）$(xy)z=x(yz)$，

（七）$x\cdot 1=x$，

（八）对于任意非零复数 x，有一个确定的复数 y，使 $xy=1$，

（九）$(x+y)z=xz+yz$.

说明（一）和（二）利用（5）和（6）. 说明（三）和（四）利用（2）和（3）. 现在说明其余的算律.

设 $x=\dfrac{u}{v}$，$y=\dfrac{v}{w}$，则 $xy=\dfrac{u}{w}$. 取 t 使 $\dfrac{w}{t}=\dfrac{u}{v}=x$，则 $yx=\dfrac{v}{t}$. 但 $\dfrac{u}{v}=\dfrac{w}{t}$，故 $\dfrac{u}{w}=\dfrac{v}{t}$；因而 $xy=yx$.

设 $x=\dfrac{u}{v}$，$y=\dfrac{v}{w}$，$z=\dfrac{w}{t}$. 则 $xy=\dfrac{u}{w}$，$yz=\dfrac{v}{t}$，因而 $(xy)z=\dfrac{u}{t}$，$x(yz)=\dfrac{u}{t}$，即 $(xy)z=x(yz)$.

设 $x=\frac{u}{v}$. 命实数 1 等于 $\frac{v}{v}$，故 $x1=\frac{u}{v}=x$.

设 $x=\frac{u}{v}\neq 0$. 则 $v\neq\vec{0}$，$u\neq\vec{0}$，命 $y=\frac{v}{u}$，则 $xy=\frac{u}{u}=1$. 倘又 $xz=1$，则 $zx=1$，因而 $y=(zx)\,y=z\,(xy)=z$.

设 $x=\frac{u}{w}$，$y=\frac{v}{w}$，$z=\frac{w}{t}$. 则 $x+y=\frac{u+v}{w}$，$(x+y)\,z=\frac{u+v}{t}$，$xz=\frac{u}{t}$，$yz=\frac{v}{t}$，$xz+yz=\frac{u+v}{t}$，故 $(x+y)\,z=xz+yz$.

（四）中的 y 称为 x 的负，记为 $-x$.（八）中的 y 称为 x 的逆，记为 x^{-1}. 界说 $x-y=x+(-y)$，而 $y\neq 0$ 时，$\frac{x}{y}=xy^{-1}$. x 自乘 n 次记为 x^n.

设 u，v 是如此两个非零向量：$|u|=|v|$，而 $\angle(u,v)=90°$. 复数 $\frac{u}{v}$ 叫做虚数单位，记为 i. 命 $w=-v$，则 $|w|=|u|$，而 $\angle(w,u)=90°$，故 $\frac{w}{u}=i$. 令 $\frac{w}{u}\frac{u}{v}=\frac{w}{v}=-1$，故

（十）$i^2=-1$.

设 $x=\frac{u}{v}$，而 $u=\overrightarrow{AC}$，$v=\overrightarrow{AB}$. 取 $w=\overrightarrow{AD}$ 使 $AD=AB$ 而 $\angle(\overrightarrow{AD},\overrightarrow{AB})=90°$，则 $\frac{w}{v}=i$. 由 C 作 CM 平行 AD 交 AB 或其延长线于 M，作 CN 平行 AB 交 AD 或其延长线于 N，则 $\overrightarrow{AM}+\overrightarrow{AN}=\overrightarrow{AC}$. 命 $\overrightarrow{AM}=a\overrightarrow{AB}$，$\overrightarrow{AN}=b\overrightarrow{AD}$，得 $x=\frac{u}{v}=\frac{\overrightarrow{AC}}{\overrightarrow{AB}}=\frac{\overrightarrow{AM}}{\overrightarrow{AB}}+\frac{\overrightarrow{AN}}{\overrightarrow{AB}}=\frac{\overrightarrow{AM}}{\overrightarrow{AB}}+\frac{\overrightarrow{AN}}{\overrightarrow{AD}}\frac{\overrightarrow{AD}}{\overrightarrow{AB}}=a+bi$. 所以任意复数可以表为 $a+bi$ 的形式，a，b 为实数. 设 $\angle(u,v)=\theta$，而 $\frac{|u|}{|v|}=r$，界说 $\frac{a}{r}$ 为 θ 的余弦，记为 $\cos\theta$，$\frac{b}{r}$ 为 θ 的正弦，记为 $\sin\theta$，可以说明 $\cos\theta$ 和 $\sin\theta$ 只和 θ 有关系，和 u，v 的取法无关. 这样，x 又可以表为 $r\,(\cos\theta+i\sin\theta)$ 的形式.

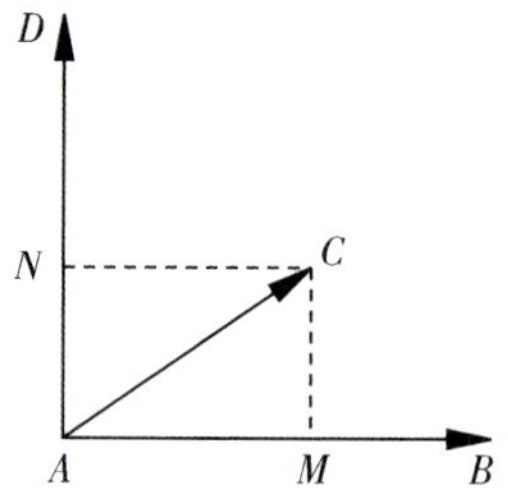

图四

倘 $a+b\,i=0$，则 $-a=bi$. 设 $bi=\frac{u}{v}$. 若 $b\neq 0$，则$\angle(u,v)=90°$，因而 bi 不是实数，不能等于 $-a$. 故 $b=0$，因而 $a=0$.

若 $a+bi=c+di$，则 $a-c=(d-b)\,i$，故 $a-c=0$，$d-b=0$，因而 $a=c$，$b=d$. 所以任意复数恰有一法表为 $a+bi$ 的形式.

根据这种表法和十个算律，可以推出复数的普通算法.

三、应　用

复数是有实数意义的. 实数是纯量的比或一条直线上向量的比，复数是平面上向量的比，虚数既然不是『虚』数，所以必有应用. 现在我们举两个例子，说明复数在数学本身的应用，其他方面的应用则超出了本文的范围.

设 u，v，w 是非零向量，$|u|=|v|=|w|$. 命 $\theta=\angle(u,v)$，$\varphi=\angle(v,w)$，则$\angle(u,w)=\theta+\varphi$. 因之，$\frac{u}{v}=\cos\theta+i\sin\theta$，$\frac{v}{w}=\cos\varphi+i\sin\varphi$，$\frac{u}{w}=\cos(\theta+\varphi)+i\sin(\theta+\varphi)$，但 $\frac{u}{v}\frac{v}{w}=\frac{u}{w}$，而 $(\cos\theta+i\sin\theta)(\cos\varphi+i\sin\varphi)=\cos\theta\cos\varphi-\sin\theta\sin\varphi+(\sin\theta\cos\varphi+\cos\theta\sin\varphi)\,i$，故

$$\cos(\theta+\varphi)=\cos\theta\cos\varphi-\sin\theta\sin\varphi,$$

$$\sin(\theta+\varphi)=\sin\theta\cos\varphi+\cos\theta\sin\varphi.$$

在平面上取定一个单位长度. 长度等于单位的向量叫单位向量. 设 O 是任意点，称为原点，而 e 是任意单位向量. 若 P 是任意点，命 $\frac{\overrightarrow{OP}}{e}=a+bi$，则 a，b 称为 P 点

的坐标，或 P 点对于 e 的坐标. 设 e' 是另外一个单位向量，而 $\angle(e', e)=\theta$. 命 a'，b' 为 P 对于 e' 的坐标，则 $\dfrac{\overrightarrow{OP}}{e'}=a'+b'i$. 令 $\dfrac{\overrightarrow{OP}}{e'}=\dfrac{\overrightarrow{OP}}{e}\dfrac{e}{e'}$，而 $\dfrac{e}{e'}=\cos(-\theta)+i\sin(-\theta)=\cos\theta-i\sin\theta$，故 $a'+b'i=(a+bi)(\cos\theta-i\sin\theta)=a\cos\theta+b\sin\theta+(-a\sin\theta+b\cos\theta)i$. 因之，

$$\begin{cases} a'=a\cos\theta+b\sin\theta, \\ b'=-a\sin\theta+b\cos\theta. \end{cases}$$

这样，对于解析几何的转轴公式和三角的和角公式我们有了一种新的理解，它们可以利用复数的计算推出来.

由于复数和向量的密切关系，不难想到我们可以利用复数解力学的问题. 譬如一个点在平面上运动，利用复数可以非常简捷地推出各种形式的关于速度和加速度的公式.

电子计算机浅说连载（上）*

高中二年级数学教材新增加了“电子计算机简介”一节。为了帮助教师掌握新教材，我们邀请王湘浩教授写了这篇文章，将分三期发表，供同志们参考。

——编者

从 1946 年第一台电子计算机制成到现在这短短的三十余年中，计算机科学技术的发展经历了电子元件为真空管、晶体管、集成电路、大规模化和超规模化集成电路这四个时代。今天，计算机软硬件的研制水平和应用的广泛性已经成为科学技术现代化的重要标志之一。

在此，就电子计算机知识分六个问题介绍：一、信息及编码；二、介绍一台模型机；三、指令系统；四、程序；五、软件；六、计算机的应用。

一、信息及编码

什么叫信息呢？自然界和人类社会中有大量的信息。例如，人说的话，就是信息。书上的字也是信息。书里的文章就是信息的汇集，它无非是把作者要说的话写下来，让读者知道。在自然界里，地质学家如何探矿呢？就是根据地层里面储存的信息，比如，化石、地质构造等等来推断地底下究竟有什么物质。要发生地震时，一些鸟兽能够感到一些信息，比人还灵敏。如，我们知道的，老鼠在震前就满街跑。气象预报就是通过收到许多信息，来推断明天或者一个时期内天气如何。从大方面来说，在宇宙空间里，如何知道一些星星的情况呢？就是通过光所传来的信息来推断。从微观来说，在原子核里存有大量信息，将这些信息收集到之后，来了解原子核的内部构造。生物遗传是通过细胞中存放的一些信息来起作用。小孩象父母，是父母的生殖细胞把一些信息传给小孩的缘故。人和外界怎样发生联系呢？人通过感觉器官，感受到

---

* 本文原载于《辽宁教育》1980 年第 3 期。

外界的信息，然后由大脑对这些信息进行处理、加工，之后再传达给执行器官，如腿脚、胳膊、手等。因此，人与外界联系也是通过信息的。收音机、电视机接收的是无线电波运载来的信息，测量仪器也无非是收集信息。

因此，信息在自然界，人类社会里是无所不在的。很难设想如果没有信息，事物将如何发生联系。物理学过去主要研究能量。信息是一个新的概念，其重要性不亚于能量。

通过以上许多例子，我们就会清楚，为什么计算机的用途如此之大。原因就在于计算机能够快速地处理大量、复杂的信息，所以它的用途非常广泛。这是人类发明的快速地处理大量复杂的信息的工具。

计算机处理信息是通过“算法”。因此，计算机科学基本的概念是信息和算法。算法的具体化就是计算机上的程序，对此我们将在后面介绍。

因此，我们可看到，计算机科学是自然科学的一门新学科。从基本概念，研究的对象来看，它不同于数学、物理学、化学、天文学、地学、生物学，而有它本身的对象，是一门独立的学科。它既不是数学的一部分，也不是物理学的一部分。从某种意义上来说，它是一门技术学科，技术性很强。而另一方面，又是一门基础学科。

当我们使用计算机的时候，怎样把信息送到计算机里，让计算机处理，然后再将结果送出来给人看呢？这就要以某种形式将信息送到计算机里。目前送到计算机里面去的信息主要是文字或数字。当然也有让计算机直接识图和听人的语言，但目前尚处于研究之中。

那么究竟用什么形式和方法将信息送入计算机呢？比如，处理一篇文章或要作某种计算时，怎样将文字或数字送入计算机呢？回答是用编码形式将它们送入计算机的。

编码没什么稀奇的。这如同打电报一样。打电报无论是表示数字或字母，都是用“滴”“答”来代表。“答”代表一条线，就是“—”，“滴”代表一个点，就是“·”。用它们编成不同的码，来表示数字或字母。计算机中所用的编码，也基本上一样。就是用两个符号：“1”和“0”。“1”类似“答”，“0”类似“滴”。将“1”、“0”编起来，用以代表文字和数字，然后送进计算机里。如，101001就是一个编码，我们可以规定它代表某个字母或数字。一串 0 和 1 组成的编码我们称之为一个“字”。

在计算机里面，0 和 1 要用具体的物质的东西来代表。例如用低电平和高电平分别代表 0 和 1，或用无电流和有电流分别代表 0 和 1。存放一个 0 或 1 的电子器件叫触发器。它的输出端有低电平和高电平这样两个状态。高电平代表 1，低电平代表 0。我们可以通过在输入端加信号，使输出端输出 1 或 0。如何用电子元件构成触发器，在这就不讲了。

一个触发器能够代表“1”或“0”，那么一串触发器就能代表一个编码，就是说，代表一个字。一串触发器组成的器件叫寄存器。如，十六个触发器组成一个 16 位寄存器。这样一个寄存器就可以存放一个十六位的编码，就是说，存放一个十六位的字。

这样，在有信息的时候，就可把它们存到寄存器中。

另一种为计算机所使用的贮存一个 0 或 1 的元件是磁芯。它是一种磁性元件，形同一个环。它可以很小，如同一个小米粒那么大。计算机用来大量存储信息的部分，叫作存储器，一般是用磁芯构成。磁芯代表“1”或“0”的道理是什么呢？这就是磁芯可以有两种不同的磁化方向。对这样两种方向，我们就可以规定一种方向代表“1”，另一种方向代表“0”。这也类似触发器。我们把磁芯串上金属导线，通过电磁感应，就可改变其磁化方向。一串磁芯可以起到寄存器的作用。如，串成一串的十六个磁芯就组成了十六位的代码。

于是我们可看到，信息可通过二进编码来代表，然后送进计算机里。而计算机处理信息也就使用二进编码。那么数是怎样实行二进编码的呢？下面就介绍二进制数。

普通我们常用的数是用 0、1、2、3、4、5、6、7、8、9 共十个符号组成的十进制数。这是人人都熟悉的。人为什么使用十进数呢？其实十进数并没有什么特殊含义。之所以使用，就是因为人有十个手指头。在古代的时候，人打鸟或打兽就是用手指头或脚指头来计算。现在小孩刚开始学算术，不也常常用手指头吗？如果人人都是六指，那就会使用十二进制数。因此十进制数并非是天经地义的，也完全可以用其它进制。最简单的就是二进制。我们可结合十进法来想二进法，就是用 0 和 1 来组合，代表所有的数。十进制数用 0 到 9 共十个符号，二进制数仅用两个符号：0 和 1。现在我们就举个具体例子来看，如图1。

```
   0
   1
  10
  11
 100
 101
 110
 111
1000
```

图1　二进制数

在十进制数算法中，够10就进位。第一位称为个位，第一位的1代表1个。第二位的1代表十，第三位的1代表十个十，等等。在二进制数的算法中，够2就进位。个位的1就是1个。第二位的1代表2。右数第三位的1代表二个2，即代表2×2=4。第四位代表2×4=8。第五位代表2×8=16。依此类推。

写成101001的二进制数是41。原因是32+8+1=41。

二进制数是简单的。它的加、减、乘、除法也简单。只要结合十进制数的加、减、乘、除法来看就容易明白。

加　法、如：
```
     1010
+）  1011
    10101
```

在十进法中，是够十进1。而这里的二进法是够2进1。

减　法、如：
```
     1100
—）  1011
     0001
```

在十进法中，是借1来十。而这里的二进法是借1来2。

乘　法、如：
```
       101
  ×）  101
       101
      000
+）  101
     11001
```

道理与十进法类似。

除　法、如：

```
           1010
      ________
  101 )110110
       -)101
       ______
          111
       -) 101
       ______
          100
```

道理与十进制数的除法一样。这里商是1010，化成十进数就是8+2=10。余数是100化成十进数就是4。

看来小孩子们学这个会很快，也不用背“小九九”。学珠算也用不着背口诀。

以上就是二进制数的加、减、乘、除法。

我们知道，在代数里有正数和负数。在二进制数里，和十进制数一样，也有正、负数。在计算机里是怎样表示的呢？比如一串触发器代表一个二进制数，那么符号如何表示以便区别正、负数呢？这只要在寄存器前加一个触发器就行了，这个触发器叫作符号触发器。“0”状态为“正”，“1”状态为“负”。例如，16个触发器，让第一个代表符号，其余15个代表数。因此用触发器做的寄存器可以代表正、负数。对于磁芯也如此，加一个磁芯来代表符号就行了。

二、介绍一台模型机

我们这里所谓的模型机，就是尽量简化的、理想化的计算机。比如，在物理上讲发电机，讲得比较简单，而实际上却很复杂。原因就是讲时为了讲原理，于是就抽出主要的部分，舍掉次要的部分，尽量简化，使之成为模型，这样就便于讲原理了。我们这里也是如此，把计算机尽量简化，使之成为一台模型机，以便我们介绍原理。

计算机有五大部件。

运 算 器：计算机用它来处理信息，如进行加、减、乘、除及其它操作。顾名思义，可以为它就是被用来算。可把它想象为一个算盘，或台式的手摇计算机。

存 储 器：它是用来存储大量数据或大量信息的。可想象它是纸，被用来记录运算过程的中间结果及最后结果。原始数据，中间结果，最后结果都放在这里面。运算器里没有这么多地方放这些东西，因此须有存储器。

输入装置：这里特殊用的是光电输入机。人用它把信息送入计算机。

输出装置：这里用的是穿孔机，计算机用它将结果送出来。

控 制 器：它控制整个计算机进行工作。

以上为计算机的五大部件。还有控制板，人用它来控制计算机的运行。

现在，一一具体地说明。

运 算 器：它里面的核心部分是一个十六位的寄存器。在此，它被称为累加器，用字母 A 来表示，也可写为“累加器 A”。若存一个二进制数，那么最左边那个触发器就是符号触发器，也称为“符号位”。其余 15 个触发器，或称 15 位，就代表二进制数的数值。为什么称它为累加器呢？因为计算结果总是放在它里面。它是运算器里最主要的部件。

存 储 器：用许多排磁芯组成。为了具体化一些，我们在此就说 16 个磁芯组成一排，用来放二进制数，或其它信息。存储器共有 4096 个地方，亦即 2^{12} 个存储单元。对于每个存储单元，我们给它一个号码。这道理就如同人在旅馆里住的房间号，如某同志住 xx 号房间。可设想一个存储单元如同一个房间，里面放二进制数或其它信息。但还要给房间号，在存储器里就叫作存储单元的地址，也叫“地址码”，它共有 12 位。我们可用电磁感应的方法，按照存储单元的地址，把一个数或其它信息存到存储器里的某个单元；也可按照存储单元的地址，把存储器里那个单元里面存的数或其它信息取出来，送到运算器里。

我们可以用十进数表示存储单元的地址。在计算机中要用二进制数表示，总之，每个存储单元要给一个号码，称为存储单元地址，共有 4096 个地方，亦即 4096 个单元，每个可存一个二进制数或其它信息。

光 电 机：它通过光电效应可将穿孔纸带上的信息送入计算机。穿孔纸带有好多种，这里仅介绍八单位的穿孔纸带。每排可以穿 8 个孔，一排一排的。穿孔就表示 1，不穿孔就表示 0。两排合起来是 16 位。两排表示一个 16 位的二进制数或其它信息。第一排表示右边八位，第二排表示左边八位。用两排来表示一个数或其它信息。使用的时候，先将二进制数或其它信息穿在纸带上，然后将纸带安置在光电机上，纸带在光电管上面通过时，有孔的就通电，无孔的地方就不通电。这样就将编码通过光电效应送入计算机。

穿 孔 机：它是穿纸带的，把计算机对数或其它信息处理后的结果穿到纸带上，输出来。这里纸带上的孔与排的规定与输入时一个样。

控 制 器：控制整个计算机来进行工作的。对它就不详细讲了。

控 制 板：控制板是人用来控制计算机进行工作的。它上面有下列一些东西：电源开关，运行灯，地址开关，引导钮，启动钮。

电源开关拧上去就使电源接通，拧下来就将电源关闭。运行灯用以标志计算机是否在运行中，计算机在运行时此灯便亮，否则便灭。地址开关是一排十二个开关，用以安排一个地址，每个开关扳上去代表 1，扳下来代表 0。引导钮用以把穿孔纸带上的数据或其他信息送入存储器。先把纸带安置在光电机上，再利用地址开关拨一个地址，然后按一下引导钮，于是纸带上的数据或其他信息便从所拨的那个地址开始进入存储器中去了。启动钮用以启动计算机从某个地方开始执行程序（后再详细介绍）。

电子计算机浅说连载（中）*

三、指令系统

计算机是用来处理信息的非常复杂的工具，它可以处理很多很复杂的工作。但是它最基本的功能并没有多少。所谓指令系统，就是一台计算机所具有的最基本的功能，或者说，它所能作的最基本的操作。它们总共不过几十种或几百种。那些复杂的工作要化成这些指令，由这些指令配合起来完成。

在设计一台计算机时，要事先规划好计算机能作哪些基本操作。即是说，这台计算机要具有多少条指令。我们假定这台模型机可以作 16 种基本操作，也就是说，它的指令系统有 16 条指令。它们是加、减、乘、除、取、存、读、穿、转、比、停等等。它们具体的功能是这样子的（左边的 0000，0001 等是操作码，其意义下面再讲）：

0000 加：把存储器某单元中的数，加到累加器里的数上；

0001 减：把存储器某单元中的数，从累加器里的数上减去。

0010 乘：把存储器某单元中的数，乘到累加器里的数上；

0011 除：把存储器某单元中的数，去除累加器里的数；

0100 取：把存储器某单元中的数取到累加器里，代替原有的数；

0101 存：把累加器里现有的数存到存储器里的某个单元，代替这个单元中原有的数；

0110 读：从光电机纸带上读一个数到累加器里，并存到存储器里某个单元中去；

0111 穿：把存储器里某个单元中的数用穿孔机穿到纸带上；

1000 转：在执行程序时，当执行到此处就转移到程序中的别的地方去执行（这在下面再详细说明）；

---

* 本文原载于《辽宁教育》1980 年第 4 期。

1001 比：按累加器里的数是否小于存储器中某个单元里的数，决定是否跳过下一条指令（这也在下面再详细说明）；

1010 停：让计算机停止工作。此时，控制板上的运行灯就熄灭了。

至此共 11 条，还有 5 条，就不说了。

四、程　序

现在我们举几个简单的程序设计的例子来说明。

例 1　设有 a、b、c 三个数分别存在 1、2、3 这三个单元里，求 a+b−c。

现在我们编个程序，让计算机执行，以便把 a+b −c 算出来。1、2、3 是单元的地址。

我们编了一个有 4 条指令的程序。现在解释一下："取 1" 就是从 1 单元中取出数放到运算器里的累加器中，这也就是把 1 单元中的 a 取出来放到累加器中了。"加 2"，把 2 单元中的数，也就是 b 取出来加到累加器中，这样就算出 a+b。"减 3"，把 3 单元中的数，也就是 C 取出来，去减到累加器中的数上，这样就得到了 a+b−c。"停"，就是让计算机停下。

取	1
加	2
减	3
停	

这是个最简单的程序。编好程序之后，如何通知计算机让计算机执行呢？这就要把这几条指令编码。指令就是让计算机作的事，它也是信息，用编码的形式通知计算机。编码当中，既要有操作，又要和存储器某个单元地址有关系。

因此，指令编码时，有操作码和地址码，合起来刚好是 16 位的一个字，我们现在就用这样的字来代表指令。前 4 位是操作码，代表操作。这是把操作编码，它用 4 位二进制数代表，这在上面已经列在每条指令的左边了。4 位操作码对于 16 条指令刚好够用。后 12 位是地址码，代表地址。这 12 位刚好够 4096 个单元。

现在我们把上面那 4 条指令编码：

取 1　$01000\underbrace{\cdots\cdots}_{11个0}01$

加 2　$00000\underbrace{\cdots\cdots}_{10个0}010$

减 3　$00010\underbrace{\cdots\cdots}_{10个0}011$

停　$10100\underbrace{\cdots\cdots}_{12个0}0$

其中“11 个 0”或“10 个 0”或“12 个 0”，是为清楚起见，我们加的说明。实际编码时，要把所有这些 0 写上。另外，“停”与存储器无关系，因此地址码写任何数码都可以，只要操作码写对就可以。我们这里写的是 0。

编码后的四条指令要送到计算机的存储器中去，以便让计算机执行。假定我们要把它们送到 4、5、6、7 四个单元里面，并假定 a、b、c 三个数还没有送入 1、2、3 三个单元。我们可以把这三个数连同四条指令同时送入计算机。

于是我们写出下面的程序：

1	数 a
2	数 b
3	数 c
4	01000…001
5	00000…010
6	00010…011
7	10100…000

这个程序有三个原始数据，有 4 条指令，三个数据 a、b、c 当然要是具体的二进数。左边从 1 到 7 是这 7 个字要送入的单元。为了往计算机里送，以便让计算机来执行，就要先把这 7 个字穿孔在纸带上。共有 7 个字，那么在纸带上就共有 14 排。纸带穿好孔，再安置在光电机上。这时把控制板上的地址开关拨成 1，也就是前 11 个开关扳下来，都是 0。最后一个开关扳上去，是 1。再按“导引钮”，计算机就将纸带上的信息读入存储器，而且是从单元 1 开始存放这 7 个字。

把程序送进计算机以后，就该让计算机执行程序了。那么，从哪儿开始执行程序呢？单元1、2、3是三个数，而从第4单元开始存放的是指令，因此要从第4单元开始执行。在控制板上将地址开关扳成4，然后按“启动钮”。此时“运行灯”亮，计算机就从第4条指令开始逐条执行，最后停机。这时“运行灯”就灭了。这样计算机就执行了这个程序。

为什么计算机能一条条地执行指令呢？它一是靠运算器执行，一是靠控制器控制执行。这在设计计算机时就已经设计好了，所以计算机能够这样逐条执行指令。

下面的例子是求斐波纳奇数，我们先说一下什么是斐波纳奇数。斐波纳奇有一道有趣的算术题；假定新生的一对兔子过一个月还不能生产，从第二个月起便每月生一对兔子，那么，新生的一对兔子一年后孳生成多少对？

因为过一月还不能生产，所以过一个月仍是1对，过两个月变成2对，3个月后成为3对，4个月后成为5对，如此类推。实际上，我们有下面的公式：

上月数＋本月数＝下月数。

利用这个公式可以列出下面的表：

月数	兔子对数
1	1
2	2
3	3
4	5
5	8
6	13
7	21
8	34
9	55
10	89
11	144
12	233

所以一年后孳生成233对。当然我们还可以接着往下算。这样一直往下算得到的一串数1，2，3，5，8，……。就叫斐波纳奇数。斐波纳奇数和“优选法”有关系，

在程序设计中也有用处。

例 2 求小于 20000 的所有斐波纳奇数。

我们写出下面的程序然后再加以解释：

单　元	内　容
0	穿 10
1	取 10
2	加 9
3	比 11
4	停
5	存 10
6	减 9
7	存 9
8	转 0
9	1
10	1
11	20000

这个程序如上面那一例所示是要存到 0 单元到 11 单元中去的。0 单元到 8 单元的内容是指令，10 单元存本月数，9 单元存上月数，11 单元存常数 20000。

这个程序要从 0 单元开始执行。执行了“穿 10”，就把单元 10 中的 1 穿到纸带上去了，这是一串斐波纳奇数中的第 1 个。接着执行“取 10”，把单元 10 的内容 1 取到累加器。A 执行了“加 9”，就算出第二个斐波纳奇数 2。然后执行“比 11”，以累加器 A 的内容和单元 11 的内容相比较，若小就跳过下一条指令，否则不跳过而接着执行下一条。现在 A 的内容 2 小于单元 11 的内容 20000，所以跳过下一条指令而去执行“存 10”，这就把方才算出的第 2 个斐波纳奇数 2 存入了单元 10。接着执行“减 9”重新算出第 1 个斐波纳奇数 1，而执行“存 9”后便把此数存入了单元 9。然后执行“转 0”，这就是转到 0 单元去执行那里的指令“穿 10”。于是重复上面那些步骤，不过并不是简单的重复，而是在一个新阶段上的重复，这次穿出去的是第 2 个斐波纳奇数；

而执行到单元 7 中的指令后，单元 10 中存入了第 3 个斐波纳奇数，单元 9 中存入了第 2 个斐波纳奇数。然后执行“转 0”穿出第三个斐波纳奇数，如此类推。直到 A 中算出的斐波纳奇数已经不小于 20000，这时执行“比 11”便不再跳过下一条指令，于是执行“停”使计算机停止工作，而小于 20000 的所有斐波纳奇数都已经穿在纸带上了。

这个程序里面有两点需要特别指出：指令“比 11”叫做“条件转移”。这条指令使得计算机能够按照当时的不同情况自动地作不同的处理，换句话说，使得计算机有“随机应变”的能力。这是计算机科学的一大发明。第二点需要特别指出的是“循环”的技巧。这里只有 9 条指令，但计算机执行时要循环多次。实际算一算可以看出计算机要循环 21 次，执行这 9 条指令共 165 次。如果没有这种“循环”的技巧，这个题目编程序就要写出一百多条指令。

例 3 一些款项已以分为单位穿在纸带上最后穿的是 0 表示已穿完，求总和。

我们编出程序如下：

单　元	内　容
1	读 10
2	读 0
3	比 11
4	转 7
5	穿 10
6	停
7	加 10
8	存 10
9	转 2
10	0
11	1

这个程序就是逐步从纸带上读数加入单元 10。直到读进来的数是 0。这表示各笔款项已经读完，于是把单元 10 中的总和穿出去。程序从 1 单元开始执行。首先，读

进纸带上的第一个数送入单元10。接着，读第二个数到累加器A。此数还要送入单元0，但这对于我们的题目是不起作用的。接着，以读进的数和单元11中的1相比较。若小于1，那就必然是0，这表示纸带上的各笔款项已读完，于是跳过下一条指令把单元10中的总和穿出去而停机。若不小于1，便执行“转7”，把读进来的数和单元10中的数相加再送入单元10，然后“转2”再去读纸带上的数。如此类推。

电子计算机浅说连载（下）*

五、软　件

这个名字听起来特别，它是国际上通用的名词。其实软件就是比较复杂的程序，没有什么特别难于理解的。相对来说，计算机本身，也就是机器，叫作硬件。计算机是人类制造的机器，它有个特点，不同于其它机器，就是要用软件。光有计算机本身的话，它只能执行基本操作，很简单。要解决复杂的问题，就一定要有程序。对于一个复杂的问题，人要事先设计解决问题的方法，然后根据方法编程序。一个复杂的问题，无非是把许多指令编排起来进行解决的。程序有的是很大的。上面举的几个例子很简单。复杂的程序可有几百条，几千条，几万条，几十万条。

软件特别指计算机本身要带的程序。计算机出厂时，一方面要有硬件，另一方面还要带大量的程序，穿在纸带上或记录在磁盘、纸带等上面，这就是计算机所带的软件。这是一些用得多，用得普遍的基本的程序。它的作用是为了方便用户，省得用户还要设计那些基本的程序。

软件非常重要，它可以把计算机的功能增强许多。下面举几个例子。

汇编程序　上节例 1 中的三个数 a、b、c 是具体的数。假定 a=23　b=14　c=18 于是，那里的程序具体写出来是：

1　0000000000010111

2　0000000000001110

3　0000000000010010

4　0100000000000001

5　0000000000000010

6　0001000000000011

7　1010000000000000

*　本文原载于《辽宁教育》1980 年第 5 期。

这样的程序很容易写错而且很难看懂。但是，为了让计算机执行，只好写成这样，以便送到存储器中去。后来，人们想了一种办法：人编程序时，可以用符号来写，然后让计算机把这种用符号写出来的程序自动翻译成如上的“码子”程序。例如，上面算 a＋b－c 的题目可以由人写出程序如下：

	始址	1
第一数		23
第二数		14
第三数		18
启　动	取	第一数
	加	第二数
	减	第三数
	停	
	完	启　动

说明一下。“始址 1”表示要从单元 1 开始安排地址。“完”表示程序已经写完，后面接写的“启动”表示要在左边注有“启动”二字的那条指令开始执行程序。那些方块字就是我们所说的“符号”。用方块字，计算机的外部设备中就要有中文打字机。没有中文打字机时，可以把方块字改成汉语拼音文字而使用普通的电传打字机。

这种用符号写的程序不容易写错，又容易看懂。这叫做用“汇编语言”写的程序。用“码子”写的程序说是用“机器语言”写的程序。为了把汇编语言翻译成机器语言要使用一种软件，称为“汇编程序”。计算机出厂时应该带有这种软件。

编译程序　用汇编语言写程序比用机器语言当然好得多。但是，计算机多种多样，指令系统五花八门。用于某台计算机的程序在另一台计算机上就不能用。一个人在不同的时间地点可能要使用多种计算机，因而要学习多种指令系统和汇编语言，这对于一般用户来说，当然是很不方便的。于是，人们又想了一种办法：创立不依赖于计算机指令系统而接近自然语言的所谓“高级语言”。著名的高级语言有 ALGOL，FORTRAN，BASIC，等等。把高级语言翻译成机器语言的软件叫做“编译程序”。一般计算机都配有上述几种高级语言的编译程序。用户只要学会一种高级语言，例如

ALGOL，不论使用什么计算机，根本不用管它的指令系统，只要用自己掌握的那种高级语言编程序就行了。

上节例 1 的题目用 BASIC 语言编程序只用两句话：

1 穿　23＋14－18

2 完

操作系统　计算机有各种“资源”，例如处理器（包括运算器和控制器），存储器，外部设备，汇编程序，各种编译程序，各种档案，等等。人使用计算机就要使用它的各种资源，为此，有时需要编很复杂的程序。一台计算机经常有许多用户争先恐后地要使用，如何安排和调度以便最有效地使用计算机的资源就更是一个非常重要的问题。处理这个问题的办法有两种：成批处理和分时。成批处理是由用户把自己编好的解题程序随时提交给计算机管理人员，管理人员把不断送来的程序根据某种标准排队，成批地送入计算机处理。分时是用户各有自己的终端设备，随时可以把自己的解题程序送入计算机，计算机把时间分成小段，让各用户轮流使用。不论成批处理还是分时，都需要一种极其复杂的软件进行安排和调度。这种软件叫“操作系统”。

应用软件　计算机有各种各样的应用，针对某种应用编出的软件叫“应用软件”。例如，科技情报管理、工厂管理、生产控制、空防系统管理、计算机辅助设计、辅助教学、问题解答，等等，都需要专用的应用软件。解一个简单的问题所用的程序，例如上节的三个程序，那就只是简单的程序，不算是软件了。

现代计算机不能只有硬件没有软件。只有硬件的“裸机”好比是赤膊上阵的李逵，配上软件就好比是顶盔贯甲跨马提刀，这才可以大显身手。今天的计算机体系，特别是大型、巨型计算机和许多计算机连成的网络都是软硬件交互作用的。设计一台计算机必须软硬件一起全面考虑。“计算机”这一概念本身就包括软件和硬件在内。

六、计算机的应用

计算机问世之初，只在大学和科研部门使用，社会公众对它并不注意，甚至不知道有这种东西。今天，计算机已经广泛应用于国民经济甚至日常生活的各个方面。据 1974 年的统计，计算机的应用达 2670 种。事实上，现在不需要问计算机能够应用于哪些方面而只要问究竟还有哪些方面用不上计算机了。

计算机的应用主要可以分为三大类：科学计算，大量信息处理，实时自动控制。

科学计算 以前，数学是“科学的皇后”，其应用于各种科学技术，主要是作为说理的工具，真正能够进行计算的只是一些极其简单的理想化了的东西。自从有了电子计算机，人们才能够在短时间内进行大量的数值计算，数学才真正成为进行科学研究，进行设计施工和指导生产的有力工具。以前，一种新产品在投入生产前必须进行多次的耗费大量人力物力的实验，现在，使用计算机进行计算可以代替许多实验。有的产品在使用前根本无法做实验，例如人造卫星就是这样，这就必须事先利用计算机进行精确计算，然后施工和使用。在这方面值得特别提一下的是计算机辅助设计。例如设计一个建筑物，工程师把数据送入计算机，计算机经过计算，立即把预想的那个建筑物的图形在电视屏上显示出来，工程师可以让计算机把图形旋转各种角度，以便看自己的设计是否合意，他可以用光笔修改图形，然后让计算机绘出设计图纸。

大量信息处理 属于这方面应用的主要是情报检索和业务管理。情报也就是信息，这是多种多样的，例如图书资料，科技情报，经济情报，人事档案，航空和卫星摄影图片，等等。这些情报可以用计算机储存起来。人们需要某种情报时，可以按照一定的方式向计算机提问，计算机检索储存的情报，经过某些处理和推理回答人们的问题。这就是情报检索。至于业务管理，现在已经发展为一门新兴学科，叫做“管理科学”。这门科学大量使用计算机，并且和情报检索相结合。例如工资计算，生产计划，成本计算，库房管理，银行业务，等等。此外，象一项大型的科学研究，必须组织成千上万的人员，组织许多科研机关、学校，工厂分工协作，不利用计算机进行周密的管理，势必乱成一团。例如制定研究计划，确定研究课题，协调各方面的研究工作，物质条件保证等等，不用计算机是完全不可能的。

这里必须特别提一下计算机辅助教学。由在某门课程上教学经验丰富的教师编好教学程序送入计算机。学生要学这门课程的某个章节时，只要打开计算机，以某种方式选择这一章节，计算机便把本章节的第一段课文在电视上显示出来。学生认为已经看懂后，以某种方式通知计算机，计算机立即显示几个问题让学生回答，然后按照他答题的情况转到另一段课文。如果他答得好些转到新课文，否则转到进一步讲解或纠正其错误的课文。以后仍用这种讲解加问答的方式对学生进行“因材施教”，直到他学会这一章节为止。这种办法大大有利于提高教学质量和节省教师力量，并且可以大量用于普及教育。

实时自动控制 生产全盘自动化是人们世世代代的理想。现在有了计算机，这已经不是遥远的幻想了。例如日本有一个钢铁厂过去用十五万人，使用计算机控制生产后，只用四千人就行了。除了生产，象交通运输、战略武器系统等等，现在都使用计算机进行实时控制。一个现代化的物理实验室或化学实验室必须大量使用计算机进行控制，甚至实验仪器上都带有微型计算机。

除了上述三类重要应用，计算机日益进入人们的日常生活，例如订飞机票，到商店买东西开账单，到医院治病计算机辅助医疗，等等。计算机还用以代替某些家务劳动，以后电话机、打字机、照相机、缝纫机都要装上计算机。计算机将成为家庭必需品。

常言说得好：“戏法人人会变，各有巧妙不同”。在计算机的各种应用中都要编程序，有的人编得笨一些，有的人编得巧一些。编得巧就使得计算机的“智能”因素多一些。计算机代替人的部分智能活动并没有什么神秘，对此用不着大惊小怪。计算本来就是一种智能活动，十七世纪发明的机械台式计算机能自动作加减乘除，这就是初步代替人的一点智能活动。计算机科学中的“智能模拟”研究，其目的就是要使计算机更有效地代替人们更多的智能活动。这种研究包括数学定理证明和公式推演、学习、概念形成、逻辑推理、模式识别、自然语言理解、问题解答、判定决策、博奕、绘画、作曲、机器人，等等。这种研究目前还处于初级阶段。但是，有迹象表明本世纪内这方面将有重大突破。例如，古老的数学难题“画地图只用四种颜色分别相邻区域够不够?”终于在前年利用计算机进行大量判断得到正面解决，就令人信服地说明计算机的应用将进入一个带有更多智能因素的新阶段。人类发明了计算机，但还没有充分理解它的作用。到本世纪末，计算机的体系设计及其应用将达到怎样的高度，实在是我们今天所难以想象的。

三段论的一般规则*

三段论是普通逻辑书中演绎推理的主要部分，但一般书恰在这一部分说理不够严格和充分。例如关于名词的规则“中词至少要周延一次”。既然说是规则，那就不同于公理，就必须证明。但一般只是举一个例子，或针对这个例子说明一下。这显然不符合“充足理由律”。中词在一个前提中不周延有四种可能情形，在两个前提中都不周延可以搭配出十六种情形。再考虑到结论可有 A、E、I、O 四种可能情形，这样就共有六十四种情形。如果要一一举例，那就要举六十四个例子才算充分进行了说理。

另外有的书不用举例法而证明如下：“中词在前提中起媒介作用，大词和小词的联系是通过中词而实现的。中词在前提中至少周延一次，大词和小词才有必然的联系；如果中词在前提中一次也不周延，大词和小词的联系就不确定，就得不出确定的结论。”这些话有的不知其确切含义是什么，其中的用语究竟作何解释。有的话绕来绕去只不过是原规则的同语反复。所以，这些话实际上什么也没有证明。

关于前提的那几个规则也有类似情况。逻辑要求人们在从事思考和推理时要严谨。如果逻辑书自身就不严谨，那就不好要求读者了。本书尝试较严格地证明三段论的一般规则，即关于名词和关于前提的规则。

一些基本概念和简单事实

首先，需要明确所谓一个三段论正确究竟是什么意思，然后论证才有所依据。

有的书只谈“判断”而不谈“命题”，这似乎没有太多的理由，本文仍使用“命题”这个词。试看下面的一些语句：

凡 S 是 P，

凡 S 非 P，

有 S 是 P，

有 S 非 P。

---

* 本文原载于《吉林大学社会科学学报》1987 年第 4 期。

若S和P是取定的名词，这些语句当然就是四种类型的直言命题了。如果S和P看作是未定的名词，这些语句的真假就是未定的，因而就不是命题，我们称之为直言式。以后，未定名词用S、P、M、X、Y、Z等代表，取定的名词用A、B、C等代表，直言式用F、G、H等代表。直言式F中的两个未定名词是X和Y时，F可以写成F（X，Y）。以取定的名词A、B分别代替F中的X、Y所得的命题记为F（A、B）。

设F（M，P），G（S，M），H（S，P）是三个直言式。下面的推理

$$
\begin{array}{l}
F(M, P), \\
G(S, M), \\
H(S, P)
\end{array} \tag{1}
$$

称为一个三段式。分别以取定的名词A、B、C代替S、P、M所得的推理

$$
\begin{array}{l}
F(C, B), \\
G(A, C), \\
H(A, B)
\end{array} \tag{2}
$$

是三段式（1）的一个实例，称为一个三段论。

定义一：三段式（1）说是正确的。如果不论怎样取A、B、C所得的实例（2）中，只要两个前提F（C，B），G（A，C）为真，结果H（A，B）必为真。反之，若有一个实例（2），其中两个前提为真而结论却为假，则三段式（1）说是不正确的。

定义二：三段论（2）说是正确的三段论，那么以之为实例的三段式（1）是正确的。这两个定义中有一点需要说明：一个名词所指称的事物有可能不存在。换句话说，名词的外延可能是空的，这样的名词称为空名词。对一个三段式（1）可以加非空限制：其中有的未定名词，甚至所有三个未定名词，只允许以非空名词代入。例如，可以限制三个直言式中的主词只能以非空名词代入。如果有非空限制，定义一中的和定义二中的实例（2）必须符合所加的限制。易见，不论有没有非空限制，下面的简单事实总是成立的：

事实一：若能取非空名词A、B、C使（2）的两个前提为真而结论为假，则三段式（1）是不正确的。

含A、B的直言命题共八个，全称特称各四，肯定否定各四，A在其中周延与不周延者亦各四，如下表所示：

表一

<table>
<tr><td></td><td>全称</td><td>特称</td><td></td></tr>
<tr><td rowspan="2">肯定</td><td>凡B是A</td><td>有B是A</td><td rowspan="3">A在其中
不周延</td></tr>
<tr><td>凡A是B</td><td>有A是B</td></tr>
<tr><td rowspan="2">否定</td><td>凡A非B</td><td>有A非B</td></tr>
<tr><td>凡B非A</td><td>有B非A</td><td rowspan="2"></td></tr>
<tr><td></td><td colspan="2">A在其中周延</td></tr>
</table>

定义三： 名词A和B说是互相重合，如果A和B的外延相同。

定义四： 名词A和B说是互相分离，如果A和B的外延无公共部分。

定义五： 名词A和B说是互相交叉，如果A和B的外延有公共部分并各有一部分在此公共部分之外。

定义六： 名词A说是从属于B，果如A的外延是B的外延的一部分，那么A和B非重合。

以下举例将用到一些具体名词。设丁大学有中文、历史、哲学等系，每系都有男生和女生。我们将用到下列五个名词：丁校学生、丁校男生、丁校中文系学生、丁校历史系学生、丁校哲学系学生：这五个名词将依次简记为校，男、文、史、哲。它们的相互关系如下表所示：

表二

	校	男	文	史	哲
校	重合				
男	从属	重合			
文	从属	交叉	重合		
史	从属	交叉	分离	重合	
哲	从属	交叉	分离	分离	重合

参照表一不难验证下面的四个简单事实是成立的：

事实二： 设A和B重合且非空。含A和B的四个肯定命题为真，四个否定命题为假。

事实三： 设 A 和 B 分离且非空。含 A 和 B 的四个否定命题为真，四个肯定命题为假。

事实四： 设 A 和 B 交叉。含 A 和 B 的四个特称命题为真，四个全称命题为假。

事实五： 设 B 非空且从属于 A。含 A 和 B 而 A 在其中不周延的四个命题为真，A 在其中周延的四个命题为假。

关于名词的规则

普通逻辑书关于名词有三个规则，其第一个是：三段论只能有三个名词。这实际上不必列为规则，因为，三段论当然只能有大中小三个名词，这是三段论的定义中所规定的。只是由于普通语言中的名词往往有不同含义，因而使用三段论时可能犯所谓四名词的错误，而诡辩家也可以利用名词的歧义钻空子而已。四名词的错误可以放到专门讨论证明中错误的章节中去，和三段论的一般规则摆在一起是不伦不类的。

现在证明关于周延性的两个规则：

规则一： 中词至少要周延一次。

证明：设有三段式如下：

$$F(M,X),\quad G(M,Y),\quad H(X,Y), \tag{3}$$

其中中词 M 在两个前提内都不周延。求证此三段式（3）不正确。分两种情形来看。首先，设 H（X，Y）是肯定的。以校、文、史分别代 M、X、Y 得三段论如下：

F（校，文），
G（校，史），　　（4）
H（文，史）。

由事实五，两个前提为真；由事实三，结论为假。因之，由事实一，（3）不正确。其次，设 H（X，Y）是否定的。以校、文、文分别代 M、X、Y 得三段论如下：

F（校，文），
G（校，文），　　（5）
H（文，文）。

由事实五，两个前提为真；由事实二，结论为假。因之，由事实一，(3) 不正确。

规则二： 在结论中周延的名词在前提亦必周延。

证明：设 (3) 中的 X 在 H (X, Y) 中周延而在 F (M, X) 中不周延。求证 (3) 不正确。先设 G (M, Y) 是肯定的。以文、校、文分别代 M、X、Y 得三段论如下：

$$
\begin{gathered}
\text{F（文，校）},\\
\text{G（文，文）},\\
\text{H（校，文）}。
\end{gathered}
\tag{6}
$$

由事实五与二，两个前提为真；由事实五，结论为假。因之，(3) 不正确。次设 G (M, Y) 是否定的。以文、校、史代 M、X、Y 得三段论如下：

$$
\begin{gathered}
\text{F（文，校）},\\
\text{G（文，史）},\\
\text{H（校，史）}。
\end{gathered}
\tag{7}
$$

由事实五与三，两个前提为真；由事实五，结论为假。因之，(3) 不正确。

关于前提的规则

规则一： 两个前提至少有一个是肯定的。

证明：设三段式 (3) 的两个前提都是否定的。求证 (3) 不正确。先设 H (X, Y) 是肯定的。以文、史、哲分别代 M、X、Y 所得的两个前提 F (文，史) 和 G (文，哲)，由事实三皆为真，而结论 H (史，哲) 由事实三却为假，故 (3) 不正确。次设 H (X, Y) 是否定的。以文、史、史分别代 M、X、Y，所得的两个前提，F (文，史) 和 G (文，史) 皆为真，而结论 H (史，史) 由事实二却为假，故 (3) 不正确。

规则二： 若两个前提一是否定的，一是肯定的，则结论必是否定的。

证明：设 (3) 中的 F (M, X) 是否定的，G (M, Y) 是肯定的，H (X、Y) 也是肯定的。求证 (3) 不正确。以文、史、文分别代 M、X、Y，两个前提由事实三与二为真，而结论由事实三为假，故 (3) 不正确。

规则三： 若两个前提都是肯定的，则结论也必是肯定的。

证明：设（3）的两个前提都是肯定的而结论是否定的。把 M、X、Y 都代为文，所得的两个前提文为真而结论为假，故（3）不正确。

规则四： 两个前提至少有一个是全称的。

证明：设（3）的两个前提都是特称的。若 H（X，Y）是肯定的，以男、文、史分别代 M、X、Y；若 H（X，Y）是否定的，以男、文、文分别代 M、X、Y。这样所得的两个前提总是真的而结论总是假的，故（3）不正确。

规则五： 若两个前提一是特称的，一是全称的，则结论必是特称的。

证明：设（3）中的 F（M，X）是特称的，G（M，Y）是全称的，H（X，Y）也是全称的。若 G（M，Y）是肯定的，以文、男、文分别代 M、X、Y；若 G（M，Y）是否定的，以文、男、史分别代 M、X、Y。这样所得的两个前提总是真的而结论总是假的，故（3）不正确。

若两个前提都是全称的怎样呢？对此笔者将在《关于三段论法》一文里予以探讨。

关于三段论法*

如所周知，三段论式有四个格，每格可以有 64 式。普通逻辑在这 256 式中沙里淘金筛出 19 式，或利用主宾词式命题的所谓“差等关系”再添上 5 式共得 24 式，这些被认为是正确的三段论式。由于这种筛选过分繁重，一般逻辑书只是举一举例，并不认真把论证进行到底。这里面还有一个重要问题：主词指称的事物有可能不存在，换句话说，主词的外延可能是空集。据此，许多数理逻辑学家认为，不但 5 式不能添，19 式也要再淘汰 4 式，剩下的 15 式才是真金不怕火炼的。象罗素所举的第三格 Darapti 的例子“金山是金的，金山是山，所以，有的山是金的”，由于金山事实上不存在，虽然大小前提皆真，结论却是假的。L. C. Franklin 利用集合论的语言和所谓“反三段式”把 15 式归结为唯一的一种形式。这样一来，判断三段论式是否正确就不需要现查书或硬记那 15 式了。

但是，我们不妨把事情再考虑一下。虽然罗素所举的那个三段论法不正确，但象下面的例子“鲸是用肺呼吸的，鲸是水生动物，所以，有的水生动物是用肺呼吸的”，由于鲸的确是有的，这一推理却是完全正确的。数理逻辑所讨论的命题形式固然比主宾词式命题要丰富得多，但主宾词式仍不失为最常见的命题形式，而主词外延非空是主宾词式命题的一般情形。由于 9 式在主词外延为空集这种特殊情形下不正确就把 9 式统统不要，这不象俗话说的倒洗澡水把孩子也一起倒掉了吗？这就难怪普通逻辑书宁愿列举那 24 式或 19 式连同它们的绰号 Barbara，Celarent 等等而置 Franklin 的反三段式于不顾了。为此，本文把 Franklin 的反三段式定理试加扩充而证明了下面的定理，这样就恢复了普通逻辑的 24 式，而用这一定理代替 24 式仍象 Fnanklin 定理那样有记忆和应用的方便。

首先，我们明确一些概念和用语，把一个三段论法中的三个名词换成以符号表示

---

* 本文原载于《哲学研究》1987 年第 6 期。

的“变名词”所得到的推理格式称为一个**三段论式**，而原三段论法是这一三段论式的实例，例如，上述关于金山和关于鲸的三段论法都是下列三段论式的实例：

$$\left.\begin{array}{r}\text{凡}\,M\,\text{是}\,P,\\ \text{凡}\,M\,\text{是}\,S,\\ \hline \text{所以，有}\,S\,\text{是}\,P\end{array}\right\}\qquad(1)$$

设 U 代表论域。以 $\overline{X}$ 表集合 X 在论域 U 中的**余集**，$X+Y$ 表 X 和 Y 的**并集**，XY 表 X 和 Y 的**交集**，O 表**空集**。个体名词的外延看作是单独这一个个体作成的集合，而单称命题看作是全称命题的特例。

命 A，B，C 分别为 S，P，M 的外延。于是，三段论式（1）可用集合论的语言写成下面的形式：

$$\left.\begin{array}{r}C\overline{B}=O,\\ C\overline{A}=O,\\ \hline \text{所以，}AB\neq O\end{array}\right\}\qquad(2)$$

以下我们凡说三段论式就假定它是用集合论的语言表示的。一个三段论式说是正确的，或说是恒真的，当且仅当不论以怎样的三个具体集合代替其中的三个“变集合”，只要两个前提皆真，结论便为真。若三段论式中有的变集合限定为非空，则代替它的具体集合便取非空集合。当一个三段论式是正确的，而且也只有这时，作为其实例的三段论法才算是正确的。

把一个三段论式的结论加以否定，以之与两个前提联立，这样三个命题的合取称为一个**反三段式**。例如，由三段论式（2）导出的反三段式可以写成：

$$\left.\begin{array}{l}C\overline{B}=O,\\ C\overline{A}=O,\\ \hline AB=O\text{。}\end{array}\right\}\qquad(3)$$

一个反三段式说是恒假的，如果不论以怎样的三个具体集合代替其中的三个变集合，三个命题总不能都成立。关于非空的限制同三段论式。易见，一个三段论式是正确的，当且仅当其所导出的反三段式是假的。

反三段式中的一个字母说是**偶出现**，如果一次出现带横一次出现不带横，否则说是**单出现**。例如（3）中，A 和 B 都是偶出现的，C 是单出现的。

下面在定理的证明中以辽、吉、黑、长、城、男分别表示所有辽宁省人、吉林省人、黑龙江省人、长春城里人、东北三省城里人、东北三省男人组成的六个集合。用到这些集合时，论域由所有东北三省人组成。

试看下列四种不等式：

$XY\neq O$，$X\overline{Y}\neq O$，$\overline{X}Y\neq O$，$\overline{XY}\neq O$

若取 $X=$吉，$Y=$城，则此四个不等式都成立。事实上，吉林省有城里人也有非城里人，故前二式成立；黑龙江省也是这两种人都有，故后二式成立。

定理　一个三段论式是正确的，当且仅当其所导出的反三段式具有如下形式：

i）三个命题中，一个是不等式，两个是等式，

ii）不在不等式中出现的那个字母是偶出现的，

iii）在不等式中出现的那两个字母都是单出现的；

或反三段式具有如下形式：

1）三个命题都是等式，

2）三个字母中，一个是单出现的，两个是偶出现的，

而名词外延有下列限制：

3）单出现的那个字母代表非空集合。

证明：设对反三段式 i）ii）iii）成立，于是，反三段式可写成如下形式：

$$\left.\begin{array}{l} EC=O, \\ B\overline{C}=O, \\ \hline \widetilde{E}B\neq O. \end{array}\right\} \qquad (4)$$

其中 $\widetilde{E}$ 或是 E 或是 $\overline{E}$，做代换 $A=\widetilde{E}$，（4）就成为

$$\left.\begin{array}{l} AC=O, \\ B\overline{C}=O, \\ \hline AB\neq O. \end{array}\right\} \qquad (5)$$

以 B 乘（5）中第一式的两边，以 A 乘第二式的两边有

$$ABC = O, \tag{6}$$

$$AB\overline{C} = O, \tag{7}$$

此二式相加得

$$O = ABC + AB\overline{C} = AB(C + \overline{C}) = AB \tag{8}$$

故由（5）中的前二式可推出

$$AB = O, \tag{9}$$

此与（5）中第三式矛盾。可见，（5）中的三个命题不能同时成立，此即反三段式恒假。因之，原三段论式是正确的。

今设 1）2）3）成立，反三段式中的任意等式或不等式中最多有一个字母带横，因而整个反三段式中字母带横最多出现三次。设 2）中所说单出现的那个字母为 A，其余两个字母为 E 和 F，由 2)，$\overline{E}$ 和 $\overline{F}$ 都在反三段式中出现。若 A 的两次出现都是 $\overline{A}$，则反三段式中字母带横将出现四次，此不可能。所以，A 的两次出现都不带横，作必要的代换，反三段式可写成

$$\left.\begin{array}{l} AC = O, \\ B\overline{C} = O, \\ \hline A\overline{B} = O。\end{array}\right\} \tag{10}$$

如上，由（10）中的前二式可推出 $AB=O$，此式与第三式相加得

$$O = AB + A\overline{B} = A(B + \overline{B}) = A \tag{11}$$

即

$$A = O \tag{12}$$

此式与 3）矛盾。因之，（10）中三式不能同时成立，即反三段式恒假。从而原三段式是正确的。

现在反过来设原三段式是正确的。于是，反三段式必恒假。试证反三段式或满足 i）ii）iii）或满足 1）2）3）。

先证反三段式或满足 i）或满足 1）。假若不然，三个命题必然都是不等式或两个是不等式一个是等式。若三个都是不等式，则取其中三个集合为吉、城、男，三个不等式便都成立，此与反三段式恒假矛盾。若两个是不等式一个是等式，作必要的代换

可使等式成为

$$AB=O \tag{13}$$

的形式。设其余一个字母是 C，取 A、B、C 分别为辽、吉、男，则三个命题都成立，矛盾。这就证明了反三段式必满足 i）或满足 1），以下分这两种情形讨论。

第一种情形：反三段式满足 i）。假定不满足 ii），即不出现在不等式中的那个字母是单出现的。作必要的代换，反三段式可写成：

$$\left.\begin{array}{l}\overline{A}C=O,\\ \overline{B}C=O,\\ \hline \overline{A}\widetilde{B}\neq O。\end{array}\right\} \tag{14}$$

取 A、B、C 分别为吉、城、长，则三式都成立，矛盾。可见 ii）必满足。

今设 iii）不满足，即不等式中有一个字母是偶出现的。作必要的代换，反三段式可写成

$$\left.\begin{array}{l}AC=O,\\ B\overline{C}=O,\\ \hline \overline{A}\widetilde{B}\neq O。\end{array}\right\} \tag{15}$$

取 A、B、C 分别为辽、长、吉，则三式成立，矛盾。所以，在第一种情形下，i）ii）iii）都满足。

第二种情形：反三段式满足 1）。于是，三个命题都是等式。假定 2）不满足。这时，或三个字母都单出现，或三个字母都偶出现，或两个单出现一个偶出现。若三个字母都单出现，作必要的代换，反三段式可写成

$$\left.\begin{array}{l}AC=O,\\ BC=O,\\ \hline AB=O。\end{array}\right\} \tag{16}$$

取 A、B、C 为辽、吉、黑，则三式成立，矛盾。若三个字母都偶出现，反三段式必可写成如下形式：

$$\left.\begin{array}{l}A\overline{C}=O,\\ \overline{B}C=O,\\ \hline \overline{A}B=O。\end{array}\right\} \tag{17}$$

取三个集合都为吉，则三式成立，矛盾。若两个字母单出现一个字母偶出现，作必要的代换，反三段式可写成

$$\left.\begin{array}{l}AC=O,\\ B\overline{C}=O,\\ \hline AB=O。\end{array}\right\} \tag{18}$$

取 A、B、C 分别为黑、长、吉，则三式成立，矛盾。可见，2）必满足。

以上取具体集合时一直没有取为空集（也没有取为全论域），所以，即使限制有的名词外延须非空甚至三个名词外延都要非空，上面的所有论证都是有效的。现在证明在此第二种情形下，必须限制单出现的那个字母，比方 A，非空。A 的出现必不带横，否则反三段式中字母带横将出现四次。可见，作必要的代换，反三段式可写成

$$\left.\begin{array}{l}A\overline{C}=O,\\ BC=O,\\ \hline A\overline{B}=O。\end{array}\right\} \tag{19}$$

若没有上述限制，则取 A、B、C 分别为 O，辽，吉，三式便成立，矛盾。因之，在第二种情形下，1）2）3）都满足，定理证毕。

近世代数数系章内容草稿*

编者按： 著名数学家王湘浩先生，近年十分关心高师数学教育的改革。他曾为此事在全国人民代表大会上提出过专题提案。1987 年 12 月在大连召开第一次"高师数学教育研讨会"时，他曾到会两次讲话。在召开第二次"高师数学教育研讨会"时，他因故没能到会，但将自己对高师数学教育改革的一些思考，写成两篇文章交给了大会。这些，都表现了一位老数学家对数学教育事业的高度热忱，使人十分敬佩。

这里发表的一篇，是一份专题提纲。另外一篇，是论文《初等几何中几个需要深入理解的问题》，本刊将于下期全文刊登。

1. 自然数——自然数系以加法为本之公理系统，直观地说明公理成立。自然数的基本性质。记数法。定义有限集合并推证其性质。* 直接定义有限集合并以此为基础建立自然数系。* 以 Peano 公理系建立自然数系。

2. 量及分数——以公理定义量。分数作为可通约量之比，定义相等、加法、乘法。据此推出相等、加法、乘法之数字定义并推证算律。小数作为分数之特殊情形。

3. 正实数——正实数作为同类量之比，展开为无尽小数。分数等于循环小数。定义加法、乘法。推出加法、乘法表为极限之定义，证明算律。开方。* 以几何法建立正实数理论。* 试拟中学代数第一章。

4. 实数——实数作为直线向量之比。定义加法、乘法。推出加法、乘法之数字定义并推证算律。* 以公理法建立实数系。

5. 复数——复数作为平面向量之比。定义加法、乘法。推出 a＋bi 表示法及加法、乘法之数字定义并推证算律。复数无大小。

*6. 论数——算术的基本定理。最高公因，最低公倍。同余。Felirlct-Eulel 定理。循环小数性质。一些无理数。代数数及超越数。Ziouville 定理。代数数及超越数

---

* 本文原载于《齐齐哈尔师范学院学报（自然科学版）》1989 年第 1 期。

俱不可数无穷多。

*7. 一些同构问题——实数域之同构变换。复数域之同构变换。实数加法群到正实数乘法群之同构映射，指数函数及对数函数。对数函数利用反比曲线下之面积定义。复数加法群到非零复数乘法群之同态映射，复变指数函数及对数函数。

注：带*者为数系理论课增加内容。

初等几何中几个需要深入理解的问题*

摘　要

本文以初等几何中一些看似粗浅实际上需要深入分析的问题为例来说明，如何利用高等数学学到的知识，来研究中学数学等有关高师数学教育改革应引起大家思考的几个问题。

关键词　初等几何。

一个中学教师，如果不具备必要的高等数学知识，他就不能居高临下地、深入地掌握中学数学教材。但是，学了象如今综合大学数学系的那些必修课和选修课，未见得就自然而然加深了对中学数学的理解。所以，一般师范院校数学系的学生有必要学习一定份量的利用高等数学研究初等数学的课程，只有这样，他所学的那些高等数学课才对理解中学教材真正起作用。本文是想以初等几何中一些看似粗浅实际上需要深入分析的问题为例来说明我们的这一观点。

1　两点之间线段最短

这在中学几何课本中是作为公理提出来的，可说是再初等不过的了。但仔细想一想，这事并不象表面上那样简单。线段最短，是和哪些东西比较起来最短呢？回答是：和连结那两点的所有曲线比较。什么叫曲线？曲线和线段怎样比较长短？看来。严格的几何体系中，把“两点之间，线段最短”作为公理是不合适的。删掉这条公理行不行呢？课本上是根据它推出“三角形两边之和大于第三边”这条定理的。是否连这条定理也一起删掉呢？学了如今一般大学的那些数学，未必自然而然就能圆满地回答这些问题。

下面，为了简便，我们在平面上讨论这些问题，其实在空间也是一样的。先看“三角形两边之和大于第三边”这一定理怎么办。这个问题其实在欧几里得《几何原

*　本文原载于《齐齐哈尔师范学院学报（自然科学版）》1989 年第 2 期。

本》中早已解决了，原来，这一定理是可以直接证明的。

先证三角形中大角对大边。设有三角形 ABD 如图 1 所示。若 $AD=BD$，则$\angle B=\angle A$。设 $AD>BD$。取 E 使 $ED=BD$，于是，$\angle DBE=\angle DEB$，但三角形的外角大于不相邻的内角，故$\angle B>\angle DBE=\angle DEB>\angle A$。同样，若 $BD>AD$，则$\angle A>\angle B$. 由以上所证可以推得：$\angle B=\angle A$，$\angle B>\angle A$，$\angle B<\angle A$ 时，分别有 $AD=BD$，$AD>BD$，$AD<BD$。

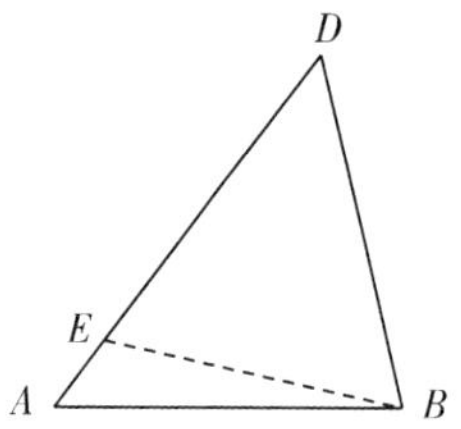

图 1

今证$\triangle ABD$ 中，$AD+BD>AB$。如图 2，取 E 使 $DE=BD$。于是，$\angle ABE>\angle DBE=\angle AEB$。但三角形中大角对大边，故 $AD+BD=AE>AB$。这就证明了“三角形两边之和大于第三边”。

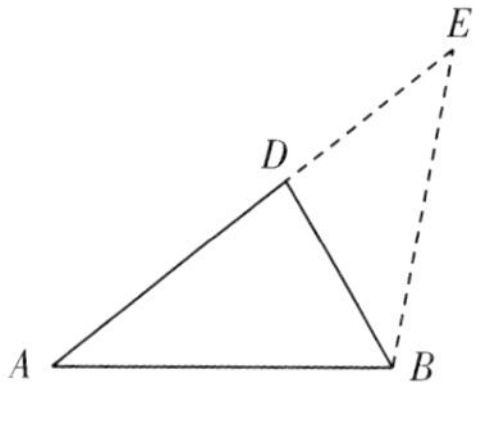

图 2

顺便提到，证明“三角形的外角大于不相邻的内角”可以不用平行公理，因而以上的证明也适用于罗氏几何。

既然在严格的几何体系中，“两点之间，线段最短”不宜作为公理，那么，“三角形两边之和大于第三边”就只能象上面那样证明。所以，这一证明虽毫无深奥之处却是非常基本的，师范院校毕业科班出身的教师无论如何是应当知道的。

现在讨论“两点之间，线段最短”这一命题。

关于曲线的定义和曲线比较长短的定义，我们基本上采用数学分析中或拓扑学中

“可直曲线”的讲法而稍加变通。综合几何较前面的部分不需要引进实数和长度的概念，线段的相等、大小、相加减都是直接规定的，所以，此处我们也不假定实数和长度为已知。

设 PQ 是一条线段，φ 是 PQ 到平面内的一个映射，其影象集合 C 不只含一个点而且具有下列性质：任意直线两侧都有 C 的点时，直线上也必有 C 的点。于是 C 与 φ 相结合而形成的概念 $\{C, \varphi\}$ 称为一条曲线。映射是一种对应关系，所以只是 φ 不能说是一条曲线。只有点集 C 也不行，因为没有说它的那些点是怎样“排列”的。所以要说得啰嗦一些把曲线说成是 $\{C, \varphi\}$，这就是说，点集 C 由 φ 规定了一种“排列”才成为一条曲线。当然，只要不忘记那个 φ，曲线 $\{C, \varphi\}$ 也不妨简称为 C。曲线的上述定义似嫌太泛。一般可直曲线的定义中要求 φ 具有连续性。可以证明，不只含一个点的点集 C 是线段的一个连续影象。当且仅当 C 是一个局部连通的连续统。例如，适当取 φ 可使其影象填满一个正方形面。可见，即使对 φ 加上连续性要求，定义仍是太泛。不过，把曲线概念限制得太窄也不好。此处，为了使定义尽量有初等性，我们索性降低了连续性要求。

下面定义曲线的等长和长短。设 $\{C, \varphi\}$ 是一条直线，其中 φ 是线段 PQ 到点集 C 上的映射。命 $A=\varphi(P)$，$B=\varphi(Q)$。在 PQ 上顺从 P 到 Q 的方向取 $n+1$ 个点如下：

$$P_0, P_1, \cdots, P_{n-1}, P_n \tag{1}$$

其中 $P_0=P$，$P_n=Q$。命

$$A_i=\varphi(P_i), \quad i=0, 1, \cdots, n-1, n \tag{2}$$

n 个线段

$$A_{i-1}A_i, \quad i=1, \cdots, n \tag{3}$$

连成一条折线

$$A_0A_1\cdots A_{n-1}A_n \tag{4}$$

称为曲线 C 的一条内接折线。$A_{i-1}=A_i$ 时，$A_{i-1}A_i$ 说是一条零线段，对任意折线 l，把组成 l 的非零线段相加，设所得的线段为 a。线段 $x=a$，$<a$，$>a$ 时，x 分别说是等长于 l，短于 l，长于 l，设 C，C'是两条曲线。我们说 C，短于 C'如果可取

线段 x 使短于 C' 的某条内接折线但长于或等长于 C 的任意内接折线。C 说是等长于 C' 如果对任意线段 x，x 短于 C 的某条内接折线当且仅当 x 短于 C' 的某条内接折线。

现在证明"两点之间，线段最短"。设 A，B 是任意两个不同的点。设 $\{C, \varphi\}$ 是一条曲线，其中 φ 是线段 PQ 到平面内的一个映射，$\varphi(P)=A$，$\varphi(Q)=B$。这样一条曲线说是连结 A，B 的一条曲线。若 C 含有直线 AB 之外的点，则 C 上说是有弯路。为了方便，我们说 P 在 Q 左，A 在 B 左。若 C 全在直线 AB 上，但 PQ 上有 X、Y，X 在 Y 左面 $\varphi(Y)$ 在 $\varphi(X)$ 左，则 C 说是有回头路。若 C 上既无弯路也无回头路，则 C 看作和线段 AB 相同。这样的曲线 $\{C, \varphi\}$ 是存在的，只要取 PQ 为 AB，取 φ 为把 PQ 上任意点映到它自己的映射即可。设 $\{C, \varphi\}$ 为按上述规定和 AB 看作相同的曲线，而 $\{C', \varphi'\}$ 为按上述规定不同于 AB 的曲线，其中 φ' 为线段 $P'Q'$ 到平面内的映射，$\varphi'(P')=A$，$\varphi'(Q')=B$，我们要证明的是：C 短于 C'。既然 C 上无弯路也无回头路，易见 C 的任意内接折线等长于 AB。C' 既不同于 AB。则 C' 或有弯路或有回头路。设 C' 含有直线 AB 之外的点 D。因为"三角形两边之和大于第三边"，故 AB 短于 C' 的内接折线 ADB，取 $x=AB$ 便据曲线长短的定义知 C 短于 C'。设 C' 全在直线 AB 上，而 $P'Q'$ 上有 X、Y，X 在 Y 左而 $\varphi'(Y)$ 在 $\varphi'(X)$ 左。于是，C' 的内接折线 $A\varphi'(X)\varphi'(Y)B$ 长于 $AB+\varphi'(X)\varphi'(Y)>AB$。取 $x=AB$ 便知 C 短于 C'。这就证明了"两点之间，线段最短"。

2 角的度量

这也见于中学几何课本第一章。书上说：把周角分成 360 等份，每一份叫 1 度；把 1 度分成 60 等份，每一份叫 1 分；把 1 分分成 60 等份，每一份叫 1 秒。问题是：周角能不能分成 360 等份呢？怎样证明 1 度的角存在呢？

证明存在定理有两种方法：一是纯理论性的，只在理论上证明其存在，并不给出求法；一是构造性的，说明怎样把要证明其存在的东西具体求出来，既然能求出来，当然就存在了，初等几何中的直尺圆规作图实质上就是存在定理的构造性证明。例如用直尺圆规二等分任意角。这就是用构造法证明任意角必有二等分线存在。为什么限于用直尺圆规呢？这是由于我们有关于直线存在和圆存在的公理："两点定一直线"，

"以任意点为圆心，任意线段为半径，可作一圆"。用量角器行不行呢？不行，因为我们没有关于量角器的公理。

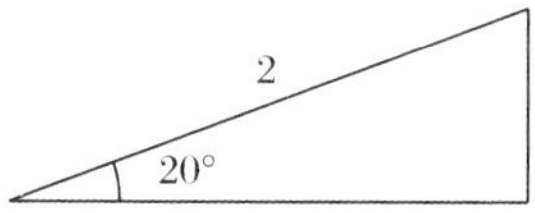

图 3

现在先看，能不能用直尺圆规作出1°的角从而证明其存在呢？20个1°的角并起来就成为一个20°的角。所以，若1°的角可用尺规作图，则20°的角也可用尺规作图。命 $a=2\cos 20^{\circ}$。取定一个长度单位。作直角三角形如图3可见，1°的角可用尺规作图时，长 a 的线段也可用尺规作图。在公式 $\cos 3\theta=4\cos^{3}\theta-3\cos\theta$ 中取 $\theta=20^{\circ}$ 得 $1/2=4\cos^{3}20^{\circ}-3\cos 20^{\circ}$，因而 $1=a^{3}-3a$，所以 a 是方程

$$x^{3}-3x-1=0 \tag{5}$$

的根，容易说明多项式 $x^{3}-3x-1$ 在有理域 R 上不可约，故扩域 $R(a)$ 对 R 的次数 $(R(a):R)=3$。用 Galoie 理论可以证明：从单位线段出发，长度为 ξ 的线段可用尺规作图，当且仅当 ξ 包含在次数为2的若干方的 R 的一个正规扩域 K 之内。因之，若长 α 的线段可用尺规作图，R 应有正规扩张 K 存在使 $\alpha\in K$ 而 $(K:R)=2$ 的若干方，比方 2^{n}. 因为

$$(K:R)=(R(\alpha):R)(K:R(\alpha)) \tag{6}$$

命 $m=(K:R(a))$ 有

$$2^{n}=3m \tag{7}$$

但 2^{n} 不是3的倍数，此不可能。可见，1°的角不能用尺规作图。人们学过泛代拓，懂得 Galoie 理论和以尺规三等分任意角不可能，但不一定自然而然就知道这同1°的角是否存在有关系。三等分任意角也许不重要，1°的角是否存在总是很重要的吧？

1°的角存在我们既然没有能够给出构造性证明，只有退而求其次给一个纯理论证明。师范院校的毕业生，如果学高等数学时有联系中学数学的习惯，大概已经想到这要用到实数系的连续性了。

在几何上，我们有关于直线的 Dedekind 公理，这与实数系的连续性等价。为了

简便，我们用左右描述直线上点的排列。Dedekind 公理可叙述如下：设把直线 l 上所有的点分为两个互无公共元素的非空集合 U 和 V，使 U 全在 V 的左边。于是，U 和 V 之间恰有一个分界点 P，P 或是 U 的最右元素或是 V 的最左元素.

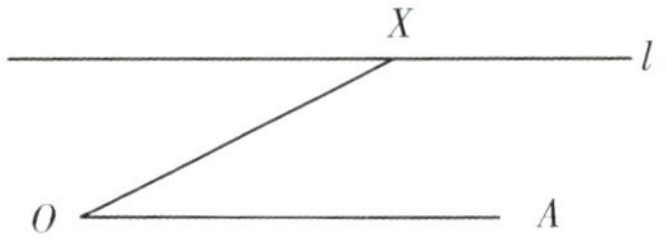

图 4

如图 4，取线段 OA 及与之平行的直线 l。把上的点分为 U、V 两个点集如下：

$$U=\{X \mid \angle AOX \text{ 的 } 90 \text{ 倍} \geqslant \text{直角}\}. \tag{8}$$

$$V=\{X \mid \angle AOX \text{ 的 } 90 \text{ 倍} < \text{直角}\}. \tag{9}$$

易见，U 和 V 互无公共元素而且 U 全在 V 的左边。若$\angle AOX$ 为直角，则 $X\in U$，故 U 非空。因为任意角可以二等分，可取 X 使$\angle AOX$ 等于直角的 123 分之一，从而 $X\in V$，故 V 非空。命 P 为 U 与 V 的分界点。P 不可能在 V 内。假如在 V 内，在 P 的左边取 X 使$\angle POX$ 等于（直角$-90\angle AOP$）的 128 分之一，易见 $90\angle AOX<$直角，从而 X 仍在 V 内，此与 P 为分界点矛盾。可见 P 在 U 内。不可能 $90\angle AOP>$直角，否则在 P 的右边取 X 使$\angle POX$ 等于（$90\angle AOP-$直角）的 128 分之一，易见 $90\angle AOX>$直角，从而 X 仍在 U 内，与 P 为分界点矛盾。因之，$90\angle AOP=$直角，故 1°的角存在。

3　面积和体积

多边形的面积和多面体的体积在小学里就部分地学过，中学课本里并没有多少提高。本节主要讨论多边形的面积问题，多面体的体积只在最后略微涉及。所谓多边形的面积就是多边形所围的那块面的面积。这里有两个问题：一是多边形所围的面是什么意思，二是怎样定义那块面的面积。

设平面上有 n 条线段，$n\geqslant 3$，线段的端点各是两条线段的公共端点，而内点只在一条线段上。这样 n 条线段说是组成一个简单多边形，我们这里所说的多边形皆指这种简单多边形而言。要说明什么是多边形所围的面，就要证明任意多边形分平面为两部分，一个有限部分和一个无限部分。这在凸多边形情形下容易证，对于可能凹进去的一般多边形就要困难得多。我们知道，证明 Jordan 曲线定理不是很容易的，有一个证明就是先证任意多边形分平面为两部分，在此基础上再证任意 Jordan 曲线分平面为两部分。所谓多边形围成的面就是分出来的那块有限部分再把多边形本身作为边界添上去，我们称之为多边面。任意多边面可以分为一些三角面，这证明起来也要费点事的。

取定一个线段 u 作为长度单位。为了定义多边面的面积需要证明下列定理：任意多边面可以分为许多三角面重新组合凑成宽度为 u 的一个长方面。证明这一定理当然也要费些词说，但并不算难。比较难证的是：上述宽度为 u 的长方面由原多边面唯一确定，与多边面分为三角面的不同分法无关。有了这些准备工作，我们定义多边面的面积如下：若多边面变成的长方面之长度为 au，则原多边面的面积定义为 au^2。最后还要证明：取另一长度单位 v 而设 $u=kv^2$，则原长方面的面积为 ak^2v^2。至此，多边面的面积理论才算是建立了。

Gouss 曾提出，立体几何讲多面体体积能否不用 Cavajieri 原理（实际上应称为祖日恒原理）而像平面几何那样只用分割的方法？Hilbert 23 个问题中的第三个问题也就是问：能否仿照多边形的面积理论来建立多面体体积理论，比方说，证明这样一个定理：取一个单位长度 u，任意多面体可以分为许多四面体重新组成一个长方体，其宽度与厚度皆为 u。1900 年，Dehn 举反例证明了像上面那样的定理是不正确的，从而解决了这个 Hilbert 问题。所以，讲多面体积还不得不用祖日恒原理。

多边形的面积理论以及体积和面积的上述本质上的不同所揭示的宇宙奥秘是非常深刻和引人入胜的。懂得这些理论对于教师提高自己的教学素养和数学鉴赏力是很有好处的。

4　几个例子

从某种意义上说，中学数学教材比大学数学教材难懂，中学数学教师比大学数学教师难当。为什么呢？大学教材可以不太考虑可接受性，定义定理清清楚楚写出来，只要下点功夫就能够弄懂。中学课本就不然，只要注意逻辑性又不能写得太严格，说理往往要靠直观，有时吞吞吐吐不敢直言无隐，有时甚至不得不用点骗招。从前三节的几个例子可以看出，补上那些不严格的打折扣的地方并不容易。教师如果不求甚解，知其然而不知其所以然，就不能针对学生的具体情况，灵活地、因材施教地进行教学。只有把问题彻底弄明白了，才能恰当掌握分寸，知道什么地方要细讲，什么地方要诉诸直观，什么地方必须一带而过，什么东西千万不能提等等。

下面再举几个例子供讨论问题时参考：

例 1. 中学几何课本在讲什么是二线段之比时说：在同一单位下，两条线段长度的比叫做这两条线段的比，两条线段的比值与所采用的长度单位没有关系。那么，什么是线段在某一单位下的长度呢？书上在第一章说过：在小学时，我们曾使用刻度尺来度量线段的长度。这些话里有哪些不严格的地方？其中有没有骗招？试在正实数理论已知的假定下，建立线段长度的理论，证明二线段之和的长度等于二线段长度之和，并证明换单位公式。上引课本上的话，在讲课时，是指出其中有打折扣的地方好呢，还是混过去好呢？

例 2. 几何课本讲比例线段和相似三角形时，以“平行线分线段成比例”为基本定理，其他定理由此定理推出。但基本定理的证明是不严格的。这本来是写初等几何课本的难点，但教师自己应当能够弄清楚这些问题。试以两种方法严格建立三角形的相似理论：一是假定正实数理论和线段长度理论已知，在此基础上讲三角形相似；一是只假定自然数理论已知，以几何方法建立三角形的相似理论，并在此基础上建立正实数理论。

例 3. 讲长方形的面积时，书上在整数和有理数的情形下，举例证明了“长方形的面积等于长乘宽”，然后把这一公式列为一条公理这显然是打了折扣。试如下把书上的讲法予以严格化：取一个长度单位 u。对长方形面积提出下面几条显然应当满足的要求：1. 任意长方形的面积是一个正实数；2. 如图 5，长方形 W 分为长方形 U 与 V，则 W 的面积等于 U 与 V 的面积之和；3. 边长为 u 的正方形面积为 1，据此推出长方形面积公式。然后以此公式定义长方形面积，证明此定义满足上面三条要求，并证明换单位公式。

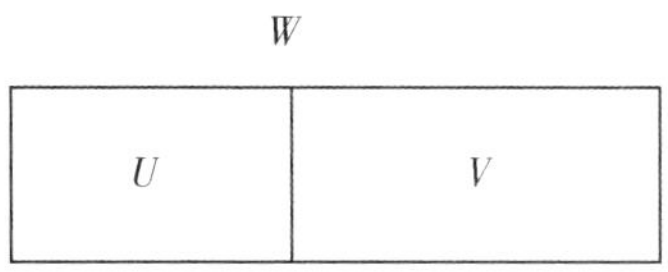

图 5

例 4. 书上以“能够完全重合”定义两个图形全等。这样，自然就应当用叠置法证明两个三角形全等的 SAS 判定方法：“有两边和它们的夹角对应相等的两个三角形全等”。但是，书上不用叠置法证明，却让学生做一个实验，然后把 SAS 列为一条公理。书上为什么这样处理 SAS？教师应怎样讲这段课文？怎样定义全等和处理 SAS 才恰当？

例 5. 圆与直线的位置关系有三种情形：圆心与直线的距离 $d>$半径 r 时相离（相离定理）。$d=r$ 时相切（相切定理），$d<r$ 时相交于两点（相交定理）。前两种情形比较简单，问题是怎样证明相交定理。有的课本诉诸直观，有的课本证明如下：如图 6，过圆心 O 作直线 l 的垂线 OD. 设 $d<r$。在 l 上 D 的两边取 A 与 B 使 DA 和 DB 的长度等于 $\sqrt{r^2-d^2}$。易见 A 与 B 都是圆与直线的交点，而且只能有这两个交点。这一证明的缺点是：圆与直线的三种关系开始讲尺规作图时就要用到，使用勾股定理来讨论就太晚了。有一种实验教材在讨论了相离和相切两种情形后说：经过半径 OA 的端点 A 作与 OA 不垂直的直线 l。这条直线和圆不能只有一个公共点，还必须有一个交点 B。然后才下定义说：如果一条直线和一个圆有两个公共点，我们就说，这条直线和这个圆相交。似乎这样就算是把三种关系讨论清楚了。但是，这里面的骗招太明显，教师真要这样讲，恐怕同学要挑出毛病。那么，相交定理究竟应当怎样严格证明呢？当然，最好不要动用 Dedekind 公理。证明相交定理能否做到这一点呢？如果回答能，须给出证明。如果回答不能，并以 Dedekind 公理证明相交定理。

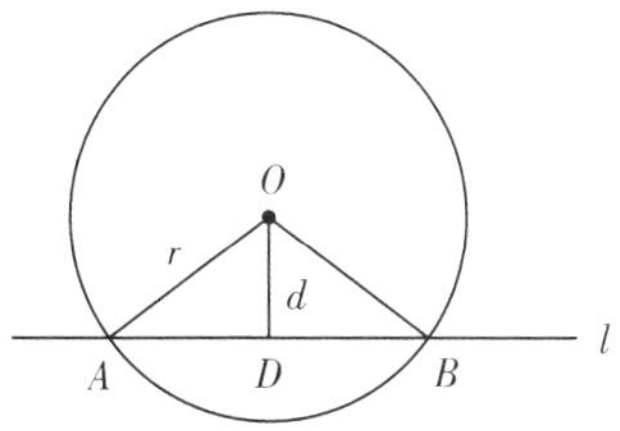

图 6

讨论二圆的位置关系问题。

例 6. 中学几何课本在讲到圆周率 π 时说：我们知道，圆周长 C 与半径 R 之间有下面的关系 $C=2\pi R$。这指的是在小学里学过，似乎这样就把 π 的问题交代过去了。在小学里是怎样学的呢？小学课本是要求学生在纸上画一个圆，量出直径长，剪下圆来在尺子上滚一周就量出圆周长，除以直径长得 3 多一点。再画一个圆，同样度量和计算仍得 3 多一点。于是书上说：圆周比直径是一个常数 π，$\pi=3.1416$ 我们不是说中学和小学里不能这样讲 π，但老师总该知道得多一点吧？高中毕业时关于 π 的知识也还是小学里那些。进了大学，学的是泛代拓，大学毕业后，关于 π 也仍然限于小学里那点知识。可能辩论说：圆周长可以用积分求出来怎能说大学毕业后还只知道“在尺子上滚”呢？但用积分求圆周长，积出来有 arcsin，这要知道 $\sin x$ 的导数才行，而证明（$\sin x$）$=\cos x$ 要用到圆周长或圆面积，其实角的弧度单位已经涉及圆周长。因此，所谓“用积分求”其根据仍是“在尺子上滚”。有的课本加了个附录，但也不敢直言无隐，只说是讲 π 的算法，不说是证明圆周比直径等于常数。

能否不用极限严格地讲圆周率 π？讲课时，能不能讲得让学生相信而又不难？

Several Problems Needing Comprehension in the Elementary Geometry

Abstract

In this paper several problems in the elementary geometry, which seem simple but in fact need thorough analysis, are taken for example to explain how to make use of the knowledge learnt from the higher meths so as to study the problems concerning the eductional reform of the higher maths, which should lead to our consideration.

Key words elementary geometry.

其他文章

书 评*

YUEH-LIN CHIN. *Lo chi* (Logic). Shanghai 1937. Third edition, The Commercial Press, Shanghai 1948, 362 pp.

This book (written in Chinese) is a text-book for the use of university students. It consists of four parts.

Part one deals with traditional formal logic, omitting all unnecessary epistemological and psychological discussions.

Part two is a criticism of the same. In traditional logic, the meanings of the four kinds of propositions, A, E, I, O, are not clearly defined. According to different points of view toward the question of existential import, the author gives four different interpretations:

(1) A_n *as* $S\overline{P}=0$. E_n *as* $SP=0$. I_n *as* $(SP\neq 0)\vee(S=0)$. O_n *as* $(S\overline{P}\neq 0)\vee(S=0)$.

(2) A_c *as* $(S\overline{P}=0)\cdot(SP\neq 0)$. E_c *as* $(SP=0)\cdot(S\neq 0)$. I_c *as* $SP\neq 0$. O_c *as* $S\overline{P}\neq 0$.

(3) A_n, E_n, I_c, O_c.

(4) A_h, E_h, I_h, O_h: same as A_n, E_n, I_c, O_c, with presupposed existence of S.

From the standpoint of each of these four different interpretations, the rules for immediate inferences and (categorical) syllogisms are examined. There are also discussions about hypothetical and disjunctive syllogisms and dilemmas, and criticisms as to the narrowness of scope of traditional logic.

---

* 本文原载于《符号逻辑杂志》(*Journal of Symbolic Logic*) 1949 年第 2 期。

In the third part, the author reproduces a part of *Principia mathematica* (Part I of the frst book), including the calculi of propositions, propositional functions, classes, and relations. Since this part of the book is intended to be an introduction to mathematical logic, the elaborate theory of types and material directly bearing on it are omitted. It seems to the reviewer that the distinction between a formal primitive proposition and a rule of inference should be emphasized.

In the fourth part, the author outlines a theory of logical systems. First, the nature of deductive systems is explained. A deductive system consists of two parts: the basic part and the derivative part. Primitive ideas and propositions constitute the basic part. The primitive propositions are premises in the system, and are tools for deciding what is to be retained and what is to be rejected. All that is to be retained belongs to the derivative part. Sometimes, the primitive propositions are also tools of inference; if this is the case, the system is self-sufficient.

The logical systems are distinguished among deductive systems in that the "things" they retain are necessities (tautologies) and those they reject are contradictions. Suppose all "possibilities" are divided into n categories. Then, a proposition asserting separately these n possibilities is a necessity in this n-valued system (p. 277), and one denying all these n possibilities is a contradiction in the same (p. 278).

The author is of the opinion that, while a tautology may assume different "forms" in different logical systems, its "substance" is one and the same. Every logical system represents logic, but it is not true that logic has to be represented by one particular system (p. 282).

In Chapter II of this part, while explaining the nature of logical systems in greater detail, the author discusses the three so-called Laws (Principles) of Thought.

The Principle of Identity is what makes a possibility a possibility and a term a term of meaning. In the authors opinion, it should be expressed in the following way: "If x is A, then it is A", where x denotes concrete things and A denotes names.

A contradiction, on the negative side, denies all possibilities; on the positive side, it asserts all possibilities simultaneously (p. 309). The Principle of Contradiction expresses that possibilities cannot be asserted simultaneously (p. 308). A necessity, on the positive side, asserts the possibilities separately; on the negative side, it rules out omissions of possibilities (p. 308). The Principle of Excluded Middle expresses that possibilities cannot be omitted (p. 308).

In a particular system, the first two principles may be expressed in particular forms, and can be proved. This is not so with the Principle of Excluded Middle. All necessary propositions in a system express this principle, and the process in developing the whole system may be regarded as its proof (p. 307).

The principle of construction of the four functions on p. 295 is not sufficiently explained. One of them is a disjunction and the others are conjunctions.

In Chapter III, the primitive ideas and propositions of a logical system are discussed.

The book is carefully written. Apart from a few misprints, the reviewer found only an insignificant mistake in p. 199, where the negations of two propositions are incorrectly stated.

SHIANGHAW WANG

评“组合拓扑学基础”中文译本的两个附录*

邦德列雅金著“组合拓扑学基础”的中文译本（冯康译，中国科学院1954年7月出版）较原著增加了许多习题和两个附录，这，诚如原著者在中文版序言中所说，可以使本书更容易被接受．但是，两个附录本身却十分难读．这里，我们提出一些意见，供再版时参考．

两个附录所以难读，主要是由于文字不够流畅，定理的叙述和证明有时也写得比较难懂．例如155页命题1.16的证明中说：“组$\{f^{-1}(Fa)\}$显然是紧密空间R的一组闭集，……”实际上，这并不是十分显然的，因为，f是映R入S的一个连续映像，我们不能不加证明而把f看作映R成$f(R)$的一个连续映像，因而更不能立即得到上面的话了．

169页命题4.7把许多很少关联的事实罗列在一起，使读者不易掌握要点．再者，命题4.7第2条之后似乎该接写“群H的任意元素的完全原像是模核的一个剩余类”，但这个重要的意思却完全漏掉了，只隐含地放到定理4.9的证明中去．

159页命题2.7的证明中，文字上有些毛病，似乎不应说“必要条件不证自明．兹证充分条件：……”而应说“条件的必要性不证自明．兹证明充分性”．定理2.8证明中的提法也应当是“充分性”和“必要性”．此外，定理2.7的证明中用到点列似乎是不必要的，直接证明可能使读者更容易体会些．

154页定理1.11的证明中，“设$C(x)=C_1\cup C_2$，…；$x\in C_1$，M是一个含有点x的连通集”．第一设是实际的假设，有反证法之意；第二设是不妨假定$x\in C_1$；第三设是用M代表含有x的任意连通集．三者的逻辑意义是很不一样的．今三者不加区别地堆在一句话里，并且第一二两设之间还夹杂着一个注语，这就使读者很难体会作者所说的究竟是什么．类似这样的地方还有很多．

---

* 本文原载于《东北人民大学自然科学学报》1955年第1期。

命题和定理的区别好像很不明显. 定理似乎应比命题更重要些，但 155 页命题 1.15 和 1.16 是相当重要的，而 159 页定理 2.8 倒好像不太重要.

“紧密”是本书重要的用语，必须坚持译名的统一性，但 161 页有一处、163 页有两处、164 页有一处却无故出现“封闭”二字.

两个附录中最大的缺点是有三个地方是错的：172 页定理 4.14 的叙述，177 页命题 5.7 的证明，185 页命题 6.5 的证明. 错误的来源是同样的. 就是作者认为“若 $G=G_1+\cdots+G_n$，F 是 G 的子群，而 $F_i=F\cap G_i$，则 $F=F_1+\cdots+F_n$”这样的错误是不应该发生三次的. 因为，第一，只要随便举一个例子便看出来这话不对；第二，在交换群的理论中我们所以要讲整数矩阵等等（当然也可以避免整数矩阵而用其他方法）主要就是由于上面的话不成立，否则交换群的理论便异常简单. 比方，如果上面的话成立，则 180 页定理 6.2 的证明就十分容易，只要把 G 分为巡回群 G_1，…，G_n 的直接和. F 便相应地分为 F_1，…，F_n 的直接和. 这样，则 $A=A_1+\cdots+A_n$，其中 $A_i\cong G_iF_i$ 都是巡回群. 然后，把 A_i 中级数有限的那些再分一分并一并定理便证明了.

评儿玉哲夫君的论文“关于正规单纯代数的交换子群”*

儿玉哲夫君的这篇论文（T. Kodama, On the Commutator Group of Normal Simple Algebra, Memoris of the Faculty of Science, Kyusyu Univ., Ser. A, 10, 141, 1956）目的在于将评者关于代数数域上的单纯代数的交换子群的一个结果（王湘浩, On the commutator group of a simple algebra, Am. J. 72, 323, 1950）推广到示性数为 0 的任意单纯代数. 儿玉君曾在写给评者的一封信里，把证明的要点告诉我，最初我并没有注意到论证中的一个错误，后来发现，便写信把我的意见告诉他. 在他所发表的这篇论文里，虽然已将证明作了修改，但原来的错误之点并没有能够免除. 我认为这一个关于单纯代数的交换子群的问题是一个难而有趣的小问题，希望终将有人加以解决，并愿以此与儿玉君共勉.

儿玉君的证明归结到下面的论点（见论文 147 页）：设 D 是域 F 上面的一个正规单纯代数，其指数是一个质数 p 的若干方，比方 p^s. 设 L 是 D 的子域，并且是 F 的 p 次巡回扩张. 于是，若 $\alpha \in D^L$ 对于 F 的缩减模 $N_{DF}\alpha$ 为 1，则 $N_{D^L L}\alpha$ 是一个 p 次单位根. 这一个论断是错误的，因为如果这是对的，则 $\alpha \in L$ 而 $N_{LF}\alpha = 1$ 时上面的结论应当对. 但这时 $N_{D^L L}\alpha = \alpha^{p^{s-1}}$，因而根据上面的结论有 $\alpha^{p^s} = 1$. 这就是说，除 p^s 次单位根外，L 中没有对 F 之模为 1 的元素. 这自然不会是对的，因为不难说明 L 中有无穷多个对 F 之模为 1 的元素. 事实上，根据 Hilbert 定理，$N_{LF}\alpha = 1$ 必要而且只要 $\alpha = \beta^{1-\sigma}$，其中 σ 为 L/F 的生成同构变换. 若 $\gamma^{1-\sigma} = \beta^{1-\sigma}$，则 $(\gamma/\beta)^{\sigma} = \gamma/\beta$，因而 $\gamma/\beta \in F$. 换句话说，只有当 $\gamma = c\beta$ 而 $c \in F$ 时，$\gamma^{1-\sigma} = \beta^{1-\sigma}$ 才成立. 由于 F 中有无穷多个元素，据此易见 L 中确有无穷多个对 F 之模为 1 的元素.

---

* 本文原载于《东北人民大学自然科学学报》1957 年第 1 期。

关于王、谢编高等代数中的一处错误*

我们所编“高等代数”（人民教育出版社 1960 年版）第 149 页施斗姆定理的证明中有错误，这就是第 12 到 14 行的那句话是不正确的．那句话应修改如下：“这表示，x 经过 α 时，若 α 不是 $f(x)$ 的根，则

$$f_0(x),\ f_1(x),\ \cdots,\ f_m(x)$$

之间的变号数不变；若 α 是 $f(x)$ 的根，则

$$f_1(x),\ \cdots,\ f_m(x)$$

之间的变号数不变．”

这一错误承宁夏水电局李学记同志发现并写信告诉我们，谨在此致谢．

* 本文由王湘浩、谢邦杰共同创作，原载于《数学通报》1963 年第 10 期。

加速发展我国计算机科学*

实现“四个现代化”，关键在于实现科学技术现代化，而电子计算机的广泛应用是科学技术现代化的重要标志之一。历史上，蒸汽机的发明和电力的使用曾经在生产技术上引起了划时代的革命。今天，电子计算机的发明和发展必将引起而且正在引起一场新的重大的技术革命.

现代的通用电子计算机是能够按照人所安排的程序自动地快速地进行复杂的数值计算和处理其他信息的工具。它不同于算盘和手摇的或电动的计算器，这些计算工具或者不具有自动性，或者只能自动进行一次运算而不能执行复杂的计算程序，也不能处理数字以外的信息。现在自动计算机的发明是科学史上科学研究走在生产前面的一个突出事例。远在一八三三年，英国数学家白贝之（Babbage），在已有的各种计算工具的基础上，设计成功了一个“分析机”，能够自动地执行某些程序，具有现代计算机的五个基本部件，即运算器、存储器、控制器、输入设备和输出设备。白贝之的先进思想远远超出了他的时代。由于当时种种条件的限制，虽经多方奔走呼吁，他的设计最终还是没有能够得到支持，更谈不上进入制造。直到一百多年以后，才在美国基本上按照他的设计思想制造成功了一台机械的自动计算机。接着，一九四五年，第一台电子管的自动计算机问世，而在一九四九年制造成功了第一台程序内存电子计算机。时至今日，电子计算机的发展经历了四代：电子管，晶体管，集成电路，大规模集成。速度从每秒千次运算发展到亿次运算。主机类型达千余种，外部设备花样繁多。另一方面，发展了程序系统，即所谓“软件”，从而相应地把计算机本身称为“硬件”。由于每台计算机有了这种预先设计好的程序系统，这就大大提高了计算机的功能和效率，并且大大方便了操作人员和用户。此外，发展了终端设备与分时系统以及联结各地计算机的大型网络，因而为计算机的应用创造了更好的条件。

---

* 本文原载于《自然科学争鸣》1977 年第 5 期。

现在，电子计算机已经广泛应用于国防、科学实验和国民经济的各个部门，甚至应用到日常生活的各个方面。据统计，电子计算机的应用已经从六十年代的几百种扩大到七十年代的几千种。可以预料，到本世纪末，计算机的应用将更为广泛。那时，工业生产将在计算机控制下实现高度的自动化，农业生产的某些方面也是一样。计算机将能够识别印刷的，甚至手写的文字，能够直接听取人的语言。人工智能的研究将有重大突破，计算机将能够进行某些类似思维的活动，将直接与人进行对话，回答人们提出的问题，协助人们做某些推理和判断。以计算机为"大脑"的机器人将研制成功，能够听取人的命令自动去做某些事情。按那时的标准说，具有中等功能的计算机早已成为家庭必需品，人们手腕上带着的不是手表而是微型计算机。人造卫星上设有巨型的、其功能比现有计算机不知高多少倍的计算机系统，人们可以通过随身携带的袖珍式通讯装置与之对话，用以解决高度复杂的问题。

自从一九五六年制订十二年科学技术发展规划以来，我国的计算机事业有了很大的发展。但是，发展还不够迅速和全面。同世界上科学技术较先进的国家相比，还存在着差距。在推广应用方面也很不够。毛主席说：**"中国人民有志气，有能力，一定要在不远的将来，赶上和超过世界先进水平。"**我们一定要响应党中央加速实现科学技术现代化的号召，奋起直追，大大加快计算机科学的发展步伐。

我们的软硬件质量不高，根本原因是基础理论的研究开展不够。计算机科学是一门新兴的学科，它既是边缘学科，同时又是基础学科。目前，这门科学正在迅速发展。我们要在计算机科学上赶超世界先进水平，必须结合软、硬件的研制，大力开展计算机科学基础理论的研究。例如从软件来看，除了研制急需的系统软件和应用软件以外，语言理论、自动机理论、算法分析和复杂性理论、模式识别、人工智能等方面的研究必须迅速开展起来。与计算机科学有密切关系而过去比较薄弱的数学学科，如数理逻辑、组合数学等方面的研究也要相应开展。至于计算方法的理论研究虽然过去有比较好的基础，也急待进一步加强。不进行基础理论的研究，就根本谈不到赶超，充其量只能做到跟在人家后面"赶抄"，即"赶紧抄"，赶紧抄还来不及。

在培养计算机事业所需要的科技人员方面，除了重视数量，也要重视质量。高等学校计算机软、硬件方面的专业必须加强基础理论课和基本技能的训练。软件专业要建立软件实验室，为教师的科学研究和学生的上机实习服务。实验室要具有不断更新和扩大计算机设备的能力，要建立自己的软件体系和数据库，有方向、有目的、有计划地进行软件方面的实验，从事日积月累的记录、统计和理论分析工作。学生要有大量的上机实习，而不能认为只分担了某个软件任务中的一小部分工作就算是学了软件。

要把计算机的使用广泛推广到生产斗争、国防、科学实验、计划工作、情报资料工作等方面，甚至用到日常生活的各个方面，必须大力开展科学普及工作。计算机科学只有专业队伍是不够的，一定要有千千万万各个方面的人参加。要大力发展终端设备，建立简便易学的程序语言，使人人都能使用计算机。要举办各种短期训练班，编写普及性读物，打破计算机的神秘性。计算机使人望而生怯的原因之一是计算机软、硬件的书籍本身就有神秘性，因而成为双重神秘。有一种流行的说法，认为计算机的软、硬件本来是复杂的东西，而且具有“循环性”，不可能深入浅出地、由简到繁地讲给别人听。根据我们的教学经验，这是不确实的。计算机书籍之所以难懂，主要不是这门科学本身的问题，更多的是文风问题。

在此，让我们本着“争鸣”的精神提出一个异常紧迫的问题：要迅速发展计算机科学，解决汉字输入刻不容缓，而要实现四个现代化，文字改革势在必行。毛主席教导我们说：**“文字必须在一定条件下加以改革，言语必须接近民众，须知民众就是革命文化的无限丰富的源泉。”**从计算机科学的需要来看文字改革，我们只举一个方面的例子，事情就清楚了。要实现科学技术现代化，必须实现科技情报资料的计算机存储和检索。

我们的祖先曾经创造了算筹和算盘等在当时比较先进的计算工具。今天，在计算机科学和技术上，我们也一定能够在本世纪末赶上和超过世界先进水平。让我们在党中央的英明领导下，全国一盘棋大搞协作，在计算机科学事业上阔步前进。

和青年们谈谈计算机科学*

我国八年科技规划提出了六个影响全局的带头学科，计算机科学技术是其中之一。

今天，计算机已经广泛应用于国民经济、科学技术及日常生活的各个方面。计算机的应用主要可以分为三大类：科学计算、大量信息处理、实时自动控制。

1. 科学计算

以前，数学应用于各种科学技术，主要是作为说理的工具，真正能够进行计算的只是一些极其简单的东西。自从有了电子计算机，人们才能够在短时间内进行大量的数值计算，数学才真正成为进行科学研究、设计施工和指导生产的有力工具。现在，使用计算机进行计算，可以代替耗费大量人力物力的实验。有的产品使用前根本无法做实验，例如人造卫星必须事先利用计算机精确计算，然后施工和使用。计算机还能够帮助人们进行设计，例如设计房屋，工程师把数据送入计算机，计算机经过计算，立即把预想的那个建筑物的图形在电视屏上显示出来，工程师可以让计算机把图形旋转各种角度，可以用光笔修改图形，然后让计算机绘出设计图纸。

2. 大量信息处理

属于这方面应用的主要是情报检索和业务管理。情报也就是信息，例如图书资料、科技情报、人事档案、航空和卫星摄影图片等，这些情报可以用计算机储存起来，人们需要某种情报时，可按照一定的方式向计算机提问，计算机检索储存的情报，经过某些处理和推理回答人们的问题。至于业务管理，例如工资计算、生产计划、成本计算、库房管理、银行业务，以及科学研究的人力组织和协作，研究计划的

---

* 本文原载于科学普及出版社 1982 年出版的《选准目标，立志成材：科学家向青年介绍专业与志愿》（上册）。

制定，科研课题的确定，物资条件的保证等等，无不大量使用计算机。属于大量信息处理的还应当提到计算机辅助教学，教师将编好的某门课程的教学程序输入计算机，学生需要其中某章节时可进行选择，计算机在电视屏上显示课文和一些问题，针对学生回答问题的不同情况，计算机显示不同的新课文和新问题，这样逐步对学生进行“因材施教”。

3. 实时自动控制

生产全盘自动化，在有了计算机的今天，已经不是遥远的幻想了。除了使用计算机控制生产外，像交通运输，战略武器系统，或一个现代化的物理实验室或化学实验室等，也必须使用计算机进行实时控制。而许多实验仪器上都带有微型计算机。

此外，计算机已日益进入人们的日常生活领域，代替许多方面的家务劳动，它将成为家庭必需品。

现在计算机软、硬件的研制水平和应用的广泛程度已成为科学技术现代化的重要标志。这样就建立起了一门新学科——计算机科学。它既是一门技术科学，又是一门基础科学，具有很强的实践性和理论性，它的研究对象是“如何用算法处理信息”。例如对一些数据进行某种复杂的计算，总要把算法分解成许多基本操作，像加减乘除和某些逻辑运算。把使用这些基本操作来解决一个复杂问题的步骤列出来叫作一个“程序”。一台计算机作为一个机器，能做加减乘除和逻辑运算等基本操作，这个机器就是硬件；解题的程序就是软件。

计算机科学的内容主要如下：

1. 计算机体系结构　现代计算机是软、硬件交互作用的体系，对软、硬件全面考虑以设计高功能、稳定可靠，而又价格低廉的计算机体系，必须进行深入的理论分析。

2. 硬件理论与技术　研究硬件设计的理论和实现设计的实际技术。

3. 软件理论　研究软件设计的理论、技巧和帮助设计的工具。

4. 计算机应用的研究　针对计算机在各方向的应用进行理论和方法的研究。

5. 人工智能　研究如何提高计算机应用的“智能”因素，从而使计算机更灵巧地处理问题。

自从 1956 年开始研制电子计算机以来，我国的计算机事业是取得了很大成绩的。在党的正确科技方针的指导下，我国的计算机事业，包括计算机科学的研究工作，必将蓬勃发展，赶上世界先进水平！

庆祝吉林大学数学系建系40周年*

当此吉林大学数学系建系40周年之际，我们不由得回想起建系之初。1952年，国家对全国高等学校做了大规模的调整，原只有文科的东北人民大学改建为一所综合大学，1958年改名为吉林大学。新建的数学系当时有三个年级的学生，一年级是新招的，二、三年级从外校调来。然而，教师从教授到助教只有十四名，要开出四个系的十几门数学课程。图书设备基本上没有。人力物力条件的艰苦是异乎寻常的。

但是，在学校党政的领导下，经过十年左右的努力，我系便发展成为一个学科方向比较齐全，师资队伍比较雄厚，在国内外有一定影响的数学系。所以有这样快的发展，我认为正是由于建系时条件极其艰苦，大家团结一致，把办好系作为共同事业，因而系里逐步形成了一种良好的、正派的系风和一套办学传统，较好地处理了教学与科研、理论与实际、年长教师与青年教师这三种关系。

我系教师在教学中认真负责，严格要求学生。从建系之初我们就一直坚持教材建设，所编写的教材有的至今仍被校内外采用，有些还获得优秀教材奖。同时，我们大力开展科学研究。年长教师以身作则，经常同时开两门课并积极从事科学研究，而且通过讨论班等措施指导青年教师。对青年教师，提出既要过教学关，又要过科研关。年长教师对青年教师在教学和科研上热忱帮助，严格要求。对在教学和科研两方面都有突出表现的青年教师，系里大胆提拔，其中有几个助教，本科毕业后工作两年就提升为讲师。1955年《吉林大学自然科学学报》创刊号上，数学系发表论文二十余篇。

我系既重视基础理论，又重视联系实际。建系后不久，我们在基础学科方面建立了代数、拓扑、泛函分析、微分方程等专门化。这些方面所取得的成果在国内外受到了重视。在比较更接近实际的方面，我们在1957年就主动投入力量建立了计算数学专业，后来又建起了控制论（即计算机理论及设计）专门化，这在国内都是较早的。

---

* 本文原载于吉林大学出版社1992年出版的《数学科学学术研讨会（长春）论文集》，为该书的代序。

此外，我们建立了力学专业和概率统计专门化。这些方面除在理论上取得不少成果外，在同实际单位的协作中都发挥了应有的作用。

以上说的都是老皇历了。“文革”之末，我调到计算机科学系工作。“文革”后，数学系有了新的重大发展。如今，系里有数学、计算数学及其应用软件、应用数学、数理统计、力学等五个专业。这些专业都设有硕士点，其中的两个专业还设有博士点并被国家教委评为重点学科和教学科研人才培养基地。1978 年，学校借靠数学系建立了数学研究所，这是国家最早批准的数学研究所之一。数学系和数学研究所现有教师百余名，其中教授、博士导师占很大比例，老中青结构比较合理，学科方向已发展到二十余个。系所除继续向全国输送着大批本科毕业生和研究生外，还不断在国内外刊物上发表大量学术论文，出版了多种数学专著，不少科研成果获国家和部委奖励。

后来这些我都没能躬与其盛。但是，我们虽调到外系工作，数学系仍是我们的老家，工作上也有不少联系。我每次回到老家看看，家乡水、家乡人，总感到非常亲切，耳闻目见都使我感到系里仍具有当年艰苦奋斗的精神和朝气，保持着并发扬了以前的良好系风和办学传统。

让我们在发展祖国的数学事业中共同前进，取得更大的胜利！

恭贺我恩师江泽涵先生九十寿辰*

江泽涵先生是拓扑学专家，但在数学各方面的学识都是异常丰富的。自从他在 30 年代初学成归国执教北大并任数学系主任以后，才把当代处于发展前沿的数学引进了北大，使北大数学系进入一个新阶段，开始从事研究工作，改变了过去达而不作的状态。

先生执教六十余年以来，桃李满天下。我属于跟随先生亲受教诲时间最长的弟子之一。我 1933 年入北大学习，二年级的高等微积分是先生讲授的。1935 年选修了先生开设的拓扑学课；同年又参加了先生组织的拓扑学讨论班。这是我跟先生学习拓扑学的开始。1937 年我在北大毕业时，先生曾让我做他的研究生。恰在这时，七七事变发生了，北大南迁长沙，成为长沙临时大学和后来的昆明西南联合大学三个成员之一。我在该校做了两年助教。学校稍稳定后，先生开始招收研究生，我是他第一期两个研究生中的一个。我的毕业论文题目是“*Mapping of Graphs in Closed Surfaces*”。过去，关于“图”的一个著名定理是：一个“图”可以拓扑地画在球面（或平面）上，当且仅当它不包含一个完全五点形也不包含一个以九线互连的双三点形。上述论文题目是先生让我读了一些与此著名定理有关的文献后选定的。我的论文算是完成后，先生多次催我简化后寄往国外发表。但我觉得取得的结果很不好，只得到一个有限步骤的判定方法，而文章又不适当地太长难以简化，终于没有尝试发表。

在昆明，不论条件如何艰苦，先生总是坚持为我们组织讨论班。讨论班的题材除了拓扑学，还有近世代数、拓扑群等。拓扑群涉及数学众多的分支，例如拓扑代数、李群李代数、几乎周期函数、拓扑群上的积分和积分方程等等。这开阔了我们的眼界，加深了我们对当代数学和经典数学的理解。我出国在普林斯顿大学做研究生时，人家根本不理会我曾在国内研究生毕业，但我不到一年便取得了他们的硕士学位。我所

---

* 本文原载于北京大学出版社 1993 年出版的《数学泰斗　世代宗师》。

以能有这样的基础完全是由于先生的培养，任何一个跟先生学习的人都会是这样的。

先生待人厚道，乐于助人，对他的学生也是一样。但对学生却是严格要求和督促的。而当他们稍有进益时便极口称赞。七七事变发生，我经西安流亡到长沙，不料先生一时尚不能到校。过了一段时间，我弄得身无一文，困在长沙，向人借到两块钱暂时吃饭。幸而先生终于有一天来到长沙，立即让我做了助教，救拔我于穷途末路。我嗜好下棋和打桥牌，甚至因此放松了学习。偏偏有时正做这些事，先生来宿舍看我们，不须他出言责备，我自己先就脸红了。有一年暑假我住在西山和尚庙里，专门看维摩诘经等佛经和元杂剧选等闲书，先生使人责我说，虽然开卷有益，但放下数学，专看那些书，这不好。我做先生的研究生时，不能再拿助教的工薪，当时物价腾涨，助学金不够用，有一段时间先生让我在他家里吃饭。

上面提到先生几次催我发表毕业论文。我有一篇没多大意思的短文，在先生坚持下，寄往国外发表了。申又枨先生引进了一种各项前都冠以正号的“行列式”，他对这种“行列式”有一个猜想，是要用在他的函数逼近研究中的。我证明了这一猜想，先生对人几次称赞。我在国外完成了学位论文，先生非常欣喜，到处跟人说。

我 1949 年夏回国时，上海已经解放，那时蒋军封锁大陆各口岸，美国船不肯照原定开往上海，强使我们在香港下船。我困在香港，幸遇先生，才搬到一个不花钱的住处。过了些时，先生好不容易托人买到英国船票，携带我上船，夜间息灯偷渡渤海回国。我从上大学时起，始终是在先生培养教育和提携之下的。

“文化大革命”刚过去，我第一次来北大拜见先生。他见到我，什么也顾不得讲，第一句话就脸色和声音都显得是出了大事的样子对我说，他的一个学生的论文受到国际上的高度重视和赞誉。这表明他把这事看得比什么都重要，“文化大革命”中所遇到的那些折腾丝毫没有动摇他对数学和对拓扑学的热爱以及以全部精力为发展祖国数学事业而培育青年一代的决心。

欣逢先生九十大庆，谨书此以表我对先生的崇仰和感戴之情。愿先生永如青松不老，使我们能够时闻謦欬，长聆教导。

王湘浩诗歌

这经文，若能背，微积分，便学会；
n 次幂，算微商，乘以 n，降一方；
赛因 X，作微分，结果是，柯赛因；
柯赛因，求导数，得赛因，加个负；
指数函，更方便，微一微，它不变。

红楼梦新探*

---

*　《红楼梦新探》为王湘浩先生所著，首版于 1993 年由吉林大学出版社出版，2020 年为纪念先生诞辰 105 周年，新增“王湘浩院士的部分照片汇集”等内容再版。文集略去该书文前照片部分，并将伍卓群先生所写代序为文集所用，其余部分全文收录。

《红楼梦新探》序言

王汝梅

著名数学家、计算机科学家、中国科学院学部委员王湘浩教授撰著的《红楼梦新探》一书，将由吉林大学出版社出版。这本数学家和计算机科学家的红学专著，将会引起红学界的极大兴趣与重视，也将会给古典小说研究者以新的启示。

王湘浩教授早年毕业于北京大学数学系。他曾以无可辩驳的反例推翻了格伦瓦尔（Grunwald）定理，动摇了有理单纯代数理论的基础，引起国际代数界的震动。从1958年起，他重点从事计算机科学研究，创建了吉林大学计算机科学系。王湘浩教授在从事伟大的科学观念的推演、抽象与理解之时，相伴着伟大艺术的美学鉴赏。他酷爱优秀的传统艺术，尤其喜爱引以为我们民族骄傲的《红楼梦》。流传至今的《红楼梦》是一部伟大的未完成品，书未最后定稿竣工，雪芹泪尽而逝。因此，有红学家称曹雪芹的作品是伟大的惊叹号又是伟大的省略号。八十回以后的小说情节、人物命运、故事结局到底是什么面貌，成为中国的斯芬克斯之谜，成为中国文学史上的一桩公案。红学界对此难题的探讨意见分歧很大。王湘浩教授不畏艰难，乐此不疲，以勇于求解难题的探索精神，缜密地分析了前八十回提供的信息，梳理千头万绪的伏线，阐释脂砚斋的评语，在这些工作的基础上，提供了《红楼梦》一种可能的结局。在王湘浩教授的期待视野中，补续建构了《红楼梦》完整的艺术世界。

王湘浩教授在诸多有争议的重要问题上，提出了自己的新见解。关于八十回后的史湘云，《红楼梦新探》认为史湘云守寡后改嫁给贾宝玉，对“白首双星”“金麒麟”，联系杜甫的诗作，进行了新解。关于宝黛钗的爱情悲剧，论证了“木石前盟”与“金玉姻缘”终于两毁，而不能两全，也不能一全。关于荣国府内部之争，提出贾兰与主流派对立，是巧姐的“奸兄”。关于薛宝琴，有的红学家觉得宝琴是谜一般的人物，使读者感到困惑，但谁也没有详加研讨。《红楼梦新探》对这一问题用了一万余字篇幅加以探索，提出宝琴是尘世中的兼美的看法，并因此而提出雪芹用宝钗、黛玉二女的矛盾来写宝玉自己的思想斗争的见解。关于薛宝钗，王湘浩教授给予了充分肯定评价与鉴赏，并续补构思了薛宝钗及薛氏家族在八十回后的故事梗概。《红楼梦新探》根据前八十回的草蛇灰线，推演八十回后的情节，把未知的、佚散的给予索解，显示出一位科学家的创造性想象力。《红楼梦新探》使我们感受到科学想象力对艺术想象力通过撰著者在大脑中的泛化而产生的启发。王湘浩教授关于《红楼梦》研究本有一长远的计划。很痛惜的是，他已离我们而去。《红楼梦新探》竟成为了王湘浩教授的遗著，我们再也不能看到这位科学家继续以一种创造精神对伟大艺术的美学鉴赏。

早于《红楼梦》二百年问世的《金瓶梅》给曹雪芹创作《红楼梦》开启了新路，积累了艺术经验，从继承关系上说，没有《金瓶梅》就不可能产生《红楼梦》。《金瓶梅》的艺术之谜更有许多。在张竹坡评点《金瓶梅》校点本出版之后，笔者曾把一部精装本赠送给王湘浩教授，盼望他对《金瓶梅》《红楼梦》这两部伟大的艺术作品同时进行美学鉴赏。这时，他正客居大连海滨，来信时兴奋地谈到这方面的研究计划。但是，现在我们不但不能拜读到他的关于《红楼梦》研究的全部成果（尚遗留下一部分未完成稿），更不可能读到他关于《金瓶梅》的富有科学家的艺术想象力的论著了。王湘浩教授的数学研究贡献与红学研究贡献，长驻人间，永垂不朽！

1993 年 8 月 19 日

于吉林大学中国文化研究所

第一篇
《红楼梦》八十回后的史湘云

八十回后，曹雪芹要怎样写史湘云这个人物，这一直是红学界众说纷纭的问题。人们重视这个问题，不但由于史湘云是十二正钗中的重要人物，应当有个结局，而且由于这牵涉到小说的中心人物贾宝玉的最后命运。贾宝玉是否出家当了和尚？和他一起过着贫困生活的是薛宝钗还是史湘云？这样的问题直接关系到《红楼梦》的整体艺术构思和思想性。关于史湘云的最后结局，主要有下列五种看法：

1. 夫死立志守寡；
2. 嫁后早亡；
3. 嫁后因故与夫终生分离；
4. 守寡后改嫁贾宝玉；
5. 直接嫁给贾宝玉。

这些不同的看法，除第三种外，其余四种早在清代就已经有了。至于第三种看法则是近年才提出来的。最近播放的电视连续剧写史湘云做了妓女，可说是第六种看法。但电视剧在一些人物的结局上似乎留有不尽之意，关于史湘云好像就是这样的。作为一个《红楼梦》爱好者，我对史湘云问题一直感兴趣。我是赞成上述第四种看法的，觉得只有这样才能说明较多问题而不致产生矛盾。本文想把自己的理由写出来，以就正于专家。

一　白首双星

人们对史湘云的结局问题之所以有这样多的不同看法，主要是由于第三十一回的那句回目“因麒麟伏白首双星”与第五回册子中的判词及《乐中悲》曲子似乎有矛盾。从册子和曲子来看，湘云应当是嫁后早亡或夫死守寡。但若根据那句回目来看，既然湘云佩戴金麒麟，而宝玉得到一个金麒麟，那么，因麒麟而伏的白首双星自然就是他们，因而他们应当成为白头到老的夫妻。主张嫁后早亡或夫死守寡的人是以册子、曲子为准，不考虑那句回目；主张嫁给宝玉的是只考虑那句回目，忽视了册子和曲子。

为了解决这里所说的矛盾，朱彤先生在《释“白首双星”——关于史湘云的结局》（《红楼梦学刊》一九七九年第一辑）一文中，提出前述第三种看法。文中说：

> “双星”一词，在中国古代文学语言里，是一个专用名词，从古以来，它一直具有固定的、特有的内涵，即指牵牛、织女二星，不能另作他解。

接着，文中举了许多例子予以证明。然后做出论断说：“白首双星”四字暗伏后来史湘云嫁后与其夫因某种变故而分离，像神话传说中的牵牛、织女二星那样，一直到老不得重聚。这种说法在册子和曲子上似乎也通得过，所以矛盾就解决了。香港梅节先生《史湘云结局探索》一文持同样的论点。两文令人信服地论证了“双星”一词是专指牛女二星而言，这是研究史湘云结局的重要进展。但是，由此而做的论断却似乎理由还不够充分。首先，“白首”二字含有和好、偕老之意，以之形容因故分离到老不得重聚的夫妻，这与习惯用法不符合。“双星”一词必须按从古以来的习惯用法来理解，“白首”一词反而可以不按习惯用法来理解，这恐怕是说不过去的。另外，神话传说中的牛郎、织女并非“不得重聚”，而是每年聚会一次的。“双星”一词见于古人诗词也往往是取义于牛郎、织女七夕会合之好，并不特别重视其分离的一面。周汝昌先生在《红海微澜录》（《红楼梦研究集刊》第一辑）一文中曾引《长生殿·怂合》一折里牛女双星上场时的《二犯梧桐树》，其中说：

> 斗府星宫，岁岁今宵会。银河碧落神仙配。地久天长，岂但朝朝暮暮期。[五更转] 愿教他人世上夫妻辈，都似我和伊：永远成双作对。

这表明在《长生殿》作者洪昇笔下，牛女双星年年相会，地久天长，这非但不是“不得重会”，而是“永远成双作对”，其情好远胜世人的朝朝暮暮。一般诗词中使用“双星”一词是侧重哪一面，雪芹当然是熟悉的。况且他祖父曹寅对《长生殿》一剧赞赏备至，他当然也是知道的。他怎么可能用“双星”一词去指永远不得重聚的夫妻呢？

我认为，没有必要把事情说得离奇，应当承认湘云既然像册子和曲子说的嫁给了一个“才貌仙郎”，丈夫死了守寡；又像三十一回回目所暗示的嫁给了贾宝玉，夫妻白头到老。一女可以嫁二夫吗？这没什么不可以。只要两个丈夫是一前一后的就行了。这就是说，湘云是在先嫁的丈夫死后改嫁宝玉的。册子和曲子说的是她的前半生。三十一回回目说的是她的后半生。二者是互相补充的，本来就不存在什么矛盾。

其实，与夫不得重聚守活寡和夫死守寡并没有实质上的区别。如果湘云的结局就是这样的，这在册子和曲子中已经有了。对于前面早已说过的事情，雪芹怎么会又在三十一回当作秘密用“双星”二字去暗示，又用“麒麟”去作伏笔，并且郑重其事地标出“因麒麟伏白首双星”这句回目呢？可见，这句回目必然隐含新的重要信息，而“双星”二字也一定另有寓意。按照神话传说，牛郎、织女结为夫妇是私相配合的。天帝闻知大怒，使他们分离在银河两边（或说是王母娘娘用金簪在他们中间画了一道天河），他们却又在喜鹊搭桥的帮助下在夜里私会。这样一来，天帝和王母娘娘无计可施，只得任凭他们去“永远成双作对”了。因此，牛女双星不只是一般恩爱夫妻的象征，他们还是争取婚姻自主反抗封建礼法最终取得胜利的艺术形象。回目既然以“双星”形容日后的宝湘结合，则他们的结合应当也具有这种特点。但是，湘云这样一个侯府千金，虽然“襁褓中父母叹双亡”，她的婚姻也一定是由家庭中或近亲内的长辈包办的，婚姻自主有可能吗？只有在一种情形下才可能，这就是旧时代俗话说的“初嫁由父母，再嫁由自身”。寡妇再嫁，社会上是要非议，父母是要反对的，所以只能是当事人自己做主，经过斗争才能实现的。

在使用“双星”二字的诗词中，特别值得注意的是杜甫的《奉酬薛十二丈判官见赠》。全诗共六十句，摘抄若干如下：

> 忽忽峡中睡，悲风方一醒。……谁重断蛇剑，致君君未听。志在麒麟阁，无心云母屏。卓氏近新寡，豪家朱门扃。相如才调逸，银汉会双星。客来洗粉黛，日暮拾流萤。不是无膏火，劝郎勤六经。老夫自汲涧，野水日泠泠。我叹黑头白，君看银印青。……丈人但安坐，休辨渭与泾。……荣华贵少壮，岂食楚江萍。

这里，集中在中间部分的十三句中，“麒麟”“双星”“白头”都出现了！如果我们说雪芹在纂写“因麒麟伏白首双星”这句回目时，曾受到这首著名杜诗的启发，这大概距离事实不会太远吧？“白首”即“白头”，雪芹是由于平仄考虑才改用“白首”的。我们看，这首杜诗使用“双星”一词恰恰是用在卓文君嫁司马相如这种寡妇再嫁的事情上！《史记·司马相如列传》说：

卓王孙有女文君新寡。好音，故相如缪与令相重，而以琴心挑之。及饮卓氏弄琴，文君窃从户窥之，心悦而好之，恐不得当也。既罢，相如乃使人重赐文君侍者，通殷勤。文君夜亡奔相如。相如乃与驰归，家居徒四壁立。卓王孙大怒曰：“女至不材，我不忍杀，不分一钱也。”

寡妇再嫁本来就是违反封建礼法的，何况还夜里私奔呢？卓王孙自然要大怒了，如此不材女，杀之也不为过。织女嫁牛郎也是私奔，惹得其父天帝大怒。可见，上述杜诗使用“双星”一词是用得极为确切的。

三十二回中也有一些迹象说明雪芹的确受到这首杜诗的启发。据说薛十二丈曾有类似司马相如娶卓文君之事，写诗给杜甫为自己辩解，杜甫的诗回答他说：这类事无须分什么渭与泾，要建立功业才是正事。诗中有“客来”二字，三十二回便有客来要会见宝玉。诗中薛十二丈的那位“卓文君”劝郎读六经，三十二回宝玉的未来的“卓文君”便劝他讲“仕途经济”。杜甫劝薛十二丈说：荣华贵少壮，要志在麒麟阁建立功业，不要把心思用在云母屏边粉白黛绿的事情上去；史湘云劝宝玉说：“如今大了，……也该……谈谈讲讲些仕途经济的学问，……没见你成年家只在我们队里搅些什么？”所有这些何其相像！

“白头”二字除了在这首杜诗中出现，还有一个重要出处雪芹一定也注意到了，这就是卓文君曾写过一首《白头吟》：

皑如山上雪，皎若云间月。闻君有两意，故来相决绝。今日斗酒会，明日沟水头，躞蹀御沟上，沟水东西流。凄凄复凄凄，嫁娶不须啼。愿得一心人，白头不相离。竹竿何袅袅，鱼尾何簁簁？男儿重意气，何用钱刀为！

她为什么作这首诗呢？《西京杂记》说：

司马相如将聘茂陵人女为妾，卓文君作《白头吟》以自绝，相如乃止。

我们试猜想一下雪芹撰写三十一回回目时的艺术构思。他要在回目中暗示将来湘云改嫁宝玉，首先想到的自然是借用人所共知的卓文君改嫁司马相如的故事。但直接引用，那就太明显，不成其为暗示了。他进一步想到上述那首杜诗，诗中曾用到文君、相如之事并比之为“双星”，而“我叹黑头白”这句诗又使他联想到文君的《白头吟》。于是，他用“白首双星”四字曲折而巧妙地暗示了湘云改嫁宝玉之事。

二　金麒麟

如果“白首双星”四字的来源我们猜想得不错，那么，回目中的“麒麟”二字又是怎么来的呢？宝玉戴通灵玉，宝钗戴金锁，大概雪芹也想让湘云戴点什么东西。二宝的“金玉姻缘”在《终身误》曲子中是定了的，但宝湘也要结合，因此湘云戴的东西也以金器为宜。下面的情节雪芹已是胸有成竹了：让宝玉在外面得到一个同样的金器，却又得而复失，让湘云拾到，以此作为二人日后关系的伏笔。那么，让湘云戴什么金器呢？据说司马相如在文君家里“以琴心挑之”，除了琴音还有歌词：

凤兮凤兮归故乡，游遨四海求其凰。有一艳女在此堂，室迩人遐毒我肠，何由交接为鸳鸯！

宝玉得到的那个金器对于湘云的金器应当是像相如歌词中的凤求凰那样起“挑之”的作用的。不正是这样吗？翠缕说：“可分出阴阳来了！”湘云拿到手里以后，“只是默默不语。正自出神，忽见宝玉从那边来了，……”所以，为了暗示凤求凰，让湘云戴金凤凰最切题，只是太明显些。相如的歌词中还有“鸳鸯”一词，戴金鸳鸯也可以，但那就不合少女身份了。忽然想到杜甫诗中的“麒麟”二字，戴金麒麟最好！麟为百兽之长，凤乃百禽之王，麟凤是相连的。《礼记·礼运》：“山出器车，河出马图，凤凰麒麟，皆在郊棷。”杜甫《幽人》：“麟凤在赤霄，何当一来仪。”湘云既然戴金麒麟，三十一回的那句回目“因麒麟伏白首双星”就完全定下来了。回目中“麒麟”“白首”“双星”三个词指向杜甫的那首诗，其中的“白首”一词联系文君的《白头吟》，“麒麟”一词联系相如的《凤求凰》。雪芹的巧妙构思、用古入化，真使我们不得不叹为观止。他还怕读者由“麒麟”联想不到司马相如的那段歌词，特在三十二回黛玉来探消息时写道：

原来林黛玉知道史湘云在这里，宝玉又赶来，一定说麒麟的原故。因此心下忖度着，近日宝玉弄来的外传野史，多半才子佳人都因小巧玩物上撮合，或有鸳鸯，或有凤凰，或玉环金佩，或鲛帕鸾绦，皆因小物而遂终身。今忽见宝玉也有麒麟，便恐由此生隙，同史湘云也做出那些风流佳事来。

相如歌词中有凤凰和鸳鸯，这里就用凤凰和鸳鸯来陪衬麒麟以暗示歌词，这真是字字都有来历，笔笔皆非闲文。

现在我们看，三十一回中是怎样因麒麟而伏白首双星的呢？庚辰本在回末有一条脂批：

后数十回若兰在射圃所佩之麒麟，正此麒麟也。提纲伏于此回中，所谓草蛇灰线，在千里之外。

由于这条脂批，人们猜想湘云将来要嫁给卫若兰其人。这个问题后面再讨论，现在我只说，我同意周汝昌先生的观点：卫若兰不是湘云的丈夫，而是她和宝玉之间的绾合者。据脂批，八十回后宝玉的通灵玉曾被盗。可做如下设想：盗玉者同时也盗去了宝玉的金麒麟，经过几次转卖，金麒麟为卫若兰所得，他喜欢它，便佩戴在身上。七十五回说，贾珍请了几位世家弟兄及亲友，每日早饭后习射，贾政曾命宝玉、贾兰等跟着去练习。习射者里面可能有冯紫英、卫若兰等人。经过许多变故以后，宝钗故去，湘云守寡。宝玉漂泊归来，可能同卫若兰等人在“射圃”相遇，认出了若兰所佩的金麒麟，谈及往事，不胜欷歔。若兰听出了他对湘云的怀旧之情，慨然自任撮合。此时翠缕已嫁，但仍与湘云有来往。若兰设法同翠缕取得联系，像相如经文君侍者通殷勤那样，请她转达宝玉求凰之意，并把金麒麟交翠缕带给湘云，以明前缘已定，不可有违。这样，宝湘二人，便不顾湘云夫家、史家、贾家的反对和人言的非议，终于成为眷属。三十一回中宝玉丢掉的麒麟，经翠缕到湘云手中，这就是伏于该回中的“提纲”，也就是“因麒麟伏白首双星”这句回目的含义。

三十一回回目暗伏湘云改嫁宝玉已如上述。那么，第五回的册子和曲子对于湘云改嫁是否一点消息也没有透露呢？这倒也不是。今按庚辰本把《乐中悲》曲子抄录于下：

襁褓中，父母叹双亡。纵居那绮罗丛，谁知娇养？幸生来，英豪阔大宽宏量。从未将儿女私情，略萦心上。好一似，霁月光风耀玉堂。厮配得才貌仙郎。博得个地久天长。准折得幼年时坎坷形状。终究是云散高唐，水涸湘江。尘寰中，消长数应当，何必枉悲伤？

此曲共有十个整句，每个整句都是由三个字开始，后跟一个或两个小句。例如，第一个整句的格式是“三五”，第二个整句是“三三四”，最后一个整句是“三五五”。有的版本中，最后那个整句前面多出“这是”二字，已卯本就是如此，这就破坏了曲子的节奏。可见，“这是”二字乃是不应有的衍文。顺便说一下，现存庚辰本是否全由现存己卯本过录而来，红学界有不同的看法。今在《乐中悲》曲子中，己卯本有衍文，庚辰本倒保持了原状，过录之说就值得怀疑了。

如果我们仔细体会一下最后那个整句，就会觉察到其中颇有弦外之音。旧时代的一个年轻妇女，夫死守寡，以后的苦日子望不到边，世间哪还有比这更悲惨的遭遇，怎么能说“何必枉悲伤”呢？这话的言外之意其实就是说：何必这样想不开，白白受苦呢？“往前走一步”也未尝不可嘛！明白了这一层，最后那个整句的意思就清楚了，“云散高唐，水涸湘江”是说夫死守寡。但世界上的事情有“消”也还能有“长”，这也是自然之理，应有之数，和你说的“阳尽了就成阴，阴尽了就成阳”是一个道理，何必枉悲伤呢？不要再这样死心眼儿守下去了吧！若多出“这是”二字，“尘寰中消长数应当”就不是泛论人世沧桑而成为专指此曲前面的那些话而言，这就使人觉得湘云的一生已经由那些话盖棺论定，她只能认命守寡到底了。从曲名看，乐极可以生悲；但从最后那个整句看，消甚也能有长。这种哲学思想在三十一回由湘云论阴阳加以阐述，这也是示意读者不要只看到事情的一面，须要体会《乐中悲》曲子的全面含义。

册子和曲子对湘云的后半生只能像上面说的略有所示。

若在第五回就让读者知道或猜到湘云寡后改嫁宝玉这一结局，那就让读者过早地知道了宝玉一生的婚姻情况，读下去就没有趣味了。但瞒过这一结局，册子、曲子对湘云一生遭际的提示毕竟太不完全。因此，作者才在三十一回回目中和另外一些地方用暗示的方式加以补充。这是作者的巧妙安排。

甲戌本第一回有脂批说：

若云雪芹披阅增删，然后（则）开卷至此，这一篇楔子又系谁撰，足见作者之笔狡猾之甚。后文如此处者不少。这正是作者用画家烟云模糊处，观者万不可被作者瞒弊（蔽）了去，方是巨眼。

雪芹以三十一回回目补册子、曲子之不足，但仍不肯让读者轻易地猜透玄机，于是在三十一回和三十二回用了不少烟云模糊法以“瞒蔽”读者。先由王夫人把湘云“前日有人家来相看，眼见有婆婆家了”吵嚷得尽人皆知，然后在怡红院又让袭人当着宝玉面说：“大姑娘，听见前儿你大喜了。”这就给读者造成一种印象，似乎湘云已经有了婆家，不能嫁给宝玉了，因麒麟而伏的白首双星与宝玉无关。其实相看不等于婚事成功；即使成功，也完全可以在守寡后改嫁宝玉。又写宝玉把麒麟失掉了，这使得读者推想宝玉与湘云无缘。殊不知失而复得正暗示日后湘云嫁出而又嫁回。湘云称赞宝钗，议论黛玉，又“心直口快”地讥笑宝玉，使读者感觉宝湘二人很不和睦。特别是湘云说“仕途经济”时，宝玉撵她到别的屋里坐坐，又说黛玉从来不说这些混帐话，要说早和她生分了。这就更使读者觉得二人本来不和睦，这次又闹了严重矛盾，从此必然就生分了。其实只要读下去就会发现，二人不但没有生分，反而越来越接近。脂砚还怕所有这些不能“瞒蔽”读者的“巨眼”，又在三十一回加回前批说：

金玉姻缘已定，又写一金麒麟，是间色法也，何颦儿为其所感？故颦儿谓情情。

没想到这条批语反而泄漏了春光。既然你说写金麒麟不过是间色法，只是骗骗颦儿的，没有别的意思，怎么回目中说这麒麟很重要，是用以伏白首双星的呢？可见你的批语只是烟幕弹。

三　圃冷斜阳忆旧游

有的同志说，前八十回对史湘云未来命运的伏笔和暗示非常少，仅有的几处又过于含蓄隐晦。其实，书中关于湘云的伏笔和暗示是相当多的，而且也并不都是很隐晦的。本节要讨论的菊花诗，其含意就颇为明显。

如所周知，咏白海棠表面上是宝钗夺魁，实际上是湘云后来补作的两首诗压卷。在这两首诗中“自是霜娥偏爱冷”这一句下面，庚辰本有一双行批语：

又不脱自己将来形景。

有正本漏掉了“又”字。这个字很重要，表示湘云另外的诗句中也有不脱自己将来形景的。要找这样的诗只能往后面去找，因为白海棠诗以前，书中没有湘云的诗。在白海棠诗之后，首先见到的就是菊花诗。这次是黛玉夺魁，宝钗的螃蟹咏也被评为绝唱。湘云这次的表现似乎平平。但是，且看下面这段文字：

> 黛玉道："我那个也不好，倒底伤于纤巧些。"李纨道："巧的却好，不露堆砌生硬。"黛玉道："据我看来，头一句说的是'圃冷斜阳忆旧游'，这句背面傅粉；'抛书人对一枝秋'已经妙绝，将供菊说完，没处再说，故翻回来想到未折未供之先，意思深远！"

雪芹通过黛玉之口指出湘云的"圃冷斜阳忆旧游"是十二首菊花诗中头一句好的，这岂是泛泛之笔，分明是让读者特别注意这句诗了。

现在我们把湘云的《对菊》《供菊》两首诗抄在下面：

对菊

别圃移来贵比金，一丛浅淡一丛深。
萧疏篱畔科头坐，清冷香中抱膝吟。
数去更无君傲世，看来惟有我知音。
秋光荏苒休孤负，相对原宜惜寸阴。

供菊

弹琴酌酒喜堪俦，几案婷婷点缀幽。
隔坐香分三径露，抛书人对一枝秋。
霜清纸帐来新梦，圃冷斜阳忆旧游。
傲世也因同气味，春风桃李未淹留。

我们看，"圃冷斜阳忆旧游"这句诗连同上一句表示什么意思。"霜"者，"孀"也；"斜阳"等于"展眼吊斜晖"中的"斜晖"二字。湘云居孀清冷，梦中、回忆中的是什么呢？是大观园中欢乐的日子，是和宝玉青梅竹马、两小无猜的童年，是共同结社吟诗、一起烧鹿肉吃的往事，是拾到宝玉小心戴在身上却在雨中失落在蔷薇架下的那个金麒麟时的情景，等等，等等，所有这些都涌上心头。这说的是未改嫁之先的心情，是背面傅粉。至于改嫁和改嫁之后，那就要看这诗的其余几句和上一首《对菊》

了。“别圃移来”就是改嫁的意思，和宝玉《种菊》中的“携锄秋圃自移来”相呼应。移来者其贵如金，非比寻常，二人之结合乃是又一“金玉姻缘”。前后两个“金玉姻缘”，前者“浅淡”而后者“深”。为什么呢？因为“数去更无君傲世，看来惟有我知音”。黛玉说“抛书人对一枝秋”已经妙绝，这句连同“弹琴酌酒喜堪俦”是说宝湘结缡后的夫妇琴瑟之好。“春风桃李”和册子中的“桃李春风结子完”有关。宝玉漂泊归来，宝钗殂谢，麝月适人，他与湘云结合之时，想必贾兰“威赫赫爵禄高登”，贾府颇有家道复兴之意，其实无非是回光返照而已，不久就“昏惨惨黄泉路近”。当时那种表面的“春风桃李”未使宝玉淹留，他断然携妇离开京城，隐于山村，共同过着像文君、相如那样“家居徒四壁立”，然而却是自食其力的生活，“傲世也因同气味”。从菊花诗开始，湘云使用别号“枕霞旧友”，“枕霞”示隐居，这个别号暗示将来和宝玉成为共隐青山的旧时之友。两首菊花诗的暗示如此明白无误，为什么脂砚等人不在这里加批，却在白海棠诗上加批呢？在菊花诗上加批是不行的，那就彻底“泄露天机”了。在白海棠诗上加批不妨事，因为那两首诗只是说孀居时的景况。

菊花诗唯一的解释恐怕只能是湘云改嫁宝玉，任何另外的假定在菊花诗上都通不过，比方说，假定湘云守寡到底，不论是夫死守寡还是与夫终生分离守活寡，怎样解释“来新梦”“忆旧游”呢？若以为湘云最初嫁的就是贾宝玉，“别圃移来”“来新梦”都讲不通。避开菊花诗，只在“白首双星”上做文章是不能解决问题的。

上面说关于湘云命运的伏笔、暗示并不少，现在再举几处为例。

“寿怡红群芳开夜宴”，湘云掣的签画着一枝海棠，题首“香梦沉酣”四字，另一面刻的诗句是“只恐夜深花睡去”。这使大家自然而然想到白天的“憨湘云醉眠芍药裀”，以为此签只不过是指那件事而言。但众人抽的花名签都和本人的命运有关，湘云此签怎么会只应在那件小事上呢？签上的诗句来自何处？来自苏东坡的《海棠》一诗：

> 东风袅袅泛崇光，香雾空蒙月转廊。只恐夜深花睡去，故烧高烛照红妆。

于是我们恍然大悟：那句诗要结合下一句才能明白是说什么的。“故烧高烛照红妆”，这分明是洞房花烛夜的情景。湘云白海棠诗是说她的孀居景况，以之与东坡此诗相对照，白海棠变为红海棠，这正是守寡后再嫁之兆。“香梦沉酣”是说湘云寡居，如同沉睡。长夜漫漫，只恐名花睡去不醒，再结花烛之后，方从梦中醒来也。此签的含意其实非常明显。即使没有其他暗示和伏笔，只根据此签也可以断定：湘云的日后命运只能是夫死再嫁。雪芹惟恐读者猜破机关，让黛玉打趣湘云说：“‘夜深’二字改‘石凉’两个字倒好。”以诱使读者去想日间湘云卧石之事，读者稍一大意，就又被作者瞒蔽去了。

“史湘云偶填柳絮词”，调寄《如梦令》，其词如下：

> 岂是绣绒残吐？卷起半帘香雾。纤手自拈来，空使鹃啼燕妒。且住，且住，莫使春光别去。

此词暗伏她在守寡后有一段漂泊生活。她拿给宝钗和黛玉去看，黛玉笑道：“好！新鲜有趣，我却不能。”但照字面来看，这首词好像也没有什么特别新鲜有趣的，而且还很难讲通。一定要结合她的寡居、漂泊和改嫁才能理解。“岂是绣绒残吐”，这是说做一个寡妇像残吐的绣绒被命运抛弃。“卷起半帘香雾”是形容自己漂泊无依的生活。“纤手自拈来，空使鹃啼燕妒”是说改嫁乃自己选择的道路，让一些人去叫嚷，一些人因妒忌而非议吧。“且住，且住，莫使春光别去”是说不听这些闲言，寡居漂泊的生活必须停止了，不要使有限的春光如流水般逝去。词牌是“如梦令”，这是说自己的寡居漂泊生活像一场梦，和花名签上的“香梦沉酣”“只恐夜深花睡去”是一样的意思。

湘云与黛玉联诗，最后，湘云说了一句“寒塘渡鹤影”，黛玉对了一句“冷月葬花魂”。这时妙玉走出来，说：“好诗，好诗，果然太悲凉了。”又说：“此亦关人之气数。”这两句诗自然是暗伏她二人的结局，上句说的是湘云，下句说的是黛玉。“塘寒”如菊花诗中之“圃冷”，指孀居而言。“寒塘渡鹤影”，鹤离寒塘而另有所适也。“此亦关人之气数”，即《乐中悲》曲子中所说“尘寰中，消长数应当”之意也。

四　才貌仙郎

前面说过，我不同意有些同志说的史湘云嫁给了卫若兰。此人只在第十四回送殡的诸王孙公子名单里提了一下，除此，在全书正文中，就再也没有提到。前引三十一回后的那条脂批也只是说，后数十回卫若兰曾在射圃戴过宝玉的那个金麒麟。此外，在第二十六回写宝玉、薛蟠、冯紫英几个人谈话时有一条眉批：

惜卫若兰射圃文字迷失无稿，叹叹。

此人才华怎样，相貌如何，一概不知道，怎么能根据上述那点材料把他硬派给湘云算是她的“才貌仙郎”呢？我们曾经指出，在三十一回和三十二回中，雪芹用了许多烟云模糊法以“瞒蔽”读者，脂砚等人关于“间色法”的回前批也是烟幕弹。可见，在三十一回雪芹显然不愿意让读者猜到湘云的丈夫是谁，脂砚等人怎么反而会在回后加批把谜底偷偷告诉读者呢？这就从反面证明：脂砚等人在回后点出卫若兰的名字，说他以后戴过那金麒麟，这只不过是要使读者上当，误认为湘云要嫁给这个卫若兰。回前那个批语是实者虚之，回后这个批语是虚者实之，二批都是烟幕弹。

那么，这个“才貌仙郎”究竟是谁呢？先说我的答案，然后再讲理由。我认为，这就是和贾宝玉同名、同貌、同性格、大概也有同样才华的甄宝玉。此人说是“才貌仙郎”，是当之无愧的。

我的主要根据是第五十六回。这一回中写甄府四个女人到贾府请安，说她们那里也有个宝玉。后来见了贾宝玉唬了一跳，因为不但名字、性格一样，连相貌也一样。现在摘引此回的一些段落如下：

贾母便向李纨等道：“偏也叫作个宝玉。”李纨忙欠身笑道：“从古至今同时隔代重名的很多。”……

这里贾母喜的逢人便告诉：也有一个宝玉，也都一般行景。众人……皆不介意。独宝玉是个迂阔呆公子的性情，自为是那四人承悦贾母之词；后至蘅芜院去看湘云病去，将这话告诉，湘云说：“你放心闹吧。先是单丝不成线，独树不成林，如今有了个对子，闹急了，再打狠了，你逃走到南京，找那一个去。”宝玉道：“那里的谎话，你也信了，偏又有个宝玉了！”湘云道：“怎么列国有个蔺相如，汉朝又有个司马相如呢？”宝玉道：“这也罢了，偏又模样儿也一样，这是没有的事！”……湘云没了话答对，因笑道：“你只会胡搅，我也不和你分证。有也罢，没也罢，与我无干！”

宝玉心中便又疑惑起来：“若说必无，也似必有；若说必有，又并无目睹。”心中闷闷，回至房中榻上，默默盘算，不觉就忽忽的睡去。……一面顺步早到了一所院内……进入屋内，只见榻上那个少年叹了一声，一个丫环笑问道：“宝玉，你不睡，又叹什么？想必为你妹妹病了，你又胡愁乱恨呢。”……

宝玉听说，忙说道：“我因找宝玉，来到这里，原来你就是宝玉！”榻上的忙下来拉住：“原来你就是宝玉，这可不是梦里了！”宝玉道：“这如何是梦，真而又真了。”……

一个宝玉就走，一个宝玉便忙叫：“宝玉快回来！”袭人在旁听他梦中自唤，忙推醒他，笑问道：“宝玉在那里？”宝玉……因向门外指说：“才出去了。”袭人笑道：“那是你梦迷了，你揉眼细瞧，是镜子里照的你的影儿。”

我们看，这时湘云正因时气所感病卧在蘅芜院，全书提到史湘云患病只有这一次。恰好在这时，贾宝玉梦见甄宝玉的丫环对甄宝玉说：“想必为你妹妹病了，你又胡愁乱恨呢。”这决非巧合，显然是一种暗示，这是第一。

第二，同贾宝玉讨论甄宝玉的不是别人，恰恰是湘云。一般情形下，贾宝玉有什么事，总是先找黛玉去谈，这次不但没去找黛玉，到蘅芜院也没有去见宝钗，回到房中只是自己默默盘算，也没和袭人等谈论，只找湘云谈不找别人，这难道不是雪芹故意安排而有所示意的吗？

第三，“从古至今同时隔代重名的很多”，为什么雪芹偏偏让湘云把两个宝玉比作蔺相如和司马相如呢？将来湘云是要像卓文君嫁司马相如那样嫁给贾宝玉的。比作司马相如的贾宝玉是湘云的后夫，那么，比作蔺相如的甄宝玉又是她的什么呢？

第四，湘云说：“有也罢，没也罢，与我无干！”真的无干吗？这正是从反面暗示日后此人与湘云是大有干系的！这是雪芹涉笔成趣的妙文。读者阅至八十回后湘云嫁甄宝玉时，回想此时她所说“与我无干”的话，定当拍案叫绝也。

第五，贾宝玉盘算，那个甄宝玉是有是无呢，结果进入梦迷，自唤其名，梦中两个宝玉搅在一起，分不清谁真谁假、谁有谁无了，这真是神来之笔！但这仍然是为湘云日后命运所做的伏笔。对于湘云来说，贾作甄时甄亦贾，无为有处有还无，前后两个宝玉对于她来说是一而二、二而一的。

第六，众人皆不介意，只有贾母喜得逢人便告诉，为什么这样写？贾宝玉是老太太的心尖子，知道甄家的孩子和她的爱孙同名、同貌、同性格，自然很高兴，这不奇怪。但写她喜欢得逢人便告诉，这必是作者又在暗示什么。很可能八十回后是贾母做主把她的侄孙女许配给甄宝玉的。

甲戌本第二回有脂批说：

> 甄家之宝玉乃上半部不写者，故此处极力表明，以遥照贾家之宝玉。凡写贾宝玉之文则正为真宝玉传影。

此批中之“真”字显系“甄”字之误。由此批可见，下半部将有甄宝玉之事，而且必然是重要的情节，否则不会在第二回“极力表明”。这些情节也一定和上半部的人物有关联，决无自成体系之理。我们根据脂批知道，雪芹《石头记》原稿应共有一百一十回，第五十六回正是下半部的第一回。以上摘引的那些段落及有关的文字在这一回占了很大篇幅。这些虽不是关于甄宝玉的正面描写，但写在下半部的开始，肯定是重要提示。在此重要提示中，有一整段文字涉及湘云，则湘云和甄宝玉很有关系由此也可想见了。

第十八回元春归省，点了四出戏。庚辰本在四个剧目下都有批注：

> 第一出豪宴——一捧雪中。伏贾家之败。
>
> 第二出乞巧——长生殿中。伏元妃之死。
>
> 第三出仙缘——邯郸梦中。伏甄宝玉送玉。
>
> 第四出离魂——牡丹亭中。伏黛玉死。
>
> 所点之戏剧伏四事，乃通部书之大过节、大关键。

贾家败、元春死、黛玉死是全书之大过节大关键，这不错。甄宝玉送玉这件事本身怎么能是大过节大关键呢？这和其余三事完全不能相比。可见“甄宝玉送玉”这一批语中还隐藏着什么事。《仙缘》是汤显祖《邯郸记》传奇中的第三十出，内容是吕洞宾引卢生至仙境，八仙共度卢生的故事。“仙缘”暗示什么呢？有两种可能：一是甄宝玉真遇到了神仙；二是借指姻缘之类的事。若第一种解释是对的，可能是甄宝玉梦见警幻仙姑引他到太虚幻境。即使是这样，也必定同姻缘之类的事有关系，因为警幻仙姑是专管人间之风情月债、尘世之女怨男痴的。所以，可以肯定，“仙缘”总是暗示姻缘的。“甄宝玉送玉”自然是送玉给贾宝玉了，因此，所说的姻缘必定同甄贾二宝玉都有关系。

根据这一分析，我们可以做下面的猜想：贾宝玉丢失的通灵玉，辗转为甄宝玉购得。他不知是贾宝玉的，见它晶莹可爱，便自己戴着。这样一来，他有通灵玉，史湘云有金麒麟，二人就成了“金玉姻缘”。湘云看到甄宝玉戴着的通灵玉，当然认得是贾宝玉之物。后来甄宝玉将此玉送还了贾宝玉。甄宝玉此举意味着什么呢？这表示：他迟早是要退位让贤，把这“金玉姻缘”双手奉送给贾宝玉的。这一大过节大关键才是可与其余三事相比的。脂砚之批，为什么在其余三事上说得明明白白，而此事说得如此隐晦曲折呢？说得太明白，岂不又泄露天机了吗？

五　此系身前身后事

庚辰本第二十一回有脂批：

> 宝玉看（有）此世人莫忍为之毒，故后文方能悬崖撒手一回。若他人得宝钗之妻，麝月之婢，岂能弃而僧哉！

这不是说宝玉当了和尚吗？而且分明说他在“悬崖撒手”时有“宝钗之妻，麝月之婢”。如果说湘云嫁给他，那只能是他当了和尚以后又还俗结婚了，这不大可能吧？

但是，要注意这里有“一回”两个字。这就是说，后三十回中有宝玉“悬崖撒手”这样的“一回”，并没有说这是他的最后结局，也没有说这一回是全书的最后一回。如果宝玉真当了和尚，那就应当直截了当地说“故后文方能悬崖撒手”，为什么要拖泥带水添上“一回”两个字呢？这就好像前八十回中有“听曲文宝玉悟禅机”那样一回，说悟禅机实际上并未真悟。黛玉曾和他开玩笑说：“作了两个和尚了。我从今以后，都记着你作和尚的遭数儿。”可见宝玉说作和尚，原也可以不作的。何况他是一贯“毁僧谤道”的呢？

脂靖本六十七回有回前批说：

> 末回撒手，乃是已悟：此虽眷念，却破迷关。是何必削发？青埂峰证了前缘，仍不出士隐梦中；而前引即三姐。

这岂不正是说宝玉撒手是在全书最末一回吗？但是，此批虽录在回前，其实只是批此回开始处下列一段文字的：

> 柳湘莲见三姐身亡，痴情眷恋。却被道人数句冷言，打破迷关。竟自截发出家，跟随这疯道人飘然而去，不知何往。暂且不表。

所以，“末回撒手”说的是柳湘莲，与宝玉无关。正文说柳湘莲何往“暂且不表”，可见后文还有湘莲之事。批语的意思是：柳湘莲末回撒手，才是真悟；此时眷念三姐，尚未真悟，却也打破了迷关。若真悟，又何必削发呢？所谓“青埂峰证了前缘”，可能是警幻把“情榜”张贴在青埂峰下的那块顽石上，尤三姐引导众情鬼去看“情榜”。众情鬼造劫历世之事是第一回在甄士隐梦中交代的，所以说“仍不出士隐梦中”。为什么由尤三姐引导呢？因为六十六回说过，她是奉警幻之命掌管册子修注一干情鬼之案的。

有的同志说，像贾宝玉这类人物，经历了从富贵繁华到抄家破落的巨大政治变故以后，在绝望之余，只有遁入空门才符合他思想性格的逻辑。我们认为，这一推理的说服力是很不够的。因为，一个明显的反例就是作者曹雪芹本人。贾宝玉的思想性格无非是在一定程度上反映了曹雪芹自己的思想，而且曹雪芹也经历了类似的政治变故，但他并未遁入空门。敦诚《挽曹雪芹》诗中说“新妇飘零目岂瞑”，可见他不但没有当和尚，还有个新妇。他没有“绝望”，也未“斩断尘缘”，他走的是一条远为积极的道路。他傲视封建统治者，对他们采取一种不合作的态度，“羹调未羡青莲宠，苑召难忘立本羞”；他自食其力，“卖画钱来付酒家”，宁愿过着“举家食粥酒常赊”的穷困生活而“著书黄叶村”，写出这部具有反封建意义的文学巨著《红楼梦》。如果写后三十回的贾宝玉时，部分地反映了作者自己所走的这条道路，岂不比写他当和尚更能表现其叛逆思想性格吗？

所以，无论从脂批来看还是从“逻辑”来看，都不能证明宝玉当了和尚。另一方面，我们却有相反的材料说明他的确没有当和尚。

庚辰本第十九回有批语说：

> 宝玉自幼何等娇贵，……与下部后数十回“寒冬噎酸齑，雪夜围破毡”等处对看，可为后生过分之戒，叹叹。

由此批可见，贾府经抄家破败后，宝玉确实有一段像雪芹那样非常穷困的生活。问题是，这段生活是在他“悬崖撒手”以后还是以前呢？若在以前，则是宝钗、麝月等和他共度这段生活的；若在以后，则他并没有真当和尚可知。

雪芹的友人明义曾见《红楼梦》初稿，有《题红楼梦》七绝二十首。今录其最后的第十九首和第二十首于下：

莫问金姻与玉缘，聚如春梦散如烟。石归山下无灵气，总使能言亦枉然。

馔玉炊金未几春，王孙瘦损骨嶙峋。青娥红粉归何处？惭愧当年石季伦。

第十九首说那块顽石已归大荒山下，第二十首却又说“王孙瘦损”，可见宝玉尚在人间。这是怎么回事呢？原来，在雪芹原著中，宝玉是神瑛侍者转世，那顽石本来与他无关，只不过茫茫大士、渺渺真人为了使那蠢物去经历经历，和警幻仙姑商妥，把它夹带在情鬼历世案内，由神瑛侍者转世的宝玉衔它降生，程高为了安排那戏剧性的“调包计”，必须让宝玉失玉疯癫，因而要肯定那通灵玉的确是宝玉的命根子，于是篡改原著，把神瑛侍者说成是那顽石变的。人们往往忘记这一点，认为“石归山下”就等于贾宝玉“复归到青埂峰下的原位去了”。从明义的第十九首诗来看，“莫问金姻与玉缘，聚如春梦散如烟”表示宝玉已经“悬崖撒手”舍宝钗等而去，这时那顽石已回归青埂峰下，第二十首“王孙瘦损”乃是神瑛侍者转世的宝玉“寒冬噎酸齑，雪夜围破毡”时的情景。明义的这两首诗有力地证明了宝玉这段穷困生活是在他“悬崖撒手”以后，从而无可怀疑地否定了他出家为僧之说。

可能有的同志会提出另一个问题：宝玉当和尚不成，会不会回到宝钗身边，和她共同过那段穷苦日子呢？张之先生的《红楼梦新补》正是那样写的。我认为那种事情是不可能的，但讨论薛宝钗的问题需要另写专文，此处我想只指出一点：根据宝钗的柳絮词《临江仙》，她被宝玉遗弃后，并未委芳尘，而是置身青云之上做出一番事业，她是过不惯穷日子的。

“悬崖”意谓一种重要的抉择关头。可做如下猜想：以前贾政虽已不想让宝玉从科甲出身做官，但在贾府破败后，为使家道有复兴之望，贾政等人却逼迫宝玉像高鹗续书所说的那样下考场。这违反宝玉坚决不肯当“禄蠹”的原则，又因屡经严重打击，他产生了逃世的念头。这时茫茫大士、渺渺真人便乘虚而入来劝他出家为僧，同他商定，先送那顽石归位，然后趁他下考场之机度脱他出家。宝玉离家后，未下考场，混出长安，却不见大士和真人到约定地点。原来大士和真人带那蠢物到太虚幻境销号时，提起要度宝玉出家，警幻仙姑坚决不同意，说神瑛侍者在她所主持的那情鬼历世案内情缘未了，他们引诱他出家纯属“挖墙脚”的行为。经过一番争吵，大士和真人自知理亏，只得把那蠢物送归青埂峰下了事，不再赴宝玉之约。宝玉久等二人不来，他一向毁僧谤道，普通和尚老道他看不上眼，不肯随便跟他们出家，只得开始过着流浪生活。后来回到长安，与湘云结婚。

众人起社填柳絮词，香尽时探春才有了半首《南柯子》：

空挂纤纤缕，徒垂络络丝。也难绾系也难羁，一任东西南北各分离。

这半首词显然不只是说她自己的远嫁，而是说贾家败亡时“树倒猢狲散”的情景。宝玉本来已经交了白卷，看到这半首词，忽然有所触动，提笔续出下半阕：

落去君休惜，飞来我自知。莺愁蝶倦晚芳时，纵是明春再见隔年期。

宝玉这半首词是回答探春的，今译为白话：

树倒猢狲散你不要惋惜，我将飞回长安，这我确信无疑。分离在这莺愁蝶倦的晚芳时候，我们还会相见的吧，相见在春天到来的新时期。

宝玉经过像柳絮那样的漂泊，果然回到了长安并和史湘云相见了。他所期待的“春天”虽未到来，但他们却开始了一种有意义的新生活，什么有意义的新生活呢？这就是宝王在湘云的帮助下编写《红楼梦》。

关于《红楼梦》的作者虽有一些争论，我是确信作者是曹雪芹的，这当然指的是前八十回和失去的后三十回。但如前引第一回中的一条脂批所说，雪芹之笔“狡狯之甚”，他说自己只不过“披阅十载，增删五次，纂成目录，分出章回”而已，而真正的“作者”另有其人。且看他在第一回开始时所说：

此开卷第一回也。作者自云：因曾历过一番梦幻之后，故将真事隐去，而借通灵之说撰此石头记一书也。……但书中所记何事何人？自又云：今风尘碌碌，一事无成，忽念及当日所有之女子，一一细考较去，觉其行止见识，皆出于我之上，何我堂堂须眉诚不若此裙钗哉！……万不可因我之不肖，自护己短，一并使其泯灭也。虽今日之茅椽蓬牖，瓦灶绳床，其晨夕风露，阶柳庭花，亦未有妨我之襟怀笔墨。虽我未学，下笔无文，又何妨用假语村言，敷演出一段故事来？……

按雪芹狡狯之笔，这里所说的“作者”不是他自己，也不是那块顽石，因为，说是那蠢物所撰只不过是为了隐去真事而已。那么，这个“作者”是谁呢？此人必须亲历过《红楼梦》那番“梦幻”，必须是个须眉男子，而和书中那些女子非常接近，以致能对其行止见识细细考较，此人又必须具有堂堂须眉不如裙钗的特殊观点，这除了贾宝玉还能是别人吗？这“作者”既是贾宝玉，为何不说“宝玉自云”呢？第一回宝玉还未出场，不能点他的名。还有一件事，在此又一次得到证明：宝玉写《红楼梦》时，其居处是“茅椽蓬牖，瓦灶绳床”，这显然是俗家的陋室景象，可见他没有当和尚。猜想在末回可能要补叙此书“出则”如下：宝玉在湘云的帮助下写完《红楼梦》，本想传世。等到众情鬼归位时，警幻仙姑得知有这部书，认为不宜流传，将书收归太虚幻境，送入薄命司存档，茫茫大士、渺渺真人闻知此事，来到幻境，主张将书刻在那块顽石上，若后世偶遇有缘人抄去，就无妨问世传奇，若无此机缘，也便作罢。警幻不好拒绝，只得同意。大士与真人刻完此书，又以石头的口气吟成一偈刻在后面：无才可去补苍天，枉入红尘若许年。此系身前身后事，倩谁记去作奇传。“石归山下无灵气”，它归位后尘世间的事情，乃是它身后之事了。

六　傲世也因同气味

湘云菊花诗中说：“数去更无君傲世，看来惟有我知音。”又说，“傲世也因同气味”。这表明，她和宝玉的结合不但有深厚的感情基础，而且有共同的思想基础。但是，不知从什么时候起，史湘云被人扣上了落后帽子，人们说她具有“受封建主义濡染很深的思想性格”，说曹雪芹“现实主义地揭出她思想性格中的落后面”，“这么两个对世界和人生具有根本对立看法的人，作者怎么可能违背人物性格逻辑的制约，让他们晚年好合，结为患难中的夫妻?!”这些观点影响所及，史湘云在大观园那些女孩子中，几乎成为典型的落后分子。例如，有同志把“数去更无君傲世，看来惟有我知音”说成是黛玉的诗句。为什么会这样呢？因为，这两句诗似乎很进步，所以不用查书就知道是林黛玉的，绝不会是史湘云的。可见，为史湘云平反昭雪很有必要。

我们认为，大观园里的那些女孩子都很年轻，是比较天真和纯洁的，不宜硬性地把她们划分为进步和落后两类，她们之间时或发生这样那样的矛盾，她们对一些问题也有不同的看法。但不能把这些都夸大为“封建正统派与叛逆者之间的思想搏斗”。这些女孩子之间的矛盾不同于叛逆者与贾府上层封建势力之间的矛盾。这些女孩子各有优点，也各有缺点，宝玉也是这样。

宝玉坚决反对封建主义的几乎一切方面，这是很可贵的。但他的思想言行中也有一些幼稚的、过“左”的成分，因而周围的人往往对他不理解。例如，他认为做官的人是“禄蠹”，这就是说，封建统治者是社会的寄生虫、吸血鬼，因此他自己坚决不肯走“仕途经济”的道路，这无疑是有很高认识的。但是，那时的人很少有不赞成读书做官的，一般人不可能有他那样的认识水平。三十二回湘云劝他走“仕途经济”的道路，这也不过是常人之见，结果他“大为光火，立即斥之为混帐话，给她难堪，赶她到别屋里去坐”，这就做得有些过分了。三十六回他说：“好好的一个清净洁白女儿，也学的钓名沽誉，入了国贼禄鬼之流！”这帽子也太大了些吧？他为了表示决心，“采取革命行动”把自己所有的书都烧了，只留下四书，必要时用作挡箭牌对付他父亲，这种焚书的做法岂不有点幼稚可笑了吗？

书上说：“众人见他如此疯颠，也都不向他说正经话了。独有黛玉，自幼不曾劝他去立身扬名，所以深敬黛玉。”可见只有黛玉一个人了解他。但是，在他“大受笞挞”后，黛玉也动摇了，哭着对他说：“你从此可都改了吧！”在外面，最了解他的是秦钟，但秦钟临死来了个一百八十度的大转弯，对他说：“以前你我见识，自为高过世人，我今日才知自误了，以后还该立志功名，以荣耀显达为是。”宝玉最引为知己的黛玉和秦钟，其思想高度不过如此，为什么湘云说过一次“仕途经济”，她就具有“受封建主义濡染很深的思想性格”呢？

有同志可能反驳说，湘云对“仕途经济”缺乏认识，这就是落后；宝玉“给她难堪”，这就是和她“格格不入”。前面说过，雪芹和脂砚在三十一、三十二两回用了许多“烟云模糊法”，湘云说“仕途经济”，宝玉“给她难堪”，这正是作者“狡猾之甚”的笔法之一例，“观者万不可被作者瞒（蔽）了去，方是巨眼”。但这姑且不论吧，在

“仕途经济”这一问题上，我们只想再说明一点：有证据表明，八十回后，湘云提高了认识，在对官场的看法上和宝玉取得了完全一致的意见。二十二回众姐妹的谜语都隐着本人的命运，这是书有明文且有脂砚诸批为证的。但那一回中连惜春的谜语都有，独无湘云的谜语，这显然是雪芹有意隐瞒湘云的日后命运。直到第五十回才有湘云的一个谜语，形式是一支《点绛唇》：

溪壑分离，红尘游戏，真何趣！名利犹虚，后事终难继。

谜底是“耍的猴儿”，是宝玉猜着的。当时这一谜语以特笔出之，可以肯定地说，必与湘云的日后遭遇有关。试作如下猜想：湘云出嫁时，史家和甄家都已衰败，她与甄宝玉居于“溪壑”，这大概是指不太繁华的地方而言。湘云力劝丈夫读书应考，后来甄宝玉得中，湘云随夫离家宦游，这就是所说的“溪壑久分离，红尘游戏”。她亲眼看到官场中的事情沐猴而冠，和耍猴儿差不多，所以说“真何趣”。甄宝玉也没有做到什么大官，名未成，利未就。“后事终难继”是说他死在任上，湘云开始过着寡居漂泊的生活，尝尽人间苦味。“真何趣”三字说明湘云通过亲身经历认清了官场的尔虞我诈、卑鄙龌龊，克服了以前对“仕途经济”的天真幻想。她的谜语由宝玉猜着表示二人在这一问题上意见一致了。此谜恐怕只有如上解释才能讲得通。《点绛唇》作为一首词来看，此谜只是半阕，所以说的是湘云日后命运的一半。

人们“揭发批判”史湘云，除了“仕途经济”，还有一条罪状，这就是她在芦雪庵联诗时曾“颂圣”。中秋凹晶馆联句，黛玉说了一句“色健茂金萱”，湘云说：“不犯着替他们颂圣去。”为什么黛玉说的这句诗是颂圣，陈诏先生在《〈红楼梦〉小考（一）》（见《红楼梦研究集刊》第一辑）一文中有考证和分析。可见湘云不赞成颂圣，倒是黛玉有颂圣之嫌了。这是有书上的明文为证的，怎么反而把颂圣的帽子扣到湘云的头上呢？

说湘云在芦雪庵联诗曾“颂圣”大概是由于她说过一句“加絮念征徭”和一句“瑞释九重焦”。但是，她在“瑞释九重焦”之下接着说的是“僵卧谁相问”。“僵卧”二字的出处是东汉时的袁安在大雪积地丈余中，因考虑到人皆饿，自己忍饥僵卧不出。但诗文中用典，往往只是借用，不一定就是说出处中的那件事。事实上，袁安僵卧是有人相问的，而且还因此举了孝廉。所以，“瑞释九重焦”和“僵卧谁相问”连在一起是说：皇帝在深宫见下雪，以为瑞雪兆丰年，可以高枕无忧了，但大雪天因饥寒交迫而僵卧的人们有谁过问呢？这哪里是颂圣，分明是在挖苦皇帝老儿，是《红楼梦》中的“碍语”之一。

“加絮念征徭”的出处据说是这样的：唐开元中，宫中制棉袍赐边军。有士兵在袍中找到一个宫女的诗：“沙场征戍客，寒苦若为眠？战袍经手作，知落阿谁边？蓄意多添线，含情更著棉。今生已过也，重结后生缘！”后此兵竟因诗与这一宫女成为夫妇。不管出处是否如此，“念征徭”的总是“加絮”之人，并不是皇帝。可见，这句诗和颂圣是不相干的。

评论旧社会一个人的思想倾向主要应当看她对统治者和对下层人民的态度。上述的“碍语”清楚地表明，湘云的思想感情是较为倾向下层人民的。这是在相对意义上说的，并不是说她已经很“革命”的了。我们看，全诗共七十句，其中有三句涉及下层人民的疾苦：“加絮念征徭”“僵卧谁相问”“清贫怀箪瓢”，而这三句都是湘云说的。另外有一句“诚忘三尺冷”。有同志认为这句诗是说将士不顾三尺剑冰冷而执剑守边，但这样解释就和下雪无关了。我认为这句诗的出处是“程门立雪”之事。《宋史·道学传》中说，杨时师事程颐：“一日见颐，颐偶瞑坐，时与游酢侍立不去。颐既觉，则门外雪深一尺矣。”程门为学本于“诚”字，而一尺改为三尺则出于平仄考虑与艺术的夸张。可见这句诗是说尊师重道，不是说人民疾苦。显然，雪芹写那些女孩子联诗，是自己先大致作好全诗，然后按照人物的思想性格和日后遭遇把诗句分配给大家，在此过程中还要根据总的意图多次修改原诗和分配方案。那么，他把涉及下层人民疾苦的三句诗统统分给了湘云，这难道还不足以说明雪芹是要怎样塑造湘云这个人物和要赋予她什么样的思想感情吗？顺便说一句：这三句诗都分与湘云也使我们更加相信，日后和宝玉共同过着“寒冬噎酸齑，雪夜围破毡”的是湘云，不是宝钗。

宝玉和湘云的思想性格有很多相近之处。在对待下层人民的态度上就是如此。例如，凤姐和鸳鸯拿刘姥姥开了一个极其恶劣的玩笑，潇湘子雅谑补余香说刘姥姥是“母蝗虫”，又说惜春的画可名为“携蝗大嚼图”，宝钗也在旁捧场凑趣，宝玉和湘云虽然也跟着大家笑，他们自己却不可能讲这些谑近于虐的话。

湘云性格开朗，口快心直，“英豪阔大宽宏量”“好一似，霁月光风耀玉堂”。在这方面，宝玉的性格也是和湘云最接近而和黛玉并不接近。四十九回“脂粉香娃割腥啖膻”：湘云悄悄和宝玉计较，要一块鹿肉，拿了园里弄着又顽又吃，宝玉听了，巴不得一声儿。果然二人拿到芦雪庵烧着吃起来，吸引得一些人也跟着吃。黛玉笑道：“那里找这一群花子去！罢了，罢了，今日芦雪庵遭劫，生生被云丫头作践了。我为芦雪庵大哭。”湘云冷笑道：“你知道什么，是真名士自风流，你们都是假清高，最可厌的。我们这会子腥膻大吃大嚼，回来却是锦心绣口。”宝钗笑道：“你回来作的不好了，把那肉掏了出来，就把这雪压的芦苇子揌上些，以完此劫！”接着，争联即景诗。完了一数，参加联诗的共十二人，全诗七十句，其中宝琴十三句，黛玉十一句，宝钗只有五句，湘云独有十八句，占全诗四分之一强，又一次夺魁。众人都笑道：“这都是那块鹿肉的功劳。”《红楼梦》无一闲笔，这些文字生动地描绘出湘云的名士风流、文惊四座的风采。将来和宝玉共隐青山，傲视王侯，“萧疏篱畔科头坐，清冷香中抱膝吟”的只能是她。

以上分析了湘云的思想和性格。书中多次描写了她的出众才华和健美的体魄。三十二回着重叙述了她赶做针线活的勤苦生活和她刺绣的精美。宝玉拿她做的扇套子去和黛玉比，结果黛玉赌气把那扇套子铰了。所以，仅从前八十回看来，史湘云已经是德、智、体、美、劳全面发展的优秀女青年，这是大观园中任何另一个女孩子，连同贾宝玉在内，所难以相比的。当然，她也和其他少女一样，思想还未定型，还在发展变化。八十回后，她经历了贾史两家抄家败亡的巨大变故和夫死守寡的悲惨遭遇，接着在漂泊中历尽艰辛，但也接触到下层人民，她所经受的这些磨炼和教育必然使她在思想感情上发生深刻变化而转向人民。宝玉的情形也是一样。二人结合时，思想感情上已经不再是当年的侯府千金和贵公子。这就是“傲视也因同气味”这句诗的真实含义。“傲世”当然不是傲视人民而是“横眉冷对千夫指”、傲视封建统治者之意。

十分明显，雪芹用极大力量塑造了史湘云这个人物，决不是为了在八十回后把她塞给一个卫若兰，然后让她守寡或守活寡了事，《红楼梦》也决不会在宝玉当和尚、“落了片白茫茫大地真干净”中收场。那就是玉石俱焚，让先进人物和书中所有善良的人们统统为四大家族殉葬。那就好比写一本国际政治小说，结局是发生世界大战，许多核弹落下来，人类毁灭，地球成为齑粉，于是一切矛盾统统解决了，岂不快哉！

曹雪芹的哲学思想和艺术构思不是那样的。“阳尽了就成阴，阴尽了就成阳”，旧事物的衰落，其中孕育着新事物的萌芽。“落了片白茫茫大地真干净”的是四大家族，封建统治者的末日不等于世界末日。曹雪芹是一定要写顽石“身后”那部分文字的，他所塑造的贾宝玉和史湘云这两个人物也一定要有所作为的。他不是写什么“悲剧”“喜剧”，他只是在写《红楼梦》。

茅盾有一首诗赞美思想成熟后的曹雪芹，诗中的主要意思看来也适用于雪芹笔下共隐青山时的贾宝玉和史湘云。我们抄录于下作为本文的结束：

浩气真才耀晚年，曹侯身世展新篇。自称废艺非谦逊，鄙薄时文空纤妍。

莫怪爱憎今异昔，只缘顿悟后胜前。懋斋纪盛虽残缺，已证人生观变迁。

第二篇
金玉姻缘与木石前盟

宝黛钗的爱情婚姻纠葛，研究者的讨论已经很多了。本文主要是想弄清楚一些重要事件的真相而不过多地评论有关人物的是非曲直。如果不明真相而急于评论是非，说不定“乱判葫芦案”把官司断错了。可能有人说，小说不比历史，书上没有说的就是没有的，有什么真相可言呢？但《红楼梦》不同于一般小说，其中有许多曲笔和“不写之写”，故用画家“烟云模糊”之法，读者一不小心，就会被作者“瞒蔽了去”。所以，虽然小说不比历史，但对《红楼梦》这部特殊的小说而言，其中的重要情节确有弄清真相的必要。

一　婚事初提

百二十回本《红楼梦》中，二宝的婚事是在八十四回由凤姐提起的。所以，一般认为，八十回前这一婚事不但未定，而且也未提起过。真实情况如何呢？

第四回，薛家母子三人远离家乡来到京城，说是要入部算账并“送妹待选”。真是这样吗？薛家是半贵族半商业性家庭。其实所谓贵族不过是祖上的事了，如今不但没个做官的，而且人丁单弱，只有一母一子一女，儿子又不成材。虽“家中有百万之富”，“自薛蟠父亲死后，各省中……伙计人等……便趁时拐骗起来；京都中几处生意，渐亦消耗”。可见这个家庭已是岌岌可危。所以，离家来京的真实意图是要依傍贾家以图挽救危局，否则为什么住在贾家后就不再收拾京中原有房舍以便搬去呢？

书上说，宝钗“自父亲死后，见哥哥不能依贴母怀，他便不以书字为事，只留心针凿家事，好为母亲分忧解劳”。照书上后来的描写，她有过人的才识、坚强的性格，而其治家之能在红楼诸芳中为第一，超过凤姐和探春。五十七回薛姨妈对黛玉说：“你这姐姐就和凤哥儿在老太太跟前一样，有了正经事就和他商量，没事时幸亏他开我的心。”离家去依傍贾家这一重大决策，只能是出之于宝钗，薛姨妈无此才略，而薛蟠到后来也不知有此意图。

是否离开金陵前就已经打主意要成就那个“金玉姻缘”呢？至少要探索这种可能性。薛姨妈是王夫人的亲妹妹，自然早就知道姐姐有衔玉而生的男孩子宝玉。癞头和尚送那两句吉利话，说要錾在金器上，等有玉的才可婚配，这“有玉的”不是指宝玉而言是指谁呢？果然安顿在贾家不久，“金莺微露意”“送妹待选”的话便再也不提了。这哪里是待选入宫，分明是待选入贾家为媳。二十八回有下列一段话：

> 薛宝钗因往日母亲对王夫人等曾提过“金锁是个和尚给的，等日后有玉的方可结为婚姻”等语，所以总远着宝玉。昨儿见元春所赐的东西，独他与宝玉一样，心里越发没意思起来。幸亏宝玉被一个林黛玉缠绵住了，心里念念只记挂着林黛玉，并不理论这事。

薛姨妈对王夫人等说“金玉”之事，这岂不等于亲自为女儿向贾家提亲吗？当然，这种说法是比较婉转的，如果王夫人不同意，尽可以装作不懂，薛姨妈也不会感到太尴尬，所谓“王夫人等”大概是王夫人和凤姐。这话是“往日”说的，可见来贾家后不久，先让“金莺微露意”，然后薛姨妈便以试探的口气向贾家提出了这一亲事。

王夫人和内侄女凤姐联手早已掌握了荣府的大权。若娶甥女宝钗为媳，其权自更巩固。所以，这门亲事王夫人是求之不得的。转问贾政时，贾政从家庭利益考虑，也甚合意。贾家“生齿日繁，事务日盛，主仆上下，安富尊荣者尽多，运筹谋画者无一；其日用排场费用，又不能将就省俭，如今外面的架子虽未甚倒，内囊却也尽上来了”。若和薛家结成儿女亲家，一贵一富互相依靠，对两家都有好处。如今掌家的凤姐，总是要回到大房去的，那时二房就掌家无人了。看宝钗秉性贤淑，颇有才能，得她为媳，一者可用闺中之力规引宝玉“入正”，二者可接替凤姐持家，这门亲事实在是天造地设，尽善尽美的。宝钗知道姨父姨母已有允婚之意，自应遵礼远着宝玉一点。

二　为弟相亲

元春的端午节礼物，只有宝钗同宝玉一样，黛玉等都较少，这自然是对宝玉择配表示了明确的态度。问题是，她深居宫内，只省亲一次，怎么会忽然做出这一选择呢？

自从有旨准每月二六日椒房眷属入宫看视到元妃省亲，其间经过了一年多。在这一年多时间内，王夫人必曾几次入宫。元妃除垂询父母、祖母等的情况外，最关心要问的自然就是爱弟宝玉。听了王夫人的答对，关于宝玉择配，元春在钗黛之中已有所偏。但这事关系着家运的盛衰，非同小可，她必须亲自相看一番。因此，归省时同本家女眷相见后，便问到宝钗黛玉。召见叙礼间，元妃看宝钗时，只见：

> 唇不点而红，眉不画而翠。脸若银盆，眼如水杏。罕言寡语，人谓藏愚；安分随时，自云守拙。

好一个天然艳丽而又稳重端庄的福相。再看黛玉时，生得：

> 两弯似蹙非蹙罥烟眉，一双似喜非喜含情目。态生两靥之愁，娇袭一身之病。泪光点点，娇喘微微。闲静时如娇花照水，行动处似弱柳扶风。心较比干多一窍，病如西子胜三分。

元妃不由暗道："父亲母亲所见果然不差。祖母虽未言明，亦必选中宝钗无疑。这黛玉如此娇弱，实非福寿之相。"然仅凭外表还难遽定，乃命各题一匾一诗，以观其才而窥其内。完卷上呈，元妃看宝钗所题匾额是"凝晖钟瑞"四字，其曰：

> 芳园筑向帝城西，华日祥云笼罩奇。
> 高柳喜迁莺出谷，修篁时待凤来仪。
> 文风已著宸游夕，孝化应隆归省时。
> 睿藻仙才盈彩笔，自惭何敢再为辞。

果然文如其人，且又颂扬得体，谦抑适宜，真吾弟之好逑，贾门之贤妇也。黛玉匾额是"世外仙源"四字，其诗乃五律一首：

> 名园筑何处，仙境别红尘。
> 借得山川秀，添来景物新。
> 香融金谷酒，花媚玉堂人。
> 何幸邀恩宠，宫车过往频。

元妃暗摇头：此园专为盛世隆恩之省亲旷典而建，怎么比之为避秦的世外仙源，说是想不到幸邀恩宠宫车频过呢？我们看宝玉“试才题对额”时，有清客题匾曰“秦人旧舍”，宝玉批驳道：“秦人旧舍说避乱之意，如何使得？”黛玉之诗匾恰恰犯了此病。“原来林黛玉安心今夜大展奇才，将众人压倒，不想贾妃只命一匾一咏，倒不好违喻多作，只胡乱作一首五言律应景罢了。”其实，黛玉的诗品反映着她的思想和性情，即使认真作，也难以取得贵妃娘娘的欢心。于是，元妃回宫后，以端午赐品作了明确表示。

三　金蝉脱壳

宝钗到贾府后，很快就对宝玉产生了爱慕之情。这从“比灵通”那一回所描写的“金娃对玉郎”的旖旎风光看得很清楚。自从姨父姨母有允婚之意，她便感到和宝玉的关系已非寻常。但是宝玉却“被一个林黛玉缠绵住了，心心念念只记挂着林黛玉”，这使宝钗产生了强烈的忌妒之心。于是，二女成为情敌。黛玉对付情敌的办法是对宝玉试探哭闹，对二宝讽刺、挖苦，使得宝玉摔玉、砸玉、赌咒、发誓。宝钗由于性格上的矜持和平日言行上的庄重，受到黛玉的挖苦，不但不能还嘴，还要表现得似乎“浑然不觉”。

这天，宝钗追扑两个蝴蝶到滴翠亭边，听里面红玉和坠儿正私传手帕并要推开槅子看外面是否有人，宝钗躲不及，使了个“金蝉脱壳”之计，假作是在追寻黛玉，说自己在河那边看见黛玉在此蹲着弄水，问题是：宝钗此事只是以急智掩盖自己呢，还是有意把事情扣在黛玉头上呢？

我们说，这要通过效果检查动机，宝钗去远后，红玉拉坠儿道：“了不得了！林姑娘蹲在这里，一定听了话去了！”又说：“若是宝姑娘听见，还倒罢了。林姑娘嘴里又爱刻薄人，心里又细，他一听见了，倘或走露了风声，怎么办呢？”从这样的效果看来，应当认为，宝钗一方面固然是在掩盖自己，一方面也是要把事情扣在黛玉头上。此事之前，姑娘们祭饯花神，独不见黛玉，宝钗去找她，遥见宝玉进了潇湘馆，她为避嫌疑，转身回去，这时显然已颇含酸意。忽见一双“玉色”蝴蝶，一上一下迎风翩跹，这使她想到“二玉”正在潇湘馆嬉笑玩戏，增强了心中的忌妒。恰遇二婢之事，就借机让情敌替自己背上了黑锅。

四　两条旧帕

宝玉大承笞挞后，晚间人静时，遣开袭人，命晴雯给黛玉送去两条旧帕。黛玉先还不解其意，细心搜求，方才大悟，体贴出手帕的深意，不觉神魂驰荡。究竟有何深意，书上没有说。研究者都强调这两条手帕的重要性，但到底是什么意思，据笔者所知，似乎都没有说，我们不妨试猜一下。

书上写黛玉悟出帕子的深意后，感情激动，心想：

> 宝玉这番苦心，能领会我这番苦意，又令我可喜；我这番苦意，不知将来如何，又令我可悲；忽然好好的送两块旧帕子来，若不是领我深意，单看了这帕子，又令我可笑；再想令人私相传递与我，又可惧；我自己每每好哭，想来也无味，又令我可愧。如此左思右想，一时五内沸然炙起。

问题的关键是：黛玉的"苦意"是什么？如果我们能像宝玉那样领会黛玉的"苦意"，两条旧帕的深意就不难猜想了。

三十二回黛玉听到湘云说"仕途经济"，又听宝玉说："林妹妹不说这样混帐话，若说这话，我也和他生分了。"黛玉惊喜自己的眼力不错，宝玉果然是个知己。接着"诉肺腑心迷活宝玉"，二人不用多言，已是心意相通，默默相感而定情。宝玉"不肖种种大承笞挞"后，黛玉哭得眼睛红肿，气噎喉堵，心中虽有万句言词，只是不能说得，半日，方抽抽噎噎地说道："你从此可都改了罢！"黛玉心中的"万句言词"是什么呢？

宝玉反对封建礼教，坚决不和封建官僚同流合污，说他们都是些国贼禄鬼。黛玉由于天性、家庭和她自身的遭遇，对官场深为厌恶，因而从没劝过宝玉去立身扬名或讲究什么"仕途经济"。这就成为两人爱情的共同思想基础。但以黛玉的聪明，她自然也明白，宝玉的思想同他的家庭对他的企望大相径庭，她隐隐感觉到他们的爱情很可能是不幸的。这次宝玉被父亲狠狠责打，归根结底是为了他的那种一般人看来不长进的乖僻性情。他若坚持不改，舅舅等人是绝对不能容许的。这次打得这样重，以后还不知要怎样管教。她吃惊地意识到，她给予宝玉思想上的支持是支持了他对家庭的反抗，因而必然给宝玉造成可悲的后果。但难道也像别人一样劝他去立身扬名吗？这是违反她的天性的。她一方面痛惜宝玉的被责打，一方面为两人的爱情前途柔肠百转，这就是为什么她哭得眼睛红肿，气噎喉堵。她痛苦地下了决心，为了宝玉，必须劝他改变他的犟脾气，要违反自己本意地去劝他，甚至为此牺牲自己的爱情，为了他，她可以牺牲一切。这是爱情的最高境界，而这也就是她的“苦意”。但见了宝玉，却一句话也说不出来，半日才勉强说出那句：“你从此可都改了罢！”

宝玉当时也没有完全领会这句话所包含的“万句言词”，过后才体会出她的“苦意”。他给黛玉的两条旧帕就是重申两点：一是决不改变旧志，二是决不改变旧情。黛玉悟出了帕子的深意，两人的爱情和互相了解又加深了一层，两人的“叛逆思想”在互相支持下也更进了一步，以致宝玉有焚书之举。

“都道是金玉良姻，俺只念木石前盟。”木石前盟是什么呢？在虚幻世界中，这当然就是神瑛侍者和绛珠仙草的那段公案，这是作者的浪漫主义艺术构想。但在人间，木石前盟就是两条旧帕所代表的海誓山盟：海枯石烂，不改旧志，不改旧情。金主富，玉主贵，金玉良姻乃是符合贾薛两家家庭利益的婚姻。两条旧帕经过黛玉的题诗恰成为“金玉”的对立物。

五　钗黛修好

宝黛钗的爱情婚姻矛盾日益激化了。一方面是元妃赐礼，一方面是宝黛定情。这天，宝钗到宝玉房内，袭人出去走走，宝钗身不由己坐下来，替袭人绣那鸳鸯肚兜，形成宝玉睡在床上，宝钗坐在他身边做针线，旁边放着蝇帚子，以便替他赶着蚊蝇的“景儿”。忽听宝玉在梦中喊骂说：“和尚道士的话如何信得？什么是金玉姻缘，我偏说是木石姻缘！”宝钗听了这话，不觉怔了！这表示二玉的爱情已达到刻骨铭心的程度，真是“睡里梦里也忘不了”！她痛苦地认识到，自己在这场竞争中已经失败。

宝钗是理智型的女性。她从痛苦中逐渐冷静下来分析当时的形势：从两个家庭的意图来看，她和宝玉的婚姻已成定局，即使她自己想退出也已经不可能。宝玉这次挨打，宁死也不讨饶。看来他家里若在婚姻问题上硬逼他，他也一定至死不从。这是一个无法解开的死扣！怎么办呢？她做出了一个果断的决定：主动同黛玉改善关系，化敌为友！

四十二回“蘅芜君兰言解疑癖”，宝钗因黛玉行牙牌令时用了《牡丹亭》和《西厢记》词句，戏说要审她，接着说了一大段十分坦白诚恳的话，又说自己也看过那些杂书。一席话说得黛玉垂头吃茶，心下暗服，只有答应“是”的一字。这些话并不能改变黛玉的思想和性情，她“心下暗服”是感激宝钗无人时才向自己指出不该在大庭广众之中用那些书里的词句招人议论，而且说出她对这类事物的看法的真心话，她劝黛玉的话显然是为了黛玉好。

“金兰契互剖金兰语”，宝钗帮黛玉治病又送给她燕窝，黛玉真心感激宝钗的情分，说了许多肺腑之言。几次说“往日竟是我错了，实在误到如今”，又说燕窝“东西事小，难得你多情如此”。

自此，二人果然消除了敌意，建立了深厚的友谊！宝琴来了，黛玉不但不忌妒老太太待她好，却赶着宝琴叫妹妹，并不提名道姓，直是亲姊妹一般。

论者说，这是宝钗又在使奸，黛玉上了她的当。但“心较比干多一窍”而又疑心很重，早对宝钗怀有戒心以为她“藏奸”的黛玉那样容易上当吗？这事也要通过效果看动机，宝钗劝黛玉不要当众显出自己看过那些杂书又帮黛玉治病，这都是对黛玉有好处的，所以宝钗的居心应当认为是好的。否则，对别人不好，别人说是使坏；对人好了，别人又说是使奸，做人未免太难了吧？

不，不是的。二人的友谊是真诚的。但婚姻问题怎么解决呢？庚辰本四十二回有回前脂批如下：

> 钗玉名虽二个，人却一身，此幻笔也。今书至三十八回时，已过三分之一有余，故写是回，使二人合而为一。请看黛玉后宝钗之文字，便知余言不谬矣。

关于此批，笔者将在另一篇文字中作进一步讨论。此处只针对婚姻问题来谈一点看法。

先看宝钗。她既然和黛玉结为“金兰之好”，决不准备将来又反目成仇，而且自然想到了日后的婚姻问题。那么，事情很明显，唯一可能的方案是：二女共效英皇！我必须赶紧声明，我坚决反对多妻制！宝玉和黛玉也不会接受这样的安排，雪芹也不可能有这种主张。我只是说，雪芹在八十回后要写宝钗有这种主观设想。那时在官僚家庭中，男子三妻四妾是常事，共效英皇被视为美德，许多小说、鼓儿词用此法解决矛盾，《聊斋志异》中这样的事情多得很，《儿女英雄传》几乎全书写的就是这件事，《红楼梦》许多续书也照此办理，如果黛玉死后宝钗只是非常悲痛，那能算二人“合而为一”吗？二女共侍一夫总多少有“合而为一”的意思吧？这就是为什么脂批之人看了黛玉逝后宝钗的文字，方知“余言不谬”。

宝钗的这一方案并未立即得到她母亲的支持，薛姨妈认为这是多此一举。直到“慧紫鹃情辞试忙玉”薛姨妈才看到事情的严重性，懂得二女一夫是解决问题唯一可行的方案。五十七回“慈姨妈爱语慰痴颦”，当面对黛玉说要向老太太出主意把她定给宝玉，这其实就是试探她肯不肯与宝钗共侍宝玉，因为，谁都知道薛姨妈是早已极力宣扬“金玉”之说的，决不可能不顾宝钗单把黛玉说给宝玉。

现在看黛玉这方面。黛玉越来越看得清楚，即使她的病完全好了，她也不可能按照舅舅家的要求做宝玉之妇。她能够像凤姐那样管理家务吗？她能够劝宝玉去读八股做举业求取功名吗？她是孤高自许、一尘不染的，是“世外仙姝寂寞林”，是天生的心灵上和生活上的孤独者。只有宝玉是她的知己，而宝玉所以是她的知己，正是由于只有他理解她的孤独和寂寞。人生得一知己可以无恨，爱情的归宿不一定是婚姻。她本来有为了宝玉牺牲自己的“苦意”，逐渐她更加下定了决心：必须劝宝玉不要再和家里硬顶，即使因此会使得他们那金玉姻缘成就，她也无怨。七十九回她劝宝玉说：“又来了，我劝你把脾气改改罢。一年大二年小，……”她说出这话，内心是十分痛苦的，话未说完便咳嗽起来。

第三回写到黛玉因宝玉曾摔玉，自己淌眼抹泪地说：“今儿才来，就惹出你家哥儿的狂病，倘或摔坏了那玉，岂不是因我之过？”甲戌本批道：“所谓宝玉知己，全用体贴工夫。”有正本批道：

> 惜其石必惜其人，其人不自惜，而知己能不千方百计为之惜乎？所以绛珠之泪至死不干，万苦不怨，所谓“求仁而得仁，又何怨？”

那块玉应当爱惜，黛玉能够体贴到这一点。若宝玉和家里对抗，固执到底，则不但危及其前途，且将危及其身，黛玉能不千方百计为之惜乎？即使自己因此受尽万苦，其泪至死不干，亦将无怨。

所以，钗黛二人，一个有共效英皇之心，一个有为情舍己之志，婚姻问题上的矛盾不再存在了。当然，这不包括思想和性格。

六　贾母表态

前面说过，二宝的金玉姻缘，从贾家的利益来看，乃是无可挑剔的。况且连贾妃都已明确表态，如果贾母再一点头，这事应当早已定了。事情一直定不下来，这说明贾母迟迟没有表态。

作为最高家长，贾母当然重视家庭利益。但她不像贾政等人一样，她也要考虑其他方面。例如，黛玉是她极为爱怜的外孙女。又如，宝钗沉默寡言、外和内冷的性格，她未必十分满意。但在家庭利益之外的所有因素中，在她心中最有分量的是她的爱孙宝玉本人的意愿。宝玉是她的命根子，她看得出来，宝玉和黛玉越来越亲密，而和宝钗的性格极为不合。如果硬让宝玉丢下黛玉而娶宝钗，一定会造成夫妻不睦，这不但违反她爱孙的意愿，而且对家庭也未必有利。二十九回贾母对张道士说的话表示她要等一等看，等“大一大儿”，孩子们的性格可能有变化，黛玉的体弱多病也可能有好转。

但在等一等中她看到了什么呢？她逐步觉察到，宝黛之间的亲密关系助长着宝玉的怪脾气，他的性情越来越怪，谁劝一劝他，他就说谁是“国贼禄鬼之流”，听说还烧了书。老太太表面上似乎什么事也不关心，一天到晚只知道欢乐，听子们说说笑笑。其实，她在大事上是不糊涂的。在她的心目中，宝玉是全家的希望所在，是荣府世职的承袭者，她不赞成贾政那种过分严厉的管教方法，认为考什么举人进士也没有必要，但宝玉必须走上“正道”，必须关心经济世事，以便担当起继承祖业的大任。为此，他也必须有一个贤内助。不论她对孤苦的外孙女黛玉如何爱怜，她越来越感到，从体弱多病来看，从性情孤僻来看，尤其是从对宝玉的思想影响来看，黛玉作她的孙媳是不适当的。在外面找呢，“模样性格难得好的”。来了个宝琴，老太太非常喜爱，却又许配了梅家。宝钗虽然不无缺点，但从家庭利益来看，实在是万中选一的。于是，在选孙媳问题上，老太太就是这样逐渐偏到宝钗方面。五十四回元宵夜宴，黛玉把自己的酒杯放在宝玉唇上边，宝玉一气饮干，黛玉笑说：“多谢。”老太太看到他们在大庭广众之中的这种亲密表示，借对那些演唱才子佳人的鼓词“掰谎”，向宝黛二人提出了严重警告。

没想到事实也向老太太提出了严重警告。五十七回“慧紫鹃情辞试忙玉”，由于一句顽话，宝玉便“死了大半个”。若在婚姻问题上硬逼他，那还了得！那就连宝玉的命都保不住了，还讲什么家庭利益呢？这样一来，连王夫人都不知所措了。

七十一回贾政回京后，王夫人向他说了宝玉的情形，贾政更是束手无策。但忽然两人都来了个一百八十度的大转弯：慈母变得非常严厉，严父变得非常慈爱。王夫人抄检大观园，逼死晴雯，逐出芳官等人，又责令宝玉明春搬出大观园。

贾政却多次夸奖宝玉，不但不再逼他读四书做举业，反而时常带他到外面参加诗会。怎么会有这种奇事呢？这固然是由于承继世职之事时机紧迫（参见拙作《从荣府内争看贾兰和巧姐》），但固执异常的贾政怎么会有此突变呢？晴雯为贾母所赐，王夫人如果没有先斩后奏的上方宝剑，她怎么敢逐出晴雯呢？宝玉是按元妃的谕旨住进大观园的，王夫人敢擅自决定让宝玉搬出园外吗？事情十分明显，命宝玉搬出曾请示元妃，而贾政的转变、王夫人的施威只能是老太太指示机宜的。

贾母为人十分精明，遇到大事也能杀伐决断。三十五回宝钗说：“我来了这么几年，留神看起来，二嫂子凭他怎么巧，再巧不过老太太。”贾母道：“我的儿！我如今老了，那里还巧什么？当日我象凤丫头这么大年纪，比她还来得呢！”经过紫鹃情辞试玉，贾母知道事情决不能硬来，必须逐渐使宝玉从女孩子圈里走出来，引导他同那些为官做宦的人来往。贾政必须变往日的严厉为慈爱，以有利于宝玉接受教导。对他身边那些“狐狸精”却不能手软，但也要按不同情况分别对待。

贾政回京后，贾母把自己的意图讲给他和王夫人听，并命他们向薛姨妈定下宝钗为媳，但须绝对保密，看情况发展如何，再正式遣媒补行订婚之礼。贾政夫妇大喜过望，谨遵慈命而行。

七十八回王夫人向贾母汇报了处理“狐狸精”的经过，并说贾政带宝玉在外参加诗会得奖品，还夸奖宝玉。贾母深喜儿子和媳妇都还能体会自己的意图，并以做小结的口气对宝玉何以爱同女孩子们亲近做了分析，指出这不只是由于人大心大知道男女的事了，其中还有一种背离正道的思想在起作用。

王夫人向薛姨妈谈二家之事时，薛姨妈自然答应。当她提起宝玉因紫鹃的顽话得病及宝钗共效英皇的设想时，王夫人却不以为然，说并妻匹嫡，悖于常理，决不可行，并说以后宝玉多和外面的人来往，就不会再那样孩子气了。

七 水月镜花

二宝定亲的秘密终于泄露了。黛玉的病本已日渐加重，二宝之事虽早在她意料之中，她也并不怨恨谁，但这一消息对她的打击毕竟是太大了。

黛玉死时，大概宝玉不在家中，可能是贾政又放了外任，为了让宝玉懂得些世事人情，命他随任去了。他听到黛玉死耗时自是痛不欲生，赶回京城，奔赴潇湘馆“对景悼颦儿”即在此时。这段文字定是写得令人不忍卒读的。

跟着，各种灾祸接踵而来，元妃死去，贾府抄没，贾母死去，人亡家散。贾政和宝玉从狱中出来时，四大家族俱已衰败。贾政把贾环母子、贾兰母子分出去让他们自过，以后不许他们登门（参看拙作《从荣府内争看贾兰和巧姐》）。这时，贾政夫妇俱老病，无人主持家务，因与薛家商议，要娶宝钗过门。宝玉尚在祖母服中，只能从权行婚礼而暂不同房。他虽苦念黛玉，此时奉养无人，难以拒婚。

五十八回“杏子阴假凤泣虚凰”，芳官对宝玉说藕官和药官之事，做戏时是假夫妻，不做戏时竟也是好友。又说：

> 药官一死，他哭的死去活来，至今不忘，所以每节烧纸。后来补了蕊官，我们见他一般的温柔体贴，也曾问他得新弃旧的。他说：“这又有个大道理。比如男子丧了妻，或有必当续弦者，也必要续弦为是。便只是不把死的丢过不提，便是情深意重了。若一味因死的不续，孤守一世，妨了大节，也不是理，死者反不安了。”

书上说，“宝玉听说了这篇呆话，独合了他的呆性。”这里显然是遥遥伏下宝玉虽不忘黛玉却不拒婚宝钗之事，否则只要写藕官药官要好就行了，何必写补了蕊官仍是温柔体贴而且还讲了一篇大道理呢？雪芹难道会有这种毫无取意的蛇足文字吗？

二十回有脂批说：

> 妙极！凡宝玉宝钗正闲遇时，非黛玉来即湘云来，是恐泄漏文章之精华

也。若不如此，则宝玉久坐忘情，必被宝卿见弃，杜绝后文成其夫妇时无可谈旧之情，有何趣味哉！

此批要点在“杜绝后文成其夫妇时无可谈旧之情”这句话上。一般认为这是说：若不如此，……则成其夫妇时就不好谈旧时之情了。这样解释，既与原文的文义不合，又非合情合理。愚意以为，“无可谈旧之情”即“旧时所无之情”。二宝成其夫妇却未圆房，在此情形下，相遇同坐时，宝玉必然忘情而有亲昵表示，宝钗必然拒绝。这样才符合二人的性格。若见了面，宝钗主动找宝玉说话，宝玉把脖子一拧，理也不理，那还是贾宝玉吗？二人互相“温柔体贴”而又有分寸的情形，雪芹一定写得十分精彩，所以脂批说，若前边已有久坐忘情之事，后文成其夫妇时再写这样的事，那就不成其为“旧时所无之情”，而前文所写就泄露了文章的精华了。

他们虽然还能做到一般地互相温柔体贴，这一婚姻却仍然是不幸的。二人的思想性格不合，宝玉“空对看山中高士晶莹雪，终不忘世外仙姝寂寞林”，因而，“到底意难平”。家里越来越穷，靠宝钗和几个丫环的十指维持生活，有时还靠袭人蒋玉菡夫妇接济一点。贾政仍思恢复，重又逼宝玉读四书做举业，宝钗也“借词含讽谏”。宝玉回思两条手帕的海誓山盟，娶宝钗出于不得已，苦居然做举业当禄蠹，还何以对死去的黛玉呢！干点别的事吧，自己又一无所能。他这时才忽然发现：自己是家里的负担，这个家有自己还不如没有自己。于是他做了一个痛苦的抉择：悬崖撒手弃家而去了。

我同意周汝昌先生的观点，二宝终未同房，藕官和药官蕊官都是假凤虚凰，宝玉和黛玉宝钗也都是假凤虚凰。

《红楼梦曲》中的《终身误》是钗黛合于一曲，此曲是从宝玉的观点看二女。香港和台湾的研究者，例如高阳先生和蒋凤先生，认为《枉凝眉》也是钗黛合于一曲。此说初看觉奇怪，细想很有道理。《枉凝眉》是从二女的视角来看的。我们一般认为此曲是说宝黛二人，为什么想不到宝钗身上去呢？一是不知道宝钗的金玉姻缘也是水月镜花；二是由于最后那几句：“想眼中能有多少泪珠儿，怎禁得秋流到冬尽，春流到夏！”其实，宝玉走后，宝钗“焦首朝朝还暮暮，煎心日日复年年”，她难道就不哭吗？如果说，宝钗总不至于像黛玉那样一年四季哭个没完吧，那么，若认为此曲是说宝黛二人，宝玉一个大男人家反而会像黛玉那样哭个没完吗？

宝黛钗的金玉姻缘和木石前盟，非但不能两全，一全也不可能，终于成为两毁。这一爱情婚姻悲剧具有封建时代的特点，同时也具有超出其时代的永久性意义而在多方面发人深省。

第三篇
从荣府内争看贾兰和巧姐

《红楼梦》中，贾兰和巧姐是荣国府的最末一代儿孙。由于前八十回他们都还年幼，书中对他们的描写很少。据太虚幻境十二钗正册的判词和红楼梦曲子看来，他们在八十回后的事迹和遭遇好像比较清楚。所以，对他们似乎没有什么可研究的。但是，同他们有关的判词和曲子中也不是完全没有疑点。例如："如冰水好空相妒"究竟是什么意思？巧姐的"狠舅奸兄"到底是谁？这些都迄无确解。为了解答这些问题，我们分析了荣府的内部矛盾，试图通过这种分析找到线索。

一 自执金矛又执戈

庚辰本二十一回有回前总评如下：

有客题《红楼梦》一律，失其姓氏，惟见其诗意骇警，故录于斯：

自执金矛又执戈，自相戕戮自张罗。

茜纱公子情无限，脂砚先生恨几多。

是幻是真空历遍，闲风闲月枉吟哦。

情机转得情天破，情不情兮奈我何。

凡是书题者，不可（不以）此为绝调，诗句警拔且深知拟书底里，异乎失石（名）矣。……此曰"娇嗔箴宝玉，软语救贾琏"；后曰"薛宝钗借词含讽谏，王熙凤知命强英雄"。……何今日之玉犹可箴，他日之玉已不可箴耶？今日之琏犹可救，他日之琏已不能救耶？……今因平儿救，此日阿凤英气何如是也？他日之强，何身微运蹇，展眼何如彼耶？人世之变迁如此。……文是一样情理，景况光阴事却天壤矣。多少恨泪洒出此两回书。

题诗客"深知拟书底里"而总评盛赞其诗，足见此诗对理解《红楼梦》是极其重要的。此诗开始两句说的是什么呢？显然说的就是作为《红楼梦》主要描写对象的荣国府的内部矛盾和斗争。

荣府这个封建贵族家庭，其内部有各种各样的矛盾，这些矛盾交织在一起，非常复杂。有些矛盾是封建贵族家庭所共同具有的。例如主奴之间的矛盾就是这样。又如冷子兴说的“外面的架子虽未甚倒，内囊却也尽上来了”，以及他说的“如今的儿孙竟一代不如一代了”，这些，许多封建大家庭也都有，而且也是所有这些大家庭的必然趋势。这些矛盾只决定这类家庭的共性。《红楼梦》不是一般化的、概念化的小说。掌握这些共性固然也很重要，但只有这些还不足以理解荣国府，也就难以阐明《红楼梦》中的一些具体问题。

荣府不同于四大家族中的史王薛三家，也不同于宁府。宁府那些乌七八糟的事情，像柳湘莲对宝玉说的“你们东府里除了那两个石头狮子干净，只怕连猫儿狗儿都不干净”和焦大骂的“爬灰的爬灰，养小叔子的养小叔子”，还有后来居然以习射为名开局聚赌，这些事在荣府就没有如此严重，所以惜春才“矢孤介杜绝宁国府”。但荣府自有其极为严重的特殊问题，这就是世职承袭的矛盾。

雪芹的特殊笔法之一是对比法。他往往用对比法把问题摆出来，让读者自己做出论断。在世职承袭问题上，雪芹就是摆出了宁府和荣府的鲜明对比。

这个问题在宁府是不存在的。贾代化有两个儿子，但长子贾敷在八九岁就死了，自然由次子贾敬袭官。如今贾敬已把世职传给他的独子贾珍。贾珍也只有一子贾蓉，死后当然由他接班。

在荣府，情况就完全不一样了。对这个问题作者主要是用暗笔来写的，所以表面上好像看不出来，其实“冷子兴演说荣国府”时，问题已经摆出来了。他说：

> 如今代善早已去世，太夫人尚在。长子贾赦袭着官，次子贾政自幼酷喜读书，祖父最疼，原欲以科甲出身的。不料代善临终时遗本一上，皇上因恤先臣，即时令长子袭官外，问还有几子，立刻引见，遂额外赐了这政老爹一个主事之衔……

“次子贾政自幼酷喜读书，祖父最疼”放在“长子贾赦袭着官”之下，这又是用对比法暗示矛盾：如果传贤或者传爱，本该由贾政袭官，代善遗本中也一定委婉地表示了这种愿望，说长子表现一般，次子德才兼备。没想到遗本上去，皇上不管三七二十一呆板地照老章程办事，即时让长子袭官，只赐给贾政一个额外主事虚衔。史太君自然知道丈夫一向的心意，况且她自己也是偏爱老疙瘩不喜欢大儿子的，对皇帝的圣裁当然甚为不满。可见那时在世职承袭问题上就存在矛盾。

贾赦有两个儿子：长子贾琏，次子贾琮。贾政有三个儿子：贾珠、宝玉、贾环。贾珠和宝玉是正出，贾环是庶出。贾琏和贾琮的正庶书上没有说。七十三回邢夫人说自己一生无儿无女，说迎春“是大老爷跟前人养的”，又说“如今你娘死了”。可见冷子兴说到迎春时，庚辰等本作“赦老爷前妻所出”，此“妻”字乃“妾”字之讹。既然邢夫人无儿女而贾赦无前妻，则贾琏贾琮皆庶出可知。

只要史太君健在，分家就谈不到。姐妹和弟兄都按大小排行排序。贾珠是珠大爷，贾琏是琏二爷。宝玉等还小，暂时还不称爷；再说老太太怕她的宝贝孙子不好养活，吩咐众人都只叫他宝玉。贾珠十四岁就中了秀才，贾琏则“不肯读书”。这又是一个鲜明的对比。事情很清楚，不论传嫡传长传贤，贾赦手里的世职将来都势必要传给贾珠。贾赦眼睁睁看着世职要被二房夺去而无计可施。冷子兴说“长子贾赦袭着官”，这句话很传神。“袭着官”者，暂时袭官而未必能长保和传诸子孙之谓也。后来有的版本这句话作“长子贾赦袭了官”，想必是校对者不明原句深意，觉得“袭着官”这种说法很别扭，随手把“着”字改为“了”字。

贾珠留下一子贾兰死去后，形势有了变化。甲戌本在冷子兴谈到贾珠之死时有脂批道“略可望者即死，叹叹”。贾琏比宝玉大。宝玉“潦倒不通世务，愚顽怕读文章，行为偏僻性乖张，那管世人诽谤”。贾琏虽也不肯读书，但“于世路上好机变，言谈去的”，正所谓“世事洞明皆学问，人情练达即文章”。所以宝玉虽占了个“嫡”字，贾琏却占了“长”和“贤”两个字。况且世职毕竟在贾赦手里，看来传给儿子大有希望。不知是老太太的主意还是贾政夫妇的主意，忽然又不用大排行而改用小排行了，宝玉成了宝二爷，贾环成了环三爷。贾琏的琏二爷是早已叫惯了的，况且珠大爷虽死，珠大奶奶尚在，凤姐当然仍称琏二奶奶，贾琏只得仍称琏二爷。于是有了两个二爷。这无非是要给众人造成一种印象，似乎长幼无所谓，你看是两个二爷嘛，重要的是嫡庶之分。有的版本，因校勘者不知两个二爷的奥妙，认为按大排行宝玉应称宝三爷，按小排行贾琏应称琏大爷，无论怎样也说不通，因而把冷子兴说的“若问那赦公也有二子，长名贾琏”中的“长”字改成了“次”字，根本不顾贾琏没有哥哥而有个弟弟贾琮的明显事实。二房的司马昭之心，贾赦岂有不明白的？赶紧给贾琏捐了个同知，让他在管理荣府家务之余多在外面活动，使人人皆知琏二爷是荣府现在的掌权者，将来的接班人。看来鹿死谁手，还在未可知之数。

前者贾赦夫妇因老太太越来越偏心，和二房住在一起心情不舒畅；儿子娶媳妇时，老夫妇索性搬到由荣府花园隔断的小院中去住，把原来的住房给儿子做新房，让儿子儿媳帮二房料理家务，明里是代替贾赦夫妇侍奉老祖宗，暗里是不让二房独揽大权并可探听其动静。不料，由于和王夫人的姑侄关系，凤姐心向二房，而贾琏又有惧内的毛病，于是贾赦夫妇在荣府更加孤立。接着，两件出乎意料的事情发生了：一是宝玉得到了贾府一个重要靠山北静王水溶的赏识，二是最爱宝玉的元春入宫后被册立为妃。形势急转直下，二房完全掌握了家事的大权，而在朝中宫中的力量也已远远超过大房。一旦贾政的严厉管教把宝玉“规引入正”，在贾赦死前便由皇命立宝玉为嗣是完全可能的。当此极端不利的局面，大房采取了什么有效的对策，书上没有说，不好猜测。

二房内部也不是平静的。贾环母子缺少自知之明，也想觊觎“神器”。赵姨娘说是要夺“家私”，这其实就是要夺世职，有了世职就有了“家私”。他们的主要手段是暗害。先是贾环推翻蜡灯烫宝玉，企图烫瞎他的眼睛。接着赵姨娘勾结马道婆用魇魔法要害死宝玉和凤姐，后来贾环又在贾政前诬告宝玉强奸逼死金钏以致宝玉几乎被父亲打死。这些描写是以贾环母子的暗害活动为例说明荣府内部夺嗣斗争的残酷性，并暗示八十回后将更是你死我活的。

到了七十五回赏中秋，夺嗣斗争已接近摊牌。当时行击鼓催花之令。花传至宝玉鼓止时，他遵贾政之命作了一首诗，得了赏。贾兰一见，也作诗得赏。贾赦一见二房一子一孙出了风头；相形之下，贾琏根本不会作诗，干陪着连一句响话也没有说。这勾起了他平时的积愤，花在他手中鼓住时，他故意说了个天下父母多偏心的笑话，惹得贾母不痛快，后来还落了泪。接着，花落在贾环手里，他也作了一首诗。贾赦看了，连声赞好，说他的诗“不失咱们侯门的气概”，并把争夺世职的矛盾公然亮在桌面上，说："将来这世袭的前程定跑不了你袭呢。"这话一方面在二房的二子一孙之间进行挑拨借收渔人之利，一方面是告诉二房，休想让你们那宝贝儿子宝玉袭我大房的世职，我看连贾环也不如。

这次贾赦出去时被石头绊了一下，崴了腿。贾母说到贾敬已死了二年之久，有脂批说："不是算贾敬，却是算赦死期也。"此批一是说贾母这时已深恨贾赦；二是说贾母及众人因贾赦病弱崴腿，说话颠三倒四，估计他不久于人世。此次赏中秋，凄清之中隐含剑拔弩张之势，夺嗣斗争已临决战关头。贾赦有可能上表告病，请求将世职传给长子贾琏。

宝玉这时仍然整天和姑娘丫头们厮混做他的富贵闲人，并坚持他那些离经叛道的思想，毫无悔改表现。奇怪的是，七十五回前后，王夫夫和贾政都一反常态，慈母忽然变得非常严厉，严父忽然变得非常慈爱。邢夫人拿偶然得到的绣春囊向王夫人将了一军。王夫人借机抄检大观园，逼死晴雯，逐出芳官等人，又责令宝玉明年搬出大观园。王夫人这些行动表明，二宝婚事已经内定，以便婚后借宝钗闺中之力使宝玉走上正路，故须完全割断宝黛之间的联系；宝钗也遵礼搬回家去，婚前尽可能不与宝玉相见。前贾政逼宝玉读四书做举业，多次责打无效。今因时机紧迫，索性不再以举业逼他，反而设法帮助他以诗词在士大夫中取得令名。因此，贾政时常带宝玉和贾环、贾兰参加诗会，并"闲"征姽婳词，以便在同僚中夸耀，有机会甚至送往礼部转星御览，宝玉镇日只在情字上讨生活，哪里能体会父亲的政治意图，他作诗也只是为作诗而作诗，"茜纱公子情无限，闲风闲月枉吟哦"。这就是八十回结束时荣府夺嗣斗争的形势。

二　到头谁似一盆兰

不要忘记荣府子孙还有个贾兰。他在荣府的内争中是否置身事外呢？十二钗正册关于李纨的那一页，画的是一盆茂兰，旁有一位凤冠霞帔的美人，判词是：

桃李春风结子完，到头谁似一盆兰。

如冰水好空相妒，枉与他人作笑谈。

结合那幅画来看，判词前两句含意很清楚："李、完"隐李纨这个名字，她那桃李春风般的夫妻生活在生子贾兰后就“完”了。后来贾兰做了官，李纨得享“老来富贵”。不好理解的是判词的后两句，而第三句似乎连字面上也很难解释得通。

人民文学出版社1982年版《红楼梦》，81页有注解说：

> 后二句句意难以确定，或谓化用唐代僧人寒山《无题》诗“欲识生死譬，且将冰水比。水结即成冰，冰消返成水。”说李纨一生三从四德，晚年荣华方至，却随即死去，只留得一个诰封虚名，白白地给世人作谈资笑料。

但若“冰水”用作“生死譬”，那个“好”字就用得不恰当了。李纨“荣华方至，却随即死去”，这是坏事，怎么能给她叫好，还当作笑料呢？

吉林文史出版社1986年版《红楼梦诗词解析》，71页释词中说：

> 如冰水好：这句不好解。或者是说李纨清心寡欲，忠于故夫，恪守贞操，教子有方，像冰水一样纯洁。空相妒：不值得羡慕和嫉妒。

这样解释，“好”字是讲得通了。但“空相妒”解作“不值得羡慕和嫉妒”却在字义上不很贴切。况且清心寡欲，教子有方，终于子贵母荣，给老贾家争了一口气，这不是很值得称赞和羡慕的吗？

那么，“如冰水好空相妒”究竟是什么意思呢？我们认为，这句判词说的不是别的，正是荣府的内争。冰水同质而相妒不能相容，放在一起，不是水凝结成了冰，便是冰融化成了水。这正像曹子建《七步诗》所说：“煮豆燃豆萁，豆在釜中泣。本是同根生，相煎何太急！”也像探春说的：“咱们倒是一家子亲骨肉呢，一个个不象乌眼鸡，恨不得你吃了我，我吃了你！”三四两句判词就是说：荣府的骨肉相争如冰水有同根之好而相妒不能相容，结果家亡人散一场空，只不过让他人作为笑料来谈论罢了。判词第二句说的是贾兰，后两句却说到荣府的内争上去，可见他在内争中并不是旁观者。

贾珠死后，贾兰随着年龄的增长，自然渐渐明白，家里的那个世职本来是要由他父亲承袭的，因而有朝一日会传给他。现在他父亲虽死，但他是嫡孙，仍然有资格承袭世职。他看得出来，二叔宝玉是曾祖母的心肝宝贝，全家都像凤凰似的捧着，这世职大有被二叔袭去的可能。因此他一直和宝玉合不来，却和贾环接近。

“顽童闹学堂”，书上有下面的描写：

> 这里茗烟先一把揪住金荣，……金荣气黄了脸，说：“反了！奴才小子都敢如此，我只和你主子说。”便夺手要去抓打宝玉秦钟。……贾菌年纪虽小，志气最大，极是淘气不怕人的。他在座上冷眼看见金荣的朋友暗助金荣，飞砚……落在他桌上，……贾菌如何依得，……抓起砚砖来要打回去。贾兰是个省事的，忙按住砚，极口劝道：“好兄弟，不与咱们相干。”

众顽童打闹起来，有宝玉在内，贾兰却行若无事，袖手旁观。贾菌要参战时，他极口劝阻。有你亲叔在内，怎么说是不与你相干呢？可见他和宝玉极其不睦。

贾母设灯谜会，众人都在，贾兰不肯来。贾政偶然来参加，见他不在，问时李纨忙起身笑着回道：“他说方才老爷并没去叫他，他不肯来。”众人都笑道：“天生的牛心古怪。”其实，在这种场合“往常间只有宝玉长谈阔论”，大家不重视贾兰，所以这次他索性不来。他并不知道有贾政参加，李纨的话是明知儿子不来之故，却对老公公说谎。后来贾环去唤，贾兰听说是爷爷唤他，还派有两个婆娘，面子不小，他才来了。来后“贾母命他在身旁坐了，抓果品与他吃”，风头胜过了宝玉。

贾赦身上不好，宝玉去请安后又去见邢夫人。书上写道：

> 邢夫人拉他上炕坐了……又命人倒茶来……只见贾环贾兰小叔侄两个也来了，请过安，邢夫人便叫他两个椅子上坐了。贾环见宝玉同邢夫人坐在一个坐褥上，邢夫人又百般摸娑抚弄他，早已心中不自在了。坐不多间，便和贾兰使眼色儿要走。贾兰只得依他，一同回身告辞。宝玉见他们要走，自己也就起身要同回去。邢夫人笑道：“你且坐着，我还和你说话呢。”宝玉只得坐了。邢夫人向他两个道：“……今儿不留你们吃饭了。”贾环等答应着便出来回家去了。……宝玉道：“大娘方才说有话说，不知是什么话？”邢夫人笑道：“那里什么话，不过是叫你等着同你姐妹们吃了饭去。……”

邢夫人故意当着贾环贾兰对宝玉爱抚备至，而对他二人冷淡无情，以此来挑拨他们同宝玉不睦。此处有脂批道：“一段为五鬼魇魔法引。”可见，由于邢夫人的恶意挑拨，贾环更恨宝玉，以致他和他母亲下了毒手。这种挑拨当然也对贾兰起作用，使他对宝玉更忌恨而同贾环的关系更密切。

二十六回有这样的描写：

宝玉……顺着沁芳溪看了一回金鱼。只见那边山坡上两只小鹿箭也似的跑来。宝玉不解其意，正自纳闷，只见贾兰在后面拿着一张小弓追了下来，（脂批：此等文可是人能预料的！）一见宝玉在前面，便站住了，笑道："二叔在家里呢，我只当出门去了。"宝玉道："你又淘气了。好好的射他作什么？"贾兰笑道："这会子不念书，闲着作什么？所以演习演习骑射。"（脂批：答的何其堂皇正大，何其坦然之至！）

宝玉见贾兰拿着小弓追两只小鹿，这种描写似很平淡无奇，为什么脂批赞叹说是出人意料的？可见文中有深意。这是一种隐喻，暗示贾兰也要"逐鹿中原"，参与夺嗣之争，看看究竟鹿死谁手！他对宝玉的回答有几层意思：一是堂皇正大，标榜自己不习文便练武；二是以此讥刺宝玉不是"出门去"便是在家里"闲着"；三是坦然为自己无故杀生辩解。小小孩童，如此工于心计，其应对便捷，确有过人机智。宝玉知道他的话来得厉害，窘迫间说了句没力气的话："把牙栽了，那时才不演呢。"

现在我们看，贾兰后来做的是什么官呢？关于李纨的那支曲子名为《晚韶华》，今按甲戌本抄录于下：

镜里恩情，更那堪梦里功名！那美韶华去之何迅，再休提绣帐鸳衾。只这带珠冠，披凤袄，也抵不了无常性命。虽说是人生莫受老来贫，也须要阴骘积儿孙。气昂昂头戴簪缨，气昂昂头戴簪缨，光灿灿胸悬金印。威赫赫爵位高登，威赫赫爵位高登，昏惨惨黄泉路近。问古来将相可还存？也只是虚名儿与后人钦敬。

从"气昂昂、光灿灿、威赫赫"这三句话来看，贾兰做的是品位很高的武职官。两府的世职本来就是武官，例如第十三回中说贾珍是世袭三品爵威烈将军。可见贾兰正是袭了荣府的世职！

七十五回回目中的"新词得佳谶"意思是说作新诗得了好兆头。谁作诗得了好兆头呢？贾政评宝玉和贾环的诗说："可见是弟兄了。发言吐气总属邪派，将来都是不由规矩准绳，一起下流货。妙在古人中有二难，你两个也可以称二难了。只是你两个的难字，却是作难以教训之难字讲才好。哥哥是公然以温飞卿自居，如今兄弟又自为曹唐再世了。"对于贾兰的诗呢，"贾政看了喜不自胜，遂并讲与贾母听时，贾母也十分欢喜，也忙令贾政赏他”。从这些描写看来，作诗得了好兆头的显然是贾兰而不是贾环。贾赦的话别有用心，是不能算数的。我们知道，书中实际上没有这三首诗。庚辰本在此回回前有脂砚等所记“缺中秋诗，俟雪芹”。三首诗其实不过是三首绝句，如果很容易作，雪芹就不会暂时搁置，以致成为千古疑案。宝玉和贾环的诗由贾政的话看来没什么难拟之处，不太容易作的定是贾兰的诗。他初学诗，其诗应浅，但须含有袭世职的佳谶，而这又必须很隐晦，所以他的诗就难拟了。

贾府被抄前，贾赦、贾政、贾琏、宝玉都还在。那时，当然轮不到贾兰袭官。甄士隐《好了歌解》有两句话说“昨怜破袄寒，今嫌紫蟒长”，甲戌本对此有脂批“贾兰贾菌一干人”。可见贾兰是在贾府被抄后经过一段“破袄寒”的生活，后来才袭官的。两府被抄后是一败涂地，“落了片白茫茫大地真干净”的。那么罪臣之裔的贾兰怎么会又袭了世职呢？唯一合理的解释是：在朝内那场导致贾府败亡的政争中，贾兰属于胜利者一方。

现在我们再看一下第一节所引二十一回的回前总评。律诗的前两句说的是荣府的残酷内争已如前述。此评将荣府内争与贾家前后景况之判若天壤相联系，而题诗客又“深知拟书底里”，可见贾府之抄家败亡与此内争是有密切关系的。探春在七十四回“抄检大观园”时流泪说：“你们别忙，自然连你们抄的日子有呢！……可知这样大族人家，若从外头杀来，一时是杀不死的，这是古人曾说的‘百足之虫，死而不僵’，必须先从家里自杀自灭起来，才能一败涂地！”看来敏探春是懂得外因通过内因起作用这种哲学道理的。

内争激化到一定程度，各方就要向外部寻求支援。具体情节是怎样的只能做一些猜想。北静王水溶由于赏识宝玉，当然是支持二房和宝玉的。三十三回宝玉大承笞挞前，忠顺王府长史官因琪官的事气势汹汹来向贾政和宝玉问罪，很可能忠顺王是北静王和贾府的政敌。这一回中还有贾环“手足耽耽小动唇舌”诬陷宝玉。以这些为线索可以猜想日后贾环直接或间接同忠顺王联系上，至少红学界有些同志是有此观点的。于是贾环继续以前的暗害活动和“小动唇舌”向忠顺王一方“大动唇舌”提供了许多不利于宝玉、凤姐、贾琏、贾赦等人的情报。忠顺王正伺贾府之隙，自然乐于利用贾环。贾兰在内争中和贾环由关系密切逐渐互相配合，于是也就卷入了忠顺王一派。他倒也未必像贾环那样提供不利于自己家庭的情报，但他的聪明才智远非贾环可比，很快就得到了忠顺王的赏识。经过一番较量，忠顺王终于战胜了政敌。贾妃先已薨逝，这时北静王水溶又获罪，贾府所依靠的两座冰山俱“溶”于“水”，于是两府抄没，连宝玉都由于和北静王的密切关系而系于狱。为防物议，贾兰过了一段“破袄寒”的生活，后来先做了个小官。又过了几年，事情更冷下去，经忠顺王保奏，皇上“念贾家先世之功”，特旨将荣府世职赐还其嫡孙贾兰。

贾兰的乌纱帽不是好来的。正因为如此，判词才把他母亲的凤冠霞帔同荣府的内争相联系，并说这不过让人作为笑谈而已。

雪芹定这种夺嗣斗争不知有没有曹家的真事为背景，但肯定有影射雍正夺嫡之意，所以八十回后的稿子绝对不能外传，只有埋藏在青埂峰下，希望后世的癞头和尚或空空道人来发现，并望再有一芹一脂整理成书了。

三　劝人生济困扶穷

现在讨论巧姐在八十回后的遭遇。十二钗正册巧姐那一页的图画是一座荒村野店，有一美人在那里纺绩，判词道：

势败休云贵，家亡莫论亲。

偶因济刘氏，巧得遇恩人。

第十一支曲子《留余庆》说的也是巧姐：

> 留余庆，留余庆，忽遇恩人，幸娘亲，幸娘亲，积得阴功。劝人生，济困扶穷，休似俺那爱银钱忘骨肉的狠舅奸兄！正是乘除加减，上有苍穹。

从判词和曲子看来，日后巧姐遭到不幸，因凤姐曾周济刘姥姥，巧姐得遇恩人。再根据四十一回写大姐儿和板儿交换所拿之物及该处的脂批“小儿常情，遂成千里伏线”，可以断定巧姐嫁给了板儿，故册子有荒村野店纺绩之画。遭到的不幸是什么呢？第一回《好了歌解》中“择膏粱，谁承望流落在烟花巷”这句话，研究者都认为是指巧姐而言。另外，甲戌本第六回有脂批“老妪有忍耻之心，故后有招大姐之事”。可见巧姐是流落在烟花巷，故招她为板儿之妇须有忍耻之心。

问题是，“狠舅奸兄”是谁呢？高续把狠舅定为凤姐之兄王仁。这在前八十回似乎找不到根据，王仁称为“忘仁”是高续一百零一回贾琏说的。另一个嫌疑犯是薛蟠，这人是什么坏事都干得出来的。第八十回薛姨妈要卖香菱，宝钗说：“咱们家从来只知道买人，并不知有卖人之说，妈可是气的胡涂了。”这也未尝不可以看作一种伏笔。但薛蟠是表舅，王仁才是亲舅；从“忘骨肉”三字来看，还是暂定狠舅是王仁为宜。

本文主要讨论奸兄。先说明一点，奸兄未必是伙同狠舅一起卖巧姐的。那是从高续得来的印象。狠舅卖巧姐，所以说是“狠”；奸兄是在巧姐落难后，当管、能管、嘴里说要管而实际上不管，所以说是“奸”。

高续把奸兄定为贾芸，这是不对的。贾芸除了曾行贿走凤姐的后门，在别的事情上表现还好。脂批说“醉金刚一回文字伏芸哥仗义探庵”，可见贾芸在贾府败后有仗义之事。有的研究者主张奸兄是贾蔷或贾蓉。贾蓉自然不是好人，但用“奸”字形容他的性格并不确切。至于贾蔷，“顽童闹学堂”中有这样一段描写：这贾蔷“既和贾蓉最好，今见有人欺负秦钟，如何肯依？如今自己要挺身出来报不平，心中却忖度一番”，于是把茗烟唤到身边，如此这般挑拨他闹起来，自己却借故溜之大吉了。这可算得是近于“奸”字了，但细按似更近于“狡猾”。况且说他在巧姐的事上当了奸兄，前八十回里也缺少证据。

那么，多年来隐藏很深的这个奸兄到底是谁呢？我认为这不是别人正是“威赫赫爵位高登”为“后人钦敬”的李纨的宁馨儿贾兰！

首先，贾兰是草字辈中巧姐最近的骨肉。其次，他够不够得上这个“奸”字？我们不以上一节对八十回后的探讨为证据，因为那是推论。算不算奸只看他在前八十回中的表现。其实，只要看他在“顽童闹学堂”中那种行若无事、不动声色的神气就够了。小小孩童，其奸在贾蔷之上！再说，贾蔷因秦钟被人欺负，不肯坐视，有打抱不平之心，这总不好说他是“忘骨肉”吧？贾兰呢，亲叔要被人抓打了，却说这事不与他相干，这不正是“忘骨肉”吗？灯谜会中他和他母亲的那番表演也够奸的了。“逐鹿中原”时回答宝玉的那几句台词就更表明，其奸已达到炉火纯青的程度了。根据贾宝玉的定义，凡热衷功名讲究读书上进的人谓之“禄蠹”，而“禄蠹”一定不是好人。贾宝玉的观点未必等于曹雪芹的观点，但可以说是《红楼梦》的观点吧。贾兰是《红楼梦》中典型的“禄蠹”，八十回后决不会把他写成好人，否则岂不是自相矛盾吗？况他小时即奸，等到做了官，必然成为像贾雨村那样的奸雄。

贾兰是巧姐之“兄”中最符合“骨肉”这个条件和最够得上“奸”字的，但这还不足以证明他就是那个“奸兄”。我们来回忆一下李纨的那支曲子《晚韶华》。曲子中有这样两句：“虽说是人生莫受老来贫，也须要阴骘积儿孙。”这意思是说，李纨由于怕受老来贫，只顾攒钱，不肯救助他人。她的这种性格，前八十回书中也略有透露。姑娘们起诗社，李纨自荐掌坛为社长，说在她那里作社，她做个东道主人。但第一次做东的是探春，第二次湘云硬充大老官，以后也没见李纨做东。攒金给凤姐过生日时，连丫环们都出银两，只有李纨没出一文钱。老太太和凤姐都说替她出，她不作声算是默认了。生日过后，姊妹们来敦请凤姐做监社御史，凤姐猜到是敲她的竹杠，便给李纨细算收入账，说她舍不得拿出点钱来陪姑娘们顽顽。李纨反唇相讥，说凤姐不像个诗书名门小姐；而她自己所用语言之粗俗，实在不像是出于诗书名门小姐之口。这是因为凤姐的话点中了她的吝啬真病，使她恼羞成怒了。

曲子里说的李纨不肯拿出钱来积阴骘是在什么事情上呢？不会是在普通事上，普通事就不会写进概括她一生的曲子中去了。只要拿《晚韶华》和并排在前面的《留余庆》对照看一下就明白是在什么事情上了。两支曲子中含有下列对比鲜明的句子：一支曲子说，“虽说是人生莫受老来贫，也须要阴骘积儿孙”；另一支曲子说，“幸娘亲，积得阴功。劝人生，济困扶穷”。为了引导读者注意这种对比，作者故意在两处都嵌入了“人生”二字，并在二曲中都安排了“二重一单，二重一单”这种加强语气的句子结构。十分明显，雪芹是又用对比法暗示读者，李纨正是在巧姐遇难这个问题上，“爱银钱，忘骨肉”，损了阴德。母子一体，事情是商量着办的。母亲如此，儿子自然也是在这个问题上做了损事。

具体情节可能如下：两府被抄后，家亡人散。巧姐被王仁卖入娼家时，两府主人层已只剩贾兰母子在京。后刘姥姥闻知巧姐陷入娼家，进城拟报与贾家以便营救，先问至小红贾芸夫妇家。此时贾兰已得官，刘姥姥与小红到其官邸，赂守门人乃得进见贾兰及太夫人。贾兰与太夫人闻耗，甚表惊讶及伤感，力保或官休或私赎定能救出巧姐，对刘姥姥极口称谢，谓此事不可传扬，恐婚家闻风有异谋。刘姥姥等辞去后，久等不见动静，再去贾宅，守门人托词不肯通报，数次皆然。刘姥姥始知贾兰母子不肯援手，及与贾芸等多方筹措，又有醉金刚倪二等帮助，巧姐才得赎身，后嫁与板儿。

贾雨村忘甄士隐周济之恩而不顾被拐卖之英莲，贾兰忘骨肉而不顾陷娼门之巧姐，两个奸雄，恰是一对。

贾兰的结局如何呢？我同意梁归智先生的看法，“昏惨惨黄泉路近”是说贾兰，不是说李纨。前面同样结构的三句话都是说他，这句话自然也是说他。由此看来，他的结局是很不好的。况且有关他和贾菌的“今嫌紫蟒长”语意也不佳。宝玉评“文死谏，武死战”的那段怪论固然可以说真意是反对忠君，但他的论调实在太怪了，所以我觉得很可能是暗伏后文的什么事情。可做如下猜想：贾兰袭官后，援引其好友贾菌也做了个不小的官，贾菌因而也属于忠顺王一派。贾兰乃是文武全才，袭的世职也是武职官。后柳湘莲一干强梁造反，贾兰领兵征讨。谁知世事无常，忠顺王在这时的又一次政争中失败。贾菌为保忠顺王，对皇上犯颜极谏，皇上怒其狂妄，处以死刑。贾兰在军中闻忠顺王事败，贾菌进谏而死，惊叹之余，恐株连及己，乃冒险进军，希以战功免罪。不料全军覆没，贾兰仅以只身逃回京城。于是两罪俱发，皇上降旨赐死。正是：“正叹他人命不长，那知自己归来丧！”

十二钗，三亲四春一尼一幼三妇。《喜冤家》曲子下面有脂批：“题只十二钗，却无人不有，无事不备。”雪芹塑造了十二钗，用以代表旧时代妇女的十二种类型。她们各有不同的性格和行事，各有其薄命的遭遇。其中三妇也和另外九钗一样，都是薄命人。但三妇在品质和行事上各有严重缺点，用贾宝玉的话来说，她们都是出嫁后染上了男人气的。作者塑造了这三个艺术形象，通过凤姐写贾府这个封建贵族家庭外对人民内对婢仆的压榨迫害、敲骨吸髓，借用李纨写荣府的骨肉相争、无情无义，围绕秦可卿写宁府的生活糜烂、腐败堕落。三个形象缺一不可。畸笏叟不明此义，硬要雪芹删去天香楼的情节，雪芹虽勉强同意，但在册子、曲子和正文中保留或增加许多暗示之处。

第四篇
暖香坞的诗谜

《红楼梦》第五十回和五十一回写众人在暖香坞雅制春灯谜。宝钗、宝玉、黛玉各有一首诗谜。宝琴以十个地方的古迹为题作了十首怀古诗，既怀往事，又隐俗物十件。这十三首诗谜形式一样，每首都是七言绝句。

十首诗谜所隐十二个女子是又一种十二钗。这是在第五回的十二正钗中去掉了妙玉和巧姐，添上了尤三姐和宝琴。这个十二钗不妨称之为“暖香坞十二钗”。十首诗谜是这个暖香坞十二钗的“判词”，宝钗和黛玉的诗谜相当于第五回《红楼梦曲》的第一支《红楼梦引子》，而第一首怀古诗相当于第十四支《收尾·飞鸟各投林》。最妙的是，那里的判词黛钗合于一首，这里的“判词”黛钗湘合于一首。宝玉“爱博心劳”，但同他有爱情婚姻关系者，前半生为黛钗，故在第五回的判词中“合一”；一生中则为黛钗湘，故在暖香坞“判词”中亦“合一”也。暖香坞十二钗隐于书的深层，作者以暗笔为她们安排了“判词”“引子”和“尾声”，特犯不犯有意和第五回十二钗相对照。如此郑重其事，足见暖香坞十二钗异常重要。

现在具体分析几首诗谜。

1. 宝钗诗谜

镂檀锲梓一层层，岂系良工堆砌成？

虽是半天风雨过，何曾闻得梵铃声！

此诗是说《红楼梦》这部书。前两句说，此书镂檀锲梓，精雕细刻，层层相因，环环相扣，非仅良工堆砌而成者。后两句说的是“以假语村言”“将真事隐去”。虽然书中隐藏着半天风雨，但日常的欢声笑语掩盖了风吹梵铃之声。

2. 黛玉诗谜

騄駬何劳缚紫绳，驰城逐堑势狰狞。

主人指示风雷动，鳌背三山独立名。

此诗是赞美《红楼梦》的四个主要人物。宝玉如一匹骏马，不受羁绊。他的思想驰骋奔腾，冲击一切旧的堡垒，其势猛不可当。主人指《红楼梦》的作者而言。作者腕有风雷，语不惊人死不休。他所写的黛钗湘这三个人物，定将流传久远；但她们像传说中的那三座海外仙山，只存在于我们的幻想之中，尘世间是不会真有这样的人物的。

3. 赤壁怀古

赤壁沉埋水不流，徒留名姓载空舟。

喧阗一炬悲风冷，无限英魂在内游。

蔡先生说，此诗是写贾家“这个封建大家族在衰败过程中，死亡累累，恰如赤壁鏖兵中曹家人马之‘一败涂地’。否则，赤壁之战，可写的话正多，何至于句句说死，写得如此阴森凄惨？小说不是自传，曹操与作者同姓，这是巧合，但小说中有作者的家世感慨在，这也是不言而喻的”。我同意蔡先生的分析，尤其同意他指出“有作者的家世感慨在”。我觉得“曹操与作者同姓”还不只是巧合，作者正是要借曹操的姓做影射，第二句“徒留名姓载空舟”可为明证。

以上是“引子”和“尾声”，下面分析几首“判词”。

一　宝玉诗谜

天上人间两渺茫，琅玕节过谨提防。

鸾音鹤信须凝睇，好把唏嘘答上苍。

此诗隐含黛玉、宝钗、湘云三人的日后遭遇。“琅玕”，石之似玉者也，宝玉的那块通灵玉正是石之似玉者。当年此石经茫茫大士大施佛法变成一块美玉，由神瑛侍者转世的贾宝玉衔它降生。宝玉悬崖撒手舍宝钗等而去时，通灵玉复还顽石本质，大士真人仍将它送回青埂峰下。“琅玕节”就是书中的玉化为石这一重大关节。这一重大关节过后，黛玉已亡，宝钗独在，对于宝玉来说，两俱渺茫矣。这时，剩下的只有凄凉，提防过去那举目无亲穷愁潦倒的流浪生活吧。凹晶馆联诗，湘云有警句“寒塘渡鹤影”，故“鹤信”者，湘云之消息也。湘云终于要离开寡居之寒塘渡到宝玉身边来了，“鸾音”当然就是婚事的佳音了。“鸾音鹤信须凝睇”，湘云的消息和共结鸾俦的佳音尚须等待。到了那个时候，念及以往上赖天恩下承祖德，锦衣纨绔饫甘餍美之时，愧悔今日一事无成半生潦倒之罪，在湘云的帮助下，编述一集，以此聊“把唏嘘答上苍”吧。

二　交趾怀古

铜铸金镛振纪纲，声传海外播戎羌。

马援自是功劳大，铁笛无烦说子房。

这首诗说的是元春，雪芹每以谐音做暗示，此处马援的“援”字谐元春的“元”字。“铜铸金镛”是有的朝代宫中之事，意思是在宫内悬铜钟令宫人闻声而起。此诗前三句显然是说元春帮助皇帝整顿宫闱有很大功绩，贤声闻于海内外。第五回判词中的“榴花开处照宫闱”原也隐含元春在整顿宫闱方面有功，但这种说法过于含蓄，因而研究者过去从未言及元春之功。

第四句“铁笛无烦说子房”是什么意思呢？传说汉军围困项羽于垓下时，张良命军士以笛吹奏楚歌，瓦解楚军军心，故“铁笛”指张良的功绩而言。刘邦做皇帝后，朝内政争甚烈，张良知机，功成身退，得免于祸。《恨无常》曲子说：“望家乡，路远山高。故向爹娘梦里相寻告：儿命已入黄泉，天伦呵，须要退步抽身早！”贾家的家乡是金陵，元春死时，可能贾政放了南京或附近城市的外任。“铁笛无烦说子房”，张良有“铁笛”等大功，尚且知机退步，你贾政无功元妃又死，岂可恋栈？但对他说这些话是白费唇舌，他是不能听的。

元春是怎样死的呢？她归省时点的戏第二出是“乞巧”，脂批“长生殿中，伏元妃之死”。这暗示她像杨贵妃那样非正常死去。但杨贵妃之死是由于将士恨杨国忠等祸国殃民，“六军不发无奈何”，而由这首怀古诗看来，元妃有功且贤声远播，这与杨贵妃不同。所以我认为元春是因功招忌，被其他后妃暗害而死的，例如病中被在药内下毒之类。再说，若她是被定罪赐死的，她在梦中劝贾政早日引退就不可解。那样，贾政必立被株连，哪还容他“退步抽身早”呢？

马援死后遭诬陷，以致“名灭爵绝”，迟迟得不到平反。此诗以马援喻元春，则元春死后应亦被诬。判词中“二十年来辨是非”就是要为元春辨一辨是非功过。

甲戌本第十六回有回前批：

借省亲事写南巡，出脱心中多少忆昔感今。

既然写元春省亲是借以写康熙南巡，那么，写元春之功自然就是借以写康熙的治绩，而写元春被暗害显而易见也就是影射传说中的康熙被雍正所弑了。雪芹“忆昔感今”，他在书中以暗笔褒康熙而贬雍正是完全可以理解的。“二十年来辨是非”也就是要和雍正算算他弑父夺嫡这笔账。另外，“虎兔相逢大梦归”是隐指康熙死于寅卯二年之交，“虎兔”不应讹为“虎兕”。

三　钟山怀古

名利何曾伴汝身，无端被诏出凡尘。

牵连大抵难休绝，莫怨他人嘲笑频。

此诗说李纨。孔稚珪《北山移文》说，世有周子，隐居钟山，后来做了官，是个假隐士。这就像李纨“隐居”稻香村，后来贾兰做了官，她也穿戴上凤冠霞帔。周子隐于钟山时，身无名利，心里有没有名利呢？如果连心里也没有，怎么会“被诏出凡尘”呢？李纨住在稻香村时，真的就“心如槁木死灰”吗？姑娘们起诗社，她自己承认不大会作诗，却毛遂自荐掌坛当社长，可见她不但一天到晚盼望儿子做官，自己也要过过官瘾。后来“被诏出凡尘”，当上了荣国府“三驾马车”式掌家班子的第一把手。凤姐给她算收入账，她比别任的收入多几倍，却不肯拿出点钱来陪姑娘们顽顽。攒金给凤姐过生日时，连丫环们都出银两，李纨一个大钱也没出。曲子说她怕受老来贫，不肯积德只顾积钱。可见她名心、利心都很重。她只有住在赫赫荣府大观芳园的稻香村时，才“竹篱茅舍自甘心”的。宝玉评论这个稻香村违反天然，及“人力穿凿扭捏而成”。稻香村是假的，稻香老农也是个假隐士。她的谜语“观音未有世家传”，谜底是“虽善无征”。她表面上似很善良，像个菩萨，却无实际的善行为证。

母亲是假隐士，儿子是《红楼梦》中典型的“禄蠹”。《好了歌解》中有两句话说：“昨怜破袄寒，今嫌紫蟒长。”旁有脂批：“贾兰贾菌一干人。”所以贾兰做官是在贾府被抄没以后。从曲子中“威赫赫爵位高登”这句话看来，贾兰是袭了荣府的世职。他是罪臣之裔，怎么会又袭了世职呢？唯一合理的解释是：在朝内那场导致贾府败亡的政治斗争中，贾兰是站在贾府政敌那一方的。在荣府的内争中，贾兰和贾环是互相配合的。贾环由于利欲熏心，和贾府政敌相勾结，因而贾兰也就卷入那一派。既牵连进去，就骑虎难下，损害自己家庭也没有办法了。这就是怀古诗第三句“牵连大抵难休绝”的含意。贾府败亡后等事情渐冷，贾兰便受到奖赏，以嫡孙袭了世职。李纨的凤冠霞帔是不干净的，所以册子和怀古诗都以语气很重的贬词说：“枉与他人作笑谈”“莫怨他人嘲笑频”。

第五篇
论薛宝琴

《红楼梦》中的人物，总是一出场便写得栩栩如生，使人仿佛闻看到了她们的声音笑貌。但是有一个人物却似乎是例外，这就是薛宝琴。冯育栋先生说，宝琴是“谜一般的人物”[①]，周五纯先生说：“《红楼梦》文字有没有不得力之处？有。比如写宝琴，花了不少的功夫，但宝琴还是如纸糊的美人儿，没有什么生气。”[②] 梁归智先生则提出问题说：“有一个大家都感到困惑却谁也没有详加研讨的问题，那就是薛宝琴在《石头记》中的地位。”[③]

雪芹用了不少笔墨，通过众人之口盛赞了宝琴的才貌和性格。显然，这个姑娘必是书中的重要人物之一。不料，直至八十回之末，人们也没看到她在书中占有什么重要地位。除了同大家一起玩玩，一起作作诗，关于她也没见有什么重要情节。想必她的主要事迹是在八十回之后吧？有高鹗的续书在，接着往下看就是了。没想到看完以后，人们更坠入了云里雾中。四十回内，宝琴几乎没有露面。只在最后，由王夫人交代了几句：“那琴姑娘，梅家娶了去，听见说是丰衣足食的，很好。”

看来要揭开这位薛小妹之谜，只有从八十回前的伏线、暗示和脂批中去找线索，似乎别无他途。

一　大观园的尖子

四十九回宝琴兄妹和贾家另外一些亲戚来到荣府。宝玉见过后，回来向袭人等道：“你们还不快看人去！……你们成日家只说宝姐姐是绝色的人物，你们如今瞧瞧

---

① 冯育栋：《谜一般的人物——薛宝琴》，《红楼梦学刊》1988 年第 1 期。
② 周五纯：《借一月照万川》，《红楼梦学刊》1989 年第 3 期。
③ 梁归智：《石头记探佚》，山西人民出版社，1983。

他这妹子，更有大嫂子这两个妹子，我竟形容不出了。老天，老天，你有多少精华灵秀，生出这些人上之人来！”晴雯等早去瞧了一遍回来，对袭人说：“你快瞧瞧去！大太太的一个侄女儿，宝姑娘一个妹妹，大奶奶两个妹妹，倒象一把子四根水葱儿。”大嫂子这两个妹子和大太太的一个侄女儿，都是说来作陪衬的。后来探春称赞宝琴，才把话说明白了：“果然的话，据我看，连他姐姐并这些人总不及他。”

姐妹中生得最好的本来就是宝钗和黛玉，她两个是燕瘦环肥，一时瑜亮。今照探春说来，宝琴之美竟超过了她姐姐和大观园中包括黛玉在内的所有女孩子。她一定是肥瘦适中，增之一分则太肥，减之一分则太瘦；“其鲜艳妩媚，有似乎宝钗，风流袅娜，则又如黛玉”。

芦雪庵联诗，宝琴、黛玉共战湘云，结果诗句之多，湘云第一，宝琴第二，黛玉屈居第三。宝琴诗才之敏捷竟然超黛玉而追湘云。邢岫烟、李纨、宝琴三人的咏梅诗，众人称赞了一番，都指宝琴的一首更好。黛玉、湘云二人斟了一小杯酒，齐贺宝琴。宝钗道：“你们天天捉弄厌了我，如今捉弄他来了。”黛玉的《桃花行》，宝琴戏说是她作的，宝玉也认为她确有此才。可见，宝琴的诗才兼有钗黛湘之所长。

以性格论，“宝琴年轻心热，……又见诸姊妹都不是那轻薄脂粉，且又和姐姐皆和契，故也不肯怠慢。其中又见黛玉是个出类拔萃的，便更与黛玉亲敬异常”。宝钗对湘云说：“说你没心，却又有心；虽然有心，到底嘴太直了。我们这琴儿就有些象你。”这种写法表明，宝琴兼有湘钗黛性格上的优点：有宝钗之和而无其冷，有黛玉之高而无其僻，有湘云之快而无其憨。

但是，所有这些都还是按照评论女孩的一般标准来看的，宝琴最重要的优点作者没有用明笔来写，薛姨妈因贾母有意求宝琴为孙妇，说了下列一段话：

> “可惜这孩子没福。前年他父亲就没了。他从小儿见的世面倒多，跟他父母四山五岳都走遍了，他父亲是好乐的，各处因有买卖，带着家眷，这一省逛一年，明年又往那一省逛半年，所以天下十停走了有五六停了。那年在这里，把他许了梅翰林的儿子。偏第二年他父亲就辞世了，他母亲又是痰症……”

从这段话里可以看出：宝琴的母亲有痰症，此时必已死去，只不过没等说出来，凤姐就插嘴打断了。因父母双亡，兄妹才来到京城，以后便长期住在薛姨妈家，成为其家的成员。以前是分家各过的，宝琴的父亲是普通商人，和薛蟠是皇商不同。宝琴自幼随父天下十停走了有五六停，见的世面很多。她必然视野宽广，思想甚为解放，而且注重实际，不像宝玉颇多幻想。这些都是贾府中人远不能相比的。

例如，宝琴的十首怀古诗中有两首涉及《西厢记》和《牡丹亭》。宝钗说："不大懂得，不如另作两首为是。"黛玉说："咱们虽不曾看这些外传，不知底里，难道咱们连两本戏也没有见过不成?"探春说："这话正是了。"李纨说："况且又不是看了'西厢''牡丹'的词曲，怕看了邪书。这竟无妨，只管留着。"显然，四个人都看过这种"邪书"，否则为什么都赶紧表白说没有看过呢？宝琴呢，她既然写了那两首诗，就是认为看这类书是极平常的事，不需要装作没有看过。这就比上面那四个人高了一筹。

宝玉要起社咏水仙腊梅，宝钗戏说要邀一社咏太极图，宝琴借题发挥说："我八岁时节，跟我父亲到西海沿子上买洋货，谁知有个真真国的女孩子，……有人说他通中国的诗书，会讲五经，能作诗填词……"后来她念出了那金发女郎作的那首雄浑豪放的"昨夜朱楼梦"。这表明宝琴由于胸怀广阔，她对诗的见解有所不同。她若认真作诗，其风格必然也是雄浑豪放的。这也表明，她去过外国，接触过外国社会和文化。

像这样一个顶儿尖儿的人物，在书中必然占有重要地位。

二　尘世上的兼美

庚辰本四十二回"蘅芜君兰言解疑癖"有回前脂批如下：

> 钗玉名虽二个，人却一身，此幻笔也。今书至三十八回时，已过三分之一有余，故写是回，使二人合而为一。……

俞平伯先生非常重视这条脂批，说："这对于读《红楼梦》的是个新观点。"[①] 但是，这一观点后来基本上被否定了。

愚见以为，"钗黛合一"之说有重新加以讨论的必要，而要评论此说，必须先弄明白上述脂批说的是什么。脂砚在《红楼梦》成书过程中是起了重要作用的，他深知

---

① 俞平伯：《红楼梦研究》，人民文学出版社，1988。

拟书底里，负责整理抄写，可能不少素材是由他提供的。所以，对于一条脂批，不宜于还未弄清它的含意就轻易地予以否定。实际上，越使我们感到奇怪的脂批，其中恐怕越有深意。

我认为，“钗黛合一”的观点在雪芹的整体艺术构思中占有重要位置，甚至可以说是统率全书的。这一观点有两个方面：脂批中“钗玉名虽二个，人却一身，此幻笔也”说的是第一个方面，简单地说就是“一身分为二女”。批中其余的话说的是第二个方面，简单地说就是“二女合而为一”。

现在先看“一身分为二女”是什么意思。书的主人公贾宝玉是生于封建末世的一个贵公子。他应当走什么样的生活道路，这是《红楼梦》所要处理的最重大的问题之一。《红楼梦》是要反封建的。为要处理宝玉的生活道路问题，作者可以塑造他成为这样一个人：一方面反对封建礼教，一方面又有强烈的事业心。在封建制度下，二者是矛盾的。要想有所作为建功立业，就要顺应封建礼教；否则就难以有所作为。宝玉怎样选择他的生活道路呢？要解决这个问题，书中就要写他经常自己做思想斗争。这就非常沉闷，不像是小说了。雪芹不是那样处理的。他塑造了薛宝钗和林黛玉二女，把本来要赋予宝玉一身的两种互相对立的思想分别赋予钗黛二女，使黛玉具有叛逆思想而宝钗认为人应有所作为。二女是活生生有血有肉的人物，并不是概念化的。源于生活而高于生活，为了加强二女的对立性，他使二女在容貌、才华、性格等方面都互为对立面，各有千秋。正如俞平伯先生所说，二女“若两峰对峙，双水分流，各极其妙莫能相下，必如此方极情场之盛，必如比方尽文章之妙”①。这样，雪芹就用二女的矛盾和纠纷代替了本来要写的宝玉自己的思想斗争，为了让黛玉在贾府这个封建贵族家庭中不至于太孤立，作者使宝玉也成为一个厌弃封建礼教的叛逆者而缺少事业心那一面。

我们知道，前五回对于理解全书是非常重要的，其中第五回尤为重要。尘世之外的虚幻世界中，有三种力量在生活道路上争夺宝玉。一是警幻仙子，她要培养宝玉成为情种情痴，不管世人诽谤而坚持“意淫”二字。二是宁荣二公之灵，他们希望宝玉读书上进，克绍箕裘。三是茫茫大士和渺渺真人，他们要度脱宝玉悟道觉迷，出家当

---

① 俞平伯：《红楼梦研究》，人民文学出版社，1988。

和尚。这天，警幻遇见了宁荣二公之灵，双方会谈达成协议：由警幻出面，引宝玉梦魂到太虚幻境，设法教导，让他既做情种情痴，又留意于孔孟之间，委身于经济之道。警幻先带他到薄命司，让他看那些册子。正册第一页上就是一图二女，后面那四句话无非是让他既纳宝钗停机之谏，又怜黛玉咏絮之才。然后让他听《红楼梦曲》。[红楼梦引子] 后面第一支曲子就是一曲二女的 [终身误]，这是让他既重金玉良姻，又重木石前盟。第二支曲子 [枉凝眉]，海内外研究者中都有人主张此曲也是一曲二女，这是对宝玉说，二女都很好，都和你有缘嘛，“若说没奇缘，今生偏又遇着他”。显然，这一图和二曲都是雪芹对“一身分为二女”所做的提示。看了册子听了曲，“痴儿竟尚未悟”。于是警幻打出了王牌，将其妹“乳名兼美字可卿者”配与宝玉成婚。兼美姑娘是雪芹对“二女合而为一”所做的提示。婚后第二天，宝玉和兼美携手出去游顽，迎面一道黑溪阻路，这就是迷津。茫茫大士和渺渺真人早派木居士掌舵，灰侍者撑篙，驾一木筏埋伏在迷津中，单等宝玉到来，就要把他拉上木筏，渡到彼岸，让他当和尚。警幻从后追来喊话道：“快休前进，作速回头要紧!”话犹未了，迷津内竟有许多夜叉海鬼将宝玉拖将下去。宝玉喊道：“可卿救我!”这就是说，只有“二女合而为一”的兼美才能将宝玉救出迷津，使他找到正确的生活道路。

幻境中的事情是雪芹用浪漫主义的笔法来示意书中后面的情节的。那么，书中是否有一个“二女合而为一”的尘世上的兼美呢？

有的，但这不是秦可卿。秦可卿的容貌只是“生的袅娜纤巧”，并非“鲜艳妩媚，有似乎宝钗，风流袅娜，则又如黛玉”。她的思想性格和钗黛毫无共同之处，才华更谈不到。幻境中的兼美字可卿，只不过是作者借以暗示秦可卿和宝玉的不正当关系而已。至于程乙本说的秦可卿“又起个官名叫做兼美”，那是高鹗妄加的，因为直到程甲本还没有那句话。

这个尘世上的兼美是谁呢？这不是别人，就是我们的宝琴姑娘！上节中我们已经看到，宝琴在容貌、才华、性格等方面兼钗黛之美而过之。但更重要的还要从思想上来看。

宝琴兄妹生于普通商人家庭，其父带他们到处经商，从而使他们广泛接触社会和商业活动。他们属于康雍乾时期的新兴市民阶层。这种新兴市民的思想意识和要求在当时是具有进步意义的。他们有发展其所经营的事业的要求，有反封建的自发倾向。两个方面在他们的思想中是统一的。所以，宝琴的思想正是钗黛思想的合一。不是硬捏在一起的合一，而是经过扬弃提高到一个新层次的合一。

宝琴恰恰是雪芹有意塑造的，在容貌、才华、性格、思想上全面的，尘世上的活的兼美。所以，“钗黛合一”不是抽象概念，不是幻想，而是《红楼梦》中真有这样一个人物。既然如此，那么，我们对于“钗黛合一”之说还有什么可怀疑的呢？

宝琴作为“钗黛合一”进入大观园是起一种启示的作用，预示宝玉日后将走上正确的道路，奋志著书从事启蒙工作。[①] 宝琴进入贾府的四十九回，回目上联是“琉璃世界白雪红梅”，接着五十回有关于梅花的大量描写，下列一段文字尤其值得注意：

> 一看四面粉妆银砌，忽见宝琴披着凫靥裘站在山坡上遥等，身后一个丫环抱着一瓶红梅。……只见宝琴背后转出一个披大红猩毡的人来。贾母道：“那又是那个女孩儿?”众人笑道：“我们都在这里，那是宝玉。”

宝琴是末世冰雪中一枝报春的梅花。她站在山坡上遥遥等待，等待宝玉能够从后面跟上来。作者的寓意何其深远！

八十回前，这枝报春梅花是含苞未放的。她的主要事迹还在八十回后。我们必须对此也进行探索，才能全面地认识她在书中的重要地位。

三　暖香坞的十三首诗谜

第五十回和五十一回，众人在“暖香坞雅制春灯谜”。宝钗、宝玉、黛玉各有一首诗谜。宝琴以十个地方的古迹为题作了十首怀古诗，既怀往事，又是谜语。宝琴是“谜一般的人物”，为了探索八十回后雪芹是要怎样写这个人物的，我们就从这十三首诗谜着手吧。

暖香坞除此还有几个谜语，书上都给出了谜底。惟有这十三首诗谜没有谜底，“大家猜了一回，皆不是”。蔡义江先生[②]和梁归智先生[③]认为这些诗谜都隐寓书上人物的事迹，认为这才是真正的谜底。我非常同意两位的观点，但对所隐的人或事物和

---

① 黄鹤乡：《红楼梦八十回后的史湘云》，《吉林大学社会科学学报》1988 年第 3 期。

② 蔡义江：《红楼梦诗词曲赋评注》，北京出版社，1979。

③ 梁归智：《石头记探佚》，山西人民出版社，1983。

两位的看法稍有不同。[①] 十三首诗谜是一个整体。我认为宝钗诗谜和黛玉诗谜相当于《红楼梦曲》的［红楼梦引子］，《赤壁怀古》相当于［收尾·飞鸟各投林］。其余十首诗谜隐十二个女子如下：黛玉、宝钗、湘云、元春、李纨、凤姐、尤三姐、惜春、探春、秦可卿、迎春、宝琴。宝玉诗谜隐同他有爱情婚姻关系的前三女，宝琴其余九首诗谜依次隐其余九女。这是有别于太虚幻境正册十二钗的另一种十二钗，可称之为“暖香坞十二钗”。这是以尤三姐和宝琴代替了那里的妙玉和巧姐。十二钗本来是不固定的，例如四十九回就明文列出了又一种十二钗。

现在我们分析隐尤三姐和宝琴的那两首诗谜：

广陵怀古

蝉噪鸦栖转眼过，隋堤风景近如何？

只缘占得风流号，惹得纷纷口舌多。

这首诗用于尤三姐非常贴切。“蝉噪”指贾蓉、贾琏等对三姐的戏弄，“鸦栖”指贾珍对她的侮辱。三姐破着没脸，这些人才不敢欺负，“蝉噪鸦栖”才算转眼过去了。“隋堤”多柳，指柳湘莲。三姐思嫁柳二郎，贾琏在路上恰好遇到他，把亲事说妥。“隋堤风景近如何?”这是说，三姐思念湘莲，不知他近日景况如何，盼他“早早回来完了终身大事”。诗的后两句是说，因为宁府素有风流之名，只有门前那两个石头狮子干净，所以“惹得纷纷口舌多”：二姐三姐“与贾珍贾蓉等素有聚尘之诮”，连宝玉也说“真真一对尤物”。结果湘莲退婚，“情小妹耻情归地府，冷二郎一冷入空门”。

梅花观怀古

不在梅边在柳边，个中谁拾画婵娟？

团圆莫忆春香到，一别西风又一年。

这诗只要看题目中的“梅花”二字和第一句中的“梅”字就知道说的是宝琴，不会是别人。这是她怀古诗的最后一首，以最后这首诗说她自己，这种安排也是很得体的。“不在梅边在柳边”，这句诗的含意再明显不过：日后宝琴没有嫁给梅翰林之子，而是嫁给了一个姓柳的。遍查书中男性人物，姓柳的只有柳湘莲和第十四回送殡的理国公

---

① 黄鹤乡：《暖香坞的诗谜》，《红楼》1990 年第 1 期。

柳彪之孙现袭一等子柳芳。柳芳既袭子爵，早已使君有妇。所以宝琴只能是嫁给了柳湘莲！诗的第二句说，这位冷郎君因见宝琴的画像而生爱慕之情，正像柳梦梅见到杜丽娘的画像一样。魂寄梅花观的杜丽娘和柳梦梅是一梅一柳，比作梅花的薛小妹和柳二郎也是一梅一柳，恰相对照。“春香”这个名字是借用。“香”字谐音“湘”字，“春香”借指湘莲。三四两句说：宝琴、湘莲婚后因故分离。去秋相别，今又西风萧瑟。宝琴日夜想念湘莲，盼他到来夫妻团圆。既云“莫忆春香到”，则团圆已不可得矣。

十二钗应代表妇女中的十二种类型[①]。幻境十二钗是警幻显给宝玉看的，因而有局限性。尤三姐、宝琴、湘莲都属于贾府之外的普通市民阶层，而二女都与湘莲有婚姻关系。以二女代替幻境十二钗中的妙玉、巧姐而形成暖香坞十二钗很可能是雪芹在五次增删中较晚期的构思。

四　柳絮词和梅花诗

上节对《梅花观怀古》的分析有没有附会之嫌呢？如果只有那首诗，自然难免使人觉得孤证不足信。那么，让我们再看宝琴的一词一诗。

第七十回“史湘云偶填柳絮词”后，姑娘们以柳絮为题起社填词，限各色小调。这次起社是八十回前的最后一次，接着荣府屡有变端，此后更必大故迭起，恐怕八十回后也很难再有起社之事了。所以，这次起社所填的五首词决非无病呻吟，其中必有寓意，或预示贾家败亡后诸芳如柳絮飘零，或隐含填词者本人的日后遭遇。笔者曾分析湘云的《如梦令》和探春、宝玉的《南柯子》[②]，此处我们看宝琴的《西江月》：

> 汉苑零星有限，隋堤点缀无穷。三春事业付东风，明月梅花一梦。几处落红庭院，谁家香雪帘栊？江南江北一般同，偏是离人恨重。

此词中“明月梅花一梦”分明就是说宝琴没有嫁给梅氏子。“汉苑”植有柳树，“隋堤”植柳尤多，都指柳湘莲而言。“汉苑”借指长安。“汉苑零星有限”，这是说在京城时柳湘莲和宝琴已相遇定情或已订婚，但尚相见无多。当年隋炀帝是从北到南凿河

---

① 黄鹤乡：《从荣府内争看贾兰和巧姐》，《红楼梦学刊》1989 年第 4 期。
② 黄鹤乡：《红楼梦八十回后的史湘云》，《吉林大学社会科学学报》1988 年第 3 期。

筑堤，所以“隋堤点缀无穷”说二人成婚是在薛家由北回南以后。三四两句是回想贾家败亡时，“三春去后诸芳尽”，宝琴自己和梅氏子的姻事也成为一梦。“明月梅花一梦”还含有这样的意思：宝琴曾和梅氏子在月下相见，相见后即成永别。此词上半阕等于说“不在梅边在柳边”，只不过说得更详细些而已。

尤三姐自刎时有悼词说：“揉碎桃花红满地，玉山倾倒再难扶。”这就是“落红庭院”。“香雪”形容柳絮。此词下半阕是说江南江北发生战事。有的人血洒在庭院，有的人逃去后不知如柳絮飘落到谁家。江南江北一带尽都是如此。上节说，宝琴湘莲婚后因故分离。由此词看来，二人是在战乱中离散。“最是离人恨重”，二人分散后，深怀离恨而不能再聚。“几处落红庭院”有可能隐含宝琴有和尤三姐同样的悲惨结局。

现在看宝琴的《咏红梅花》：

疏是枝条艳是花，春妆儿女竞奢华。
闲庭曲槛无余雪，流水空山有落霞。
幽梦冷随红袖笛，游仙香泛绛河槎。
前身定是瑶台种，无复相疑色相差。

作为咏梅诗，第一句说的当然是梅枝梅花。其寓意是说贾家盛时，大观园花木繁茂。下句说园中儿女的衣妆争奇斗艳，四十九回“琉璃世界白雪红梅”恰有大段文字写众人的雪中装束。薛家是在荣府东北上一所幽静房舍内居住。“闲庭曲槛无余雪”，“雪”字谐“薛”字，贾家衰败后，薛家回南，那所房舍内就没有薛家的人了。“落霞”指落下的梅花而言，“流水空山有落霞”意谓梅氏子死去，宝琴和他的姻事成空，如流水落花般去了。

“幽梦冷随红袖笛”，“冷”字指冷二郎，这句诗说湘莲和宝琴互相入梦而在梦中相随。这正像柳梦梅和杜丽娘在梦中相遇一样。“香”字谐“湘”字，指湘莲。“游仙香泛绛河槎”则是说二人如牛女会于银河而成婚了。“前身定是瑶台种”：这样的好夫妻前身定是瑶台之金童玉女。“无复相疑色相差”：岂能由于尘世之处境不同对此良缘有所怀疑呢？处境有何不同，看下节自知。此诗未说到二人婚后之事，调子较为欢快。

五　不是冤家不聚头

以上两节是从宝琴的角度来看她和湘莲的关系的，本节再从湘莲的角度来看一下。

四十七回湘莲一出场就和薛家发生了异乎寻常的关系，正所谓“不是冤家不聚头”。他苦打了薛蟠，因而“惧祸走他乡”。薛家一家人恨透了他。薛蟠“睡在炕上痛骂柳湘莲，又命小厮们去拆他的房子，打死他，和他打官司”。薛姨妈“意欲告诉王夫人，遣人寻拿柳湘莲”，幸亏宝钗劝住。

奇怪的是，六十六回湘莲和薛蟠竟结拜了生死弟兄。后来湘莲进京，“先来拜见薛姨妈，又遇见薛蟠，……薛姨妈也不念旧事，只感新恩，母子们十分称谢。又说起亲事一节，凡一应东西皆已妥当，只等择日。柳湘莲也感激不尽”。像这样出人意料的描写显然是重要伏笔。此处写湘莲和薛氏母子亲如家人，又写薛家替他筹办亲事，这不简直像是薛家要招他入赘为婿吗？“又遇见薛蝌”五字也绝非闲文，因为他们日后是亲郎舅。

尤三姐自刎后，湘莲大哭一场，“出门无所之，昏昏默默。自想方才之事，原来尤三姐这样标致，又这等刚烈，自悔不及”。下面的几句话也很奇特：

> 正走之间，只见薛蟠的小厮寻他家去，那湘莲只管出神。那小厮带到新房中，十分齐整。

此处正写湘莲思念尤三姐，下面就要跟着道士走了。怎么忽然插入这些话写他仿佛到了薛家新房之中呢？这似乎是节外生枝，毫无必要，所以后来的版本把这些完全删去了。现在我们就懂了：这正是暗示湘莲没有做成尤家的女婿，日后却成为薛家的女婿。

下面写湘莲跟道士出家的写法也有深意：

> 柳湘莲听了，不觉冷然如寒冰侵骨，掣出那股雄剑，将万根烦恼丝一挥而尽，便随那道士，不知往那里去了。后回便见。

六十七回开始，有一段类似的话：

> 柳湘莲见尤三姐身亡，痴情眷恋，却被道人数句冷言打破迷关。竟自截

发出家，跟随疯道人飘然而去，不知何往，暂且不表。

这些话显然表明，所说的出家并不是真出家，“将万根烦恼丝一挥而尽”不过是一时激动，否则跟老道出家何必削发呢？既说“不知往那里去了，后回便见”，又说“不知何往，暂且不表”，可见以后还要写他的事迹。既然要写，就必是重要事迹，否则岂非蛇足吗？

甄士隐《好了歌解》中有一句说：“训有方，保不定日后作强梁。”旁有脂批“柳湘莲一干人”。这表明柳湘莲日后落草当了山大王。这是在什么时候呢？应是在和宝琴成婚之前。成婚时，一在绿林，一是民家女，因而说“色相差”。上节说，二人婚后在战乱中分离，这显然是湘莲等造反而被朝廷战败。

《红楼梦》主要是写“诸芳”。作者自云：“忽念及当日所有之女子，一一细考较去，觉其行止见识，皆出于我之上。……万不可因我之不肖，自护己短，一并使其泯灭也。”又借石头的话说：“竟不如我半世亲睹亲闻的这几个女子，虽不敢说强似前代书中所有之人，……亦令世人换新眼目，……”高续实际上只写了林黛玉一芳，连薛宝钗都写得疲疲塌塌，史湘云、薛宝琴更“一并使其泯灭”了。雪芹在八十回后写宝琴，虽因当时文网有难以下笔之处，但一定会用曲笔写得十分精彩动人的。

第六篇 论薛宝钗及薛氏一家

据脂批，宝钗嫁宝玉后，宝玉“悬崖撒手”弃之而去。那么宝钗此后的命运如何呢？这个问题讨论者不多。本文对此作初步探讨。然后，本文将对薛氏一家的故事试拟一个提纲。

一 宝钗所填的柳絮词

第七十回“史湘云偶填柳絮词”之后，湘云、探春、宝玉、黛玉、宝琴、宝钗六个人以柳絮为题起社填词，共填了五首词。这次填词是前八十回的最后一次，接着荣府屡有变故，恐怕八十回后也很难再有起社之事了。而填词的六个人在书中的地位如何呢？宝琴来贾府前，诗社里的骨干就是宝黛钗探湘五个人，每次起社主要是他们作诗，别人跟着吃吃东西凑凑热闹。宝琴来后，她也成为社中骨干之一。我们知道这六个人是全书所写的四百多人物中最活跃、最优秀、最重要的（关于宝琴在《红楼梦》中的地位，拙文《论薛宝琴》作了详细论述）。综上所述，可想见这五首词的重要性。因此可以说，这五首柳絮词必有深刻寓意，或预示贾府败亡后诸芳如柳絮飘零，或隐含填词者本人的日后遭遇。

拙文《红楼梦八十回后的史湘云》中曾分析湘云的《如梦令》和探春、宝玉的《南柯子》，拙文《论薛宝琴》中分析了宝琴的《西江月》，下面分析宝钗所填的词。宝钗评宝琴的词道：“终不免过于丧败。我想，柳絮原是一件轻薄无根无绊的东西，然依我的主意，偏要把他说好了，才不落套。所以我诌了一首来，未必合你们的意思。”她的词调寄《临江仙》：

白玉堂前春解舞，东风卷得均匀。蜂团蝶阵乱纷纷，几曾随逝水，岂必委芳尘！万缕千丝终不改，任他随聚随分。韶华休笑本无根，好风频借力，送我上青云。

众人看了，拍案叫绝，都说："果然翻得好气力，自然是这首为尊。"宝钗这首词难道只是为了"不落套"故作翻案文章吗？如果真的相信是那样，那就又被作者瞒过了。雪芹的特殊笔法之一是：又要做暗示和伏笔，又要瞒蔽读者，不让人轻易看出他的深意。现在分析这首词。"白玉堂前春解舞，东风卷得均匀。"这说的是贾家盛时，大观园中群芳欢乐的日子。我们看，这两句词描写的情景像不像是宝钗追扑那双"一上一下迎风翩跹"的玉色蝴蝶？"蜂团蝶阵乱纷纷"，两府忽然被抄，那些抄家的官员和卫士乱搜东西乱抓人，两府乱成一团。"几曾随逝水，岂必委芳尘！"四大家族一齐衰败，贾府家亡人散一场空，宝玉也离家出走了。宝钗在困苦中又遭遗弃，难道就"随逝水""委芳尘"吗？她的回答是"不"！

宝玉去了。"万缕千丝终不改，任他随聚随分。"宝钗对他万缕千丝般的深情是不能改变的，不论相聚时也好，分离后也好，宁君负我，弗我负君，"韶华"借指柳絮。"韶华休笑本无根"，不要笑我家业衰败，劫后余生，已像柳絮那样飘摇无根。"好风频借力，送我上青云。"她将凭借好风，改变所处的逆境而置身青云之上，此词名为《临江仙》，这表示她是回到老家金陵后有所作为的。

怎样置身青云之上呢？最简便易行的办法莫过于改嫁一个做大官的丈夫因而夫荣妻贵。不，那不是薛宝钗，何况她对宝玉"万缕千丝终不改"呢。有论者认为"送我上青云"指宝钗当了宝二奶奶而言。单看这五个字倒也未尝不可那样解释，但从全词来看，"送我上青云"显然是在贾家败亡、宝玉出走以后。

现在我们再把五首柳絮词总地看一下，五首词中说到黛玉的只有"粉堕百花洲"一句，说到探春的只有"嫁与东风春不管"一句。这表明八十回后，黛玉和探春将很快地退场：黛玉泪尽夭亡，探春远嫁不返。宝湘钗琴四人将仍是书中最重要的人物。宝玉是全书的中心人物，他的一生经历了三部曲：美梦—噩梦—梦醒。"柳絮词"这一回和"悬崖撒手"那一回是三部曲的分界点。开卷第一回所说"曾因历过一番梦幻之后"就是说宝玉经历了美梦和噩梦，在漂泊中接触到各种人物，思想上起了很大变化而梦醒黄粱。他回到京城同湘云结婚后，在她的帮助下"编述一集，以告天下人"。以此来从事启蒙工作（参看拙作《红楼梦八十回后的史湘云》）。"桃花诗社"很怪，黛玉是社主，却事事由湘云带头。先是湘云打发翠缕请宝玉来看桃花诗，然后，湘云建议改“海棠社”为“桃花社”。黛玉几次要起社作诗都未起成，直到湘云“偶填柳絮词”才起社填词，而这也是湘云建议的“桃花社”实际上只起社这一次，此诗社似只为黛湘之间交班接班而设。雪芹分明是有意要这样写，暗示日后湘云将接替黛玉成为书中第二号重要人物。前八十回黛玉的诗最多，八十回后湘云的诗将最多，例如《十独吟》必是湘云寡居漂泊中所作无疑。薛氏双姝——宝钗和宝琴将是第三号、第四号重要人物，她们将从事另一种事业。

二　薛氏一家

有人说在《红楼梦》里并没有看到四大家族，只看到贾家这一大家族。说《红楼梦》写了四个家族，这不符合事实。但若说只写了一个贾家，这也不完全对。作者还是用了不少笔墨写薛家的，可以说比对宁府的描写少不了多少。雪芹安排薛家在贾家旁边，是有深刻用意的。

一切事物都有其特殊性。不宜于笼统地、不加区别地看待四大家族，以为它们反正都是封建官僚家庭，一路货色，都应该打倒。荣府不同于史王薛三家，也不同于宁府（参看拙作《从荣府内争看贾兰和巧姐》）。同样，薛家在四大家族中的特殊性也很突出。薛家是半官僚半商业性的家庭。但所谓官僚不过是祖上的事了，现在虽“家中有百万之富”，却根本没个做官的。算是皇商，也“不过赖祖父之旧情分，户部挂虚名，支领钱粮”。人丁单弱，“主子层”原只有三口，后来增加了三口：宝琴兄妹和夏金桂。另外有薛蝌的未婚妻邢岫烟。和贾家相比，贾家后继无人，“如今的儿孙竟一代不如一代了”。薛家却不是这样，人丁虽少，除了薛蟠和夏金桂，其余五个人都不能说是没出息的。

先看薛姨妈吧，薛姨妈颇通文墨，不愧为宝钗之母。行牙牌令之前，薛姨妈谦虚得很，说：“我们如何会呢，安心要我们醉了，我们都多吃两杯就有了。”又说：“只怕行不上来，倒是笑话了。”王夫人忙笑道：“便说不上来，就便多吃一杯酒，醉了睡觉去，还有谁笑话咱们不成。”看来薛姨妈害怕，王夫人倒有把握。没想到行令时，薛姨妈对答如流，王夫人反而是让鸳鸯代说了一个。这是雪芹有意以王夫人和薛姨妈作对比。

薛姨妈懂得买卖经营之事。京中和各地的生意虽有家人、总管、伙计等料理，主家也要知其大概。薛蟠是“一应经济世事，全然不知”，这些事自然全靠薛姨妈。

在家庭中，薛姨妈并不独断专行，遇事能和家里人商量，特别是很尊重宝钗的意见。在婚姻问题上，她能够考虑到子女的愿望。她选贫家女儿邢岫烟为薛蝌之妇，这点眼光和见识就不简单。书上描写这事说：

> 因薛姨妈看见邢岫烟生得端雅稳重，且家道贫寒，是个钗荆裙布的女儿，便欲说给薛蟠为妻。因薛蟠素昔行止浮奢，又恐糟踏人家女儿。正在踌躇之际，忽然想起薛蝌未娶，看他二人恰是一对天生地设的夫妻，……蝌岫二人前次途中皆曾有一面之遇，大约二人心中也皆如意。

看她考虑得多么周到，多么照顾到儿女日后的幸福。

我们看邢夫人允婚是怎么考虑的：

> 邢夫人想了一想：薛家根基不错，且现今大富，薛蝌生得又好，且贾母硬作保山，将计就计就应了。

相形之下，其见识比薛姨妈就低得多了。此处及上面以邢王二夫人与薛姨妈作对比，这显系作者有意要突出薛姨妈。关于薛蟠娶夏金桂，香菱对宝玉说：

> “一则是天缘，二则是‘情人眼里出西施’。……所以你哥哥当时就一心看准了。……你哥哥一进门，就咕咕唧唧求我们奶奶去求亲。我们奶奶原也是见过这姑娘的，且又门当户对，也就依了。”

谁也没想到娶进来是个“搅家星”。这一点就不好深责薛姨妈了，在当时的历史条件下，难以做到对女方有更多的了解后再定亲。重要的是，在这事上她也是尊重了儿女的意愿的。宝钗嫁宝玉也不是薛姨妈硬作主张，因为那符合宝钗的心意。薛姨妈能有这些“民主作风”，不能说和她的商业性家庭没有关系吧？

现在看薛宝钗。自有《红楼梦》以来，她就是争论最多的人物。宝钗究竟是怎样一个人呢？《终身误》曲子说：“空对着山中高士晶莹雪，终不忘世外仙姝寂寞林。”我觉得“山中高士晶莹雪”七字可作为宝钗最恰当的考语。

“山中高士”：旧时妇女藏在家里不许见人，好比幽居在深山之中与世隔绝一样。

探春说："我但凡是个男人，可以出得去，我必早走了，立一番事业，那时自有我一番道理。"宝钗虽没有说过这样的话，她完全具有这样的才能与器量。她是幽居在大观园和荣国府中的高明之士。

宝钗在许多方面都高人一等。讲学问，她"无书不知"。论诗，她几次夺魁。谈画，她深明画理，还能详列作大型画所需之物。即使说容貌，她也"艳冠群芳"。莺儿说："你还不知道我们姑娘有几样世人都没有的好处呢，模样还在次。"哪几样好处呢，莺儿没说出来，咱们也猜不着。不过，所有以上这些都还不足为奇，宝钗超出大观园群芳的最重要的一点是：她有非凡的管理才能。贾家最有管事能力的是凤姐和探春，宝钗比她俩要高出许多。五十六回"敏探春兴利除宿弊，时宝钗小惠全大体"，兴利之道虽是探春提出来的，但在具体规定上宝钗作了重要修改。同宝钗的办法相比，探春的方案从表面上看来公家可以多得一些收益，但承包者缺少积极性，并且必然要瞒产，实际上公家并不能多得；宝钗的办法公私兼顾，而且除承包者外还考虑到园中其他人的劳动，使分沾些利益，故能调动大家的积极性并有利于保护产品，从而能够增加产量，使公家切实得到议定的收益。接着，宝钗同大家约法三章，不许吃酒赌钱。真乃赏罚分明，宽严并济。大家听了，皆大欢喜，信受奉行。这何止是"小惠全大体"呢，这是极为高明的管理艺术。论者说，讲来讲去不过每年收入四五百两银子，对于开支浩繁的贾府来说完全无济于事。其实作者写这事其意愿不在此些少收入。李纨嫌宝钗和探春"叫了人家来，不说正事，你们对讲学问"。宝钗道："学问中便是正事。此刻于小事上用学问一提，那小事就越发作高一层了。不拿学问提着，便都流入市俗去了。"可见作者写这件"小事"，一方面是写探春的才志，一方面是让宝钗小试牛刀，借此写她不但有高明的管理才能，而且有一套管理理论作为指导。如今一些人在管理科学上非常重视中国的古籍，薛宝钗早已这样做了。

讲管理，管理者的作风是很重要的。探春御下太严，玫瑰花有刺扎手，凤姐也有这种缺点；宝钗关心群众，待人和气。凤姐损公肥私，见利忘义；宝钗律己甚严，处事公平。探春好发怒，凤姐好施威；宝钗以身作则，不须多言，威信自立。

"寿怡红群芳开夜宴"，宝钗第一个抓花名签，签上画着一支牡丹，题着"艳冠群芳"四字，诗句是"任是无情也动人"，注着"在席共贺一杯，此为群芳之冠，……"

众人笑道："巧得很，你也原配牡丹花。"说着大家共贺了一杯。"群芳之冠"，这表明宝钗是大观园群芳事实上的领袖。她学识丰富，见解高超，说句话往往就是结论。

宝钗不但在群芳中是事实上的领袖，她见理明、行事正，在家里薛姨妈也很能听她的话。例如，柳湘莲苦打了薛蟠，宝钗劝她母亲道："这不是什么大事，……况且咱们家的无法无天的人（按：指薛蟠），也是人所共知的。妈不过是心疼的缘故。……如今妈先当件大事告诉众人，倒显得妈偏心溺爱，纵容他生事招人，今儿偶然吃了一次亏，妈就这样兴师动众，倚着亲戚之势欺压常人。"薛姨妈道："我的儿，到底是你想的到，我一时气糊涂了。"不肯倚势欺人，这一点贾家谁能做得到呢？

"晶莹雪"：宝钗是"冷美人"，"艳如桃李，冷若冰霜"。金钏儿投井，人来告诉，袭人唬了一跳，宝钗只说了一句"这也奇了"。到王夫人处见她正垂泪，说："据我看来，他并不是赌气投井。多半……失了脚掉下去的。……纵然有这样大气，也不过是个糊涂人，也不为可惜。"又说："十分过不去，不过多赏他几两银子发送他，也就尽主仆之情了。"尤三姐自尽，柳湘莲不知去向，薛姨妈告诉宝钗，宝钗听了，并不在意，说道："俗话说的好，'天有不测风云，人有旦夕祸福'。这也是他们前生命定。前日妈妈为他救了哥哥，商量着替他料理，如今已经死的死了，走的走了，依我说，也只好由他罢了。妈妈也不必为他们伤感了。倒是自从哥哥打江南回来……那同伴去的伙计们……也该请一请，酬谢酬谢才是。别叫人家看着无理似的。"金钏儿不是一个普通的丫头，她是王夫人手下最得力的骨干，好比凤姐的平儿，老太太的鸳鸯，宝玉的袭人。她死后一直没有人能代替她就是明证。宝钗当然明白，金钏儿之死极为可惜，王夫人逼死了她是自残膀臂。但对姨娘能说什么呢？只能为她开脱说非她之过，并帮助处理善后问题。柳湘莲救了薛蟠，自应感恩替他料理婚事。如今"死的死了，走的走了"，已是无法可想。下一步该请请伙计们，就应该去做下一步的事，为过去的事伤感有什么用呢？

薛宝钗是妇女中的强者，借用如今流行的词儿来说，她是一个"女强人"！

再举两件小事为例。二十二回老太太设春灯雅谜，因贾政在座，"虽是家常取乐，反见拘束不乐"，宝玉和女孩子们都不敢说笑。惟有宝钗"原不妄言轻动，便此时亦是坦然自若"。七十九回写夏金桂"先时不过挟制薛蟠，后来倚娇作媚，将及薛姨妈，又将至薛宝钗。宝钗久察其不轨之心，每随机应变，暗以言语弹压其志。金桂知其不可犯，每欲寻隙，又无隙可乘，只得曲意俯就”。从这两件小事也可以看出，宝钗的确是强者。

有人说宝钗是封建卫道者，这未免太把薛宝钗看小了。据揭发，她的第一条罪状是宣扬封建教条“女子无才便是德”。女强人薛宝钗决不是教条主义者。谁都知道她无书不读，诗作得很好而且继续在作不知悔改。她甚至不打自招承认自己“也是个淘气的”，也偷背着人看过《西厢记》等“杂书”。可见她并不相信上述教条，也不按这一教条去做，这样的“卫道者”够资格吗？她对黛玉说了那段关于“杂书”的话以后，黛玉“心下暗服”，从此就向宝钗学了个乖，在人前也就装作没看过那些书了。

第二条罪状是她“有时见机导劝”宝玉讲“仕途经济”。她是怎样劝宝玉以致被打成封建卫道者的，书上没有具体写，大概也和三十二回湘云劝宝玉，宝玉骂她说混账话的事差不多吧。“仕途经济”其实就是那时的政治。宝玉有反封建思想，坚决不肯当“禄蠹”，这是好的，但不能因此就认为他焚书、讲读书无用论、讲经济世事混帐论也都是对的。作为一个女强人，宝钗具有进取精神求实精神，不赞成讲空话不干事。照她看来，年轻人总该认真学点什么。士农工商四民中，宝玉不可能务农、做工、经商，那就该好好读书并且学习经济世事，怎么可以在大观园里“闲消日月”，做“富贵闲人”呢？书上写宝钗的哥哥薛蟠“性惰奢侈，……虽也上过学，不过略识几字，终日惟有斗鸡走马，游山玩水而已。……一应经济世事，全然不知，……”这显然是说她这位令兄没出息。那么，宝钗就不懂了：为什么“潦倒不通世务，愚顽怕读文章”不但不是说宝玉没出息，反而是似贬实褒呢？为什么谁向他提点意见，谁就被打成封建卫道者呢？

宝玉克服了思想上的片面性而梦醒黄粱后，“忽念及当日所有之女子，一一细考较去，觉其行止见识，皆出于我之上。何我堂堂须眉，诚不若此裙钗哉？”这些裙钗是哪些人呢？当然就是曾“见机导劝”过他的“宝钗辈”了。

关于宝琴，拙文《论薛宝琴》中已有详尽论述，在此不作重复了。关于宝琴之兄薛蝌作一点补充。薛蝌的父亲带着家眷各处跑不嫌累赘，一方面是要亲自督促他兄妹读书，一方面也是让薛蝌学习买卖经营，以便将来接自己。因此，薛蝌不但和宝琴一样，有才学、有志气、有新思想，而且还有经营管理的实际经验。这样的人物和贾家那些老爷少爷们相比，真有天壤之别。

所以，宝琴兄妹是新一代的有为青年。作者把对明天的希望寄托在他们身上，这就是他以特笔极力突出宝琴，在容貌、才华、性格等方面把她写成大观园的顶儿尖儿的用意所在。谁说作者只是不自觉地甚至违反自己本意地写出了封建制度的罪恶，却看不到出路，也没有提出正面的主张呢？

我们再看一下邢岫烟。书上说她“家业贫寒，二则别人之父母皆年高有德之人，独他父母偏是酒糟透之人，于儿女分中平常”。所以岫烟自幼贫寒，能够吃苦耐劳。父母酒糟透，对女儿不关心，这倒也有好的一面，使得她头脑中封建思想较少。她和妙玉“是贫贱之交”，所认的字都是妙玉所授，“又有半师之分”。她在妙玉的熏陶下，“举止言变，超然如野鹤闲云”；“虽有女儿身份”，“为人雅重”，却非“佯羞诈愧一味轻薄造作之辈”。其心性为人，竟不像邢夫人及他的父母一样，却是温柔可疼的人。大观园中的女孩子，其品性像岫烟这样为作者所高度赞美的，没有第二个人。

雪芹亲自作媒把岫烟许配给薛蝌是在五十七回。这时呆霸王在外，河东狮未来，薛家一家人本来是很和睦的，不像贾家内部存在着尖锐的矛盾。薛蟠回来，特别是七十九回“薛文龙悔娶河东狮”以后，情况就不同了。一龙一狮把薛家闹翻了天。但八十回后，这两个“搅家星”将被淘汰。贾家抄没，宝玉出走以后，宝钗无依无靠，定将回到母家，后一起返回金陵。那么，雪芹把女强人薛宝钗、新青年宝琴兄妹、品质优秀的邢岫烟和脑筋开通的薛姨妈这样五个不一般的人物集中在商业性的薛家，其意图是什么呢？如果说在困穷中这五个有才能、有志气的人什么也不做，只坐在家里互相埋怨，干等挨饿，那简直是不可想象的。

事情实际上已经很明显，雪芹是要写薛家这五人将重新从白地起家，经营一种民办工商业，而且由小到大取得成就。明末清初，市民社会力量和商品经济已有一定的发展，而金陵正是江南工商业的中心。这就构成了他们事业兴盛的外部条件，即所谓“好风频借力，送我上青云”。

最后结局却是悲惨的。“金簪雪里埋”，她们的事业正在兴旺发达之际，却在封建势力的打击下被毁灭了。

三　试拟一个提纲

以上二节及拙文《论薛宝琴》对薛氏一家特别是对薛氏两姐妹——宝钗和宝琴的事迹做了探讨。本节试补充一些细节为她们的故事拟一个提纲，既然涉及细节，那就难免带有某些猜想的成分。讨论问题可不可以作猜想呢？我认为是可以的。科学上有论断本来只是假说，后来经过证实成为定律的多得很。当然有些假说后来证明是错误的，这样的事也不少。但这就促使人们提出能够说明更多事实的假说，而研究工作就前进了。例如，下面的提纲中含有一个猜想：宝琴会武艺。这在前八十回中并非毫无迹象。宝玉反对文死谏、武死战，但在七十八回他却以长诗赞美了战死沙场的姽婳将军林四娘，而不像贾环、贾兰那样应付一下了事，可见雪芹写这事是另有深意的。这似乎是暗示薄命司诸钗中有一个会武艺而结局像林四娘的人。大观园中那些深闺弱女都不会是这样的，只有自幼随父行走江湖的宝琴才有此可能。五十二回雪芹通过宝琴的话写那个会作诗填词的金发女郎，为什么要写她披甲带刀呢？又为什么在那段话里两次提到“画儿上的”美人呢？这很像是暗示宝琴乃是文武全才，而且将有一幅画把她画成一个戎装女子。六十六回尤三姐自刎后，湘莲痛哭带雄剑而去，雌剑则留在尤家。这样写似乎是留个空子，暗伏丰城神物终当复合。剑名鸳鸯，则双剑复合与湘琴二人之姻缘有关。虽有这些迹象，宝琴会武艺之说我们仍只看作猜想。如果有理由否证此说或有更好的猜想，我们当然乐于接受，因为那就把研讨推进了一步。

下面是我们的提纲：

我们从七十九回“薛文龙悔娶河东狮，贾迎春误嫁中山狼”说起。回目中所说的两个情节看似不甚重要，实际上对八十回后故事的展开起着关键作用。这给薛家和贾家树起了两个“敌国”：桂花夏家和中山狼孙家。两条战线上都剑拔弩张，危机一触即发。第八十回戚序本的回目为“懦弱迎春肠回九曲，姣怯香菱病入膏肓”。这表示两个战场上都已交锋，更大的战斗即将到来。迎春将被孙绍祖凌虐致死。这事贾家岂能善罢干休，定然兴讼。“中山狼，无情兽，……一味的骄奢淫荡贪欢媾。”孙家定虚构许多诬陷贾家和迎春之词以还击。

另一战场上，香菱既因夏金桂虐待而“病入膏肓”，定将死去。薛蟠虽早已“软了气骨”，他也还有呆霸王的一面。金桂、宝蟾吵闹得使他时常“出门躲在外厢”。他想起香菱以前服侍他，“那一点不周到，不尽心?”这时不但没人服侍他，还弄得有家难归，他必对金桂恨极。最自然的情节发展是：有一天他酒醉回家，同金桂吵闹起来，两人越吵越凶，以至于拿刀动杖。金桂见他拿刀，像往常一样，大骂着“伸与他脖项”。呆霸王醉中怒极“仗着酒胆”接连几刀，竟把夏金桂杀死在地！这一来可就闯下了塌天大祸。桂花夏家有钱有势，正是薛家的对手，非金陵冯渊家可比。于是兴起大讼，双方各以大量钱财行贿，各仗政治靠山相对抗。早在薛家进京前，“各省中所有的买卖承局、总管、伙计人等”已是“趁机拐骗起来。京都中几处生意，渐亦消耗”。此时兴讼，趁火打劫之事就更多了。

朝内的政治斗争日渐加剧。双方阵容逐步明朗化。一方以北静王为首，下有四大家族，还有七十八回提到的庆国公、杨侍郎、梅翰林；一方以忠顺王为首，下有桂花夏家，中山狼孙家，而贾雨村、贾环、贾兰也投靠到这方面。双方当然都还有另外一些人。书上说“连宫里一应陈设盆景”亦是夏家贡奉，可见这一方在宫中也有内线。上述两起讼事和政治斗争搅在一起，促使斗争迅速进入决战。结果北静王获罪，四大家族一败涂地，薛家完全破产，薛蟠在狱中只待判刑。在此期间，黛玉、元春、贾母死去，宝钗嫁出，岫烟嫁来。

第一回写葫芦庙引起的大火道：

> 不想这日三月十五，葫芦庙中炸供，那些和尚不加小心，致使油锅火逸，便烧着窗纸。……大抵也因劫数，于是接二连三，牵五挂四，将一条街烧得如火焰山一般。……只可怜甄家在隔壁，早已烧成一片瓦砾场了。

此处有脂批道：“写出南直召祸之实病。”作者把一个具体而微的薛家安排在贾家旁边，就是把它比作甄士隐家隔壁引起大火的那个葫芦庙。

贾家人散家亡。宝玉也“悬崖撒手”，离家而去。后贾政夫妇相继病故，宝钗回到母家。梅翰林父子俱发配边疆，行前准其探视家人。时当十月之望，梅氏子和宝琴在月下相见，痛哭而别。不料政敌早已买通解差，途中经过一道山岭，行至僻处，父子二人惨遭杀害。

薛家为在狱中使费，家中器物已当尽卖光，回思自家开当铺时真如隔世。这天薛姨妈取出箱中仅余的几轴字画，要拣一轴去卖。忽见薛蝌带柳湘莲进来拜见。前者湘莲出家后，又在江湖上行走，已是蓄发复了俗装。近闻薛贾二家衰败，进京来探视。叙谈间，湘莲随手取桌上一轴画打开看时，只见画中池塘边几株垂柳，一个绝色美女武装打扮在月色微风中舞剑。薛姨妈见湘莲出神，问时说是画中宝剑似是他家鸳鸯剑。原来当时尤三姐将雄剑连鞘还给湘莲，用那股雌锋自刎。尤二姐不忍丢弃此剑，另配一鞘，收在箱内，故梦中三姐劝她“将此剑斩了那妒妇”。二姐死后，凤姐趁无人时将她箱柜中物尽皆搜去，此剑亦在其中。后凤姐闻宝琴会武艺，将剑送给她，薛家的人却不知此乃湘莲之物。一天，薛姨妈见宝琴使剑，以为比那次雪中她披着凫靥裘，后随一个丫环抱着一瓶红梅还要好看。偶同惜春谈起，惜春答应照这样给宝琴画一张像，商量配何景色。薛姨妈因自己曾住梨香院，忽然想到宋人诗句：“梨花院落溶溶月，柳絮池塘淡淡风。”便请惜春照这两句诗的意境来画。这便是湘莲所见那幅画的来历。薛姨妈取剑出来，湘莲看时果然是那股雌锋。回到住处，取出雄剑抚视，多年已冷之心，此时难以平静矣。恰值冯紫英、卫若兰来访，见他看剑，问知薛家之画，二人以为天缘，愿为月老。湘莲连称不可，二人不由分说，将雄剑连原鞘带到薛家，陈明来意，并请以剑与那幅画交换作为信物。薛家欢喜答应。

这天，湘莲正在薛家商议择日迎娶之事，忽报薛蟠已判死罪，于是全家复又转喜为悲。湘莲力任设法营救。他早已与绿林豪客相交，乃上山借取金银，贿通狱卒，趁夜放出薛蟠。湘莲与几名喽啰在外接应，混出城去，带薛蟠及狱卒一起落草去了。六十大回写湘莲与薛蟠结拜为生死弟兄，正为此事伏线于千里之外也。薛蟠越狱后，薛家幸有湘莲留下的财物贿赂官府，方得免祸。然因官吏敲诈不已，贫困愈甚。因思金陵尚略余微产，乃待事冷后，勉凑盘费，一同回南。

到了金陵，省吃俭用，百计经营，不数年，办起一个民营纺织工场。场事由宝钗主之，薛蝌与宝琴为副，薛姨妈退居二线任顾问。此时家事已繁，一切由岫烟掌管。薛氏一家各展其才，亦可见八十回前之精心安排矣。其事业青云直上，交易及于各省，规模甲于金陵。

又过几年，湘莲等一干强梁造反下山，各地守军望风而逃。不久，义军进占金陵，薛蟠回家，全家悲喜交集。薛蟠欲为湘莲、宝琴完婚，薛姨妈、薛蝌俱踌躇。薛蟠曰："义弟两次相救，薛家岂有负义悔婚之理？况我既在义军中，即无此婚，亦已与朝廷为敌，有何不能决者！"问宝琴时，答以婚否均无不可，婚则为柳家妇，不婚则削发为尼耳。宝钗乃谓："琴妹不嫁则已；若嫁，除柳家郎谁堪配她？"于是婚事乃定。花烛之夜，牛女会于鹊桥之上，游仙香泛绛河之槎，其情好难以尽述。婚后，湘莲每与宝琴兄妹及宝钗论及世事，深服三人之远见卓识，欲荐薛蝌在义军中任职，薛蝌以工场事繁辞谢。

朝廷闻金陵失陷大惊。前贾兰已袭世职，朝廷命其率军讨贼。适忠顺王在又一次政争中获罪，贾兰惧株连，冀以军功免罪，乃冒险进军而全军覆没，回朝被赐死，贾雨村前升大司马，虽曾党于忠顺王，乃是不倒翁，今又有新靠山，仍掌兵部。朝廷命其为帅，统大军进剿反寇。官军势大，义军占地渐失，后仅余金陵、扬州二城，俱被围困。湘莲困守扬州，与宝琴东西相隔，不通音信。金陵先破，宝琴率女兵与官军巷战，众寡悬殊，宝琴退入一庭院内，官军涌至，宝琴身受重伤，又奋力斩杀数人，横鸳鸯剑自刎而死。女兵死伤殆尽。官军入城，烧杀淫掠，百姓死者不计其数。乱军破门入薛家，宝钗已从容仰药而亡。薛氏一门皆遇难，家中财物抢劫一空，工场焚为白地。五日后，城中稍定，雨村出榜安民。时在严冬，天气奇寒，降下一场多年未有之大雪来。

扬州不久亦陷，湘莲突围后，身边仅余数人。后改装入金陵，薛家仆人幸存者带他到丛葬处，只见一片白茫茫大地上新坟万千，薛家之墓葬时虽有记认，被雪掩盖不可辨识。湘莲伏地大哭，痛不欲生。出城去，披发入山，后云游各地，不知所终。今试杜撰一回目如下：

薛氏女埋香雪冢下，冷郎君撒手冷云间。

跋

刘大有

为纪念尊敬的王湘浩院士诞辰105周年，为满足王湘浩院士的学生、吉林大学众多校友对《红楼梦新探》[①] 一书和《王湘浩院士的部分照片汇集》的渴望，吉林大学计算机学院经过大约一年的准备，《红楼梦新探》（附“王湘浩院士的部分照片汇集”）的出版日期，指日可待。

王湘浩院士撰写的《红楼梦新探》一书，于1993年12月由吉林大学出版社出版，发行两千册。出版后，深受广大读者的欢迎和好评。

著名红学家周汝昌先生[②]对该书给予了极高的评价，他以《君书动我心》[③] 为题对《红楼梦新探》做出评论。他认为，王湘浩先生深明曹雪芹天才的笔法匠心，且以超常的思力、悟力推论了与之相应的八十回后情节的发展变化，具有“慧眼”“灵心”。他表明《红楼梦新探》是令他心折、学术品格很高、思力和识力很深的著作。他指出，《红楼梦新探》思路新颖，与一般“红学家”显得不尽相同，令人刮目：从红学分支来看，《红楼梦新探》应属于探佚学范畴，它关系到中华文化中巨大高深的灵智问题，也是对是非真妄的思辨、观照、领悟、升华的精神世界与文化造诣的证明问题。周汝昌先生认为，王湘浩先生了不起的功绩在于：他敢于违世俗而讲真话、敢于犯权威而护真理、敢于硬翻历史评论“铁案”，为书中人物鸣不平；他敢于直言不讳地为雪芹本旨所受的歪曲雪洗污浊，使《石头记》的原本精神意旨境界大白于天下。

---

① 王湘浩．红楼梦新探［M］．长春：吉林大学出版社，1993．

② 梁归智．红学泰斗周汝昌传：红楼风雨梦中人［M］．桂林：漓江出版社，2006．

③ 绳海文．记王湘浩院士［J］．往事．2015（04）：29－32．

王湘浩院士的中国古典文学修养很深。[①] 在他潜心研究数学、计算机科学和智能科学之余，他还对文学、艺术和美学进行了探究与鉴赏，展现了逻辑思维与形象思维的完美结合，以及一代学术大师的超人才智与风采，他用严谨的数学精神研究《红楼梦》搞得非常好。[②]

在“王湘浩院士的部分照片汇集”中，收集、整理了38张珍贵照片。这些照片不仅有助于读者了解王湘浩院士在教书育人、科学研究等方面的一些片段，而且也能帮助王老师的学生、吉大校友缅怀当年与王老师在一起的情景：有时王老师像一位满脸笑容的慈祥母亲，有时王老师又像一位板起严肃面孔、令人望而生畏的父亲。

王湘浩老师与学生之间是师父和徒弟的关系，王老师不仅教授学生学识，而且还教导学生做人。一方面，王老师办事、做学问非常严谨，对学生的要求非常严格；另一方面，他又把学生当作自己的孩子，对学生关怀备至。作为王老师的学生，我们都把“一日为师终身为父”当作做人的原则。王老师其学为人师，其行为世范。

20世纪50年代，计算机科学在国际上是一个刚刚出现的新事物，但一下子就被王老师捕捉到了。他从自己熟悉并做出杰出成果的代数领域毅然决然地转到了一个崭新的领域——计算机科学，这充分展现了王老师在学术上的高瞻远瞩和远见卓识，以及他一切从国家需要出发的崇高思想品格。

20世纪60年代，人工智能在国际上正处于低谷时期，许多专家、学者从人工智能领域转向其他研究领域，有的西方发达国家已不再资助人工智能项目，人工智能这一提法会不会被别有用心的人搞成政治问题，许多人对此仍心有余悸，正是在这种情况下，他以巨大的勇气在国内率先举起了人工智能的大旗。在时隔四十多年的今天，全世界众多科学家和政界人士都把人工智能看作最重要、最具战略意义的研究领域，无限感慨在我们心中油然而生。

20世纪80年代初，管纪文教授招收了很多知识工程方向的博士研究生。1987年，管纪文教授的三位博士生郑方青、马志方和黄祥喜面临导师不在国内无法进行博

① 绳海文. 记王湘浩院士［J］. 往事. 2015（04）：29－32.

② 丁石孙（口述），袁向东，郭金海（访问整理）. 有话可说：丁石孙访谈录［M］. 长沙：湖南教育出版社，2013：49－50.

士论文答辩的难题。当时这三位博士生都跟着我做项目。为此，我向王老师汇报了此事。王老师跟我说了四点意见：其一是，三位博士生的答辩是大事，不解决他们的问题，他们很可能会选择出国继续学习，这不仅对吉林大学计算机学科的建设不利，而且也会对吉林大学计算机学科的声誉造成负面影响；其二是，三位博士生的答辩问题本是管纪文教授的事，但他不回国，我是他的老师，他的事我不管谁来管；其三是，办事要按规矩，要名正言顺，你去联系吉林大学研究生处看可否将这三位博士生转到我的名下，由我指导他们完成博士论文；其四是，三人的研究方向与我不同且我又不在长春这增加了指导的困难，我需要三位教授作为我的助手协助我指导并定期带他们来大连汇报论文进展情况。学校同意王老师的意见，王老师确定了刘叙华、庞云阶和刘大有三位教授作为助手。1989 年，这三位博士生顺利通过了“高规格”的博士答辩（答辩委员会包括三位中科院院士）。在 1987 年末至 1989 年期间，我们一行 6 人多次去大连市专程向王湘浩老师汇报博士论文进展情况，同时看望王老师和王师母。此后，去大连看望王老师和王师母已成为王老师的学生们的一种责任，一种义务，一种亲情。

王湘浩院士是杰出的数学家、计算机科学家、人工智能科学家和教育家，中国人工智能研究的奠基人①②，他具有强烈的爱国心，将毕生的精力都献给了祖国的教育事业③，为吉林大学的发展和建设作出了卓越贡献，作为数学学科、计算机科学与技术学科、软件工程学科和人工智能学科④的奠基人和学术带头人，以其高瞻远瞩和远见卓识，为这四个学科的发展运筹帷幄，呕心沥血。1952 年以来，他的成就、他的快乐、他曾经的磨难以及他最后的眷恋，都与这四个学科融为一体。

他的名字、崇高品格和精神将成为吉林大学师生心中的一座丰碑，成为一个时代的象征。

---

① 伍卓群．追忆中国人工智能奠基人——王湘浩院士［J］．往事．2015：26－28．

② 高文．国家新一代人工智能发展规划［R］．ICIC 第 26 届媒体融合技术国际研讨会，2018．

③ 伍卓群．追忆中国人工智能奠基人——王湘浩院士［J］．往事．2015：26－28．

④ 2011 年软件工程被批准为一级学科，从计算机科学与技术学科中独立出来，与计算机科学与技术一级学科并列。第一届国际人工智能联合会议于 1969 年召开至今已经历 50 多个年头；从全国人工智能研究会成立起，中国人工智能学会已走过了 40 多个年头；2011 年人工智能学科（智能科学学科）也进行了申报但未获批准，目前国内已有许多大学都建立了人工智能学院（有的称智能科学学院）。

后　记

我敬爱的爸爸王湘浩离去了，他把他对祖国、对科学、对教育事业以及对学生、对家人的深沉而无私的爱留在了人世。作为他的女儿，我或许比别人多知道一些的，是他对中华民族文学艺术的挚爱。我爸爸最喜爱的还属古典文学。特别，对于那中华古典文学的骄傲，那世界文学宝库中的明珠——《红楼梦》尤其爱不释手。他对这部只剩下八十回的宝书所遗失的部分深感痛惜，就以非常认真的态度搞起了《红楼梦》探佚的研究。

多年来，尽管我爸爸工作很忙，但还是尽量抽出他有限的业余时间，凭着他在古典文学方面的造诣，当然，也凭着他惯于解数学难题的脑力，写了很多探佚方面的文章。我爸爸早就打算先在他的第一本《红楼梦新探》专辑中登载本书所搜集的这六篇文章，而其他几篇已经基本成形的文章放在以后的专辑中出版。万万没想到的是，他走得如此匆忙，以至于这第一本专辑就成了逝世以后才出版的遗作。

本书中的第一篇、第三篇和第四篇文章，我爸爸曾以黄鹤乡为笔名分别在《吉林大学社会科学学报》《红楼梦学刊》和《红楼》杂志上发表过（参见本书文章《论薛宝琴》的注释）。之所以使用笔名，是因为我爸爸不愿意借他在理科方面的名气来提携他在红学方面的工作。他的上述文章发表后，曾主动把抽印本寄给了几位红学界的同志，随后收到了不少寄给黄鹤乡的信，其中包括红学界知名人士写来的信。信中对于我爸爸文章的水平和见解给予非常高的评价，而且写信者全都理所当然地把黄鹤乡当成了文学界的某个“后起之秀”和“奇才勇士”。另外三篇没有发表过的文章，我爸爸已经写好了完整的底稿，我一字不差地抄录一遍，寄给了出版社。凭我的水平，真的不敢替他老人家改动一个字，也确实找不到任何一处需要改动的字句。

如果我爸爸奉献给读者的这本遗作，能够为红学的研究工作提供一些有新意并且有参考价值的线索，那么若他的英灵有知，定会感到欣慰的。

王坤健

1993 年 6 月 27 日于大连

致　　谢

国内外知名的数学家和计算机科学家、中国科学院学部委员、博士导师王湘浩教授离我们而去了，1993年5月12日，我们——他培养的学生们，在和王老师最后告别时，谈起了我们的老师：王老师，凭着他敏锐、严谨的思维不仅在自然科学上作出了出色的研究工作，而且在晚年的业余时间里，使用他那令人佩服的逻辑思维能力，对《红楼梦》中一些疑难问题，取得了一些独到的新颖的研究成果。他的学生们觉得将这些成果奉献给红学界可能是有好处的，一个自然科学家在人文科学领域中思考出来的结果，也许更具有独特的创造性。

这一想法得到了王老师子女的极大支持，因此，在本书问世时，应该对促成本书的出版作出关键贡献的他的学生们表示谢意。

首先应该对他的学生——伍卓群教授、陈维钧教授表示感谢，他们的支持与努力是本书得以出版的关键。

其次应该对他的学生——奚涌江高级工程师表示感谢，奚涌江是王老师在“文革”前培养的最后一名研究生，他怀着师恩如海的感情，极想为老师最后再贡献一次自己的力量，他承担了本书的排版、校对、印刷的全部工作，并负责筹措了这些工作的全部费用。

还要感谢他的学生——李志林副教授的帮助，他承担了在出版过程中的全部联系、协调的工作。

最后应该感谢王汝梅教授，他虽然不是王湘浩教授的直接学生，但是王汝梅教授愿意执弟子礼，欣然接受了为本书写序的任务，使得本书有了一个完整的形式。

他的所有学生们，祝愿他们敬爱的老师——王湘浩教授在天国安息！

学生刘叙华

1993年11月5日

附录

王湘浩年谱

- 1915 年 5 月 5 日，出生于河北安平槐林庄（现名大同新村）。
- 1928 年，从槐林庄小学毕业，考入直隶省第十中学（现为深州市旧州中学）。
- 1931 年，初中毕业，考入北洋工学院附属高中。
- 1933 年，以优异的成绩考入了北京大学算学系（数学系）。
- 1934 年，大学二年级遇到人生第一位导师江泽涵。
- 1935 年，参加江泽涵组织的拓扑学讨论班，开始系统学习拓扑学。
- 1937 年，大学毕业，几经辗转，一路奔波，追随江泽涵先生的脚步，到达长沙临时大学做助教。
- 1938 年春，随长沙临时大学（后成立为西南联合大学）南迁至昆明。
- 1939 年，结束助教生涯，成为江泽涵在西南联合大学招收的第一批研究生。
- 1941 年，在江泽涵的指导下，完成硕士论文 *Mapping of Graphs in Closed Surfaces*，研究生毕业后，成为西南联合大学数学系的一名专任讲师。
- 1943 年，公开发表的第一篇论文 *A System of Completely Independent Axioms for the Sequence of Natural Numbers* 刊登在符号逻辑的国际权威期刊 *Journal of Symbolic Logic* 上，是在这个久负盛名的国际学术期刊上发表论文的第一位华人学者。
- 1946 年，以优异的成绩脱颖而出，被北京大学选送到美国普林斯顿大学留学，成为数学大师阿廷在普林斯顿招收的第一批研究生。
- 1947 年，提前拿到了普林斯顿的硕士学位。
- 1948 年，推导出格伦瓦尔德定理的反例；同年在数学界顶级期刊《数学年刊》(*Annals of Mathematics*) 上发表了只有一页半的著名论文《关于格伦瓦尔德定理的反例》；11 月，在他的博士论文中，纠正格伦瓦尔德定理的错误，将该定理做了推广，重新证明了迪克逊猜想。
- 1949 年夏天，从普林斯顿大学博士毕业；6 月，从普林斯顿出发，乘火车到达旧金山，乘船至上海未果到达香港并和同样滞留在香港的江泽涵先生重逢，辗转漂泊两月，8 月，终于返回祖国的怀抱；回到北京后，被北京大学聘为副教授。
- 1950 年，升为教授，还在清华、北师大和辅仁大学兼课；继续围绕格伦瓦尔德定理做研究，对一般的 Abel 扩张给出了格伦瓦尔德定理成立的充分和必要条件；后来，他还得到了柯特半单循环的亚直接和表示，并讨论了与此相关的拟赋值环问题；博士论文成果发表在《数学年刊》上，格伦瓦尔德－王定理（Grunwald-Wang Theorem）创造了历史，成为国际数学界由中国数学家发现问题并修正的为数不多的著名定理之一。

- 1952 年，全国高校进行院系调整，积极响应国家号召，到祖国最需要的地方去，投身东北人民大学（1958 年改名为吉林大学）的建设工作中，组建东北人民大学数学系。
- 1954 年，正式加入中国民主同盟，以数学家盟员的身份为国家的科技和教育发展建言献策，后来相继担任吉林省民盟副主任委员、长春市民盟主任委员，民盟中央委员和参议委员；1954 年从东北工学院聘请了王柔怀；1954 年，阿廷出版的 *Class Field Theory* 讲义中专门有一章讲述格伦瓦尔德－王定理。
- 1955 年 6 月 3 日，成为中国科学院首批学部委员。周恩来总理签发了国务院令，公布首批 233 位学部委员名单，年仅 40 岁因为数学成就卓著而当选中国科学院首批学部委员，同时当选的学部委员还有陈建功、苏步青、江泽函、柯召、许宝騄、华罗庚、李国平、段学复等其他 8 位数学家，这也标志着王湘浩已经进入中国一流数学家的行列；此后，有更多的机会参与国家科技发展的规划。
- 1956 年 3 月 14 日，成为国务院科学规划委员会下设计算技术和数学规划组的 26 位委员之一；同年 5 月，作为主要参加人参与制定了《东北人民大学 1956—1967 年发展规划》草案，计划成立物质结构和特殊材料性能、计算数学等六个理科研究室，在计算数学方向背后，有一个更宏大的计划，研究当时绝对的高精尖设备——电子计算机；同年，从中国科学院聘请了学识丰富的孙以丰。
- 1958 年，带领一部分青年师生开始了电子计算机和控制论方面的研究；请来苏联梅索夫斯基赫（Mysovskih）教授帮助在我国最早建立了计算数学专业；经过 6 年的努力，把吉林大学数学系办成了一个初具规模、在国内外有一定影响的多学科系；时至今日，吉林大学数学学科在国内外都具有重要的地位，这也是王湘浩当年在吉林大学数学系的工作成果和贡献；带领团队聚焦人工智能基础理论研究。

- 20 世纪 60 年代初，提出了解决多值逻辑的函数完备性问题的重要思想，即利用“保 *n* 项关系”的方法来研究 *n* 值逻辑的完备性问题。
- 1961 年，与谢邦杰合作编写《高等代数》教材，由高等教育出版社出版。
- 1963 年，提出了多值逻辑中缺值函数的结构问题，并取得一些成果，这一问题后来也由他的学生罗铸楷完全解决了。
- 1964 年，指导学生解决了利用“保 *n* 项关系”的方法来研究 *n* 值逻辑的完备性问题；在计算数学专业内开设控制论专门化，开始培养本科学生。
- 1973 年，在他带领下，经过近半年的辛苦劳作，吉林大学与通化无线厂合作研发了第一台函数可编程计算机；在吉林大学数学系成立“控制论”专业，招收了一个班共 22 名的学生，这也是这个专业在计算机科学系建系前唯一的一次招生。
- 1976 年 5 月，吉林大学计算机科学系正式成立，担任首任系主任，并带领师生大力开展计算机教学和研究工作。
- 1976 年，计算机系采购了一批当时流行的 Nova 计算机，带领老师们研习使用 BASIC 语言编写程序。
- 1977 年，中国科学院主持召开了全国自然科学学科规划会议，以规划未来十年科技发展；王湘浩受邀主持计算机方向的十年规划，在会议期间明确提出要开展人工智能研究，并将“人工智能”写入十年科技发展规划。
- 1978 年，指导苏运霖担任组长，和管纪文、周长林共同完成“Nova 多用户 BASIC 解释程序的分析和注释”的研究，并在全国科学大会上汇报。
- 1979 年 7 月 23 日—30 日，中国电子学会计算机学会（中国计算机学会的前身）在吉林大学召开了“计算机科学暑期讨论会”，担任会议领导小组组长；和他的学生刘叙华研究了归结方法中的取因子问题并提出了广义归结方法*；这次讨论会的一个重要专题是人工智能，因此后来被称为“中国的达特茅斯会议”；8 月，王湘浩被任命为国家教委代表团团长，与清华大学林尧瑞、华中工业大学彭嘉雄参加 8 月 20—23 日在日本东京举办的 IJCAI 学术会，以了解国际人工智能研究的新动向。

---

* 广义归结方法最早于 1979 年提出，1982 年正式发表。这一创新性工作不仅包含了鲁宾逊在 1965 年的普通归结方法，还融入了默里 1982 年的非子句归结方法，为证明定理机器提供了更为全面的策略。

- 1980 年，受教育部委托在吉林大学举办了人工智能研讨班并亲自编写讲义，有清华大学、北京航空航天大学、中国科学技术大学、复旦大学、中山大学、国防科技大学、西北工业大学等共计 16 所高校的教师来到吉林大学进修、学习，不少人成为中国人工智能研究的中坚力量；创立了全国高校人工智能研究会，任会长；以后每年举行一次研讨班，是国内最早的人工智能学术研讨活动，对全国高校的人工智能研究起到了重要推动作用。
- 1981 年 2 月 10 日，全国各地从事人工智能研究的部分科技工作者近 50 人聚首北京，倡议尽快成立中国人工智能学会；会上成立了由秦元勋、王湘浩等 16 人组成的常务筹备小组，并以 48 人的名义起草了《关于成立中国人工智能学会的申请报告》，呈交中国科学技术协会和中国社会科学院等单位；6 月 13 日，国务院学位委员会第二次会议通过了国务院学位委员会学科评议组成员名单，当选计算机学科评议组组长，并被国务院遴选为第一批 2 位计算机软件方向的博士生导师之一（另一位是南京大学的徐家福先生）；10 月，中国人工智能学会在长沙正式成立，王湘浩在筹备和初创过程中发挥了重要的作用。
- 1982 年 4 月 21 日，中国计算机学会人工智能学组成立大会及学术报告会在浙江大学举行，作为学组组长亲自到场并致开幕词。
- 1983 年 5 月，带领学生管纪文、刘叙华撰写的《离散数学》教材由高等教育出版社出版，这也是国内最早一批《离散数学》教材，在国家教委第二届高等学校优秀教材评选中获得“国家级优秀奖”。
- 1985 年 6 月 1 日，中国计算机学会在北京科学会堂隆重举行成立大会，标志中国计算机学科进入了一个新纪元；再一次以副理事长的身份参加（1980 年已当选学会副理事长），在中国计算机学会中，继续带领中国计算机科学和人工智能事业高歌猛进；同年，与杨荫华合作编写《线性代数》教材，由华中工学院出版社出版。
- 1987 年六届全国人大期间，呼吁要“面向中学改革师范院校数学专业的课程设置”；9 月，指导的第一个博士生陈建华博士通过答辩。
- 1988 年夏天，回到吉林大学参加研究生答辩并与毕业生合影。
- 1989 年夏天，邀请陆汝钤、董韫美等专家到吉林大学参加郑方青、马志方和黄祥喜博士答辩，并与师生合影留念。

- 1990 年 12 月，从事高校科技工作四十年受到国家教委表彰。
- 1991 年 4 月，在北京参加全国人民代表大会，并探望导师江泽涵院士；10 月，获得国务院政府特殊津贴。
- 1993 年 5 月 4 日，因病医治无效，在大连与世长辞，享年 78 岁。

王湘浩

部分著述目录

《高等代数》*

目录

* 《高等代数》为王湘浩、谢邦杰合著的教学用书，1960年由人民教育出版社出版。文集附录仅收录该书目录。

《高等代数（1964年修订本）》*
目录

* 《高等代数》（1964年修订本）为王湘浩、谢邦杰合编的教学用书，1964年由人民教育出版社出版。文章附录仅收录该书目录。

《离散数学》*
目录

* 《离散数学》为王湘浩、管纪文、刘叙华合编的教材，1983 年由高等教育出版社出版。文集附录仅收录该书目录。

《离散数学》*

目录

* 《离散数学》为王湘浩、管纪文、刘叙华、姜云飞、王钲旋合编的教材，1988 年由吉林大学出版社出版。文集附录仅收录该书目录。

《线性代数》*
目录

* 《线性代数》为王湘浩、杨荫华合编的教材，1985 年由华中工学院出版社出版。文集附录仅收录该书目录。

后　　记

受学校和学院的委托，我代表学院负责了这次《王湘浩文集》的编撰工作，经过7年多的收集和整理，在很多熟悉的和不熟悉的朋友的帮助下，同时得到了王湘浩院士家人的信任和认可，历尽很多波折和困难，在2025年年初终于如愿以偿收集到了王湘浩院士在各个时期发表的论文、撰写的图书章节以及为报纸杂志写的科普文章和书评等，这些弥足珍贵的文字资料为文集的出版打下了重要的基础。如今，整个文集的文字校对已接近尾声，为此谨向所有为文集出版付出努力和辛勤汗水的领导、老师、同学和各界人士表示深深谢忱。

《王湘浩文集》的出版是2025年吉林大学纪念王湘浩诞辰110周年系列活动的重要事件。本书的编纂与出版，离不开张希校长的深切关怀与鼎力支持。谨向张希校长致以崇高的敬意与由衷的感谢，校长的文字与行动始终是吉大传承先生精神、践行“教育报国”使命的重要引领。

特别感谢陆汝钤院士拨冗为文集作序。陆院士的序言高屋建瓴，不仅深刻阐述了王湘浩先生在学术上的卓越贡献，更以同行视角勾勒出王先生治学为人的精神风貌，为读者理解王先生的学术遗产提供了重要指引。

衷心感谢张景中院士以手书寄语的方式表达对王湘浩先生的深切缅怀。张院士的墨宝饱含深情，字里行间既见对前辈学人的敬重，亦透露出两代学者在学术追求上的精神共鸣，为文集增添了厚重的纪念意义。

同时，我们向汤涛院士致以谢忱。汤院士对王湘浩先生的评价文字精辟而中肯，既是对前辈学人科学精神的致敬，亦为后辈提供了理解数学史脉络的钥匙。

感谢学校宣传部王庆丰部长的关心和帮助，感谢学院杨博院长给予此项工作的特别重视和鼎力支持，感谢学院宫皓宇书记对这项工作的完成给予的关心和支持，感谢李昕副院长对这项工作的大力支持和付出的辛劳。同时，文集的编辑也得到数学学院王春朋院长和韩月才副院长的帮助，在此表示感谢。部分图片来自北京大学数学学院，向陈大岳院长表示感谢！

在文集编撰过程中，得到了计算机学院刘大有教授、计算机学院欧阳丹彤教授、数学学院张树功教授以及吉大校友、中国科学院数学与系统科学院刘卓军研究员的悉心指导，在此向他们表示感谢。

感谢数学学院马晶、侯秉喆两位老师为文稿校对提供了数学专业的指导；感谢计算机学院吴春国、赖永、崔佳旭三位老师为文稿校对提供了计算机专业的指导。也感谢欧阳丹彤、吴春国、赖永、时小虎、叶育鑫、崔佳旭、王生生等老师为文稿的录入和校对推荐了众多优秀的研究生同学，保证了文集出版工作的顺利推进。尤其感谢吉林大学出版社的领导和编辑，对文集出版工作的重视和关心，为保障进度，社里为我们提供了重要的技术保障，诚挚感谢责任编辑周婷老师，从文集规划、具体实施，到文字校对，周老师付出了难以想象的辛苦。

王湘浩院士是著名的数学家、教育家、计算机科学家和中国人工智能的奠基人，他的研究极为广泛，而且各个领域均成就卓著，他发表的文章和撰写的文字涉及拓扑学、代数学、逻辑学、计算机科学、人工智能等领域；同时王湘浩院士还是一位文理兼修的奇才，是著名“红学家”，喜欢京剧、元杂剧，爱打桥牌，能看懂佛经，善填词，写楹联，晚年还续写了金庸的武侠小说《射雕英雄传》，从某种意义上说他更是一个懂生活、爱生活、有品位的先生。整个文集文稿的收集和文字录入过程中，我们始终怀着崇敬和敬仰的心情，先生的文章不但涉猎广泛，而且涉及中文简体字、繁体字、英语、法语、俄语和希伯来语等，对参与录入工作的师生是一项巨大的挑战，从拿到文稿到录入校对完成，学院组织召开了多次面对面的研讨会和文稿校对讨论会，毫不夸张地说，这次文稿的录入和校对工作对每个人都是一次最好的学科历史和精神传承教育，也让大家由衷感受到王湘浩先生的伟大精神，用颜回对孔子“仰之弥高，钻之弥坚，瞻之在前，忽焉在后”赞美来形容，再合适不过了。

在此，向在文集文稿录入和校对工作中，出色完成任务的各位同学表示衷心感谢，他们是（以姓名首字母排序）：陈曦，陈钊煌，房劭同，耿之阳，郭凯旋，江金普，姜雨辰，李得志，李东哲，李锦茹，李子恒，刘家文，马昊，欧思思，彭家骥，石岩，史梓慧，孙世昂，王斌，王继葵，王天威，邢哲哲，杨君豪，叶媛媛，余家暄，余泽骏，张雪晨，赵一铭，赵志敏。文稿录入时，你们快速且精准地将文字输入系统，使任务高效推进；文稿校对时，凭借认真负责的态度，把一个个错误纠正过来。同学们的表现充分展现了吉林大学计算机学科学生良好的精神风貌和专业素养。

文集出版还得到了旅居海外的吉大校友、王湘浩院士培养的第一位博士陈建华老师的关心和帮助，尤其感谢陈老师对文集校对提出的专业意见；同时，文稿的编撰也得到了旅居海外的吉大校友、王湘浩院士的儿子王强和王康老师的帮助，王康老师对文集的编排、图片和文字都提出了很多有建设性的和专业的意见，同时，两位老师提供了很多珍贵的史料和图片。特别感谢王湘浩院士的孙女王亦平和王亦方女士提供了很多珍贵的照片和书稿图片，很多照片和书稿都是第一次面世。为此，向王湘浩院士的家人表示衷心感谢。

在文集即将付梓印刷时，我们向远在大洋彼岸的普林斯顿大学表示感谢，在获悉吉林大学要举办王湘浩院士诞辰 110 周年纪念活动的消息后，普林斯顿大学图书馆 Sylvia Swain 女士提供了王湘浩院士 1948 年 11 月提交到普林斯顿大学的异常珍贵的博士论文全文扫描件。特别要感谢北京大学曹永知教授费尽心思帮助联系北京大学档案馆，为文集出版提供了强有力的帮助和支持。衷心感谢北京大学档案馆同志们的大力支持，尤其感谢陈沫老师精心为我们扫描了王湘浩院士在西南联合大学的硕士论文、《集合论》讲义和珍贵的书信手稿。

此外，河北省安平县各位领导对王湘浩院士诞辰 110 周年纪念活动的开展做了大量卓有成效的工作，对文集的出版工作赋予了力量，其中包括河北省安平县前县委书记曹向东、河北省安平县委书记赵东钊、河北省安平县统战部部长崔长友、河北省安平县东黄城镇党委书记赵营、河北省东黄城镇宣传委员王成芳、河北省安平县东黄城镇大同新村党支部书记王小同和中共第一个农村党支部纪念馆前馆长王彦芹老师，一并对大家表示感谢。

还有很多人对文集的出版提供了默默的支持和帮助，恕不能一一提及名字，在此一并表示感谢!

我们希望通过《王湘浩文集》的出版，能让众多后辈和学子进一步了解王湘浩院士为吉林大学计算机学科、数学学科的发展作出的杰出贡献。但由于年代久远，先生的很多手稿已难以找到，这也是一种遗憾。同时，由于先生对名利看得很淡，他很多学生发表的论文都是他亲手指导的，甚至很多主要创新思想都源自先生，但他要求这些论文不允许署他的名字，就造成很多学术论文的署名中都没有先生的名字，为了充分表达对王湘浩院士的尊重，这些文章没有收入文集。同时，由于我们的水平所限，先生的数学、计算机、人工智能专业水平非常人所能比，很多文字字迹模糊、难以辨认，我们只能按照自己的理解来录入，有可能望文生义，难免有所纰漏和瑕疵，希望读者予以谅解。

《王湘浩文集》难以全方位展现先生的所有成就，但王湘浩院士在科学和教育事业领域建树卓越，早已是不争的事实。他不仅在相关领域取得杰出成就，立下卓越功勋，创造光辉业绩，作出彪炳史册的贡献，其崇高的思想品格以及严谨求实、富于创新的治学精神，也必将垂范后世。

计算机科学与技术学院　张永刚
2025 年 4 月 6 日